AF754025

[redacted]

[redacted]

Frontiers in Bioactive Lipids

GWUMC Department of Biochemistry and Molecular Biology Annual Spring Symposia

Series Editors:

Allan L. Goldstein, Ajit Kumar, and J. Martyn Bailey

The George Washington University Medical Center

Recent volumes in this series:

ADVANCES IN MOLECULAR BIOLOGY AND TARGETED TREATMENT FOR AIDS
Edited by Ajit Kumar

BIOLOGY OF CELLULAR TRANSDUCING SIGNALS
Edited by Jack Y. Vanderhoek

BIOMEDICAL ADVANCES IN AGING
Edited by Allan L. Goldstein

CARDIOVASCULAR DISEASE
Molecular and Cellular Mechanisms, Prevention, and Treatment
Edited by Linda L. Gallo

CARDIOVASCULAR DISEASE 2
Cellular and Molecular Mechanisms, Prevention, and Treatment
Edited by Linda L. Gallo

CELL CALCIUM METABOLISM
Physiology, Biochemistry, Pharmacology, and Clinical Implications
Edited by Gary Fiskum

THE CELL CYCLE
Regulators, Targets, and Clinical Applications
Edited by Valerie W. Hu

FRONTIERS IN BIOACTIVE LIPIDS
Edited by Jack Y. Vanderhoek

GROWTH FACTORS, PEPTIDES, AND RECEPTORS
Edited by Terry W. Moody

NEURAL AND ENDOCRINE PEPTIDES AND RECEPTORS
Edited by Terry W. Moody

NEURODEGENERATIVE DISEASES
Molecular and Cellular Mechanisms and Therapeutic Advances
Edited by Gary Fiskum

PROSTAGLANDINS, LEUKOTRIENES, LIPOXINS, AND PAF
Mechanism of Action, Molecular Biology, and Clinical Applications
Edited by J. Martyn Bailey

A Continuation Order Plan is available for this series. A continuation order will bring delivery of each new volume immediately upon publication. Volumes are billed only upon actual shipment. For further information please contact the publisher.

Frontiers in Bioactive Lipids

Edited by

Jack Y. Vanderhoek
The George Washington University Medical Center
Washington, D.C.

[redacted]

Plenum Press • New York and London

Library of Congress Cataloging-in-Publication Data

On file

Proceedings of the Sixteenth Washington International Spring Symposium at The George Washington University, held May 6–9, 1996, in Washington, D.C.

ISBN 0-306-45505-6

© 1996 Plenum Press, New York
A Division of Plenum Publishing Corporation
233 Spring Street, New York, N. Y. 10013

10 9 8 7 6 5 4 3 2 1

All rights reserved

No part of this book [redacted]ny form or by any means, e[redacted]erwise, without written per[redacted] from the Publisher

Printed in the United States of America

PREFACE

Recent advances in our understanding of the fundamental role of bioactive lipids in normal and abnormal physiology was the theme of the XVIth International Spring Symposium on Health Sciences held on May 6–9, 1996 at the George Washington University School of Medicine in Washington, D.C. Over 250 participants shared their latest findings on fundamental mechanisms in lipid metabolism, transport, and signal transduction. Most of the papers presented at the plenary sessions have been collected in this volume, which is divided into seven parts.

The focus of Part I is on the new roles that fatty acids and esters play in cellular function. These activities include regulation of gene expression, control of eicosanoid-mediated responses, and intracellular calcium sequestration. Most biological effects of fatty acids, esters, and phospholipids are transmitted via the interaction of these lipid molecules with specific lipid binding proteins. The chapters in Part II detail the involvement of these proteins in the transport of fatty acids, fatty acyl CoA esters, and phospholipids, and in the modulation of the fatty acid-induced activation of the peroxisome proliferator-activated receptors.

The study of arachidonic acid metabolism continues to be the subject of intense interest and research. A number of studies in Part III deal with the substantial differences between the constitutive prostaglandin H synthase isozyme (PGHS-1 or COX-1) and the inducible form PGHS-2 (COX-2) regarding compartmentation, control, and expression. Novel information is presented on the use of lipoxygenase-deficient mice in the elucidation of this pathway's function and on the involvement of the cyclooxygenase and lipoxygenase pathways in cell survival and apoptosis. The involvement of several, more recently discovered, eicosanoids in signal transduction is described in Part IV. Results with the 5-oxo-eicosanoids and anandamide indicate that they are new players in the signaling networks present in cells. New data is also presented on the alteration of functional responses of the prostaglandin E and thromboxane receptors.

A new and exciting area in cellular signaling involves signal-transducing molecules derived from sphingolipids. Chapters on ceramide and sphingosine-1-phosphate as well as the use of labeled sphingolipids to study membrane trafficking are presented in Part V. Part VI deals with messengers originating from glycerophosphates. Signal transduction pathways involving lysophosphatidic acid and glycerophosphoinositol-4-phosphate are described, as well as recent results on controlling platelet-activating-factor metabolism.

The stimulation and regulation of cellular phospholipases is of obvious importance in signal transduction processes. Although numerous reports have been published on the mammalian phospholipases A_2 and C, less attention has been given to phospholipase D. The major focus of Part VII is on phospholipase D with additional contributions on the A_2 and C enzymes.

These chapters on the novel biological actions of lipid signaling molecules, their metabolisms, as well as their interactions with proteins, affirm that this is an active and fruitful area of research.

Finally, on behalf of the organizing committee, I would like to thank the contributors listed below for their generous support of this symposium

Jack Y. Vanderhoek
Washington, D.C.

Financial support for this symposium was received from:

- I.T.T. Continental Baking Company, Inc.
- Merck Research Laboratories and Merck Frosst Canada, Inc.
- Searle Technical Support and Administration
- Avanti Polar Lipids, Inc.
- Biomol Research Laboratories, Inc.
- Ciba-Geigy Corporation Pharmaceutical Division
- The Council for Tobacco Research—USA, Inc.
- Quest International Scientific and Regulatory Affairs
- Pfizer, Inc. Central Research Division
- Alza Corporation
- Assay Designs, Inc.
- Cascade Biochem Ltd.
- Matreya, Inc.

CONTENTS

PART I—FATTY ACIDS AND ESTERS

PART II—LIPID-BINDING PROTEINS

PART III—EICOSANOID BIOSYNTHETIC ENZYMES

PART IV—EICOSANOIDS AND RECEPTORS

PART V—SPHINGOLIPIDS

PART VI—PHOSPHOLIPIDS

PART VII—PHOSPHOLIPASES/KINASES

MOLECULAR MECHANISMS INVOLVED IN THE ADIPOGENIC ACTION OF FATTY ACIDS

Paul A. Grimaldi, Danielle Gaillard, Hidekuni Inadera, Lydia Teboul, Gérard Ailhaud, and Ez-Zoubir Amri

UMR 134 CNRS, Centre de Biochimie, UNSA, UFR Sciences, Parc Valrose, 06108 Nice cedex 2, France

INTRODUCTION

Fatty acids are important energy storage substrates and precursors of triglycerides, phospholipids and prostaglandins. In vertebrates, adipose tissue is the major depot for energy storage. Fatty acids are stored in adipose cells as triacylglycerol and released in the blood circulation during periods of nutritional deprivation.

In addition to their role as substrates, fatty acids can affect many aspects of cell function. Some effects of fatty acids are indirect and involve arachidonic acid metabolites, such as prostaglandins and leukotrienes. Furthermore, fatty acids have been shown to directly modulate a range of biological processes. For example, fatty acids activate guanylate cyclase or protein kinase C and modulate ion channel function[1, 2, 3].

In vivo studies have suggested that adipose tissue development is regulated by nutritional status. High-fat and high-carbohydrate diets lead to hypertrophy and hyperplasia of adipose tissue in adult rats[4]. More recently, it has been found that feeding rats a diet containing saturated fat lead to a 3-fold increase in the adipocyte number of retroperitoneal fat pad compared to the number of adipocytes in animals receiving a diet containing polyunsaturated fat[5]. In order to gain some insights in the possible links between high-fat diets and adipose tissue development, we have investigated the effects of fatty acids on the process of differentiation of Ob1771 preadipose cells.

Under appropriate culture conditions these cells, established from genetically obese mouse adipocytes, differentiate into adipocytes and undergo a large shift of protein biosynthesis reflecting transcriptional activation of adipose related genes. The differentiation process of Ob1771 is characterized by a sequence of events during which adipoblasts divide until confluence and are commited to preadipocytes with activation of early genes such as lipoprotein lipase and α2 chain of collagen type VI. Preadipocytes undergo several rounds of mitosis, *i.e.* post-confluent mitosis, which are critical for terminal differentiation. The last step of differentiation is characterized by the activation of a large set of genes including those involved in fatty acid synthesis and utilization for triacylglycerol metabolism such as adipocyte lipid binding protein (ALBP, also called aP2), fatty acid transporter (FAT), acyl CoA synthetase (ACS) and glycerophosphate dehydrogenase (GPDH). It is worth noting that the adipose conversion of precursor cells of established cell lines or that of precursor cells in primary cell culture does involve a limited proportion of cells. This process leads to the formation of colonies of fat cells separated from each other by cells resting in preadipose state. Such "dormant" preadipose cells are also present in adipose tissue of adult animals and it has been shown that under some conditions they are able to proliferate and terminally differentiate. Several adipogenic factors have been identified so far. Among circulating hormones, growth hormone, triiodothyronine (T_3), glucocorticoids, insulin, insulin growth factor-I and retinoids are positive effectors of the expression of adipose related genes and activate the subsequent lipid accumulation[6, 7].

During the last years, it has been demonstrated that fatty acids play an important role in the control of gene expression during adipose differentiation and that fatty acids have to be considered as adipogenic hormones. This chapter presents findings demonstrating this new action of fatty acids and the more recent evidences showing that nuclear receptors of the steroid/thyroid hormone receptor gene family mediate the transcriptional effects of fatty acids.

ADIPOGENIC EFFECTS OF FATTY ACIDS

To investigate the effects of fatty acids on the adipose conversion process, Ob1771 cells have been maintained for two weeks after confluence in standard differentiation medium, *i.e.* containing fetal calf serum, insulin and T_3, and chronically exposed from confluence to fatty acids. Fatty acid treatment led to a strong increase of morphological adipose differentiation. When maintained in standard medium, differentiation occured in about 40% of the cells, whereas chronic exposure to native fatty acids (100 μM) or to 2-bromopalmitate (25 μM) resulted in a net increase in the differentiated cell number which was estimated to be more than 90%. This phenomenon was not a passive filling-up of the cells but was related to an enhancement of the expression of adipose conversion program. Chronic treatment of cells with saturated or unsaturated long-chain fatty acids resulted in a net induction (8- to 10-fold increase when compared to cells maintained in standard medium) of the expression of adipose-related genes, whereas middle-chain fatty acids were not effective. More interestingly, 2-bromopalmitate, as compared to palmitate, exerted its adipogenic effect at lower concentrations (half-maximal effect observed at 10 μM versus 75 μM). Keeping in mind that the bromylated derivative was shown to be less metabolized than the natural fatty acid[8], this observation suggested that the adipogenic action of fatty acids was likely not related to an increase in substrate availability for triacylglycerol synthesis. Additional experiments have shown that chronic exposure to fatty acids was not required for enhancement of the adipose conversion as treatment from confluence to day 3, *i.e.* during the preadipose state, exerted similar effects on gene expression. By opposition, exposure to fatty acids beginning at day 5, *i.e.* during terminal differentiation, was shown to be ineffective. These observations suggested that fatty acids exerted their adipogenic effects by acting at the early stage of the differentiation process which, in turn, control terminal events[9].

Taken together, these observations suggested that, at the preadipose state, an increase in the intracellular concentration of unprocessed fatty acids plays the role of a signal triggering the activation of expression of the adipose differentiation program. The effects of fatty acid on gene expression in preadipose cells were next investigated.

TRANSCRIPTIONAL EFFECTS OF FATTY ACIDS IN PREADIPOCYTES

The effects of short-term exposure to various fatty acids on the pattern of gene expression have been investigated using Ob1771 cells just after confluence. At this stage, cells are still non-terminally differentiated and do not synthesize fatty acids and triglycerides. Long-chain fatty acids exerted potent effects on the transcriptional rates of genes encoding proteins directly implicated in fatty acid metabolism, such as FAT, ALBP or ACS, and have no effects on expression of other adipose-related genes, such as GPDH or adipsin which remained transcriptionally inactive. Relationships between the nature of the fatty acid, *i.e.* length and unsaturation degree of the fatty acyl chain, and its potency in inducing FAT gene expression were also studied showing that short- and middle-chain fatty acids were ineffective or less active than long-chain fatty acids. The influence of the unsaturation degree of the fatty acyl chain appeared to be less important, as oleic, linoleic and linolenic acids exerted similar effects than stearic acid[10]. The transcriptional activation of fatty acid-responsive genes was found to be a relatively slow process occuring within hours and fully and rapidly reversible upon withdrawal of the inducer from culture medium[11].

Interestingly, the nonmetabolized derivative 2-bromopalmitate was a potent inducer of FAT gene transcription in preadipose cells. As it has been demonstrated that this fatty acid derivative is not activated to acyl CoA in preadipocytes[8], this indicated that the unprocessed fatty acid is the actual inducer of gene expression. Additional experiments have shown that 2-bromopalmitate was more potent than palmitate in inducing FAT gene. This reflected bromopalmitate accumulation in the unprocessed form inside the cell, and explained why high concentrations of natural fatty acids, which are metabolized by preadipocytes, were needed to transcriptional effects.

These results demonstrated that fatty acids act as transducing molecules and activate in preadipocytes the expression of proteins directly implicated in their own metabolism. This process remains reversible upon removal of the inducer from medium in preadipocytes, but becomes irreversible after induction of terminal differentiation.

NUCLEAR RECEPTORS MEDIATE TRANSCRIPTIONAL EFFECTS OF FATTY ACIDS

Recent evidences suggest that members of the peroxisome proliferator-activated receptor (PPAR) family are implicated as mediators of transcriptional and adipogenic effects of fatty acids. The first member of this family has been cloned from a cDNA library of mouse liver and called PPARα. This nuclear receptor is activated by peroxisome proliferators (PP), hence the term PPAR, and stimulates the transcription of genes implicated in fatty acid oxidation. PPARα binds to specific responsive elements, called PPREs, which have been located in the PP-responsive gene promoters[12]. PPARα can be activated by a diversity of molecules including polyunsaturated fatty acids. As PPARα is expressed at a very low level in adipose tissue, these findings have conducted some groups to investigate whether a member of the PPAR family mediates the transcriptional effects of fatty acids in preadipose and adipose cells.

Using a DNA fragment containing highly conserved sequence in the steroid/thyroid receptor gene family, two new nuclear receptors have been cloned from adipose cDNA libraries. PPARγ[13] and fatty acid-activated receptor (FAAR)[14] are clearly related to PPARα as confirmed by their sequence and their ability to bind to PPREs. Tissue distribution of the various PPAR isoforms is different. In rodents, it has been shown that PPARα is mainly expressed in liver, kidney and brown adipose tissue. FAAR is expressed in various fat-metabolizing tissues, such as brown and white fat, small intestine, heart, skeletal muscle and brain, and is undetectable in liver. PPARγ is mainly expressed in brown and white adipose tissues and has to be considered, at least in rodents, as specific for adipose cells. PPAR isoforms are activated by a broad diversity of molecules that include natural compounds, such as fatty acids or prostaglandins and synthetic molecules, such as fibrates or thiazolidinediones. Peroxisome proliferators are potent activators of PPARα and weak activators of PPARγ and FAAR, whereas saturated long-chain fatty acids are potent activators of FAAR and almost not effective in PPARα or γ activation. Recent evidences suggested that the physiological ligands of PPAR isoforms are different. For example, it has been shown that the prostaglandin J_2 dehydration product 15-deoxy-Δ12,14-PGJ_2 is a ligand for PPARγ and not for PPARα or FAAR[15, 16]. PPAR-binding sites have been located in the promoters of ALBP[13], phosphoenolpyruvate carboxykinase[17] and ACS[18].

FAAR and PPARγ, which are both expressed at high levels in adipose tissue, play important, and probably different, roles in adipose cell biology. Both of them are induced during the differentiation process of preadipocytes. In Ob1771 cells, FAAR mRNA which is undetectable before confluence increases rapidly thereafter reaching its maximal level during the preadipose state. Expression of PPARγ mRNA is clearly delayed and is induced just before activation of terminal differentiation and lipid accumulation[14]. It has been concluded from these observations that FAAR is a very early marker of adipose differentiation and that its appearance at confluence coincides with acquisition of the response to fatty acids[10]. By contrast, PPARγ, which is expressed at a very low level in non-terminally differentiated cells, does not appear as a good candidate to mediate transcriptional effects of fatty acids in preadipocytes.

The roles of PPARs in the process of adipose differentiation and fatty acid responsivity have been investigated more directly by forcing their expression in fibroblast cells which are not responsive to fatty acids and not able to undergo adipose conversion. Quite recently, we have shown that ectopic expression of FAAR in 3T3-C_2 fibroblasts was sufficient to confer fatty acid-responsivity to the endogenous FAT and ALBP genes. Transcriptional induction of these genes occured within hours after addition of fatty acids to the medium. Furthermore, the response observed in FAAR expressing 3T3-C_2 cells exhibited the same specificity previously demonstrated in Ob1771 preadipocytes. In both cases, 2-bromopalmitate was more effective than native fatty acids. Both saturated and unsaturated fatty acids were effective, whereas middle-chain fatty acids and fibrates were weaker inducers[14].

A role for PPARγ in the control of terminal adipose differentiation has been recently proposed. It has been found that retroviral expression of PPARγ stimulated adipose conversion of various fibroblast cell lines. Long-term exposure of PPARγ-transfected cells to activators, such as unsaturated fatty acids or PGJ_2 metabolites, induced the entire program of adipocyte gene expression[15, 16, 19].

Taken together, these observations strongly suggest a direct role of these two lipid-activated nuclear receptors in the transcriptional induction of genes implicated in fatty acid metabolism and in the control of adipose differentiation by fatty acids and related molecules.

CONCLUSION

Fatty acids appear to act as signal transducing molecules in addition to playing the role of an energy source and of being essential membrane components and eicosanoid precursors. In preadipose cells, fatty acids behave as hormones and stimulate gene transcription by means of nuclear receptors of the steroid/thyroid receptor family. The temporal expression of these lipid-activated nuclear factors during the differentiation process of preadipose cells suggests that they could act in a sequential manner. FAAR which is induced at confluence in Ob1771 preadipose cells is likely involved in fatty acid activation of lipid metabolism-related genes and in the adipogenic action of fatty acids which occurs during the first days of the confluent phase. PPARγ, which is expressed at a later stage, might mediate regulatory effects of fatty acids or fatty acid metabolites occuring in partially to fully differentiated cells. Keeping in mind that exposure of preadipocytes to fatty acids induces expression of proteins implicated in acyl CoA synthesis, it can be proposed that these inductions lead to an enhancement of arachidonic acid metabolism and to accumulation of some prostaglandin metabolites including the PPARγ ligand 15d-Δ12,14-PGJ_2. Concommitant expression of PPARγ and accumulation of its activator could be the signals for terminal differentiation and triacylglycerol synthesis.

Activation of these nuclear receptors could, *in vivo*, constitute an important part of the molecular mechanism behind the adipogenic effects of overfeeding. Such a new role for fatty acids remains to be assessed *in vivo*, but it can be assumed that the flux of fatty acids entering dormant preadipose cells should be significantly increased by feeding high-fat diet and that this could be the signal for the terminal differentiation of new adipocytes leading to obesity. The characterization of fatty acid-activated nuclear receptors could also be helpful to define new pharmacological approaches to the treatment of obesity and metabolic disorders related to obesity.

REFERENCES

1. J.M. Braughler, C.K. Mittal, and F. Murad, Purification of soluble guanylate cyclase, *Proc. Natl. Acad. Sci. USA* 76:219 (1979).
2. L.C. McPhail, C.C. Clayton, and R. Snyderman, A potential second messenger role for unsaturated fatty acids: activation of Ca-dependent protein kinase, *Science* 224:622 (1984).

3. R.N. Ordway, J.J. Singer, and J.V. Walsh, Direct regulation of ion channels by fatty acids, *Trends Neurosci.* 14:96 (1991).
4. I.M. Faust, P.R. Johnson, J.S. Stren, and J. Hirsch, Diet-induced adipocyte number increase in adult rats: a new model of obesity, *Am. J. Physiol.* 235:E279 (1978).
5. G. Shillabeer, and D.C. Lau, Regulation of new fat cell formation in rats: the role of dietary fats, *J. Lipid Res.* 35:592 (1994).
6. G. Ailhaud, P. Grimaldi, and R. Négrel, Cellular and molecular aspects of adipose tissue development, *Annu. Rev. Nutr.* 12:207 (1992).
7. O.A. MacDouglas, and M.D. Lane, Transcriptional regulation of gene expression during adipose differentiation, *Annu. Rev. Biochem.* 64:345 (1995).
8. P. Grimaldi, S.M. Knobel, R.R. Whitesell, and N.A. Abumrad, Induction of the aP2 gene by nonmetabolized long-chain fatty acids, *Proc. Natl. Acad. Sci. USA*, 89:10930 (1992).
9. E.-Z. Amri, G. Ailhaud, and P.A. Grimaldi, Fatty acids as signal transducing molecules: involvment in the differentiation of preadipose to adipose cells, *J. Lipid Res.* 35:930 (1994).
10. E.-Z. Amri, B. Bertrand, G. Ailhaud, and P. Grimaldi, Regulation of adipose cell differentiation: I. Fatty acids are inducers of the aP2 gene expression, *J. Lipid Res.* 32:1449 (1991).
11. E.-Z. Amri, G. Ailhaud, and P. Grimaldi, Regulation of adipose cell differentiation: II: Kinetics of induction of the aP2 gene by fatty acids and modulation by dexamethasone, *J. Lipid Res.* 32:1457 (1991).
12. H. Keller, and W. Wahli, PPAR, a link between endocrinology and nutrition, *Trends Endocrinol. Metab.* 4:291 (1993).
13. P. Tontonoz, E. Hu, R.A. Graves, A.I. Budavari, and B.M. Spiegelman, PPARγ2: Tissue-specific regulator of an adipocyte enhancer, *Genes Dev.* 8:1224 (1994).
14. E.-Z. Amri, F. Bonino, G. Ailhaud, N.A. Abumrad, and P.A. Grimaldi, Cloning of a protein that mediates transcriptional effects of fatty acids in preadipocytes, *J. Biol. Chem.* 270:2367 (1995).
15. B.M. Forman, P. Tontonoz, J. Chen, R.P. Brun, B.M. Spiegelman, and R.M. Evans, R.M., 15-deoxy-Δ12,14-prostaglandin J2 is a ligand for the adipocyte determination factor PPARγ, *Cell* 83:803 (1995).
16. S.A. Kliewer, J.M. Lenhard, T.M. Willson, I. Patel, D.C. Morris, and J.M. Lehmann, A prostaglandin J2 metabolite binds to PPARγ and promotes adipocyte differentiation, *Cell* 83:813 (1995).
17. P. Tontonoz, E. Hu, J. Devine, E.G. Beale, and B.M. Spiegelman, PPARγ2 regulates adipose expression of the phosphoenolpyruvate carboxykinase gene, *Mol. Cell. Biol.* 15:351 (1995).
18. K. Schoonjans, M. Watanabe, H. Suzuki, A. Mahfoudi, G. Krey, W. Wahli, P. Grimaldi, B. Staels, T. Yamamoto, and J. Auwerx, Induction of the acyl-coenzyme A synthetase gene by fibrates and fatty acids is mediated by a peroxisome proliferator response element in the C promoter, *J. Biol. Chem.* 270:19269 (1995).
19. P. Tontonoz, E. Hu, and B.M. Spiegelman, Stimulation of adipogenesis in fibroblasts by PPARγ2, a lipid-activated transcription factor, *Cell* 79:1147 (1994).

ACYL COENZYME A SYNTHETASE AND THE TRANSPORT OF LONG-CHAIN FATTY ACIDS

Paul N. Black

Department of Biochemistry
University of Tennessee College of Medicine
Memphis, TN 38163

INTRODUCTION

Long-chain fatty acids and their derivatives represent essential components of membranes, importance sources of metabolic energy and important effector molecules that regulate metabolism. Prior to any type of metabolic transformation, exogenous long-chain fatty acids (C_{12} - C_{18}) must pass through the cell membrane. A saturable long-chain fatty acid uptake process that is effective at nanomolar concentrations has been described in a number of different cells type suggesting these compounds enter the cell by a facilitated mechanism. This suggestion is supported by work showing that this process is also rapid, blocked by fatty acid analogues and protease sensitive[1-6]. Once long-chain fatty acids are taken up by the cell, they become bound to intracellular fatty acid binding proteins (FABPs) or are activated into CoA thioesters. Fatty acyl CoA thioesters serve as substrates for both β-oxidation and phospholipid synthesis.

The uptake of long-chain fatty acids in the gram negative bacterium *Escherichia coli* has been the focus of our work over the past ten years. In *E. coli* these compounds are taken up by a concentrative process that is tightly coupled to metabolic utilization and requires at least the outer membrane protein FadL (product of the *fadL* gene) and the inner membrane associated acyl CoA synthetase (product of the *fadD* gene)[7-26]. FadL binds exogenous long-chain fatty acids with high affinity and facilitates their transport across the outer membrane[17,24]. Acyl CoA synthetase activates long-chain fatty acids concomitant with transport across the inner membrane by a process that proceeds through the pyrophosphorolysis of ATP[7,8,27]. An oleic acid binding protein postulated to be a H^+/fatty acid cotransporter has been purified and partially characterized that might represent an inner membrane-bound component of this transport system[28,29]. The structural gene encoding this protein has not been identified.

We and others have demonstrated that the process of fatty acid uptake in *E. coli* requires the expenditure metabolic energy[8,10,30,31]. Central to this requirement for energy is acyl CoA synthetase (fatty acid:CoA ligase [AMP forming] (EC 6.2.3.1)). This enzyme catalyzes the formation of an acyl adenylate intermediate from ATP and fatty acid that is subsequently converted to acyl CoA with the release of AMP. Several studies support the hypothesis that acyl CoA synthetase becomes membrane associated under certain metabolic conditions. Of particular note is the study by Mangroo and Gerber[30] which demonstrates this enzyme is recruited to the inner (cytoplasmic) membrane in the presence of D-lactate. It is interesting to speculate whether the proton derived electrochemical gradient across the inner membrane contributes to this process. If this is experimentally verified, this would explain the role of this enzyme in the context of fatty acid transport and would broaden our understanding of the vectorial thioesterification of exogenous long-chain fatty acids. We

Frontiers in Bioactive Lipids
Edited by Vanderhoek, Plenum Press, New York, 1996

have demonstrated that this enzyme is activated by phospholipids supporting the notion that it interacts with the membrane (Weimar and Black, unpublished) and have demonstrated that the proton motive force is required for optimal levels of long-chain fatty acid transport[31].

In the context of fatty acid transport, the *E. coli* acyl CoA synthetase appears to be functionally analogous to the arachidonate-specific acyl CoA synthetase in fibrosarcoma cells (uptake of arachidonic acid), the plasma membrane-associated acyl CoA synthetase found in rat hepatocytes (uptake of C_{18} saturated and unsaturated fatty acids), and the mouse adipocyte acyl CoA synthetase[6,32,33]. In the yeast *Saccharomyces cerevisiae* two acyl CoA synthetases (FAA1 and FAA4) have been identified that are responsible for activating exogenous fatty acids[34,35] that may comprise part of a fatty acid transport apparatus. The adipocyte acyl CoA synthetase appears to be linked to a membrane-bound fatty acid transport protein (FATP) to facilitate the transport of exogenous fatty acids[6]. To date, a fatty acid transport protein partner to the acyl CoA synthetases in fibrosarcoma cells, rat hepatocytes and yeast have not been identified.

ACYL CoA AS A BIOACTIVE COMPOUND

Our interest in acyl CoA synthetase is two fold: [1] This enzyme represents an essential component required for the transport of exogenous long-chain fatty acids destined for β-oxidation and [2] the product its enzymatic reaction, fatty acyl CoA, represents an important effector molecule in the regulation of fatty acid metabolism. This later observation is of particular interest as the transcription factor FadR responds to intracellular levels of long-chain fatty acyl CoA in *E. coli* and thus differentially regulates genes involved in fatty acid biosynthesis and fatty acid degredation[24,36-38] (DiRusso, this volume). We have recently demonstrated that long-chain fatty acyl CoA compounds accumulate in *E. coli* strains blocked for β-oxidation (*fadE*) when grown on acetate. The apparent accumulation of long chain acyl-CoA in *fadE* strains in the absence of exogenously supplied fatty acids indicates that the acyl moiety must be derived either from *de novo* synthesis or turnover of membrane lipids. In *E. coli* the primary products of fatty acid biosynthesis are $C_{16:0}$, $C_{16:1}$ and $C_{18:1}$ thioesterified to acyl carrier protein (ACP). The acyl-ACPs rather than acyl-CoA are believed to serve as substrates for phospholipid and lipoprotein biosyntheses. It has been previously demonstrated that when fatty acid biosynthesis is uncoupled from phospholipid biosynthesis and β-oxidation, long chain acyl-ACP accumulates[39]. Our work suggests, that acyl CoA synthetase also functions in the natural state to control fatty acid levels by activating "excess" fatty acids derived both from exogenous sources and from *de novo* synthesis. It is also probable, as suggested by the earlier work of Silbert and co-workers, that there is a direct interaction between the acyl CoA synthetase and acyltransferase which results in modification of membrane phospholipids[40,41]. Taken together these studies indicate that the regulatory loop controlling fatty acid and phospholipid metabolism in *E. coli* must be extended to include acyl-CoA perhaps both through effects on enzyme activities including the acyl-transferases[39,41] and through FadR-mediated effects on transcription[36-38].

CLONING, SEQUENCING AND CONTROL OF THE *FADD* GENE

We have cloned and sequenced the acyl CoA synthetase structural gene (*fadD*)[21,22]. The derived amino acid sequence of this enzyme demonstrates that it is remarkably similar to the family of AMP binding proteins and more specifically the family of acyl CoA synthetases[22,23]. Two regions within the enzyme are notable: The first lies between amino acids 205 and 230 and the second between 431 and 455. The first region of similarity is common to AMP binding proteins while the second is more restricted to acyl CoA synthetases (Figure 1). We are using both affinity labeling of purified acyl CoA synthetase coupled with site-directed mutagenesis of the *fadD* gene to address the functional importance of these regions within the *E. coli* enzyme.

The sequence of the acyl CoA synthetase structural gene led us to postulate that it is tightly controlled at both transcriptional and translational levels[21]. The *fadD* promoter contains two FadR binding sites (O_{D1},-13 to -29 and O_{D2}, -99 to -115) that may interact with each other during FadR binding to promote DNA bending and transcriptional repression.

Just upstream from -35, we identified a cAMP-CPR consensus site suggesting this gene is also controlled at the level of catabolite repression. This observation was validated using a *fadD-lacZ* reporter strain grown in the presence of glucose as opposed to the long-chain fatty acid oleate as a sole carbon and energy source[42]. Data generated using *fadD-lacZ* protein and operon fusions are consistent with this gene also being down regulated at the level of translation (β-galactosidase levels are 2- to 3-fold higher in operon fusions as opposed to protein fusions). The *fadD* gene has an alternative initiation codon, UUG, preceded by a perfect Shine-Dalgarno (ribosome binding site) sequence that appears to be important in this down regulation. Using directed mutagenesis and *fadD-lacZ* reporter strains, we have recently demonstrated that these regulatory sequences are essential for the controlled expression of *fadD*. The disruption of the Shine-Dalgarno sequence nearly abolishes acyl CoA synthetase activities in *fadD* strains harboring plasmids containing the regulatory mutation. The conversion of the UUG initiation codon to AUG increases acyl CoA synthetase activity to 122% those measured for the wild-type. The double mutant has activity levels that are intermediate indicating that these mutations are compensatory for each other. These data support the hypothesis that translational down regulation occurs at the level of initiation[42].

	1	DGWLHTGDIGXWXPXGXLKIIDRKK	25
		********* * ********	
ECACS	431	NGWLHTGDIAVMDEEGFLRIVDRKK	455
FAA1	527	DGWFKTGDIGEWEANGHLKIIDRKK	551
FAA2	531	DGWFSTGDVAFIDGKGRISVIDRVK	555
FAA3	567	DGWFRTGDIVEWTPKGQLKIIDRRK	591
FAA4	527	DGWFRTGDIAEWTPKGQVKIIDRKK	551
HIACS	433	DGWMATGDIVIMDESYSLRIVDRKK	457
POACS	410	DGWFSTGDVATIDSDGFMTICDRAK	434
M1ACS	534	DGWLHTGDIGKWLPNGTLKIIDRKK	558
R1ACS	534	DGWLHTGDIGKWLPNGTLKIIDRKK	558
R2ACS	533	DGWLHTGDIGKWLPEGTLKIIDRKK	557
H1ACS	534	DGWLHTGDIGKWLPNGTLKIIDRKK	558
H2ACS	533	DGWLHTGDIGKWLPNGTLKIIDRKK	557

Figure 1. Region of *E. coli* acyl CoA synthetase (ECACS; sw P29212) with sequence similarities found *only* in the family of acyl CoA synthetases. FAA1 (sw P30624), FAA2 (sw P39518), FAA3 (sw P39002), and FAA4 (sw P47912) are from the yeast *Saccharomyces cerevisiae*; HIACS (sw P46450) is the presumed acyl CoA synthetase from *Haemophilus influenzae*; POACS (sw Q00594) is from *Pseudomonas oleovarans*; M1ACS (sw P41216) is from mouse adipocytes; R1ACS (sw P18163) and R2ACS (sw P33124) are isozymes found in the rat liver and brain respectively; and H1ACS (sw P41215) and H2ACS (sw P33121) are isozymes found in the human liver and brain respectively. The sequence on the top represents the consensus for this 25 amino acid residue segement within the family of acyl CoA synthetases. The asterisks represent residues within the *E. coli* enzyme subjected to alanine-scanning mutagenesis.

On the basis of our studies on the regulation of the *fadD* gene, we conclude that this gene is highly regulated in order to sense fatty acids that are available from exogenous sources or accumulate from *de novo* biosynthesis. The net result is a highly controlled regulatory circuit for long-chain fatty acyl CoA production and the subsequent transcriptional control of genes involved in fatty acid biosynthesis and long-chain fatty acid transport, activation and β-oxidation.

PURIFICATION OF ACYL COA SYNTHETASE AND SITE-SPECIFIC LABELING WITH THE LONG-CHAIN FATTY ACID ANALOGUE 9-*p*-AZIDOPHENOXY NONANOIC ACID

We have constructed an acyl CoA synthetase expression system whereby mg quantities of hexameric-histidine tagged enzyme can be purified that maintains the same kinetic parameters as the native enzyme. The enzyme is purified following growth at 37°C of the *E. coli* strain BL21 (λDE3)/pLysS harboring the expression plasmid pN3576 and induction with isopropylthiogalactoside. Following growth and induction, cells are lysed by sonication and the enzyme purified using Ni^{+2} chelate affinity chromatography. Hexameric-histidine tagged acyl CoA synthetase is eluted using a step gradient at 250 mM imidazole. On the basis of HPLC profiles, this enzyme preparation is >95% pure.

Using the long-chain fatty acid analogue, 9-*p*-azidophenoxy nonanoic acid (APNA)[43,44], the His-tagged enzyme has been covalently modified in an effort to specifically define amino acid residues within the fatty acid binding pocket. ^{3}H-APNA labeling is specific and can be blocked by increasing concentrations of palmitic acid. From this work, the apparent K_D for APNA binding to the enzyme is estimated to be 12.5 μM. The modified enzyme has been purified using HPLC and subjected to limited proteolysis using trypsin. Although we have identified a labeled peptide, the data is too premature to assign the specific region of the enzyme that is labeled with APNA. Our prediction is the peptide that is labeled will include amino acid residues that are common to the family of acyl CoA synthetases (see below).

SITE-DIRECTED MUTAGENESIS OF *FADD* IDENTIFIES THE FATTY ACID BINDING DOMAIN OF ACYL COA SYNTHETASE

As noted above, we and others have demonstrated that the *E. coli* acyl CoA synthetase shares considerable similarities to other acyl CoA synthetases and more broadly, the super family of adenylate-forming enzymes[21,23]. A twenty-five amino acid segment within the *E. coli* enzyme has features common to all of the acyl CoA synthetases for which sequence information is available (Figure 1). This region contains three invariant glycine residues (at positions 2, 7, and 15) that suggests a common pattern of folding. Therefore, it is reasonable to predict that this region of all acyl CoA synthetases adopts a similar tertiary structure. The bacterial enzyme contains an aspartic acid residue at position while the others contain an asparagine indicating the presence of the carboxylate is not crucial for activity. Our mutagenesis data on this region of the enzyme bear this out. This region contains five residues in addition to the three glycine residues that are invariant in the family acyl CoA synthetases: Trp (at position 3), Asp (at position 8), Asp (at position 22), Arg (at position 23) and Lys (at position 25). The residue in the fourth position is hydrophobic and is either a leucine or phenylalanine. Interestingly, only the phenylalanine-containing enzymes are found in yeast. The tyrosine in the fifth position is invariant except in the case of the yeast FAA4. This region of the enzyme contains hydrophobic residues (either leucine, isoleucine or valine) at positions 9, 17, 19, and 20. We predict that these residues in addition to the hydrophobic residues at positions 3, and 4 may comprise part of a fatty acid binding pocket. And [7] there is a preference for basic residues at positions 18 and 24 in addition to those at positions 22 and 25.

Using these similarities as a guide, we have generated a series of site-directed mutations within *fadD* followed by a sequence of analyses that have led us to propose that this region of the acyl CoA synthetase is essential for catalytic activity and may function in fatty acid binding. We initially focused our attention on the conserved charged residues within this region of the enzyme. Alanine substitutions were generated to Asn^{431}, Asp^{438}, Asp^{452}, Arg^{453}, Lys^{454}, and Lys^{455}. Our analyses indicated that these substitutions, except N431A, reduced the growth rate on oleate minimal plates which translated into lowered oleoyl CoA activities (*ca.* 5 % wild-type levels). Somewhat surprising, and of particular interest to us was the finding that the alanine substitutions in Lys^{454} and Lys^{455} resulted in increasing the growth rate on decanoate when compared to the *fadD*+ *fadR* strain. The increased capacity of the *fadD fadR* strain harboring either the *fadD*K454A or *fadD*K455A

Table 1. Acyl CoA synthetase activities using decanoate (C_{10}) and oleate ($C_{18:1}$) as substrates in the Δ*fadD* strain RD2 harboring plasmids containing specific *fadD* mutations.

Mutation	Acyl CoA Synthetase Activity±S.E.M. (N=3) (nmol/min/mg protein)	
	C10:0	C18:1
N431A	11.63 (1.51)	10.47 (0.95)
G432A	5.30 (0.22)	4.96 (0.30)
W433A	0.06 (0.06)	0.14 (0.07)
L434A	0.66 (0.10)	1.29 (0.25)
H435A	7.59 (0.18)	6.41 (0.51)
T436A	0.35 (0.18)	0.49 (0.11)
G437A	5.40 (0.09)	4.29 (0.54)
D438A	0.52 (0.17)	0.14 (0.07)
I439A	0.96 (0.17)	3.37 (0.29)
G446A	1.23 (0.03)	0.34 (0.02)
L448A	0.40 (0.06)	3.57 (0.20)
R449A	4.90 (0.36)	10.47 (0.95)
I450A	3.78 (1.08)	3.53 (0.88)
V451A	16.90 (0.31)	5.93 (0.09)
D452A	11.43 (0.65)	1.09 (0.26)
R453A	0.00 (0.00)	0.23 (0.03)
K454A	14.37 (0.19)	4.47 (0.18)
K455A	7.62 (0.60)	0.85 (0.20)
Wild-Type	13.24 (0.78)	16.10 (0.75)
*fadD*30	0.69 (0.22)	0.12 (0.05)

mutations on plasmids to grow on decanoate translated into higher levels of decanoyl CoA synthetase activity relative to oleoyl CoA activities (Table 1). A second *fadD* allele (D452A) also resulted in high decanoyl CoA synthetase activity relative to oleoyl CoA. Interestingly, the alanine substitution in Arg^{453} resulted in eliminating acyl CoA synthetase activity on both substrates. We chose to generate an additional series of alanine-scanning, site-directed *fadD* mutations within this region to include Gly^{432}, Trp^{433}, Leu^{434}, His^{435}, Thr^{436}, Gly^{437}, Ile^{439}, Gly^{446}, Leu^{448}, Arg^{449}, Ile^{450} and Val^{451}. Substitution of Gly^{432} or Gly^{437} with Ala lowered acyl CoA synthetase activities but did not alter the specificity profiles. If these glycine residues are important in maintaining a specific structural element within this region of the enzyme, substitutions with alanine were insufficient to disrupt structure and severely alter activity. In addition, the substitution of His^{435} with Ala also resulted in acyl CoA synthetase profiles while different than wild-type were consistent with the notion that this residue was not essential to the activity of the enzyme. Substitutions in Trp^{433}, Leu^{434}, and Tyr^{436} significantly decreased acyl CoA synthetase activities using both fatty acid substrates and thus are predicted to be essential for catalytic activity. Asp^{452}, Lys^{454} and Lys^{455} appear to be involved in determining fatty acid chain-length specificity. Likewise Val^{451} appears to contribute to fatty acid chain length specificity as the V451A allele, like the D452A, K455A, and K455A alleles has a preference for C_{10} substrates. These data are of particular interest when compared to the family of acyl CoA synthetases. Two enzymes, FAA2 from yeast and acyl CoA synthetase from *P. oleovarans* have specificity towards medium-chain fatty acids. These enzymes contain a valine or alanine, respectively, at the position equivalent to Lys^{454} in the *E. coli* enzyme in support of the notion that this region is involved in determining fatty acid chain-length specificity. While we favor the interpretation that this region of acyl CoA synthetase is involved in fatty acid binding, it is premature to assign a definitive function. Our working model is that the conserved hydrophobic amino acids, including Val^{451} comprise part of a fatty acid binding pocket. Our data argue that this fatty acid binding domain also requires Trp^{433}, Leu^{434}, and Tyr^{436}. The potential roles of the charged residues are less clear with respect to fatty acid binding, especially as substitutions of Asp^{452}, Lys^{454} and Lys^{455} confer differential fatty acid chain length specificity. Perhaps these charged residues maintain a fatty acid binding pocket that is altered in the alanine substitutions that results in a "better fit" for medium chain fatty acids as opposed to long-chain fatty acids. The aspartic acid residue at position 438 and the arginine at 453 appear to be crucial for function. Perhaps these two residues are crucial for the formation of the adenylate intermediate. Analyses of hexameric-histidine tagged mutant enzyme forms confirm the data generated in whole cell extracts. To date, our data show that purified enzymes containing the D452A and K455A substitutions have high decanoyl CoA synthetase activity relative to oleoyl CoA synthetase activity. In addition purified enzymes from the *fadD*W433A and *fadD*D438A alleles have no acyl CoA synthetase activity using either substrate arguing that Trp^{433} and Asp^{438} are essential for catalytic activity.

PROSPECTIVE - LONG-CHAIN FATTY ACID TRANSPORT AND ACYL COA SYNTHETASE

Our interests in the acyl CoA synthetase of *E. coli* originate from its role in the transport and activation of exogenous long-chain fatty acids. This enzyme is predicted to interact with the membrane to bind fatty acids during its catalytic cycle. This membrane association may simply involve binding fatty acids that are dissolved in the membrane or may involve a transient interaction with a postulated H^+/fatty acid cotransporter. In mouse adipocytes and the yeast *Saccharomyces cerevisiae* there appears to be isoforms of acyl CoA synthetase that function in the transport of exogenous fatty acids[6,34,35]. In adipocytes, acyl CoA synthetase is predicted to act in concert with a fatty acid transport protein (FATP) to facilitate the uptake of long-chain fatty acids[6]. As has been documented in *E. coli*, the esterification of the imported fatty acid in adipocytes prevents fatty acid efflux making this transport process unidirectional. The H^+/fatty acid binding protein identified from membrane preparations of *E. coli* may represent the functional counterpart of FATP[28,29]. The *E. coli* acyl CoA synthetase has been shown to partition into the inner membrane under certain metabolic conditions[30] Therefore, it is plausible that this enzyme interacts with a specific fatty acid transport protein within the bacterial membrane to facilitate long-chain fatty acid transport. If there exists such a protein, its association with acyl CoA synthetase would be temporal and simply act to deliver free fatty acids to the enzyme for activation.

A thorough understanding of the role of acyl CoA synthetase in *E. coli* will establish not only its role in the uptake of long-chain fatty acids, but also define its role outside of long-chain fatty acid transport. We have recently demonstrated that long-chain acyl CoA pools accumulate in strains defective in β-oxidation when grown in the presence of acetate arguing that acyl CoA pools are formed from *de novo* synthesized fatty acids. Other studies have demonstrated that in stationary phase cultures of *E. coli* the enzymes of β-oxidation are induced inferring that sufficient pools of long-chain acyl CoA are present which in turn bind FadR and derepress the fatty acid degradative genes. As in the case where β-oxidation is blocked by mutation in acetate grown cells, the increased acyl CoA pools must necessarily be generated through acyl CoA synthetase. The central question that is unresolved at this point is the source of the free fatty acids that serve as substrates for the *E. coli* acyl CoA synthetase under these two conditions.

Exogenous long-chain fatty acids are important sources of metabolic energy and along with long-chain fatty acyl CoAs represent important effector molecules that are involved in a wide variety of cellular processes. In the context of long-chain fatty acid transport, the interaction between the *E. coli* acyl CoA synthetase and other components of the fatty acid transport apparatus is of great interest. Cardiac muscle uses long-chain fatty acids as the preferred energy source in the production of ATP. Therefore central to cardiac physiology is the uptake and activation of long-chain fatty acids. Our work describing the process of fatty acid uptake in a system that can be genetically manipulated has allowed us (and will continue to allow us) to specifically identify regions within the *E. coli* acyl CoA synthetase that are functionally important and will elucidate the role of this enzyme in the transport of long-chain fatty acids across the membrane.

ACKNOWLEDGMENTS

The author wishes to acknowledge members of his laboratory, Qing Zhang and James D. Weimar, and collaborators, Concetta C. DiRusso and Jens Knudsen, who have contributed to this work. This work is supported by grants from the National Science Foundation and the American Heart Association. P. N. Black is an Established Investigator of the American Heart Association.

REFERENCES

1. Harmon, C. M., Luce, P., Beth, A. H. and Abumrad, N. A. Labeling of adipocyte membranes by sulfo-*N*-succinimidyl derivatives of long-chain fatty acids: Inhibition of fatty acid transport, *J. Membr. Biol.* 121: 261 (1991).
2. Abumrad, N. A., el-Maghrabi, M. R., Amri, E. Z., Lopez, E., and Grimaldi, P. A. Cloning of a rat adipocyte membrane protein implicated in binding or transport of long-chain fatty acids that is induced during preadipocyte differentiation. Homology with human CD36, *J. Biol. Chem.* 268:17665 (1993).
3. Schwieterman, W., Sorrentino, D., Potter, B. J., Rand, J., Kiang, C. L., Stump, D. and Berk, P. D. Uptake of oleate by isolated rat adipocytes is mediated by a 40-kDa plasma membrane fatty acid binding protein closely related to that in liver and gut, *Proc. Natl. Acad. Sci USA* 85: 359 (1988).
4. Stump, D. D., Nunes, R. M., Sorrentino, D., Isola, L. M., and Berk, P. D. Characteristics of oleate binding to liver plasma membranes and its uptake by isolated hepatocytes, *J. Hepatol.* 16: 301 (1992).
5. Potter, B. J., Sorrentino, D., and Berk, P. D. Mechanism of Cellular Uptake of Fatty Acids, *Ann. Rev. Nutr.* 9:253 (1989).
6. Schaffer, J. E. and Lodish, H. L. Expression cloning and characterization of a novel adipocyte long chain fatty acid transport protein, *Cell* 79:427 (1994).
7. Klein, K., Steinberg, R., Feither, B. and Overath, P. Fatty acid degradation in *Escherichia coli*: an inducible system for the uptake of fatty acids and further characterization of *old* mutants. *Eur. J. Biochem.* 19:442 (1971).
8. Frerman, F. E. and Bennett, W. Studies on the uptake of fatty acids by *Escherichia coli*, *Arch. Biochem. Biophys.* 159:434 (1973).
9. Nunn, W. D. and Simons, R. W. Transport of long-chain fatty acid by *Escherichia coli*: Mapping and characterization of mutants in the *fadL* gene, *Proc. Natl. Acad. Sci. USA* 75:3377 (1978).
10. Maloy, S. R., Ginsburgh, C. L., Simons R. W. and Nunn, W. D. Transport of long and medium chain fatty acids by *Escherichia coli* K12. *J. Biol. Chem.* 256:3735 (1981).
11. Nunn, W. D., Colburn, R. W.and Black, P. N. Transport of long-chain fatty acids in *Escherichia coli*. Evidence for role of *fadL* gene product as long-chain fatty acid receptor, *J. Biol. Chem.* 26:167 (1986).
12. Ginsburgh, C. L., Black, P. N. and Nunn, W. D. Transport of long-chain fatty acids in *Escherichia coli*. Identification of a membrane protein associated with the *fadL* gene, *J. Biol. Chem.* 259:8437 (1984).
13. Kameda, K., and Nunn, W. D. Purification and characterization of acyl coenzyme A synthetase from *Escherichia coli*, *J. Biol. Chem.* 256:5702 (1981).
14. Black, P. N., Said, B., Ghosn, C. R., Beach, J. V. and Nunn, W. D. Purification and characterization of an outer membrane-bound protein involved in long-chain fatty acid transport in *Escherichia coli*, *J. Biol. Chem.* 262:1412 (1987).
15. Black, P. N., Kianian, S. F., DiRusso, C. C. and Nunn, W. D. Long-chain fatty acid transport in *Escherichia coli*. Cloning, mapping, and expression of the *fadL* gene, *J. Biol. Chem.* 26:1780 (1985).
16. Black, P. N. The *fadL* gene product of *Escherichia coli* is an outer membrane protein required for uptake of long-chain fatty acids and involved in sensitivity to bacteriophage T2, *J. Bacteriol.* 170:2850 (1988).
17. Black, P. N. Characterization of *fadL*-specific fatty acid binding in *Escherichia coli*, *Biochim. Biophys. Acta* 1046:97 (1990).
18. Black, P. N. Primary sequence of the *Escherichia coli fadL* gene encoding an outer membrane protein required for long-chain fatty acid transport, *J. Bacteriol.* 173:435 (1991).
19. Kumar, G. B. and Black, P. N. Linker mutagenesis of a bacterial fatty acid transport protein, *J. Biol. Chem.* 266:1348 (1991).
20. Kumar, G. B., and Black, P. N. Bacterial long-chain fatty acid transport: Identification of amino acid residues within the outer membrane protein FadL required for activity, *J. Biol. Chem.* 268:15469 (1993).

21. Black, P. N., DiRusso, C. C., Metzger, A. K. and Heimert, T. L. Cloning, sequencing, and expression of the *fadD* gene of *Escherichia coli* encoding acyl coenzyme A synthetase, *J. Biol. Chem.* 267:25513 (1992).
22. Black, P. N. Physical map location of the *Escherichia coli* gene encoding acyl coenzyme A synthetase, *J. Bacteriol.* 174:7875 (1992).
23. Fluda, M., Heinz, E., and Wolter, F. P. The *fadD* gene of *Escherichia coli* K12 is located close to rnd at 39.6 min on the chromosomal map and is a new member of the AMP-binding protein family, *Mol. Gen. Genet.* 242:241 (1994).
24. Black, P. N. and DiRusso, C. C. Molecular and biochemical analyses of fatty acid transport, metabolism, and gene regulation in *Escherichia coli*, *Biochim. Biophys. Acta* 1210:123 (1994).
25. Azizan, A., and Black, P. N. The use of transposon Tn*phoA* to identify genes for cell envelope proteins of *Escherichia coli* required for long-chain fatty acid transport. The periplasmic protein Tsp potentiates long-chain fatty acid transport, *J. Bacteriol.* 176:665 (1994).
26. Black, P. N. and Zhang, Q. Evidence that His^{110} of FadL in the outer membrane of *Escherichia coli* is involved in the binding and uptake of long-chain fatty acids: possible role of this residue in carboxylate binding,*Biochem.J.* 310:389 (1995).
27. Groot, P. H. E., Scholte, H. R. and Hulsmann, W. C. Fatty acid activation: Specificity, localization and function, *Adv. Lipid Res.* 14:75 (1976).
28. Kameda, K. Partial purification and characterization of fatty acid binding protein(s) in *Escherichia coli* membranes and reconstitution of fatty acid transport system, *Biochem. Intl.* 13:343 (1986).
29. Kameda, K., Suzuki, L. K. and Imai, Y. Transport of fatty acid is obligatory coupled with H^+ entry in spheroplasts of *Escherichia coli* K12, *Biochem. Intl.* 14:227 (1987).
30. Mangroo, D. and Gerber, G. E. Fatty acid uptake in *Escherichia coli*: Regulation by recruitment of fatty acyl-CoA synthetase to the plasma membrane, *Biochem. Cell Biol.* 71:51 (1993).
31. Azizan, A. and Black, P. N. Energetics of long-chain fatty acid uptake in *Escherichia coli,* Submitted.
32. Davidson, B. C. and Cantrill, R. C. Rat hepatocyte plasma membrane acyl:CoA synthetase activity, *Lipids* 21:571 (1986).
33. Neufeld, E. J., Bross, T. E. and Majerus, P. W. A mutant HSDM1C1 fibrosarcoma line selected for defective precursor uptake lacks arachidonate-specific acyl CoA synthetase, *J. Biol. Chem.* 259:1986 (1984).
34. Duronio, R. J., Knoll, L. J. and Gordon, J. I. Isolation of a *Saccharomyces cerevisiae* long-chain fatty acyl:CoA synthetase gene (FAA1) and assessment of its role in protein N-myristoylation, *J. Cell Biol.* 117:531 (1992).
35. Johnson, D. R., Knoll, L. J., Levin, D. E., and Gordon, J. I. *Saccharomyces cerevisiae* contains four fatty acid activation (FAA) genes: An assessment of their role in regulating protein N-myristoylation and cellular lipid metabolism, *J. Cell Biol.* 127:751 (1994).
36. DiRusso, C. C., Heimert, T. L., and Metzger, A. K. Characterization of FadR, a global transcriptional regulator of fatty acid metabolism in *Escherichia coli.* Interaction with the *fadB* promoter is prevented by long-chain acyl-CoAs, *J. Biol .Chem.* 267:8685 (1992).
37. DiRusso, C. C., Metzger, A. K. and Heimert, T. L. Regulation of transcription of genes required for fatty acid transport and unsaturated fatty acid biosynthesis in *Escherichia coli* by FadR, *Mol. Microbiol.* 7:311 (1993).
38. Raman, N. and DiRusso, C. C. Analysis of acyl coenzymeA binding to the transcription factor FadR and identification of amino acid residues in the carboxyl terminus required for ligand binding, *J Biol Chem.* 270:1092 (1995).
39. Heath, R. J. and Rock, C.O. Regulation of malonyl-CoA metabolism by acyl-acyl carrier protein and β-ketoacyl-acyl carrier protein synthases in *Escherichia coli*, *J. Biol.Chem.* 270:15531 (1995).
40. Lau, C., Greenway, D. L. A., Freter, C. E., Neungton, N., Wattenberg, B. and Silbert, D. F. Conditions influencing formation of 16:0/16:0 molecular species in membrane phospholipids of *Escherichia coli*, *J. Biol .Chem.* 258:13027 (1983).
41. Greenway, D. L. A. and Silbert, D. F. Altered acyltransferase activity in *Escherichia coli* associated with mutations in acyl-coenzyme A synthetase, *J. Biol . Chem.* 258:13034 (1983).
42. Black, P. N., Knudsen, J. and DiRusso, C. C. Multi-Level Regulation of *fadD* Encoding Acyl-CoA Synthetase in *Escherichia coli.* Submitted.
43. Pederson S. O.Anvendelse af fotoaffinitetsmærkning til undersøgelse af fedtsyreoptagelse og intracellulær fedtsyretransport i 3T3-L1 muse fibroblaster. M.Sc. Thesis (Specialerapport), Biokemisk Institut, Odense Universitet, Danmark (1992) .
44. Jørgensen, T. K. Analysis of proteins involved in the cellular uptake and transport of fatty acids using a photactivatable fatty acid analog. M.Sc. Thesis (Specialerapport), Biokemisk Institut, Odense Universitet, Danmark (1992).

THE ROLE OF LONG CHAIN ACYL-COA ESTERS IN THE REGULATION OF GENE EXPRESSION AND METABOLISM IN *ESCHERICHIA COLI*

Concetta C. DiRusso

Department of Biochemistry
University of Tennessee
Memphis, TN 38163

INTRODUCTION

The pathways of lipid synthesis and degradation are co-ordinately regulated in most cells at the level of transcription. A number of studies have been directed at elucidating a role for acyl-CoA compounds in the regulation of the activity of lipid-responsive transcription factors[24,25,28,32]. However, there has been only indirect evidence that acyl-CoA compounds regulate eucaryotic gene activity by interaction with transcription factors[1,7,8,15,36]. *Escherichia coli* FadR is the only transcription factor for which there is substantial and convincing evidence that direct binding of LCACoA to the protein prevents DNA binding, transcription activation or repression[11,12,17]. Among the genes directly regulated by FadR are those encoding a specific membrane associated fatty acid transport protein (FadL), acyl-CoA synthetase, all the enzymes required for the ß-oxidation of fatty acids, two enzymes essential for unsaturated fatty acid biosynthesis and a protein of undetermined function which is expressed during a number of stress reponses[2,14]. In this article we review our current understanding of FadR-dependent gene regulation and regulation of FadR by long chain acyl-CoA.

CHARACTERIZATION OF FadR-DEPENDENT TRANSCRIPTIONAL REGULATION AND DNA BINDING

E. coli can utilize long- chain fatty acids (C14-C18) or phosphatidylserine as a sole carbon and energy source[21,22]. When grown on long chain fatty acids, all of the genes encoding proteins required for transport, ß-oxidation and growth on fatty acids are co-ordinately induced (fatty acid degradative genes, *fad*). Medium chain fatty acids can not induce *fad* genes and can only act as carbon and energy sources when *E. coli* are deficient in FadR. In contrast, at least two genes required for unsaturated fatty acid biosynthesis, *fabA* and *fabB*, have reduced expression in response to fatty acids and are positively controlled by FadR. Negative regulation of *fadL, fadD and fadBA* and positive regulation of *fabA* as been extensively characterized *in vivo* using gene fusions to the *lacZ* gene[3,13,18]. A list of the genes currently known to be regulated by FadR are given in Table 1. Our current preliminary evidence suggests that the FadR regulon is extensive and includes additional genes to those listed in table 1. For example putative FadR binding sites have also been identified upstream of the acyl carrier protein gene[2] and the *fabI* gene encoding enoyl-ACP reductase (DiRusso, unpublished). In addition to these genes required for fatty acid biosynthesis, the genes for at least two catabolic enzymes, 2,4,-dienoyl-CoA reductase and long chain acyl-CoA hydratase, have not been cloned or their promoters characterized.

Table 1. FadR-Responsive Genes Currently identified in *Esherichia coli*

Gene		Map Position[a] Reference	Protein Product
fadL	52	Outer membrane long-chain fatty acid transporter	(5)
fadD	40	Fatty acyl CoA synthetase	(31)
fadE	7	Electron transferring flavoprotein	(2)
fadF	5	Long-chain acyl CoA dehydrogenase	(2)
fadG	5	Short-chain acyl CoA dehydrogenase	(11)
fadB	80	Short-chain enoyl CoA hydratase	
		3-hydroxyacyl CoA dehydrogenase	
		3-hydroxyacyl CoA epimerase	
		Δ^3 *cis*, Δ^2 *trans* enoyl CoA isomerase	
fadA	80	3-ketoacyl CoA thiolase (thiolase I)	(11)
fadH	72	2,4, dienoyl CoA reductase	(44)
fadR	25	Tanscriptional regulator	(14)
fabA	21	ß-OH -Decanoyl thioester dehydrase/isomerase	(13)
fabB	52	ß-Keto-acyl-ACP synthase I	(20)
uspA	78	Universal Stress Protein A	(30)

The *fadR* gene was cloned and sequenced and the 239 amino acid residue protein was purified and characterized by the DiRusso laboratory[10-12]. Using native polyacrylamide gel electrophoresis and glycerol density gradient electrophoresis we have recently determined that the protein is a homodimer dimer in solution[35]. At this time, we have identified nine FadR binding site by DNaseI footprinting analyses in the promoter regions of seven FadR-responsive genes (Table 2). We used purified FadR and DNA fragments containing the FadR-responsive promoters in standard gel shift assays to determine an equilibrium binding constant (Table 2). Additionally, *in vitro* transcription assays demonstrate that FadR exerts its effect at the level of transcription by directly binding to DNA to repress the expression of the *fad* genes and to activate the expression of *fabA*[13]. Maximal *in vitro* transcription of the *fabA* promoter requires only FadR and *E. coli* RNA polymerase. Maximal transcription *in vitro* of *fadBA* requires cAMP and Crp in addition to RNA polymerase. Purified *fadR* inhibits basal and cAMP-CRP stimulated transcription of *fadBA*. Transcription of FadL is independent of cAMP-CRP both *in vivo* and *in vitro* [13].

Table 2. FadR binding sites identified by DNaseI footprinting and equilibrium constants determined using protein-DNA gel shift assays.

Gene Name	DNA Sequence	Common With fadB	Common With fabA	K_{eq} X 10^{-9}M
fadB	A T C T G G T A C G A C C A G A T_{+17}	(17)	10	0.2
fadL1	A G C T G G T C C G A C C T A T A_{-9}	11	11	0.4
fadL2	C A C T G G T C T G A T T T C T A_{+16}	7	11	
fadD1	A G C T G G T A T G A T G A G T T_{-29}	12	9	1.0
fadD2	G G C T G G T C C G C T G T T T C_{-115}	7	8	0.5
fadE	A A G T G G T C A G A C C T C C T	10	11	not tested
fabA	A A C T G A T C G G A C T T G T T_{-31}	10	(17)	0.6
fabB	G G C T G A T C G G A C T T G T T_{-31}	9	15	10
uspA	A G C T G G C C A G T C A T C G A	7	8	50
CONSENSUS	A N C T G G T C N G A C N T N T T			

ANALYSES OF ACYL-CoA BINDING

De-repression of genes negatively controlled by FadR and de-activation of genes positively controlled by FadR are chain length specific[12,21]. A concentration of 1mM C_{14}-C_{18} fatty acids fully de-represses the FadR-responsive *fad* genes while medium chain fatty acids such as C_{10} are essentially ineffective[12]. It was suggested in early work by Peter Overath and co-workers that acyl-CoA were the specific ligands which bind and inactivate FadR because mutants in *fadD* encoding acyl-CoA synthetase were not derepressed when long chain fatty acids were supplied exogenously[31,43]. However, it was subsequently determined that strains carrying mutations in *fadD* were defective in transport of long chain fatty acids[5]. Additionally, the substrate specificity of the *E. coli* acyl-CoA synthetase includes medium as well as long chain fatty acids, yet medium chain fatty acids do not increase *fad* enzyme activity[31]. Therefore it was not clear that the products of the synthetase were the ligands of the transcriptional regulator.

The first evidence that acyl-CoA compounds were the natural ligands of FadR came from analysis of FadR-DNA binding using the standard gel shift assay[12]. In these experiments we showed that binding of purified FadR to DNA containing the FadR binding site within *fadB* is specifically prevented by nanomolar concentrations of C_{16}-CoA or C_{18}-CoA but C_{10}-CoA and free fatty acids are only effective at micromolar concentrations[12]. The high amounts required for the medium chain acyl-CoA compounds and fatty acids are attributed to non-specific detergent effects. Binding of long chain acyl-CoA to FadR results in the alteration of the apparent pI of the protein from 6.9 to 6.2 while decanoyl-CoA, free coenzyme A and fatty acids have no effect on pI[34]. Purified FadR is also specifically labelled with two different long chain acyl-CoA ligands, 2-azido [^{32}P]-palmitoyl CoA[39] and [^{3}H]-9-*p*-azidophenoxy-nonanoyl-CoA (Tszvetnitsky, DiRusso and Knudsen, unpublished).

IDENTIFICATION OF AMINO ACID RESIDUES IN FadR SPECIFICALLY REQUIRED FOR ACYL-CoA BINDING

As a genetic approach to identify amino acid residues required for acyl-CoA binding, we selected super-repressor mutations in plasmid encoded FadR gene, called $FadR^S$ [34,35]. Cells carrying $FadR^S$ are unable to use long or medium chain fatty acids as a carbon and energy source. In contrast, cells which carry wild type FadR can use long but not medium chain fatty acids while cells carrying mutations in FadR can grow on medium as well as long chain fatty acids. The super-repressor phenotype is expected to result from mutations that: decrease affinity for the effector; increase DNA binding affinity; or inhibit the allosteric transition which occurs upon ligand binding and that is required to prevent DNA binding. One stable $FadR^S$ allele was identified after hydroxylamine mutagenesis of plasmid DNA. The new allele had a single amino acid change of Ser219 to Asn ($FadR_{S219N}$). This amino acid substitution has no apparent affect on repression or activation of transcription as assayed using *fadB-lacZ* as a reporter for repression and *fabA-lacZ* as a reporter for activation. $FadR_{S219N}$ was purified and it was determined that the protein retained the ability to bind DNA with a slightly reduced affinity as compared to the wild-type protein, so an increase in DNA binding affinity does not account for the altered phenotype. Purified $FadR_{S219N}$ protein has a reduced affinity for long chain acyl-coenzyme A as determined by isoelectric focussing. We also estimated by quenching of intrinsic fluorescence that the K_d of the wild-type protein for oleoyl-CoA is 12nM while $FadR_{S219N}$ is 257 nM. Therefore, it was concluded that the super-repressor phenotype is specifically due to a reduced affinity of FadR for long chain acyl-CoA.

The previous experiments suggest that Ser219 in FadR is specifically required for binding acyl-CoA. Therefore we hypothesized that additional amino acids in this region of FadR would also be required for binding the ligand[35]. To test this prediction, we generated substitution of amino acid residues 215 through 230 with alanine and assayed each for the ability to repress ß-oxidation of [^{14}C]oleate *in vivo*. Those altered alleles which were functional (*i.e.* had low basal levels of ß-oxidation) were then tested for de-repression by long chain fatty acids. These experiments identified Gly216, Glu218, Trp223 and Lys228 in addition to Ser219 as required for maximal de-repression by long chain fatty acids. This region of FadR shares amino acid identities and similarities with the coenzyme A binding site of *Clostridium thermoaceticum* CO dehydrogenase/acetyl-coenzyme A synthase (CODH)[29]. Most notably Trp223 in FadR which, when substituted with alanine has a strong FadR^S phenotype, is identical to Trp418 in CODH. Trp418 was protected from chemical modification by coenzyme A thus identifying this Trp residue as part of the CoA binding site within the enzyme. Due to the alteration in binding affinity of the purified S219N protein, the non-inducible phenotype of several proteins carrying alanine substitutions and similarities to CODH we propose this region of FadR forms part of the acyl-CoA binding domain and in particular is required for binding the CoA moiety.

ACYL-COA ESTERS AS REGULATORY INTERMEDIATES IN *E. COLI* LIPID METABOLISM

We have developed a model of FadR-dependent regulation of gene expression which hypothesizes that FadR senses the long chain acyl-CoA concentrations within the cell[2]. Inhibition of FadR-dependent DNA binding by acyl-CoA results in a cascade of changes in gene expression. The regulation of these genes encoding proteins of fatty acid metabolism in *E. coli* by FadR allows the cell to slow the rate of fatty acid biosynthesis and rapidly increase the rate of fatty acid degradation under conditions of fatty acid excess. Correspondingly, when alternative carbon sources are available, FadR potentiates fatty acid biosynthesis to allow maximal growth and division in response to the nutritional and environmental stimuli. This model emphasizes a regulatory role for acyl-CoA in cellular metabolism.

There have been several studies estimating the coenzyme A pools in *E. coli* [26,41]. These were limited in that they focused primarily on estimating CoASH, acetyl-CoA and succinyl-CoA. Coenzyme A thioesters with acyl chain lengths of C6 or longer were not evaluated since the higher molecular weight, more hydrophobic compounds were in low relative concentration and would not be detected by the chromatographic methods used. Additionally, the cells used in those studies were grown in either complex media or glucose minimal media, growth conditions which are expected to result in low intracellular long chain acyl-CoA concentrations because the gene encoding acyl-CoA synthetase is subject to catabolite repression and repression by FadR[3,4]. Therefore we have quantified acyl-CoA levels in *E. coli* using a recently modified technique which involves the extraction of long-chain acyl-CoA esters by trichloroacetic acid treatment of whole cells. The long-chain acyl-CoA esters were hydrolyzed to free coenzyme A under alkaline conditions and the coenzyme A levels quantified by an enzymatic cycling assay[37]. The results presented below in table 3, demonstrated that *E. coli* grown under a variety of conditions accumulate long-chain acyl-coenzyme A esters.
Strain RD1 which is deficient in acyl-CoA synthetase did not accumulate these compounds while strain K19 which is deficient in acyl-CoA dehydrogenase activity had levels 2- to 5-fold

Table 3. Estimation of Long-chain acyl coenzyme A pools in *Escherichia coli*.

Strain	Genotype	Enzymatic Defect[1]	Carbon Source[2]	Acyl-CoA pmol/5X10^8 cells
K12	Wild Type	None	Acetate	5.3
			Oleate	46.9
RW11	*fadR*	None	Acetate	3.4
			Oleate	30.5
K19	*fadR fadE*	ACD	Acetate	17.8
RD1	fadR fadD	ACS	Acetate	[0.7][3]

[1]ACD, acyl-CoA dehydrogenase; ACS, acyl-CoA synthetase
[2]*E. coli* cells were grown to 0.4-0.5 OD_{600} in Vogels' minimal medium E containing the carbon source indicated. Long chain acyl-CoA was extracted from cells using 1ml 0.6N TCA. The values presented are the averages of 2 samples per experiment for each strain.
[3]This value is at the limits of detection for this assay. In a separate experiment it was determined that acyl-CoA levels for RD1 did not increase with an increase in cell number while for RW11, acyl-CoA values increased linearly with cell number. Therefore we concluded that RD1 does not accumulate long chain acyl-CoA esters as expected due to the deficiency in acyl-CoA synthetase activity.

higher than wild-type or *fadR* strains. The elevated acyl-CoA levels observed in the *fadE* strains correlated with a de-repression of *fad* gene activity indicating that the accumulation of acyl-CoA was sufficent to inhibit FadR-DNA binding.

The apparent accumulation of long chain acyl-CoA in *fadE* strains in the absence of exogenously supplied fatty acids also indicates that the acyl moiety must be derived either from *de novo* synthesis or turnover of membrane lipids since *E. coli* has no storage form of lipid. In *E. coli* the primary products of fatty acid biosynthesis are $C_{16:0}$, $C_{16:1}$ and $C_{18:1}$ thioesterified to acyl carrier protein (ACP). The acyl-ACPs rather than acyl-CoA are believed to serve as substrates for phospholipid and lipoprotein biosyntheses[33]. It has been previously demonstrated that when fatty acid biosynthesis is uncoupled from phospholipid biosynthesis and ß-oxidation, long chain acyl-ACP accumulates[16]. Our work suggests, that the acyl-CoA synthetase may also function in the natural state to control fatty acid levels by activating "excess" fatty acids derived both exogenously and by *de novo* synthesis. It is also probable, as suggested by the earlier work of Silbert and co-workers, that there is a direct interaction between the acyl-CoA synthetase and acyltransferase which results in modification of membrane phospholipids by acyl-CoA under some growth conditions[17,23].

FADR-DEPENDENT REGULATION OF THE UNIVERSAL STRESS PROTEIN

The Universal Stress Protein A, UspA, is a 13.5 kDa *E. coli* protein whose synthesis is induced under most stress conditions with the exception of cold shock[30]. The exact function of UspA is unknown but deletion of the gene reduces the viablilty of cells during stress response. Transcription of UspA is repressed by FadR and induced in a growth phase dependent manner[15]. Cells carrying $FadR_{S219N}$, which has a reduced affinity for long-chain acyl-CoA as detailed above, have a reduced response to growth phase-dependent expression of UspA. The effect of *fadR* mutations on UspA expression suggests that UspA may have a role in fatty acid metabolism. Since UspA expression is not induced during cold shock, we hypothesize that the specific effect of UspA with regards to fatty acid metabolism, is to regulate thermal stability of the membrane perhaps by decreasing the ratio of unsaturated fatty acids to saturated fatty acids, an effect which would be detrimental to survival as a response to cold shock. However, a role for UspA in fatty acid metabolism has yet to be clearly demonstrated. It is also possible that UspA is not specifically involved in fatty acid metabolism. If correct, the latter possibility would point to a role for FadR in processes other than fatty acid metabolism.

GROWTH RATE-DEPENDENT CONTROL OF THE FAD AND FAB GENES

We recently observed that in addition to *uspA*, expression of *fadB-lacZ*, *fadL-lacZ*, and *fadD-lacZ* became de-repressed in stationary phase in the absence of exogenously added fatty acids[15]. In contrast, ß-galactosidase levels decreased during stationary phase in a *fabA-lacZ* strain. The growth-phase dependent increased and decreased transcription of *uspA*, the *fad* genes and *fabA*, respectively, was dependent on *fadR* and consistant with the conclusion that acyl-CoA concentrations increase in stationary phase. This was evidenced by the fact that cells carrying the non-inducible *fadR* allele $FadR_{S219N}$ which has a reduced affinity for LCACoA exhibited a weak stationary phase induction of *uspA* and *fad* genes and survived long-term stasis poorly. These observations suggest that the FadR repressor is inactivated during entry of cells into stationary phase, perhaps as a result of increasing acyl-CoA pools within the cell, and that inactivation of FadR provides a selective advantage necessary to survival.

Apart from a role in stasis survival detailed above, expression of the *fad* genes appears to be critical to pathogenesis. Using a newly devised technique called *in vivo* expression technology (IVET), Mekalanos and co-workers identified promoters specifically expressed during the infection of BALB/c mice by *Salmonella typhimurium*[27]. Among the gene fusions identified, was a fusion to *fadB*. This suggests that catabolism of fatty acids is advantageous and perhaps required during the infection process.

CONCLUSIONS AND PROSPECTIVES

These recent results demonstrating the importance of FadR-dependent gene expression in survival and pathogenesis emphasize the role of FadR and acyl-CoA as a mediators of cellular homeostasis. Our current and future work investigates the mechanism of acyl-CoA interaction with FadR through structural studies and regulation of gene expression by FadR using classical and molecular genetic techniques.

ACKNOWLEDGEMENTS

The author would like to thank laboratory members Narayan Raman, Helen Jordan and Qing Zhang and collaborators Paul Black, Thomas Nyström, Jens Knudsen, and Earl Shrago each of whom contributed substantially to this work. CCD was supported by a Research Career Advancement Award from the National Science Foundation (MCB-9407220).

1. Amri, E-Z., Aihaud, G. and Grimaldi, P-A. (1994) Fatty acids as signal transducing molecules: Involvement in the differentiation of preadipose to adipose cells. *Journal of Lipid Research* 35: 930-937.
2. Black, P.N. and DiRusso (1994) Molecular and biochemical analyses of fatty acid transport, metabolism and gene regulation in *Escherichia coli*. *Biochimica Biophysica Acta* 1210: 123-145.
3. Black,P.N., Knudsen, J. and DiRusso, C.C. Multi-Level Regulation of *fadD* encoding Acyl-CoA Synthetase in *Escherichia coli*, In Submission.
4. Black, P.N., DiRusso, C.C., Metzger, A.K. and Heimert, T.L. (1992) Cloning, sequencing, and expression of the *fadD* gene of *Escherichia coli* encoding acyl coenzyme A synthetase. *Journal of Biological Chemistry* 267:25513-25520.
5. Black, P.N., Said, B., Ghosn, C.R., Beach, J.V. and Nunn, W.D. (1987) Purification and characterization of an outer membrane-bound protein involved in fatty acid transport in *Escherichia coli*. *Journal of Biological Chemistry* 262: 1412-1419.
6. Chirala, S.S. (1992) Coordinated regulation and inositol-mediated and fatty acid-mediated repression of fatty acid synthase genes in *Saccharomyces cerevisiae*. *Proceedings of the National Academy of Sciences* USA 89: 10232-10236.
7. Clarke, S.D. and Jump, D.B. (1994) Dietary unsaturated fatty acid regulation of gene transcription. *Annual Review of Nutrition* 14: 83-98.
8. Cornelius, P., MacDougal, O.A. and Lane, M.D. (1994) Regulation of adipocyte development. *Annual Review of Nutrition* 14: 99-129.
9. Heath, R.J. and Rock, C.O. (1995) Ragulation of Malonyl-CoA Metabolism by acyl-acyl carrier protein and ß-ketoacyl-acyl carrier protein synthases in *Escherichia coli*. Journal of Biological Chemistry 270: 15531-15538.
10. DiRusso, C.C. (1988) Nucleotide sequence of the *fadR* gene, a multifunctional regulator of fatty acid metabolism in *Escherichia coli*. *Nucleic Acids Research* 16: 7995-8009.
11. DiRusso, C.C. (1990) Primary sequence of the *Escherichia coli fadBA* operon encoding the fatty acid oxidizing multienzyme complex, indicates a high degree of homology to eucaryotic enzymes. *Journal of Bacteriology* 172: 6459-6468.
12. DiRusso, C.C., Heimert, T.L. and Metzger, A.K. (1992) Characterization of FadR, a global transcriptional regulator of fatty acid metabolism in *Escherichia coli*: Interaction with the *fadB* promoter is prevented by long chain fatty acyl coenzyme As. *Journal of Biological Chemistry* 267: 8685-8691.
13. DiRusso, C.C., Heimert, T.L. and Metzger, A.K. (1993) Regulation of transcription of genes required for fatty acid transport and fatty acid biosynthesis in *Escherichia coli*. *Molecular Microbiology* 7: 311-322.
14. DiRusso, C.C. and Nunn, W.D. (1985) Cloning and characterization of a gene (*fadR*) Involved in regulation of fatty acid metabolism in *Escherichia coli*. *Journal of Bacteriology* 161: 583-588.
15. Farewell, A., Diez, A.A., DiRusso, C.C., and Nyström, T. Role of the *Escherichia coli* FadR regulator in stasis survival and growth phase-dependent expression of the *uspA*, *fad* and *fab* genes. In submission.

16. Heath, R.J. and Rock, C.O. (1995) Regulation of malonyl-CoA metabolism by acyl-acyl carrier protein and ß-ketoacyl-acyl carrier protein synthases in *Escherichia coli*. *Journal of Biological Chemistry* 270: 15531-15538.
17. Greenway, D.L.A. and Silbert, D.F. (1983) Altered acyltransferase activity in *Escherichia coli* associated with mutations in acyl-coenzyme A synthetase. *Journal of Biological Chemistry* 258: 13034-13042
18. Henry, M.F. and Cronan, J.E., Jr. (1992) A new mechanism of transcriptional regulation: Release of an activator triggered by a small molecule binding. *Cell* 70: 671-679.
19. Joseleau-Petit, D., Kepes, F. and Kepes, A. (1984) Cyclic changes in the rate of phospholipid synthesis during synchronous growth of *Escherichia coli*. *Journal of Biochemistry* 139: 605-611.
20. Kauppinen, S., Siggard-Andersen, M. and von Wettstein-Knowles, P. (1988) ß-Ketoacyl-ACP synthase I of *Escherichia coli*: Nucleotide sequence of the *fabB* gene and identification of the cerulenin binding residue. *Carlsberg Research Communication* 53: 357-370.
21. Klein, K., Steinberg, R., Fiethen, B. and Overath, P. (1971) Fatty acid degradation in *Escherichia coli:* an inducible system for the uptake of fatty acids and further characterization of *old* mutants. *European Journal of Biochemistry* 19: 442-450.
22. Kriven, H.C., Franklin, D.P., Wang, W., Laux, D.C. and Cohen, P.S. (1992) Phosphatidyl serine found in intestinal mucus serves as a sole source of carbon and nitrogen for *Salmonellae* and *Eshcherichia coli*. *Infection and Immunity* 60: 3943-3946.
23. Lau, C., Greenway, D.L.A., Freter, C.E., Neungton, N., Wattenberg, B., and Silbert, D.F. (1983) Conditions influencing formation of 16:0/16:0 molecular species in membrane phospholipids of *Escherichia coli*. *J Biol Chem* 258: 13027-13033.
24. Li, Q., Yamamoto, N., Inoue, A. and Morisawa, S. (1993) Fatty acyl-CoA binding activity of the nuclear thyroid receptor. *Journal of Cellular Biochemistry* 51: 458-464.
25. Li, Q., Yamamoto, N., Morisawa, S. and Inoue, A. (1990) Fatty acyl-CoAs are potent inhibitors of the nuclear thyroid receptor *in vitro*. *Journal of Biochemistry* 107: 699-702.
26. Magnuson, K., Jackowski, S., Rock, C.O., and Cronan, J.E.,Jr. (1993) Regulation of fatty acid biosynthesis in *Escherichia coli*. Microbiological Reviews 57: 522-542.
27. Mahan, M.J., Tobias, J.W., Slauch, J.M., Hanna, P.C., Collier, J.R. and Mekalanos, J.J. (1995) Antibiotic based selection for bacterial genes that are specifically induced during infection of a host. *Proceedings of the National Academy of Sciences U.S.A.* 92: 669-673.
28. McDonough, V.M., Stuckey, J.E. and Martin, C.E. (1992) Specificity of unsaturated fatty acid-regulated expression of the *Saccharomyces cerevisiae* OLE1 gene. *Journal of Biological Chemistry* 267: 5931-5936.
29. Morton, T.A., Runquist, J.A., Ragsdale, S.W., Shanmugasundaram, T., Wood, H.G. and Ljungdahl, L.G. (1991) The primary structure of the subunits of carbon monoxide dehydrogenase/acetyl CoA synthase from *Clostridium thermoaceticum*. *Journal of Biological Chemistry* 266: 23824-23828.
30. Nyström, T. and Neidhart, F.C. (1992) Cloning, mapping and nucleotide sequence of a gene encoding a universal stress protein in *Escherichia coli*. Molecular Microbiology 6: 3187-3198.
31. Overath, P., Pauli, G. and Schairer, H.U. (1969) Fatty acid degradation in *Escherichia coli*. An inducible acyl-CoA synthetase, the mapping of *old* mutants, and isolation of regulatory mutants. *European Journal of Biochemistry* 7: 559-574.
32. Powell, G.L., Tippett, P.S., Kiorpes, T.C., McMillin-Wood, J., Coll, K.E., Schulz, H., Tanaka, K. and Kang, E. (1985) Fatty acyl-CoA as an effector molecule in metabolism. *Federation Proceedings* 44:81-84.
33. Raitz, C.R.H. (1986) Molecular genetics of membrane phospholipid synthesis. *Annual Review of Genetics* 20:253-295.
34. Raman, N. and DiRusso, C.C. (1995) Analysis of acyl coenzyme A binding to the transcription factor FadR and identification of amino acid residues required for ligand binding. *Journal of Biological Chemistry* 270: 1092-1097.

35. Raman, N. (1996) Characterization of long chain acyl-CoA interaction with the transcription factor FadR in *Escherichia coli* Doctoral Dissertation, University of Tennessee, Memphis, TN.
36. Roncero, C., and Goodridge, A.G. (1992) Hexanoate and octanoate inhibit transcription of the malic enzyme and fatty acid synthase genes in chick embryo hepatocytes in culture. *Journal of Biological Chemistry* 267: 14918-14927.
37. Rosendal, J. and Knudsen, J. (1992) A fast and versatile method for extraction and quantitation of long-chain acyl-CoA esters from tissue: Content of individual long-chain acyl-CoA esters in various tissues from fed rat. *Analytical Biochemistry* 207: 63-67.
38. Roszak, D.B. and Colwell, R.R. (1987) Survival strategies of bacteria in the natural environment. *Microbiological Reviews* 51: 367-379.
39. Shrago, E., Woldegiorgis,G., Ruoho, A.E. and DiRusso, C.C. (1995) Fatty acids as regulators of cell metabolism. *Prostaglandins, Leukotrienes and Essential Fatty Acids* 52: 163-166.
40. Spydevold, O. and Bremer, J. (1989) Induction of peroxisomal ß-oxidation in 7800 C1 Morris hepatoma cells in steady state by fatty acids and fatty acid analogues. *Biochemica Biophysica Acta* 1003: 72-79.
41. Vallari, D.S., Jackowski, S. and Rock, C.C. (1987) Regulation of pantothenate kinase by coenzyme A and its thioesters. *Journal of Biological Chemistry* 262: 2468-2471.
42. Vanderwinkel, E., De Vliegher, M., Fontaine, M., Charles, D., Denamur, F., Vandervoorde, D. and De Kegel, D. (1976) Septation deficiency and phospholipid perturbation in *Escherichia coli* genetically constitutive for the beta-oxidation pathway. *Journal of Bacteriology* 127: 1389-1399.
43. Weeks, G., Shapiro, M., Burns, R.O., and Wakil, S.J. (1969) Control of fatty acid metabolism I. Induction of enzymes of fatty acid oxidation in *Escherichia coli. Journal of Bacteriology* 97: 827-836.
44. You, S-H., Cosloy, S. and Schulz, H. (1989) Evidence for Function of 2,4-dienoyl-coenzyme A reductase in the ß-oxidation of unsaturated fatty acids *in vivo. Journal of Biological Chemistry* 264: 16489-16495.

CONTROL OF EICOSANOID RESPONSE INTENSITY

William E.M. Lands

National Institute on Alcohol Abuse and Alcoholism
NIH, 6000 Executive Blvd., MSC 7003
Bethesda, MD, USA 20892-7003

INTRODUCTION: DIVERSITY IN RESPONSE

Tissues do not have eicosanoids continuously present, but form them rapidly from essential fatty acids in response to stimuli. The intensity of an eicosanoid-mediated response depends upon accumulating sufficient eicosanoid at a receptor site to induce the tissue response. An intense tissue response requires accumulating sufficient eicosanoid by a rapid burst of synthesis to overcome the continual rapid diffusional dilution and catabolic removal of the active autacoid. That rapid synthesis depends, in turn, upon an essential fatty acid substrate accessing the oxygenase catalytic site during a time of sufficient peroxide tone (reviewed by Lands, 1979).

Tissues have diverse essential fatty acid substrates, diverse oxygenases, and diverse peroxide activators, all of which influence the intensity of the response to a stimulus. In promoting such responses, inflammatory cells release extra-cellular hydrogen peroxide (HOOH), which activates prostaglandin H synthase (PGHS-1) to form more rapidly the intra-cellular auto-activating hydroperoxide activator, PGG, and thereby enhance the rate of synthesis (Lands, 1979). Also, cytokines activate transcription and synthesis of the isoform, PGHS-2, which has a 10-fold more sensitive response to activating lipid hydroperoxide than does PGHS-1 (reviewed by R.J. Kulmacz in this book), making amplification of intra-cellular hydroperoxides less limiting as the proportion of PGHS-2 to PGHS-1 increases. All of the fatty acid oxygenases that form eicosanoids (both PGHS isoforms and the lipoxygenases) can act with both n-3 and n-6 types of essential fatty acid, forming a diverse set of eicosanoids that mediate a body's reaction (and over-reaction) to various stimuli. This report focuses upon the manner in which the diverse set of essential fatty acids maintained in tissues influence the intensity of eicosanoid responses, and addresses the biochemical and clinical consequences of this often-neglected diversity.

Some serious chronic disorders have excessive signalling by n-6 eicosanoids which can be diminished by dietary n-3 fatty acids (Lands, 1986; Lands et al., 1990a). This situation develops because the rate of eicosanoid synthesis catalyzed by oxygenases (e.g., Kulmacz et al., 1994) and the intensity of signalling at eicosanoid receptors (e.g., Needleman et al., 1979; Lee et al., 1984) is often less for n-3 than n-6 eicosanoids. Thus, although n-3 and n-6 essential fatty acids in the diet are similarly assimilated into and mobilized from tissue lipids, they differ in the intensity of formation and function of tissue eicosanoids. As a result, the n-3 acids are partial agonists that act as competitive antagonists of n-6 eicosanoid formation and function, and they diminish the intensity of eicosanoid-mediated responses.

This view of diversity in eicosanoid signalling helps interpret the different prevalence of some diseases among populations with diverse essential fatty acid intakes (Lands, 1986; Lands et al., 1990a), and it provides the basis for a nutritional strategy to diminish the incidence and severity of many chronic human diseases and disorders. For example, results from 10,000 participants at 55 centers in 24 countries (Yamori et al,1994) showed that stroke mortalities were positively related to 20:4n-6 in serum phospholipids ($r=0.575, p<.05$) and that coronary heart disease mortalities were inversely related with n-3 fatty acids in serum phospholipids ($r=-0.475$, $p<0.05$). Also, the incidence of rheumatoid arthritis in Japan is much less than predicted on the basis of HLA genotypes (Shichikawa et al., 1981), compatible with the influence of a lower proportion of n-6 eicosanoid precursors among the 20 and 22-carbon highly unsaturated fatty acids (HUFA) of tissue phospholipids. The proportion of plasma phospholipid HUFA as 20:3+20:4n-6 has a predictable relationship to the abundance of n-3 and n-6 acids as a percent of daily dietary energy (en%; Lands et al, 1992), as well as a relationship to the severity of n-6 eicosanoid-mediated pathophysiology among experimental animals and many different human populations.

Americans have about 75% of their plasma phospholipid HUFA as 20:3+20:4n-6 and an ischemic heart disease mortality rate about five times greater than that for Japanese (for whom the value is about 50%), whereas Greenlanders with a value near 20% have a negligible rate. In fact, Americans in the upper quintile of dietary n-3HUFA intake (averaging 0.66 g n-3HUFA/day compared to the population median of 0.05 g/day) showed a 40% lower relative risk of cardiovascular mortality (discussed in Lands, 1993). The Japanese remain as one of the largest populations maintaining diversity in their dietary essential fatty acids (averaging about 3 g n-3HUFA/day). The wide diversity in Japan for total dietary fat and its proportion as n-3 provides a valuable opportunity for significant clinical studies to teach the biomedical community the benefits of maintaining a diversity in dietary essential fatty acids (Lands, 1993). Unfortunately, the average Japanese diet is now changing rapidly, and the diversity is being lost (Lands et al., 1990a).

MAINTAINING DIVERSITY

The lipid metabolizing enzymes that maintain the essential eicosanoid precursors in tissue glycerolipids discriminate little between n-3 and n-6 acids (e.g., Lands et al., 1982), permitting both types of essential fatty acid to compete in forming lipid esters. This competition has been characterized in considerable detail (Mohrhauer and Holman, 1963b; Lands et al., 1990b). To facilitate better use of the thirty years of accumulated metabolic information in designing and interpreting new clinical trials with n-3 dietary fats, we developed equations that describe quantitatively the competitive interactions of essential fatty acids (Lands et al., 1990b; Lands et al, 1992). These empirical equations summarize many metabolic interactions in a way that predicts likely changes in fatty acid composition and that indicates the probable intensity of evoked eicosanoid responses. The equations employ the proportion of 20:3+20:4n-6 present among the 20 and 22-carbon highly unsaturated fatty acids (HUFA) of tissue phospholipids as a useful index of the intensity expected for eicosanoid-mediated responses. Extensive metabolic similarities between rats and humans permit transfer of many results from the more detailed intervention studies with rats (Lands et al., 1990b) to help design and interpret the more expensive and logistically difficult studies with humans (Lands et al., 1992).

$$\begin{array}{c}20{:}3+20{:}4n\text{-}6\\ \text{as \% Plipid HUFA}\end{array} = \frac{100}{1+\frac{PC6}{en\%P6}\left(1+\frac{en\%P3}{PC3}+\frac{en\%H3}{HI3}+\frac{en\%O}{CO}+\frac{en\%P6}{KS}\right)} + \frac{100}{1+\frac{HC6}{en\%H6}\left(1+\frac{en\%H3}{HC3}\right)}$$

Figure 1. Quantitative relationships among essential fatty acids in maintaining eicosanoid precursors in tissue phospholipids (Lands et al., 1992).

The constants fitting 18:2n-6 and 18:3n-3 intakes to the various proportions of n-3/n-6 HUFA in tissue lipids are analogous to Michaelis constants, reflecting relative efficacies in supporting the enzymatic events involved. Thus, values of about 0.05 en% for the constants fitting the major essential fatty acids (18:2n-6, PC_6=0.0441; and 18:3n-3, PC_3=0.0555) indicate that both essential fatty acids are comparable substrates for forming the elongated, desaturated HUFA derivatives. The values mean that a near-maximal response can be expected at 5 to 10 times that level (0.25-0.5 en%), a result confirmed long ago (Mohrhauer and Holman, 1963a; Mohrhauer and Holman, 1963b; Cuthbertson, 1976). Unfortunately, those results have not been effectively emphasized in recent decades (reviewed in Lands, 1991a and 1991b). The fact that 0.3-0.5 en% 18:2n-6 seems biologically adequate for rats or humans and that the value is similar for both 18:3n-3 and 18:2n-6 in rats (and probably other mammals) is clear evidence of a diverse supply of essential fatty acids. An unfortunate indifference to this diverse supply causes many nutritionists, clinicians and food marketers to miss the opportunity to distribute nutritional advice that probably could help diminish a wide variety of serious chronic eicosanoid-mediated disorders (reviewed in Lands, 1986).

As a result of the accumulated knowledge regarding lipid metabolism and the maintenance of fatty acid compositions in tissue lipids, estimates of the probable eicosanoid response intensity can now be facilitated with a computer spreadsheet containing imbedded equations and constants, so that entered dietary intake data for an individual will predict the probable proportions of n-6 precursors maintained in that person's tissue HUFA. Such estimates are useful in estimating the diet-related tendency of an individual for various thrombotic or inflammatory disorders mediated by n-6 eicosanoids. Alternatively, entering analytical data on the fatty acids in an individual's plasma phospholipids and triacylglycerols predicts the probable average daily intake of n-3/n-6 fatty acids by that individual. These calculations may give more precise and reliable values for the average daily dietary intakes of essential fatty acids than the results of extensive and expensive food questionnaires (Lands et al.,1992; Lands, 1995).

THE LOSS OF DIVERSITY

One striking result from our studies of individuals in Chicago (Lands et al., 1992) was evidence for very low intakes of n-3 essential fatty acids and the almost complete dominance of n-6 essential fatty acids. Because overall metabolic discrimination between the two types of essential fatty acid is slight, the loss of diversity was due primarily to food selections consciously or sub-consciously made by those individuals. The essential fatty acid status of these people resembled that seen with rats raised on commercial laboratory chow, a diet that gives near-maximal tissue contents of n-6 eicosanoid precursors so that adding more n-6 dietary acid has essentially no effect on tissue HUFA composition (Lands et al., 1990b).

As for all experimental animals housed by humans, laboratory rats can only eat what is determined by human choices in providing the food; choices that may be indifferent to the n-3/n-6 diversity in Nature. An interesting scholarly challenge lies in examining the various cultural and market forces that led Americans to move so far away from a balanced mixture in creating the average American diet that is currently ingested. Such a study is not trivial since it might help in predicting what forces could restore a more diverse and healthy future dietary balance.

Many vegetables contain similar amounts of 18:3n-3 and 18:2n-6 expressed as g/100 g food (18:3n-3/18:2n-6 ratios are: squash, 0.04/0.02; snap beans 0.04/0.02; pinto beans, 0.19/0.11; USDA tables), whereas seeds and grains contain much greater amounts of n-6 (corn, 0.02/0.54; squash seeds, 0.18/20.7), and the processed seed oils obtained from various seeds provide massive supplies of essential fatty acids - - far above the physiologically sufficient 0.3 en%. In fact, there are almost no foods available that contain less than 0.3 en% of essential fatty acid. Even corn starch used to provide "fat-free" carbohydrate in experimental diets contains enough essential fatty acid (about 0.4 en%) to prevent the development of an essential fatty acid deficiency (Christensen, 1975; Lands et al., 1990b). As a result of a readily sufficient supply of essential fatty acid in the human food chain, the more serious concern for public health lies in preventing ingestion of harmful excesses of essential fatty acids from being ingested. Harm associated with such excesses can be mediated by eicosanoids and be more easily manifested with high levels of dietary n-6 than n-3 fatty acids. Pathology is clearly evident following an over-expression of active oxygenated metabolites of vitamin D, and health advice has been issued consistently to warn against ingesting unnecessarily high doses of this essential nutrient. In the same manner, we recognize pathological events mediated by an over-expression of active oxygenated metabolites of the n-6 essential fatty acids, and some public warning to avoid ingesting unnecessarily high doses seems appropriate.

A dilemma evolves regarding whether scientists should recommend decreasing n-6 intakes to levels that are only 5 to 10 times that needed to give a half-maximal response or whether, in the face of certain resistance to loss of sales by n-6 marketers, scientists should recommend raising the intake of n-3 to attain a more balanced diversity among the essential fatty acids. Actions taken thirty years ago in a historic parallel to this dilemma suggest that adding n-3 acids may succeed more easily than decreasing n-6 acids. In the 1960's, there was an apparent public health need to decrease over-ingestion of saturated fats, but dairy and meat marketers resisted reductions in sales. Scientists then recommended greatly increased intakes of 18:2n-6 to increase dietary P/S ratios. This recommendation was vigorously supported by n-6 marketers, and dietary P/S ratios in the USA rose significantly. Now, with about 15-20% of body fat being 18:2n-6, and with 20 to 40% of human body mass being fat, most adult Americans have essential fatty acid reserves of more than 2 kg of 18:2n-6! This reserve supply seems more than sufficient, especially when coupled with the fact that nearly all foods available contain more than the 0.3 en% of essential fatty acid shown to prevent physiological signs of deficiency!! Gradually during the past decades, dietary recommendations have slowly shifted to emphasize replacing saturated fats with unsaturated fats, ingesting no more than 10 en% saturated fat, and even regarding monounsaturated fat (18:1n-9) to be as effective as polyunsaturated fat for displacing saturated fat from the diet. Now, there is a clear public health need to address the over-ingestion of the n-6 essential fatty acids (Lands, 1994).

Medical researchers attempt various interventions to decrease eicosanoid-mediated disorders, and Enders et al. (1995) reviewed current clinical efforts to use supplemental n-3 acids. Unfortunately, the review examined clinical consequences with little attention to the amount of dietary linoleate against which the n-3 supplements were competing. The current heavy "pressure" of dietary linoleate in the USA seems likely to diminish the success of such expensive and time-consuming clinical studies. Investment of increasingly scarce clinical research funds needs more careful prospective assessment of the consequence of an imprecisely determined balance in the essential fatty acids. In approaching the clinical problems of rheumatoid arthritis and coronary heart disease, Mantzioris et al. (1995) decreased the dietary linoleate supply to provide better intervention success; a simple tactic of replacing salad oils rich in n-6 with olive or flaxseed oil helps dietary n-3 HUFA give significant benefits. With further planning, much more might be done. The results in Figure 2 illustrate the dynamics of diversity which can be created and which must be handled deliberately and rationally if public funds are to be employed successfully and efficiently.

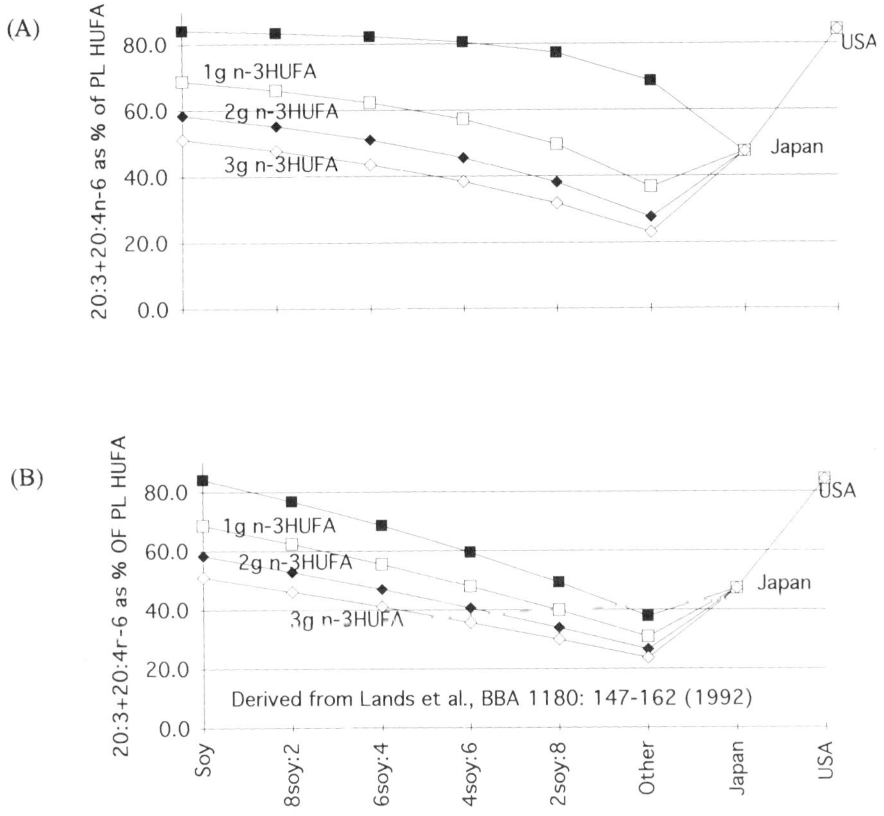

Figure 2. Predicted eicosanoid response intensity with shifted dietary oils. The ordinate value for 20:3+20:4n-6 as % of HUFA in serum phospholipids reflects the predicted eicosanoid response intensity. Average values for Japan and the USA are shown at the right of the figure. Shifting total dietary fatty acids equi-calorically from a soy-based pattern to a pattern resembling mixtures of other dietary oils is indicated on the abscissa, with 4soy:6 representing a mix of 40% soy and 60% other oil. (A) Shifting to a pattern like that of olive oil decreases the intensity value slightly, but the resultant low dietary 18:2n-6 makes the system more responsive to inclusion of dietary n-3 HUFA (such as are in seafoods). (B) Shifting to a flax-based pattern decreases the eicosanoid response intensity appreciably, making further response by the system to the inclusion of dietary n-3 HUFA less dramatic. Readers wishing more information on the spreadsheet for making these calculations can contact the author at the address above or at wlands@willco.niaaa.nih.gov.

Review of various past decisions helps clarify how Americans acquired their current status of little diversity in their dietary essential fatty acids, and some of these decisions need re-evaluation to decide whether or not they should be continued. For example, what rationale supports requiring that n-3 acids be regarded as essential only if they exhibit some unique physiological feature not seen with n-6 fatty acids? Also, what rationale supports the claim that the mere presence of some 20:3n-9 in tissue lipids is evidence of a physiological deficiency? Finally, what rationale supports regarding as desirable 80% of phospholipid HUFA in the form of n-6 acids while pharmaceutical and health care personnel are attempting to decrease the formation and function of many n-6 eicosanoids? Attention to these concerns could allow a more compassionate and effective community response to the occurrence of certain diseases and disorders in a large fraction of our population.

It hardly seems compassionate (or economical) to feed only n-6 fatty acids to rainbow trout, who then suffer fatal traumatic shock when disturbed (Castell et al., 1972). In contrast to exhibiting the normal response seen with (n-3)-rich diets, trout have a vigorous, fatal response to stress after being fed high proportions of n-6 fatty acids. That species, like many marine animals, evolved in an environment with a comparatively low n-6 essential fatty acid dietary supply. Now, trout hatchery operators control fully the essential fatty acids provided to "domestic" trout. Their decisions regarding the proportions of n-3/n-6 essential acids to be fed have important health consequences for the fish as well as for the humans who include the fish and their essential fatty acids in their diet.

It hardly seems compassionate to feed only n-6 fatty acids to companion animals (such as dogs and cats) who will then be kept through an increasingly debilitating old age filled with an increasing incidence of chronic inflammatory/thrombotic disorders. Such species have evolved in association with humans in an environment of diverse supplies of n-3/n-6 based upon the food chain available in human communities. Some tendency for an over-reaction with n-6 eicosanoids might have diminished during the centuries of human-directed selection pressure, but the onset of many disorders after the reproductive age probably left such tendencies undiminished.

It hardly seems compassionate to feed only n-6 fatty acids to institutionalized people or to family members who will be maintained through an increasingly debilitating old age filled with increased severity of chronic inflammatory/thrombotic disorders. Humans, themselves, also evolved in an environment that originally contained diverse essential fatty acids, and the current health statistics seem to indicate that retaining diversity among the various essential fatty acids may be healthier than destroying the diversity.

As we seek to regain the n-3/n-6 balance that was lost through ignorance and inattention, we should examine carefully the intended and unintended choices that brought us to restrict our current supply of n-3 essential fatty acid while ingesting massive amounts of the n-6 form. The current status may not be intended nor desirable for the future.

SUMMARY

Some serious chronic disorders have excessive signalling by n-6 eicosanoids which can be diminished by dietary n-3 fatty acids. Although n-3 and n-6 essential fatty acids are similar in their assimilation into and mobilization from tissue lipids, they differ in the intensity of formation and function of tissue eicosanoids. A polarized view of the role that the two types of essential fatty acids play in eicosanoid-linked events is: the n-3 acids can be moderate and cannot be vigorous, whereas the n-6 acids can be vigorous and cannot be moderate. As a result, the n-3 acids are partial agonists that act as competitive antagonists of n-6 eicosanoid formation and function, and they diminish the intensity of eicosanoid-mediated responses.

The proportion of plasma phospholipid HUFA as 20:3+20:4n-6 has a predictable relationship to the dietary abundance of n-3 and n-6 acids, and it has a demonstrable and rational relationship to the severity of n-6 eicosanoid-mediated pathophysiology among experimental animals and many different human populations. Americans have about 75% of their plasma phospholipid HUFA as 20:3+20:4n-6 and an ischemic heart disease mortality rate about five times greater than that for Japanese (for whom the value is about 50% HUFA), and Greenlanders with a value about 20% HUFA have a very low rate. The accumulated knowledge regarding maintenance of diverse fatty acids in tissue lipids can now be summarized in a computer spreadsheet containing the equations and appropriate constants, so that entered dietary intake data for an individual will predict the probable proportions of n-6 precursors maintained in that person's HUFA. Such estimates are useful in estimating the diet-related tendency of an individual for various thrombotic or inflammatory disorders mediated by n-6 eicosanoids. Alternatively, entering analytical data on the fatty acids in an individual's plasma phospholipids and triacylglycerols predicts the probable average daily intake of n-3/n-6 fatty acids by that individual. These calculations may give more precise and reliable values for the average daily dietary intakes of essential fatty acids than the results of extensive and expensive food questionnaires. Careful use of these calculations can ensure that planned dietary interventions attain the wide diversity that is possible with humans and obtain the statistically significant differences needed for credible conclusions.

Acknowledgements

Development of the equations and constants described in this review was supported by a Pfizer Biomedical Research Award. Scientific and editorial advice from Dr. R.J. Kulmacz was very helpful.

REFERENCES:

Castell, J.D., Sinnhuber, R.O., Wales, J.H., and Lee, D.J., 1972, Esential fatty acids in the diet of rainbow trout (Salmo gairdnerii): Growth, feed conversion, and some gross deficiency symptoms, J. Nutr. 102: 77-86.

Christensen, K., 1975, Determination of linoleic acid requirements in slayghter pigs. Report from the National Institute of Animal Science. Copenhagen, Denmark.

Cuthbertson, W.F.J., 1976, Essential fatty acid requirements in infancy, Am. J. Clin. Nutr. 29: 559-568.

Enders, S., R.DeCaterina, E.B. Schmidt and S.D. Kristensen, 1995, n-3 Polyunsaturated fatty acids: update 1995, Eur. J. Clin. Inv. 25: 629-638.

Kulmacz, R.J., Pendleton, R.B., and Lands, W.E.M., 1994, Interaction between peroxidase and cyclooxygenase activities in prostaglandin-endoperoxide synthase. Interpretation of reaction kinetics, J Biol Chem. 269: 5527-5536.

Lands, W.E.M., 1979, The Biosynthesis and Metabolism of Prostaglandins, Ann. Rev. Physiol. 41, 633-652.

Lands, W.E.M., 1986, Fish and Human Health, Academic Press, Orlando

Lands, W.E.M., 1991a, Polyunsaturated Fatty Acid Effects on Cellular Interactions, in Preventive Nutrition: The Role of Micronutrients in Health and Disease. (eds., A. Bendich and C.E. Butterworth) Marcel Dekker, Inc., N.Y.

Lands, W.E.M., 1991b, Biosynthesis of prostaglandins, Ann. Rev. Nutr. 11: 41-60.

Lands, W.E.M., 1992, Biochemistry and physiology of n-3 fatty acids, FASEB J. 6: 2530-2536.

Lands, W.E.M., 1993, Eicosanoids and health, Ann. N.Y. Acad. Sci. 676: 46-59.

Lands, W.E.M., 1994, State of the art: Where are we and where are we going?, Keynote lecture in the Proceedings from the American Heart Association Scientific Conference on Omega-3 Fatty Acids in Nutrition, Vascular Biology and Medicine. pp. 1-8.

Lands, W.E.M., 1995, Long-term fat intake and biomarkers, Am.J.Clin.Nutr. 61(suppl): 721S-725S.

Lands, W.E.M., Hamazaki, T., Yamazaki, K., Okuyama, H., Sakai, K., Goto, Y., and Hubbard, V.S., 1990a, Changing dietary patterns, Am. J. Clin. Nutr. 51: 991-993.

Lands, W.E.M., Inoue, M., Sugiura, Y., and Okuyama, H., 1982, Selective incorporation of polyunsaturated fatty acids into phosphatidylcholine by rat liver microsomes, J. Biol. Chem., 257, 14968-14972.

Lands, W.E.M., Libelt, B., Morris, A.J., Kramer, N.C., Prewitt, T.E., Bowen, P., Schmeisser, D. Davidson, M.H., and Burns, J. H., 1992, Maintenance of lower proportions of n-6 eicosanoid precursors in phospholipids of human plasma in response to added dietary n-3 fatty acids, Biochim. Biophys. Acta 1180: 147-162.

Lands, W.E.M., Morris, A.J., and Libelt, B., 1990b, The Influence of Dietary Polyunsaturated Fats on the Composition of Fatty Acids in Rat Tissues, Lipids, 25, 505-516.

Lee,T.H., Mencia-Huerta, J-M., Shih, C., Corey, E.J., Lewis, R.A., and Austen, K.F., 1984, Effects of exogenous arachidonic, eicosapentaenoic, and docosanoic acids on the generation of 5-lipoxygenase pathway products by ionophore-activated human neutrophils, J. lin. Invest. 74: 1922-1933.

Mantzioris, E., James, M.J., Gibson, R.A., and Cleland, L.G., 1995, Differences exist in the relationships between dietary linoleic and alpha-linolenic acids and their respective long-chain metabolites, Am. J. Clin. Nutr. 61: 320-324.

Mohrhauer, H. and R.T.Holman, 1963a, The effect of dose level of essential fatty acids upon fatty acid composition of the rat liver, J. Lipid Res. 4, 151-159.

Mohrhauer, H. and R.T.Holman, 1963b, Effect of linolenic acid upon the metabolism of linoleic acid, J. Nutr. 81: 67-74.

Needleman, P., Raz, A., Minkes, M., Ferrendelli, J.A., and Sprecher, H., 1979, Triene prostaglandins: prostacyclin and thromboxane biosynthesis and unique biological properties, Proc. Natl. Acad. Sci. USA 76: 944-948.

Shichikawa, K., Takenaka, Y., Maeda, N., Yoshino, R., Tsujimoto,M., Ota,H., Kashiwade,T. and Hongo, I., 1981, A longitudinal population survey of rheumatoid arthritis in a rural district in Wakayama, The Ryumachi 21(Suppl.) 35-43.

U.S. Department of Agriculture, 1984, Composition of Foods; Agriculture Handbooks No. 8-11 and 8-12. Superintendent of Documents, U.S. Government Printing Office, Washington, D.C. 20402

Yamori, Y., Nara, Y., Mizushima, S., Sawamura, M. and Horie, R., 1994, Nutritional factors for stroke and major cardiovascular diseases: international epidemiological comparison of dietary prevention, Statistics Canada, Health Reports 6(1): 22-27.

THE ROLE OF FATTY ACIDS WITHIN ENDOPLASMIC RETICULUM CALCIUM POOLS

Krystyna E. Rys-Sikora and Donald L. Gill

Department of Biochemistry and Molecular Biology
University of Maryland School of Medicine
Baltimore, Maryland 21201

INTRODUCTION

Fatty acids have a considerable range of effects on the movements and storage of Ca^{2+} within cells. These actions are dependent on a variety of factors including concentration, acyl chain length, degree of saturation, the presence of the carboxyl moiety, and orientation of double bonds.[1-6] Fatty acids, especially long-chain, unsaturated fatty acids such as arachidonic acid and linoleic acid, are effective in modulating cytosolic Ca^{2+} concentrations, $[Ca^{2+}{}_i]$, through diverse mechanisms including activation of Ca^{2+} influx[1-3] and alteration of Ca^{2+} efflux from Ca^{2+} stores.[4-8] Conversely, Ca^{2+} sequestration is *enhanced* mostly by saturated fatty acids. Using isolated sarcoplasmic reticulum vesicles, it was first reported that palmitic acid (C16:0) specifically promoted Ca^{2+} sequestration.[9] This action of palmitate appeared to be similar to that of other Ca^{2+}-complexing agents such as oxalate and phosphate used in earlier sarcoplasmic reticulum (SR) Ca^{2+} transport studies[10-12] which have continued to be in use in exploring the cytology of intracellular Ca^{2+} stores.[13]

Intracellular Ca^{2+} pools are essential to Ca^{2+} signaling events. It appears that endoplasmic reticulum (ER) or subcompartments thereof comprises a majority of Ca^{2+} pools and is considered to be the major Ca^{2+} storage organelle within cells.[14-16] Evidence indicates that intracellular Ca^{2+} pools are functionally heterogeneous; one or more of these pools are implicated in Ca^{2+} signaling events meditated by the second messenger, inositol 1,4,5,-trisphosphate ($InsP_3$).[14,15-19] In previous studies, we distinguished between $InsP_3$-sensitive and -insensitive Ca^{2+} pools.[15,17-21] Also, we have characterized a highly specific and sensitive GTP-dependent mechanism which appears to mediate the translocation of Ca^{2+} between calcium-sequestering organelles, permitting the movement of Ca^{2+} between $InsP_3$-sensitive and -insensitive pools. This GTP-mediated process is quite distinct from $InsP_3$-activated Ca^{2+} movements;[22,23] it requires the hydrolysis of GTP and is specific for guanine nucleotides.[24-26] Although little is known about the protein involved in this GTP-dependent process, it appears to reflect the action of the class of small GTP-binding proteins also called as monomeric G proteins, which have been identified as regulators of secretion and mediators of membrane trafficking in the ER and throughout the cell.[27-29] We have proposed a model in which GTP-activated Ca^{2+} movements involve the formation of a junctional communication between intracellular Ca^{2+} pools.[17,21,26,30,31] This model is subtly different from that of Dawson and coworkers[32,33] who made similar observations on the effects of GTP on Ca^{2+} movements in liver cells but considered the effects to reflect widespread fusion between ER vesicles. Our model appears confirmed by studying the effects of long chain fatty acyl-CoA esters on GTP-mediated translocation events using $^{45}Ca^{2+}$ flux assays.[34] This data revealed that the GTP-dependent Ca^{2+} translocation process could be completely *reversed* by long chain fatty acyl-CoA esters which appear to allosterically interact with a component of the junctional communication process. These results provide strong evidence that the GTP-activated translocation of Ca^{2+} was due to a pre-fusion event involving formation of a *reversible* communicating junction similar in function to pre-fusion pores that precede fusion events between membranes.[34]

In conjunction with these fatty acyl-CoA studies, the effects of fatty acids on GTP-mediated Ca^{2+} movements were also examined. In the present study, we report some important effects of fatty acids on GTP-mediated Ca^{2+} accumulation in permeabilized DDT_1MF-2 smooth muscle cells; this effect was highly chain length- and degree of saturation-specific. We provide evidence to suggest that palmitic acid can enter inside Ca^{2+} pools through an anion channel, wherein it complexes with Ca^{2+} stimulating Ca^{2+} sequestration in a manner different from that of oxalate.

MATERIALS AND METHODS

DDT_1MF-2 cells from hamster smooth muscle cell line, cultured in Dulbecco's modified Eagle's medium with 2.5% serum (Calf-Plus, Inovar, Gaithersburg, MD) and passaged every 7 days, were used in experiments after growing in culture for 4 days after passaging. Suspensions of cells (1×10^6 cells/ml) were incubated with 0.0025% saponin in intracellular-like media (ICM, comprising 140 mM KCl, 10 mM NaCl, 2.5 mM $MgCl_2$, 10 mM Hepes-KOH, pH 7.0) at 37°C until 95% permeabilization occurred. After permeabilization, cells were thoroughly washed and kept at 4°C for use in all experiments at a final cell concentration of 5×10^5 cells/ml.

Ca^{2+} flux measurements were conducted as described previously.[20,34] Briefly, accumulation of Ca^{2+} into intracellular organelles was measured using freshly permeabilized cells maintained gently stirring at 37°C in ICM containing 50 μM $CaCl_2$ (with 150 Ci/mol $^{45}Ca^{2+}$), EGTA to buffer the free calcium to exactly 0.1 μM, 3% polyethylene glycol (PEG), and 5 μM ruthenium red (to prevent mitochondrial Ca^{2+} accumulation), in a total volume of 2 ml. Effectors mentioned in the figure legends (GTP, $InsP_3$, A23187) were added at time shown by arrows. Fatty acids or oxalate, with or without GTP or DIDS (4,4'-diisothiocyanatostilbene-2,2'-disulfonic acid), when present were add ed prior to the start of the experiment. At required times, 200μl aliquots were removed from the stirring uptake medium, diluted immediately into 4 ml of ice-cold ICM containing 1 mM $LaCl_2$, then rapidly vacuum-filtered on glass fiber filters, washed, and counted. All figures show ATP-dependent Ca^{2+} uptake with that component of Ca^{2+} retained by cells and filters in the absence of ATP subtracted.

$InsP_3$ and A23187 were from Calbiochem. ATP, GTP, EGTA, PEG, saponin, ruthenium red, Hepes, DIDS, oxalate and all fatty acids were purchased from Sigma. Fatty acids were all dissolved in DMSO:MeOH (1:1) and stored at -20°C in capped vials containing N_2.

RESULTS AND DISCUSSION

The Ca^{2+}-precipitating anion, oxalate, has played an important role in discerning the mechanism by which GTP-activating Ca^{2+} movements result from formation of interorganelle junctions which allow the transfer of Ca^{2+} between distinct pools. As previously characterized[17,18,20,21,26,30,34] and as shown in Fig. 1A, in the absence of oxalate, a rapid release of Ca^{2+} accumulated by thapsigargin-sensitive Ca^{2+} pumping activity was observed upon addition of a maximally effective concentration of GTP (20 μM). GTP-dependent release of Ca^{2+} into the medium is thought to involve the formation of communicating junctions between intact and non-intact or leaky Ca^{2+}-accumulating organelles. However, in the presence of submaximal oxalate concentrations (less than 5 mM), GTP, instead of releasing Ca^{2+}, dramatically stimulates Ca^{2+} uptake. Oxalate is believed to enter one specific Ca^{2+} pool through a non-selective anion channel. ER and SR are thought to contain this anion transporter to allow the rapid transport of anions in order to dissipate charge buildup due to rapid uptake and release of Ca^{2+}.[35] Studies have shown that there are oxalate-impermeable and oxalate-permeable pools; only the latter pool can be released by $InsP_3$.[30] Upon entry, oxalate rapidly precipitates Ca^{2+}, forming a Ca^{2+} sink and greatly increasing the storage capacity of the pool for Ca^{2+}. Thus, as GTP activates translocation of Ca^{2+}, any Ca^{2+} pumped into an oxalate-impermeable pool is now allowed access to oxalate, causing the large rise in Ca^{2+} accumulation. As shown in Fig. 1B, addition of 100 μM palmitic acid prior to the start of Ca^{2+} uptake had only a small effect on Ca^{2+} accumulation, but induced a substantial stimulation of Ca^{2+} accumulation upon addition of GTP. This effect is very similar (although not identical) to that observed in the presence of 2-4 mM oxalate (see[20]). At longer times, palmitate alone did have had an effect on Ca^{2+} sequestration, an effect similar to that seen in sarcoplasmic reticulum vesicle preparations.[8,9] Thus, whereas there is a slow and modest effect of palmitate on Ca^{2+}accumulation within ER, this action is enormously potentiated by the action of GTP.

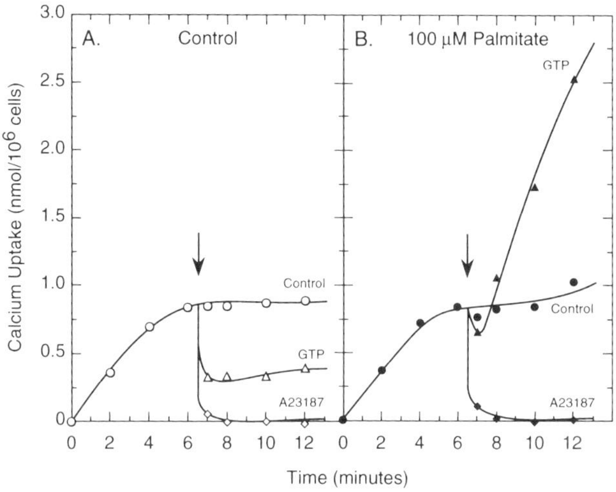

Figure 1. Action of palmitic acid on GTP-induced Ca^{2+} movements in ER. ATP-dependent uptake of Ca^{2+} into ER within permeabilized DDT_1MF-2 cells proceeded under standard conditions as described under Materials and Methods. **A.** Uptake was under either control conditions (○), or with 20 μM GTP (△) or 5 μM A23187 (◊) added at 6.5 min. **B.** Palmitic acid (100 μM) was added prior to the start of the experiment and control uptake (●) was monitored without further addition or with additions at 6.5 min of 20 μM GTP (▲) or 5 μM A23187 (◆). The experiment is representative of at least 10 similar experiments.

The EC_{50} for palmitate on Ca^{2+} accumulation in the presence or absence of GTP was approximately 25 μM (data not shown). However, even in this case, the EC_{50} for palmitate is at least 20-fold less than that we previously observed for oxalate.[15,30] A palmitate concentration of 100 μM was the observed maximally effective concentration; higher concentrations of palmitic did not further stimulate Ca^{2+} accumulation but exhibited a slight detergent-like effect presumably due to breakdown of the integrity of the ER membrane.

This effect of fatty acids on Ca^{2+} sequestration was highly specific in terms of chain length and degree of saturation (Fig. 2). In Fig. 2A, different chain lengths of saturated fatty acids were tested for their effects on Ca^{2+} accumulation in the absence and presence of GTP. Only palmitic acid (C16:0) and pentadecanoic acid (C15:0) substantially increased Ca^{2+} sequestration. In contrast, only slight effects were seen using myristic acid (C14:0) and heptadecanoic acid (C17:0). Higher chain lengths were progressively less effective. Also, no effect was observed with chain lengths lower than C14:0. All unsaturated fatty acids tested did not stimulate but instead inhibited GTP-dependent and control Ca^{2+} accumulation when added at concentrations greater than 10 μM (Fig. 2B); at 10 μM, this inhibitory effect on Ca^{2+} uptake activity was minimal. From these experiments it was clear that the action of fatty acids on Ca^{2+} sequestration in ER was highly specific for only two saturated fatty acids, palmitic acid and pentadecanoic acid.

The ensuing question was what determined the specificity of this action on Ca^{2+} sequestration.

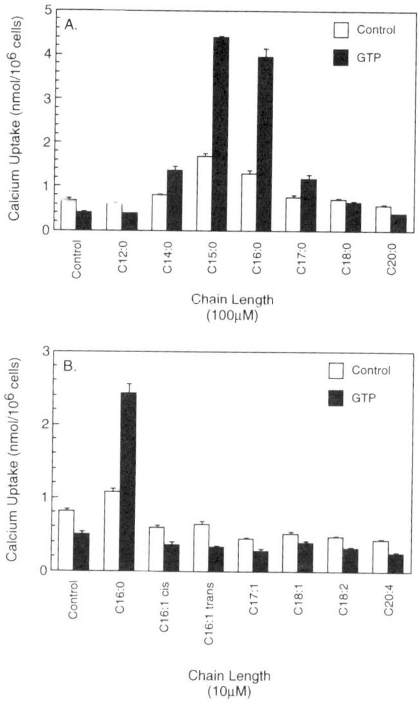

Figure 2. Specificity of chain-length and degree of saturation of fatty acid-mediated Ca^{2+} sequestration in ER. ATP-dependent Ca^{2+} accumulation was undertaken in the absence of GTP (*open bars*) or in the presence of 20 μM GTP (*closed bars*). Fatty acids and GTP were added just prior to the beginning of the experiment and Ca^{2+} loading was for 12 min, at which time three successive aliquots were rapidly taken. In panel A, saturated fatty acids tested were at 100 μM. The saturated fatty acids were: lauric acid (C12:0), myristic acid (C14:0), pentadecanoic acid (C15:0), palmitic acid (C16:0), heptadecanoic acid (C17:0), stearic acid (C18:0) and arachidic acid (C20:0). In panel B, unsaturated fatty acids were compared to palmitate (C16:0), and all were tested at 10 μM. The unsaturated fatty acids tested were: palmitoleic (C16:1 cis-9), palmitelaidic acid (C16:1 trans-9), heptadecanoic acid (C17:1, cis-10), oleic acid (C18:1, cis-9), linoleic acid (C18:2, cis-9,12), and arachidonic acid (C20:4, cis 5,8,11,14). Results are means ± SD of six determinations are representative of three identical experiments.

With respect to this question, important parameters to assess were the relative ability of fatty acids to bind Ca^{2+}, to interact with membranes, and to be transported into Ca^{2+} pools. Previous studies compared the action of a saturated fatty acid, palmitate, and an unsaturated fatty acid, oleic acid, on their ability to bind Ca^{2+},[37] to change membrane morphology of SR,[36] and to promote Ca^{2+} sequestration.[8,9] Fatty acids are capable of binding Ca^{2+} in solution[38] with an ideal ratio of 1 Ca^{2+} ion per 2 fatty acid molecules. Actual measurements showed palmitic acid binding approximately 0.4 mol of Ca^{2+}/mol of palmitate in the presence of 0.03-3 mM calcium.[37] Palmitate did not exhibit any effect on membrane morphology. In contrast, oleic acid disrupted membrane structures increasing membrane permeability and resulted in the inhibition of Ca^{2+} sequestration. Cis-unsaturated fatty acids like oleate are thought to partition into fluid domains of bilayers, whereas saturated fatty acids are considered less disrupting, preferentially partitioning in gel or solid-phase domains perhaps localizing their effects to particular sub-domains within membranes.[39] Overall, the Ca^{2+} binding activity of saturated fatty acids does not appear to be linked closely with chain-length,[37] instead, the extreme specificity of the effect of only C15:0 and C16:0 fatty acids on Ca^{2+} accumulation within ER more likely reflects specificity of the process by which fatty acids are transported into the ER.

As stated previously, oxalate is believed to enter the $InsP_3$-sensitive compartment of ER through a non-selective anion channel.[26,35] Several anion channel inhibitors including 4,4'-

diisothiocyanatostilbene-2, 2'-disulfonic acid (DIDS) have been shown to inhibit anion uptake into SR vesicles[40] and to inhibit anion-dependent Ca^{2+} uptake.[41] Our experiments reveal that oxalate-enhanced Ca^{2+} accumulation in the presence and absence of GTP was fully blocked by DIDS (Fig. 3A). Importantly, Ca^{2+} sequestration enhanced by the presence of palmitate was also completed inhibited by DIDS (Fig. 3B). This suggests that both oxalate and palmitate enter Ca^{2+} pools through similar ER-associated anion channels. Interestingly, plasma membrane fatty acid movements have been also shown to be inhibited by DIDS.[42,43] In adipocytes, the plasma membrane fatty acid transporter appears to exhibit some chain length preference.[42,43] The plasma membrane fatty acid transporter (FAT) has not only been characterized but also expressed in fibroblasts.[44] A definitive link between this activity and the one responsible for fatty acid entry into the ER has yet to be determined, and it will be important to compare the exact profile of fatty acid transport of both activities.

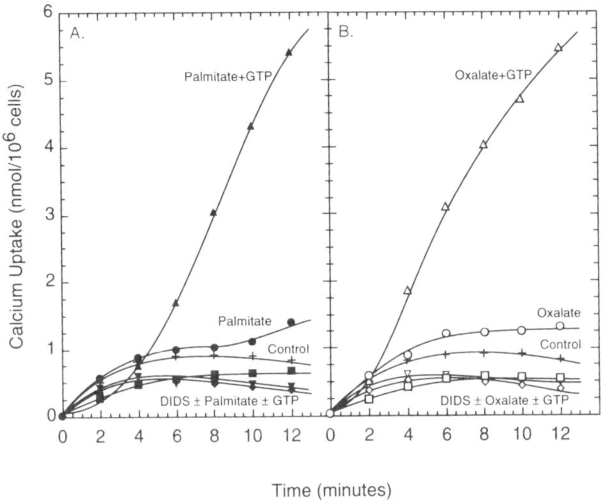

Figure 3. Action of the anion transport inhibitor, DIDS, on palmitate- and oxalate-mediated ER Ca^{2+} sequestration. Ca^{2+} accumulation was measured in the presence and absence of 100 μM palmitate (panel A) or 2.65 mM oxalate (panel B). All additions were made prior to the beginning of the experiment. In panel A, the conditions were as follows: control (+), palmitate alone (●), palmitate with 20 μM GTP(▲), 100 μM DIDS (◆), palmitate with 100 μM DIDS (▼) and palmitate with 100 μM DIDS and 20 μM GTP(■). In panel B, conditions were control (+), oxalate alone (○), oxalate with 20 μM GTP (△), 100 μM DIDS (◇), oxalate with 100 μM DIDS (▽), or oxalate with 100 μM DIDS and 20 μM GTP(□). Results are typical of five identical experiments.

Palmitate and oxalate may both enter Ca^{2+} pools through common anion channels, however, the mechanism by which they complex with Ca^{2+} may be subtly different. This was suggested by the results shown in Fig. 4. By using $InsP_3$ and the ionophore A23187, the releasability of Ca^{2+} in the presence of oxalate as opposed to palmitate were observed to be interestingly different. Upon addition of A23187, Ca^{2+} accumulated in the presence of palmitate and GTP was rapidly and completely released; however, only a small effect of $InsP_3$ was observed on Ca^{2+} accumulation under these conditions (Fig. 4A). Thus, $InsP_3$ appeared rather ineffective in releasing Ca^{2+} accumulated in the presence of palmitate and GTP. Oxalate is known to form easily observable insoluble Ca^{2+} precipitates within SR vesicles[45] which can redissolve upon dilution. As shown in Fig. 4B, A23187-induced release of Ca^{2+} was much slower than that under palmitate-induced loading conditions, whereas the actions of $InsP_3$ were if anything slight faster. Attempts by Herbette and co-workers, to visualize Ca^{2+}-palmitate structures were unsuccessful and it was concluded that Ca^{2+} bound to palmitate incorporated in the bilayer.[36] Considering this fact, it appears most likely that the Ca^{2+} ionophore, A23187, which works by incorporating into the bilayer, has enhanced effectiveness on palmitate-induced Ca^{2+} sequestration because of the close proximity of the sequestered Ca^{2+} to the

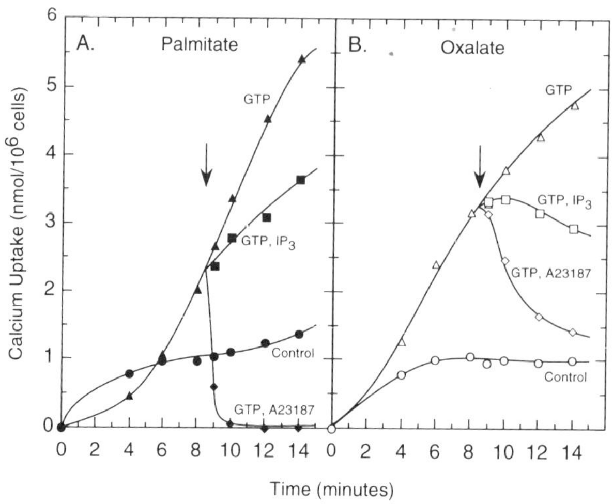

Figure 4. Relative effectiveness of ionophore A23187 and $InsP_3$ on releasing Ca^{2+} sequestered within ER in the presence of palmitate or oxalate. Ca^{2+} accumulation was examined in the presence of 100 μM palmitate (panel A) and 2.65 mM oxalate (panel B). GTP-dependent Ca^{2+} uptake was monitored for 6.5 min at which time either 10 μM $InsP_3$ or 5 μM A23187 was added. **Panel A:** the conditions were as follows: palmitate, alone (●), palmitate with 20 μM GTP (▲), palmitate with 20 μM GTP followed by 10 μM $InsP_3$ (■), and palmitate with 20 μM GTP followed by 5 μM A23187 (◆). **Panel B:** the conditions were: oxalate alone (○), oxalate with 20 μM GTP (△), oxalate with 20 μM GTP followed by 10 μM $InsP_3$ (□), and oxalate with 20 μM GTP followed by 5 μM A23187 (◊). Results are representative of five identical experiments.

membrane surface itself. In other words, the Ca^{2+} bound to palmitate may be selectively available to the action of membrane-bound ionophore. Similar reasoning may also explain the relative effectiveness of $InsP_3$ on palmitate- vs. oxalate-sequestered Ca^{2+}. In the former case, even though a significant amount of Ca^{2+} may be sequestered close to the membrane surface and hence to the mouth of $InsP_3$-sensitive Ca^{2+} channels, this Ca^{2+} may be very limited and rapidly depleted. In contrast from the action of ionophore, the channel may predominantly release Ca^{2+} that is free in solution, and therefore the release of Ca^{2+} from oxalate may be a more efficient means of delivering Ca^{2+} to the channel mouth. It is also clear that Ca^{2+} perhaps binding to a site in the vicinity of the luminal side of the $InsP_3$-sensitive Ca^{2+} channel may have a negative inhibitory effect on the action of the channel.[15] Thus, it is possible that Ca^{2+} bound to palmitate on the membrane surface may mediate or mimic this effect. Certainly, our results may be consistent with the idea that fatty acids incorporate in particular lipid domains in membrane bilayers,[39] and these may have important and differential effects on the membrane-located mechanisms that serve to transport Ca^{2+}.

A model depicting the actions of palmitate and oxalate and their effects on Ca^{2+} movements in the different pools of Ca^{2+} is shown in Fig. 5. The data presented here demonstrate that physiological concentrations of fatty acids promote intracellular Ca^{2+} sequestration in a manner which is highly chain length- and degree of saturation-specific. The transport of palmitate and its effect on Ca^{2+} sequestration appear to be similar to the effects of oxalate. However, the results examining releasability suggest subtle but fundamental differences between the state and possibly the location of oxalate- and palmitate-Ca^{2+} complexes. The membrane associated-formation of Ca^{2+}-fatty acid complexes and in turn, activation of Ca^{2+} accumulation may reflect a major physiological role for fatty acid in stabilizing Ca^{2+} within the lumen of Ca^{2+} pools as well as the means for controlling the activity of major Ca^{2+} transport mechanisms that contribute to the formation of Ca^{2+} signals.

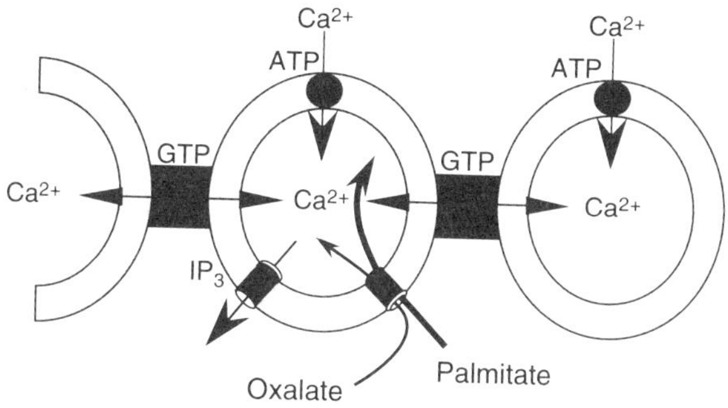

Figure 5. Model explaining the two effects of GTP on Ca^{2+} movements in ER (GTP-mediated Ca^{2+} release and GTP-activated Ca^{2+} uptake) in the absence and presence of palmitic acid or oxalate (see text for details). The DIDS-sensitive anion channel associated with the $InsP_3$-releasable Ca^{2+} pool allows influx of anions including oxalate or palmitate. Although not shown in this model, upon entering the pool, palmitic acid is believed to associate with the inner surface of the ER membrane where it complexes Ca^{2+}. In contrast, upon pool entry, oxalate reversibly precipitates Ca^{2+} and remains distributed within the ER lumen. Although, the state of Ca^{2+}-complexes and possibly its location appear to be different for palmitic acid and oxalate, both create a Ca^{2+} sink increasing the storage capacity for Ca^{2+} within these pools.

REFERENCES

1. Huang, J.M.-C., Xian, H. and Bacaner, M. Long-chain fatty acids activate calcium channels in ventricular myocytes, *Proc. Natl. Acad. Sci. USA.* 89: 6452 (1992).
2. Soliven, B., Takeda, M., Shandy, T. and Nelson, D.J. Arachidonic acid and its metabolites increase Ca_i in cultured rat oligodendrocytes, *Am. J. Physiol.* 264: C632 (1993).
3. Alonso, M.T., Sanchez, A. and Garcia-Sancho, J. Arachidonic acid-induced calcium influx in human platelets, *Biochem J.* 272: 435 (1990).
4. Cardoso, C.M. and de Meis, L. Modulation by fatty acids of Ca^{2+} fluxes in sarcoplasmic reticulum vesicles, *Biochem. J.* 296: 49 (1993).
5. Rustenbeck, I. and Lenzen, S. Effect of lysophospholipids, arachidonic acid and other fatty acids on regulation of Ca^{2+} transport in permeabilized pancreatic islets, *Cell Calcium.* 13: 193 (1992).
6. Cheah, A.M. Effect of long chain unsaturated fatty acids on the calcium transport of sarcoplasmic reticulum, *Biochim. Biophys. Acta.* 648: 113 (1981).
7. Katz, A.M., *et al.* Fatty acid effects on calcium influx and efflux in sarcoplasmic reticulum vesicles from rabbit skeletal muscle, *Biochim. Biophys. Acta.* 687: 17 (1982).
8. Messineo, F.C., Rathier, M., Favreau, C., Watras, J. and Takenaka, H. Mechanisms of fatty acid effects on sarcoplasmic reticulum III. The effects of palmitic acid and oleic acid on sarcoplasmic reticulum function-a model for fatty acid membrane interactions, *J. Biol. Chem.* 259: 1336 (1984).
9. Messineo, F.C., Pinto, P.B. and Katz, A.M. Palmitic acid enhances calcium sequestration by isolated sarcoplasmic reticulum, *J. Mol.Cell. Cardio.* 12: 725 (1980).
10. Hasselbach, W. and Makinose, M. ATP and active transport, *Biochem. Biophys. Res. Commun.* 7: 132 (1962).
11. Hasselbach, W. and Makinose, M. Uber den mechanismus des calciumtransportes reh die membranen des sarkoplasmatischen reticulums, *Biochem. Z.* 339: 94 (1963).
12. Martonosi, A. and Feretos, R. Sarcoplasmic reticulum: the uptake of Ca^{++} by sarcoplasmic reticulum fragments, *J. Biol. Chem.* 239: 648 (1964).
13. Pozzan, T., Rizzuto, R., Volpe, P. and Meldolesi, J. Molecular and cellular physiology of intracellular calcium stores, *Physiological Reviews.* 74: 595 (1994).
14. Berridge, M.J. Inositol trisphosphate and calcium signalling, *Nature.* 361: 315 (1993).
15. Gill, D.L., *et al.*Function and organization of the inositol 1,4,5-trisphosphate-sensitive calcium pool in: *Advances in Second Messenger and Phosphoprotein Research* , Putney, J.W.,ed. Raven Press, Ltd.,New York. 26: 265 (1992).
16. Rossier, M.F. and J. W. Putney, Jr. The identity of the calcium-storing, inositol 1,4,5-trisphosphate-sensitive organelle in non-muscle cells: calciosome, endoplasmic reticulum...or both?, *Trends Neurosci.* 14: 310 (1991).
17. Ghosh, T.K., Mullaney, J.M., Tarazi, F.I. and Gill, D.L. GTP-activated communication between distinct inositol 1, 4,5-triphosphate-sensitive and -insensitive calcium pools,*Nature.* 340: 236 (1989).

18. Bian, J., Ghosh, T.K., Wang, J.-C. and Gill, D.L. Identification of intracellular calcium pools,*J.Biol. Chem.* 266: 8801 (1991).
19. Short, A.D., *et al.* Intracellular Ca^{2+} pool content is linked to control of cell growth, *Proc. Natl. Acad. Sci. USA.* 90: 4986 (1993).
20. Chueh, S.-H., Mullaney, J.M., Ghosh, T.K., Zachary, A.L. and Gill, D.L. GTP-and inositol 1,4,5-trisphosphate-activated intracellular calcium movements in neuronal and smooth muscle cell lines, *J. Biol. Chem.* 262: 13857 (1987).
21. Mullaney, J.M., Chueh, S.-H., Ghosh, T.K. and Gill, D.L. Intracellular calcium uptake activated by GTP, *J. Biol. Chem.* 262: 13865 (1987).
22. Chueh, S.-H. and Gill, D.L. Inositol 1,4,5-trisphosphate and guanine nucleotide activate calcium release from endoplasmic reticulum via distinct mechanisms, *J. Biol. Chem.* 261: 13883 (1986).
23. Ghosh, T.K., Eis, P.S., Mullaney, J.M., Ebert, C.L. and Gill, D.L. Competitive, reversible, and potent antagonism of inositol 1,4,5-triphosphate-activated calcium release by heparin, *J. Biol. Chem.* 263:11075 (1988).
24. Gill, D.L., Ueda, T., Chueh, S.-H. and Noel, M.W. Ca^{2+} release from endoplasmic reticulum is mediated by a guanine nucleotide regulatory mechanism, *Nature.* 320: 461 (1986).
25. Ueda, T., Chueh, S.-H., Noel, M.W. and Gill, D.L. Influence of inositol 1,4,5-trisphosphate and guanine nucleotides on intracellular calcium release within the N1E-115 neuronal cell line, *J. Biol. Chem.* 261: 3184 (1986).
26. Gill, D.L., Ghosh, T.K., Short, A.D., Bian, J. and Waldron, R.T.GTP-mediated communication between intracellular calcium pools in: *Handbook of Experimental Pharmacology: GTPases in Biology* , Dickey, E.F. and Birnbaumer, L.,ed. Springer-Verlag,Berlin. 108: 625 (1993).
27. Ferro-Novick, S. and Novick, P. The role of GTP-binding proteins in transport along the exocytic pathway, *Ann. Rev. Cell Biol.* 9: 575 (1993).
28. Novick, P. and Brennwald, P. Friends and family: the role of the Rab GTPases in vesicular traffic, *Cell.* 75: 597 (1993).
29. Bourne, H.R., Sanders, D.A. and McCormick, F. The GTPase superfamily: a conserved switch for diverse cell functions, *Nature.* 348: 125 (1990).
30. Mullaney, J.M., Yu, M., Ghosh, T.K. and Gill, D.L. Calcium entry into the inositol 1,4,5-trisphosphate-releasable pool is mediated by a GTP-regulatory mechanism, *Proc. Natl. Acad. Sci. USA.* 85: 2499 (1988).
31. Gill, D.L., Ghosh, T.K. and Mullaney, J.M. Calcium signalling mechanisms in endoplasmic reticulum activated by 1,4,5-trisphosphate and GTP, *Cell Calcium.* 10: 363 (1989).
32. Dawson, A.P., Hills, G. and Comerford, J.G. The mechanism of the action of GTP on Ca^{2+} efflux from rat liver microsomal vesicles, *Biochem. J.* 244: 87 (1987).
33. Comerford, J.G. and Dawson, A.P. Effects of CoA and acyl-CoAs on GTP-dependent Ca^{2+} release and vesicle fusion in rat liver microsomal vesicles, *Biochem. J.* 289: 561 (1993).
34. Rys-Sikora, K.E., Ghosh, T.K. and Gill, D.L. Modification of GTP-activated calcium translocation by fatty acyl-CoA esters, *J. Biol. Chem.* 269: 31607 (1994).
35. Martonosi, A.N.Transport of calcium by sarcoplasmic reticulum in: *Calcium and Cell Function* , Cheung, W.Y.,ed. Academic Press,New York. 3: 38 (1982).
36. Herbette, L.G., Favreau, C., Segalman, K., Napolitano, C.A. and Watras, J. Mechanisms of fatty acid effects on sarcoplasmic reticulum II. Structural changes induced by oleic and palmitic acid, *J. Biol. Chem.* 259: 1325 (1984).
37. Watras, J., Messineo, F.C. and Herbette, L.G. Mechanisms of fatty acid effects on sarcoplasmic reticulum I. Calcium-fatty acid interaction, *J. Biol. Chem.* 259: 1319 (1984).
38. Markley, K.S.Salts of fatty acids in: *Fatty Acids: Their Chemistry, Properties, Production, and Uses* , Markley, K.S.,ed. Interscience Publishers, Inc.,New York. 2: 715 (1961).
39. Klausner, R.D., Kleinfield, A.M., Hoover, R.L. and Karnovsky, M.J. Lipid domains in membranes: evidence derived from structural perturbations induced by free fatty acids and lifetime heterogeneity analysis, *J. Biol. Chem.* 255: 1286 (1980).
40. Kasai, M. and Kometani, T. Inhibition of anion permeability of sarcoplasmic reticulum vesicles by 4-acetoamido-4'-isothiocyanostilbene-2,2'-disulfonate, *Biochim. Biophys. Acta.* 557: 243 (1979).
41. Chan, K.-M. and Koepnick, S.L. The mechanism of calcium uptake by liver microsomes: effect of anions and ionophores, *Biochem. Biophys. Acta.* 818: 291 (1985).
42. Abumrad, N.A., Park, J.H. and Park, C.R. Permeation of long-chain fatty acid into adipocytes, *J. Biol. Chem.* 259: 8945 (1984).
43. Storch, J., Lechene, C. and Kleinfeld, A.M. Direct determination of free fatty acid transport across adipocyte plasma membranes using quantitative fluorescence microscopy, *J. Biol. Chem.* 266: 13473 (1991).
44. Abumrad, N.A., El-Maghrabi, M.R., Amri, E.-Z., Lopez, E. and Grimaldi, P.A. Cloning of a rat adipocyte membrane protein implicated in binding or transport of long-chain fatty acids that is induced during preadipocyte differentiation, *J. Biol. Chem.* 268: 17665 (1993).
45. Carsten, M.E. and Reedy, M.K. Cardiac sarcoplasmic reticulum: chemical and electron microscope studies of calcium accumulation, *J. Ultrastructure Research.* 35: 554 (1971).

OCCURRENCE OF PREVIOUSLY UNKNOWN LPO-PRODUCTS IN BIOLOGICAL SAMPLES

Gerhard Spiteller

Lehrstuhl Organische Chemie I der Universität Bayreuth,
Universitätsstraße 30, 95440 Bayreuth - Germany

INTRODUCTION

Polyunsaturated acids (PUFAs) are easily oxidized to hydroperoxides (LOOH). These were found to be enriched in many diseased states[1,2], obviously since they decompose to toxic compounds, e.g. aldehydes. Their genesis is visualized by cleavage of LOOH in presence of Fe^{++} to $LO^{\bullet}$ radicals, these either attack another unsaturated acid by hydrogen abstraction to form a corresponding hydroxy acid (LOH) or suffer cleavage of adjacent C–C bonds to aldehydes (scheme 1)[3]:

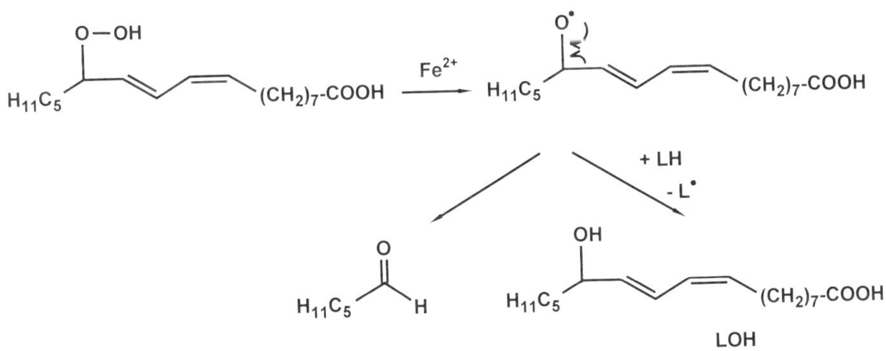

Scheme 1: Main degradation path of PUFAs exemplified with 9c-11t-13-hydroperoxy-octadecadienoic acid (13L-LOOH), one of the two main lipidperoxidation products of linoleic acid.

Frontiers in Bioactive Lipids
Edited by Vanderhoek, Plenum Press, New York, 1996

DEGRADATION MECHANISMS OF LOOHs

LOOHs are known to produce further oxidation products, e.g. malondialdehyde (MDA) or 4-hydroxynonenal (4HNE), but the mechanism of further reactions remained obscure. In order to clarify this, my collaborators Loidl-Stahlhofen[4] and Mlakar[5] subjected pure PUFAs to lipidperoxidation and investigated the thus produced compounds by GC/MS.

The results of these efforts are summarized as follows: Obviously it was previously overlooked that LOOHs and LOH (this is important) contain activated CH groups which are also prone for hydrogen abstraction by LO$^{\bullet}$ radicals. The thus formed mesomeric radicals react further with oxygen to hydroperoxyoxo - and hydroxyoxo derivatives of the original acid (scheme 2):

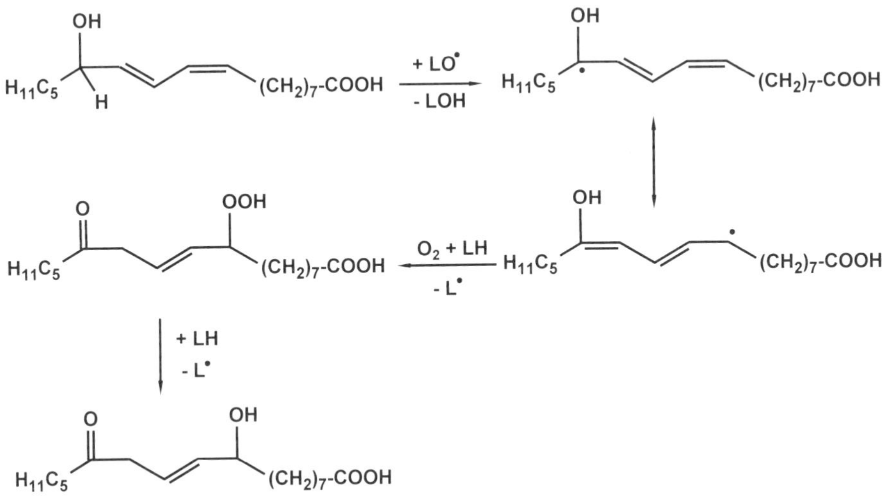

Scheme 2: Production of secondary degradation products of PUFAs, exemplified with 13L-LOOH.

Alternatively LOOHs and LOHs may react with LO$^{\bullet}$ by hydrogen abstraction from the activated CH_2 group adjacent to the diene system. The thus produced mesomeric radical reacts in similar way as discussed above with oxygen and produces another group of hydroxyhydroperoxy acids. These are cleaved to α-hydroxyaldehydes but also to (4HNE) (scheme 3):

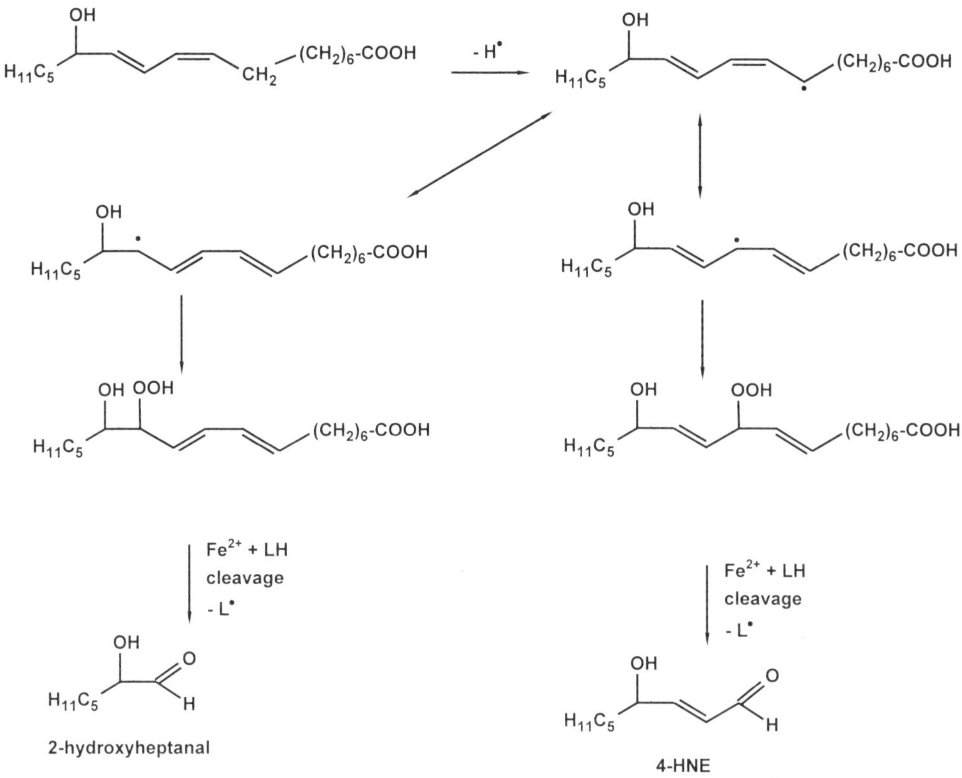

Scheme 3: Genesis of α-hydroxyaldehydes from LOHs

4HNE is known to be cytotoxic, the α-hydroxyaldehydes were found by Dr. Heinle (Tübingen, Germany) to induce an oxidative burst in stimulated macrophages[6].

REACTIONS OF LOOHs WITH OTHER MOLECULES

α-Hydroxyaldehydes are produced also in masked form from another class of compounds, plasmalogens, as shown by my collaborator Meyer[7].

In these phospholipids in position 1 of the glycerol an aldehyde is connected via an enolic bond instead an acid. In addition plasmalogens usually possess a PUFA in position 2. If the PUFA suffers lipid peroxidation a peroxylradicals is formed as intermediate. We have some evidence that this attacks the adjacent enolic double bond by epoxidation. The plasmalogen epoxides hydrolize easily to α-hydroxyaldehydes (scheme 4).

CELL DAMAGE AND ITS CONSEQUENCES

Recently my collaborator Herold observed that the above outlined reactions are initiated by cell damage[8]. Cell damage activates immediately „dormant enzymes“: Hydrolases cleave nearly instantly membrane lipids to free fatty acids. Thus formed PUFAs are attacked by lipoxygenases also liberated by cell damage to produce lipid

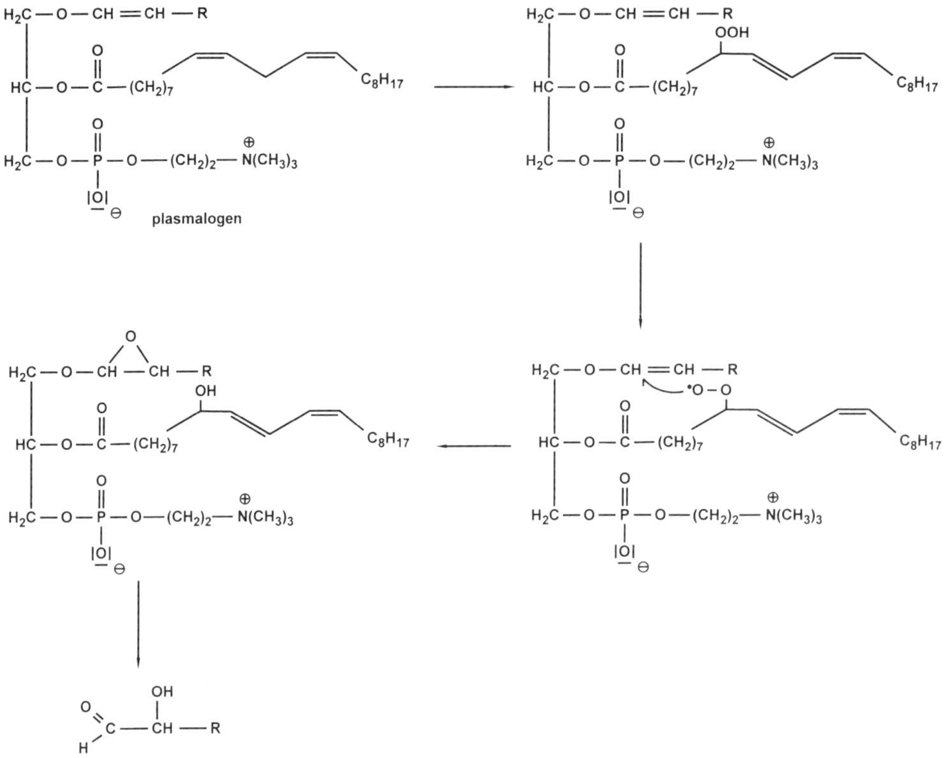

Scheme 4: Generation of α-hydroxyaldehydes from plasmalogens via lipid peroxidation

hydroperoxides. These decompose to short chain α-hydroxyaldehydes as outlined in scheme 3 in presence of Fe^{++} ions. What is the source of Fe^{++}?

We speculate that peptidases, also activated by cell injury, attack Fe^{++} containing proteins to produce Fe^{++} ions. These are able to cleave LOOH to $LO^{\bullet}$ radicals which not only initiate a chain reaction but in addition produce the above mentioned previously unknown species.

The observation that cell damage initiates lipid peroxidation explains why LPO products are generated in acute diseases, e.g. a myocardial infarction or during a transplantation. These processes are connected with severe cell injury. A gradual cell damage is also observed in chronic diseases, e.g. rheuma or atherosclerosis, and consequently also these processes should produce LOOHs.

Having this in mind my collaborator Dudda investigated if the above detected compounds are enriched during a myocardial infarction. We focussed our efforts especially to hydroxy acids, since these are the first products in the degradation cascade of LOOHs which branches more and more, so that final products, e.g. malondialdehyde (MDA), are only detected in very low yield.

Dr. Kobelt (Würzburg, Germany) induced in Göttinger mini pigs a myocardial infarction. In order to exclude individual changes, tissue from the affected and non affected heart reagion of one and the same animal was separated and each sample was investigated for its content in hydroxy acids. It must be emphazised that LOOH and derived LOH originate not only from arachidonic acid but also from linoleic acid. In fact the latter contributed even more to the hydroxy fraction than arachidonic acid. The comparison of LOHs from affected tissue and sourrounding one revealed an up to 18fold increase in hydroxy fatty acids compared to the non affected tissue of the same heart. The differences increased also considerable with time[9].

Similar, but less pronounced differences were detected in the content of secondary or tertiary oxidation products. So for instance the amount of α-hydroxyaldehydes derived from plasmalogens increased during infarction, this was connected with an analogous decrease of plasmalogens. Analogous results were obtained by a comparison of the content of α-hydroxyheptanal, derived by decomposition of oxidized Ω-6 PUFAs. Besides oxidation products of PUFAs we detected also epoxides of unsaturated fatty acids. These were neither detected in healthy hearts, nor in tissue sourrunding the infarction zone. Epoxy acids of linoleic acid had been previously detected in severe burnt tissue by Ozawa[10,11], they turned out to be cytotoxic and were called leukotoxines.

My collaborator Jira also investigated PUFAs derived from α-hydroxy acids of *low density lipoprotein* (LDL): LDL oxidation is regarded to play a dominant role in atherogenesis[12,13,14]. Our investigation confirms this: We detected a dramatic (up to 40fold) increase of hydroxy acids in LDL of patients suffering from coronar diseases[15], but recognized also that an increase of hydroxy acids occurs during the life time. Even absolutely healthy individuals aged 60 contain in their LDL about 3 times more hydroxy acids than young individuals between 20 and 30. Thus this determination might be the earliest indicator of a beginning atherosclerosis. Nevertheless although in this case other diseases must be excluded because a similar extreme high hydroxy acid level was detected to occur in LDL of young rheumatic patients or these suffering from diabetes[15].

Finally my collaborator Weisser investigated brain of young and old cattle. Plasmalogens are enriched in brain, they amount to 40% in the phospholipid fraction[16]. Therefore we investigated in this case the content of plasmalogens and aldehydes derived by hydrolysis as well as α-hydroxyaldehydes, derived by their epoxidation. The level of plasmalogens in a few days old calves and 8 year old animals turned out to be nearly constant. In contrast we observed in brain of all old animals without any exception an increase in plasmalogen derived aldehydes and hydroxy aldehydes for a factor of 40.

CONCLUSIONS

Not only lipid peroxidation products of arachidonic acid, but also of linoleic acid play a crucial role in the generation of toxic compounds. But peroxidation of lipids is only one factor which contributes to oxidative processes: Besides lipid peroxidation we have to consider that in the course of cell damaging processes obviously also epoxidases and probably also peptidases are stimulated. The action of epoxidases causes the generation of toxic epoxides, the action of peptidases liberates probably Fe^{++}. These cause then dangerous secondary oxidation reactions of PUFAs.

Acknowledgement

I am very obliged to my collaborators Angela Dudda, Michael Herold, Wolfgang Jira, Angelika Loidl-Stahlhofen, Anita Mlakar and Matthias Weisser, who contributed to the presented results not only by their skillness but also by many fruitful discussion. Some of our work was financially assisted by Deutsche Forschungsgemeinschaft.

References

1. B. Halliwell, M. Grootveld, The measurement of free radical reactions in humans. Some thoughts for future experimentation. *FEBS Lett.* 213:9 (1987).
2. K. Yagi, Lipid peroxides and human diseases. *Chem. Phys. Lipids*, 45:337 (1987).
3. G. Spiteller, Enzymic lipid peroxidation - a consequence of cell injury? *Free Rad. Biol. Med.* (1996) in press.
4. A. Loidl-Stahlhofen, G. Spiteller, α-Hydroxyaldehydes, products of lipid peroxidation. *Biochim. Biophys. Acta* 1211:156 (1994).
5. A. Mlakar, G. Spiteller, Reinvestigation of lipid peroxidation of linolenic acid. *Biochim. Biophys. Acta* 1214:209 (1994).
6. H. Heinle, N. Gugeler, R. Felde, G. Spiteller, Oxidation of plasmalogens produce highly effective modulators of macrophage function, *Free Rad. Biol. Med.*, (1995) submitted.
7. C. Meyer, A. Lutz, G. Spiteller, α-Hydroxy aldehyde derivatives as secondary products of the oxidation of plasmalogens, *Angew. Chem.,* 104:491 *(See also Angew. Chem., Int. Ed. Engl.*, 31:468 (1992)
8. M. Herold, G. Spiteller, Massive production of hydroperoxides of unsaturated fatty acids by injury of mammalian cells. *Chem. Phys. Lipids* (1996) in press.
9. A. Dudda, G. Spiteller, F. Kobelt, *Chem. Phys. Lipids* (1996) in press.
10. T. Ozawa, S. Sugiyama, M. Hayakawa, Leukocytes biosynthesize leukotoxin (9,10-epoxy-12-octadecenoate) - a novel cytotoxic linoleate epoxide. *Adv. Prostaglandin, Thromboxane, Leukotriene Res.*, 19(Taipei Conf. Prostaglandin Leukotrine Res., 1988), 164 (1989).
11. T. Ozawa, S. Sugiyama, M. Hayakawa, F. Taki, Neutrophil microsomes biosynthesize linoleate epoxide (9,10-epoxy-12-octadecenotae), a biological active substance. *Biochem. Biophys. Res. Commun. 152:*1310 (1988).
12. U.P. Steinbrecher, H. Zhang, M. Lougheed, Role of oxidatively modified LDL in atherosclerosis. *Free Radical Biol. Med.*, 9:155 (1990).
13. D. Steinberg, S. Parthasarathy, T.E. Carew, J.C. Khoo, J.L. Witztum, Beyond cholesterol. Modifications of low-density lipoprotein that increase its atherogenicity. *N. Engl. J. Med.* 320:915 (1989).
14. J.L. Witztum, D. Steinberg, Role of oxidized low density lipoprotein in atherogenesis. *J. Clin. Invest.*, 88:1785 (1991).
15. W. Jira, G. Spiteller, *Chem. Phys. Lipids* (1996) submitted.
16. M. Weisser, G. Spiteller, *Chem. Phys. Lipids* (1996) submitted.

STRUCTURAL AND FUNCTIONAL STUDIES ON THE HEART FATTY ACID-BINDING PROTEIN

Jacques H. Veerkamp and Clemens F.M. Prinsen

Department of Biochemistry
University of Nijmegen
6500 HB Nijmegen
The Netherlands

In the past we surveyed many aspects of cytoplasmic fatty acid-binding proteins (FABPs)[1-4]. These molecules with relatively low molecular masses (14-15 kDa) are supposed to play a role in fatty acid transfer and targeting, and in fatty acid metabolism. A second function is their possible modulatory role on the effect of fatty acids on gene expression, receptors, metabolic enzymes and signal transduction. An indirect involvement of FABP in modulation of cell growth and differentiation was suggested for hepatocytes, cardiomyocytes and mammary epithelial cells.

On base of structural analysis of isolated proteins and/or cDNA eight types of FABPs have been established. These types show a specific tissue and cellular distribution. The heart FABP type (H-FABP) is more generally present and is found in heart, skeletal muscle, kidney, mammary gland, lung, testes, brain and adrenals. Other types are present in only one tissue (adipose tissue, intestine and peripheral myelin types) or few tissues (e.g. liver FABP (L-FABP) in liver, kidney and intestine).

The 8 FABP types show 25-70% similarity of amino acid sequence. The heart, myelin and adipocyte FABP types show the highest similarity (about 70%) to each. other. This is reflected in the extent of cross-reactivity at comparison of the reaction of 5 human FABP types and their respective antisera in ELISA (Fig.1). A comparable pattern was observed with the rat FABP types. Molecules of a specific FABP type (H-FABP or L-FABP) from different mammalian species (human, rat, cattle, pig, rabbit) show a high immunological cross-reactivity which correlates with their high similarity (about 90%).

The ligand specificity has only been studied well for H-FABP and L-FABP. The latter FABP type binds a number of hydrophobic ligands in contrast to H-FABP[2,5,6] and also a broader range of fatty acids[5]. Recently we determined the K_d values for 5 human recombinant FABP types (Table 1). Adipocyte FABP (A-FABP) had a lower binding affinity for fatty acids than the other FABP types. The K_d value of palmitic acid was higher than of oleic and arachidonic acid. Our K_d

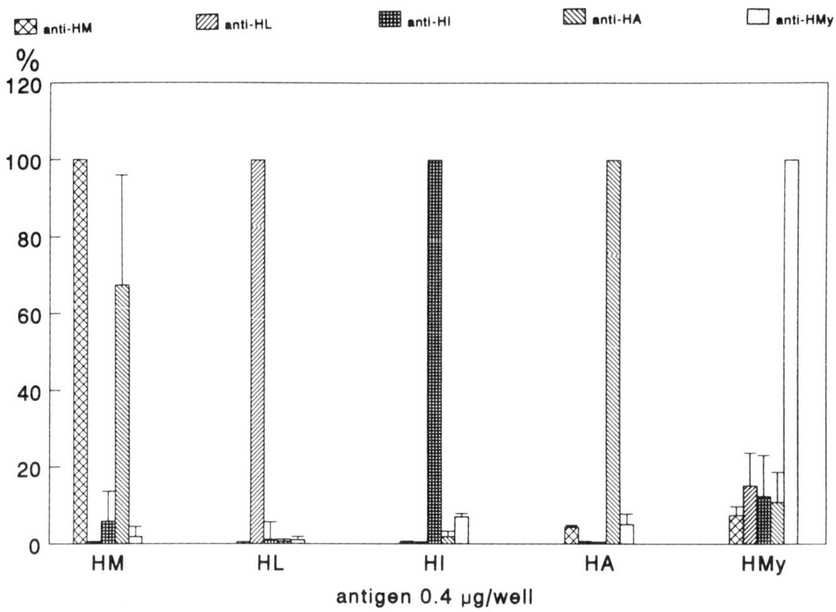

Figure 1. Crossreactivity of 5 human FABP types and their respective rabbit antisera. Antisera were applied in ELISA at a concentration giving 50% of the maximal response with their corresponding antigen.

Table 1. K_d values of human FABP types.

FABP type	Palmitic acid	Oleic acid	Arachidonic acid
Liver	4.02 ± 0.11	0.89 ± 0.03	0.44 ± 0.04
Heart	0.96 ± 0.05	0.43 ± 0.04	0.37 ± 0.02
Adipocyte	n.d.	1.56 ± 0.25	2.49 ± 0.22
Myelin	n.d.	0.31 ± 0.03	0.37 ± 0.02
Intestinal	n.d.	0.57 ± 0.04	1.49 ± 0.16

K_d values were determined with [^{14}C]fatty acid (0-3 μM) and 0.4 μM protein. Unbound fatty acid was removed with Lipidex. Values are means ± SD for 3-7 independent experiments. n.d., not determined.

values measured with the Lipidex method show the same relative differences as for those assayed with the fluorescent probe ADIFAB, although the latter values are much lower[7]. Titration calorimetry gives, however, about the same K_d values as the Lipidex procedure[6,8-10].

Displacement analysis of [^{14}C]oleic acid at different fatty acid concentrations showed a similar binding to H-FABP for oleic, elaidic and arachidonic acid, and

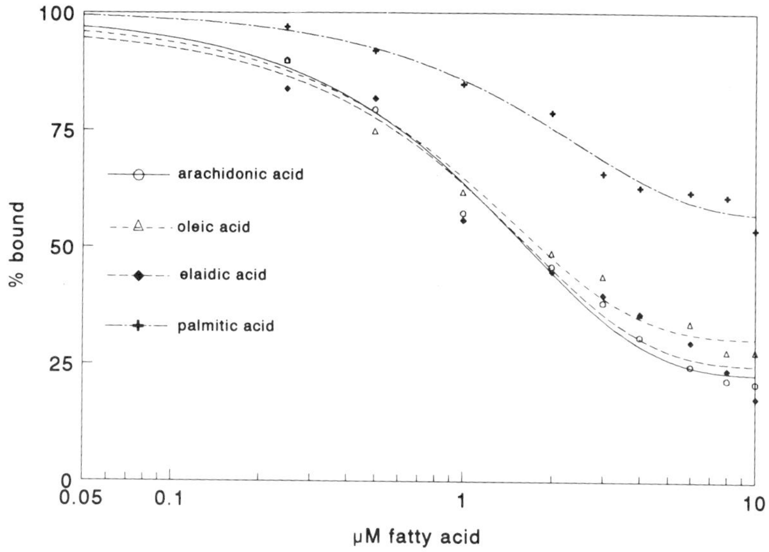

Figure 2. Displacement of [^{14}C]oleic acid from human H-FABP by various concentrations of non-labeled fatty acids. Concentrations of [1-^{14}C]oleic acid and human H-FABP were 1.0 and 0.26 μM, respectively. The Lipidex procedure was applied to remove unbound fatty acids.

somewhat less affinity for palmitic acid (Fig.2). Stearic, oleic and elaidic acid are bound in the binding cavity of H-FABP in the same arched conformation[11]. In A-FABP only arachidonic acid adopts a hairpin or U-shaped conformation[10]. The C_{16} and C_{18} fatty acids showed an extended conformation in contrast to those in H-FABP[11-13]. Crystallographic analyses[13] have indicated that H-FABP, I-FABP, A-FABP and myelin FABP bind the fatty acid with its carboxylate by interaction with Arg-106, Arg-126 and Tyr-128. Isothermal titration calorimetry[10,11] and Van 't Hoff plots for the temperature dependence of fatty acid binding[14] identified the enthalpy as the major contributor to the binding energy of fatty acids to FABP in accordance with this electrostatic interaction.

We used site-directed mutagenesis of human H-FABP to study the importance of certain amino acids for the fatty acid binding and the stability of the tertiary structure (Fig.3). The procedure and main results have been described previously[15]. In the present report we describe especially the effect of substitution of conserved phenylalanine residues by serine. At position 4 phenylalanine is present in all FABP types (except in epidermal FABP, leucine). Its ring interacts with the side chains from the C-terminal strands. Phe-16 and Phe-57 form Van der Waals contacts with the bound fatty acid. The phenyl ring of Phe-16 may form a key determinant in ligand specificity and affinity of H-FABP[12]. Phe-57 may change its orientation on ligand binding[13]. The Phe-64 is located on the βD strand in the gap region and may form one of the components of a hydrophobic patch located near the bottom of the gap between the βD and βE strands. Data for H-FABP and F16S and F57S mutants is partially obtained from previous work[15], but new mutants F4S and F64S were prepared for these studies. The F16S and F64S mutants were obtained in low yields in E. coli cultures compared to other recombinant mutant proteins. F64S was mainly present in inclusion bodies. The results of the studies on fatty acid binding and the conformational stability are given for all F→S mutants in Table 2. Replacement of Phe-64 by serine caused, like with Phe-16, the elimination of fatty acid-

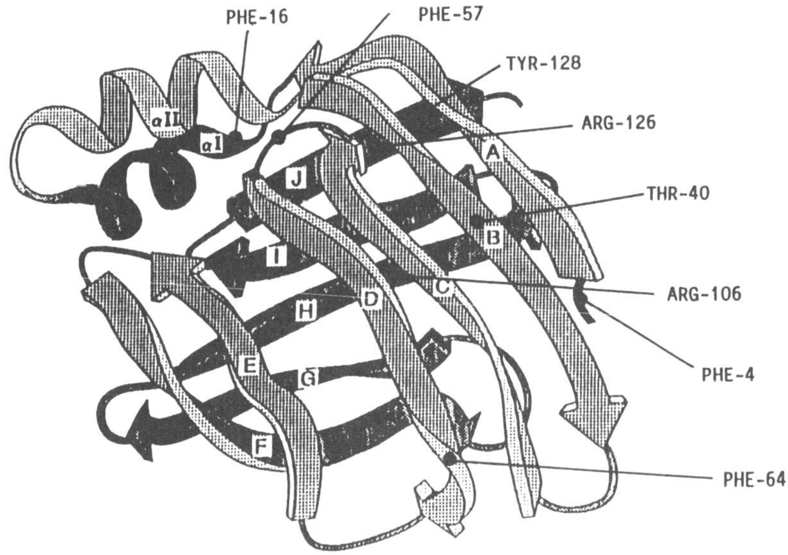

Figure 3. Model of human H-FABP with the location of the mutations applied.

binding activity. Substitution of serine for Phe-4 or Phe-57 did not markedly affect the binding of oleic acid. The K_d values of the latter mutant proteins do not significantly differ from that for the wild type H-FABP.

We tested the conformational stability by measuring the effect of urea on intrinsic fluorescence of the two internal tryptophans (position 8 and 97). This procedure could be applied on all mutants, since the tryptophans are not involved in fatty acid binding, and the presence or absence of fatty acid did not influence the fluorescence. At denaturation of the protein a red-shift of the fluorescence maximum and an increase of fluorescence intensity occur. The midpoints (the concentration of urea at which 50% of the protein is unfolded) for H-FABP and three phenylalanine mutants are given in Table 2. The midpoints for the F57V and R126H mutants were 5.3±0.1 and 4.4±0.1, respectively.

Table 3 summarizes the obtained data together with previous results[15]. We reported already that some substitutions of amino acids Thr-40 (T40E), Arg-106 (R106T) or Arg-126 (R126Q) resulted in proteins with little or no binding activity[15]. Mutants R126H, R126K and Y128F show a considerable increase of the K_d value for oleic acid[15]. These results are in accordance with the importance of the side chains of these amino acids for the electrostatic and hydrogen bonding network of the fatty acid carboxylate group. The phenyl ring of Phe-57 seems less critical for binding of the acyl chain than that of Phe-16, as well at serine as at valine substitution. Serine substitution of Phe-4 does not markedly affect the fatty acid binding and stability in contrast to that of Phe-16.

Changes in fatty acid binding and conformational stability do not always occur concomitantly. The R126Q mutant had the same stability as the wild type, but did not show binding. Other mutants as F57S and Y128F showed only a large decrease in stability. Some mutants (F4S, T40V and F57V) had a nearly similar binding activity and stability as the wild type. The results on the human H-FABP mutants are in agreement with the tertiary structure data[11,12].

Substitution of glutamine for Arg-106 in I-FABP[11,16], for Arg-126 in A-FABP[17] and for Arg-122 in L-FABP[18] caused a strong decrease of their affinity for fatty

Table 2. Oleic acid binding and urea denaturation of human H-FABP and H-FABP mutants.

Protein	Relative binding activity	K_d oleic acid (μM)	Midpoint (M urea)
H-FABP	1.00 ± 0.15	0.42 ± 0.06	6.0 ± 0.1
Phe4Ser	0.85 ± 0.12	0.67 ± 0.10	5.5 ± 0.1
Phe16Ser	0.08 ± 0.05	n.d.	2.1 ± 0.3
Phe57Ser	0.95 ± 0.14	0.41 ± 0.08	5.2 ± 0.1
Phe64Ser	0.03 ± 0.01	n.d.	n.d.

Binding activity was measured at 2 μM [1-^{14}C]oleic acid and 1.0 μM protein. K_d values are derived from Scatchard plots of binding assays at 0.1-3.0 μM oleic acid and 0.5 μM protein. The midpoint is the concentration of urea, at which the increase of maximum wave length of tryptophan fluorescence is 50%. Results are means ± SD of at least 3 independent assays.
n.d., not determined.

acids. Substitution of tryptophan for Tyr-128 had a similar effect on A-FABP[17]. In contrast to the R106Q mutant of I-FABP[19] the R106T and the R126Q mutants of human H-FABP did not show binding of retinol (data not shown). Recently some surface mutants of rat H-FABP were studied[20]. Substitution of Lys-22 by isoleucine or glutamic acid and of Lys-59 by isoleucine had no effect on fatty acid binding. The Thr28Lys mutant showed a higher K_d value for palmitic acid. Tryptophan subsitution of Phe-3, Phe-18 or Cys-69 did not change the fatty acid-binding affinity of L-FABP[21]. Mutations of rat L-FABP at Thr-102 and Arg-122 altered its structural integrity and affinity. A Thr102Arg mutant was not expressed[22]. The wild-type L-FABP had a preference for carboxylate-containing ligands as fatty acids, haem and bilirubin, whereas the Arg102Gln mutant with its less rigid structure favours more bulky ligands as oleyol-CoA, lysophospholipids and monoacylglycerol to bind with higher affinity. The affinity of the Thr102Gln mutant was intermediate between the wild-type and the Arg122Gln mutant[22].

The first and thus far the only observed natural FABP mutant was an Ala54Thr mutant of I-FABP[23]. This mutation in the tight turn between the βC and βD strands caused a twofold increase of the affinity for oleic and arachidonic acid The I-FABP threonine-encoding allele was found to be associated with insulin resistance and increased fat oxidation in vivo[23].

Tyrosine substitution[24] of Trp-6 or Trp-82 reduced the stability of I-FABP like cysteine substitution for a hydrophobic residue in two out of six cysteine mutants (V60C and L89C)[25]. The stability of all Arg-122 mutants and of the R126K mutant of L-FABP was markedly reduced compared to the wild type[18].

Studies on the effect of mutations on functional properties of human H-FABP have just been started. The growth-inhibitory and differentiation-stimulating

Table 3. Conformational stability and binding activity of human H-FABP mutants.

Stability	Binding activity	Mutants
N	N	T40V
N	↓	F4S
N	↓↓	R126Q
↓	N	T40Q, F57V
↓	↓	Y128F
↓↓	N	F57S
↓↓	↓	F16V, F16Y, R126K, R126H
↓↓	↓↓	F16S, T40E, F64S, R106T

N, normal; ↓, decreased; ↓↓, strongly decreased.

activities of human H-FABP on cultured mammary gland[26] were not affected by substitution of Arg-106 by threonine, which abolished fatty acid-binding activity. Site-specific mutations on β2 turn (K59I), helix α-I (K22I and K22E) and helix α-II (T28K) of rat H-FABP all altered the absolute rate of fatty acid transfer to membranes of small unilamellar vesicles, whereas mutations on β-strands A and I (K10I and K113I) had no effect[20]. It would be of interest to study the fatty acid transfer properties of our active recombinant proteins to mitochondria by assay of fatty acid oxidation. Previously[27] we showed that rat H-FABP was a good fatty acid donor to rat heart mitochondria. The cDNAs of these mutant proteins could be used in transfection studies with human breast cancer cells to establish the supposed tumor suppressor activity of the H-FABP gene[28] and in investigations on the effect of expression of transgenic FABP cDNA constructs[29]. The influence of mutant proteins on cardiac myocyte hypertrophy compared to H-FABP[30] is another point of interest. The aspect of the influence of fatty acid-binding on protein properties could be studied by comparison of wild-type with active and inactive stable mutants. Mutant proteins are also useful for future biophysical studies on the conformation of FABP and the energetic aspects of fatty acid binding and transfer.

We acknowledge the contributions of Mr A. Jansen and Mr H.T.B. van Moerkerk to the presented data.

REFERENCES

1. J.H. Veerkamp, R.A. Peeters, and R.G.H.J. Maatman, Structural and functional features of cytoplasmic fatty acid-binding proteins, *Biochim. Biophys. Acta* 1081:1 (1991).
2. J.H. Veerkamp, T.H.M.S.M. van Kuppevelt, R.G.H.J. Maatman, and C.F.M. Prinsen, Structural and functional aspects of cytosolic fatty acid-binding proteins, *Prostaglandins Leukot. Essent. Fatty Acids* 49:887 (1993).
3. J.H. Veerkamp and R.G.H.J. Maatman, Cytoplasmic fatty acid-binding proteins: their structure and genes, *Prog. Lipid Res.* 34:17 (1995).
4. J.H. Veerkamp, Fatty acid transport and fatty acid-binding protein, *Proc. Nutrition Soc.* 54:23 (1995).
5. R.G.H.J. Maatman, H.T.B. van Moerkerk, I.M.A. Nooren, E.J.J. van Zoelen, and J.H. Veerkamp, Expression of human liver fatty acid-binding protein in *Escherichia coli* and comparative analysis of its binding characteristics with muscle fatty acid-binding protein, *Biochim. Biophys. Acta* 1214:1 (1994).
6. B. Rolf, E. Oudenampsen-Krüger, T. Börchers, N.J. Faergeman, J. Knudsen, A. Lezius, and F. Spener, Analysis of the ligand binding properties of recombinant bovine liver-type fatty acid binding protein, *Biochim. Biophys. Acta* 1259:245 (1995).
7. G.V. Richieri, R.T. Ogata, and A.M. Kleinfeld, Equilibrium constants for the binding of fatty acids with fatty acid-binding proteins from adipocyte, intestine, heart, and liver measured with the fluorescent probe ADIFAB, *J. Biol. Chem.* 269:23918 (1994).
8. R.G.H.J. Maatman, M. Degano, H.T.B. van Moerkerk, W.J.A. van Marrewijk, D.J. van der Horst, J.C. Sacchettini, and J.H. Veerkamp, Primary structure and binding characteristics of locust and human muscle fatty-acid-binding proteins, *Eur. J. Biochem.* 221:801 (1994).
9. J.M. LaLonde, M.A. Levenson, J.J. Roe, D.A. Bernlohr, and L.J. Banaszak, Adipocyte lipid-binding protein complexed with arachidonic acid, *J. Biol. Chem.* 269:25339 (1994).
10. K.R. Miller and D.P. Cistola, Titration calorimetry as a binding assay for lipid-binding proteins, *Mol. Cell. Biochem.* 123:29 (1993).
11. A.C.M. Young, G. Scapin, A. Kromminga, S.B. Patel, J.H. Veerkamp, and J.C. Sacchettini, Structural studies on human muscle fatty acid binding protein at 1.4Å resolution: binding interactions with three C18 fatty acids, *Structure* 2:523 (1994).
12. G. Zanotti, G. Scapin, P. Spadon, J.H. Veerkamp, and J.C. Sacchettini, Three-dimensional structure of recombinant human muscle fatty acid-binding protein, *J. Biol. Chem.* 267:18541 (1992).
13. L. Banaszak, N. Winter, Z. Xu, D.A. Bernlohr, S. Cowan, and T.A. Jones, Lipid-binding proteins: a family of fatty acid and retinoid transport proteins, *Adv. Protein Chem.* 45:89 (1994).
14. G.V. Richieri, R.T. Ogata, and A.M. Kleinfeld, Thermodynamics of fatty acid binding to fatty acid-binding proteins and fatty acid partition between water and membranes measured using the fluorescent probe ADIFAB, *J. Biol. Chem.* 270:15076 (1995).
15. C.F.M. Prinsen and J.H. Veerkamp, Fatty acid binding and conformational stability of mutants of human muscle fatty acid-binding protein, *Biochem. J.* 314:253 (1996).
16. J. Eads, J.C. Sacchettini, A. Kromminga, and J.I. Gordon, *Escherichia coli*-derived rat intestinal fatty acid binding protein with bound myristate at 1.5 Å resolution and I-FABP$^{Arg106 \rightarrow Gln}$ with bound oleate at 1.74 Å resolution, *J. Biol. Chem.* 268:26375 (1993).
17. R.S. Sha, C.D. Kane, Z. Xu, L.J. Banaszak, and D.A. Bernlohr, Modulation of ligand binding affinity of the adipocyte lipid-binding protein by selective mutation, *J. Biol. Chem.* 268:7885 (1993).
18. A.E.A. Thumser, C. Evans, A.F. Worrall, and D.C. Wilton, Effect on ligand binding of arginine mutations in recombinant rat liver fatty acid-binding protein, *Biochem. J.* 297:103 (1994).
19. M.G. Jakoby, K.R. Miller, J.J. Toner, A. Bauman, L. Cheng, E. Li, and D.P. Cistola, Ligand-protein electrostatic interactions govern the specificity of retinol- and fatty acid-binding proteins, *Biochemistry* 32:872 (1993).
20. F.M. Herr, J. Aronson, and J. Storch, Role of portal region lysine residues in electrostatic interactions between heart fatty acid binding protein and phospholipid membranes, *Biochemistry* 35:1296 (1996).
21. A.E.A. Thumser and D.C. Wilton, Characterization of binding and structural properties of rat liver fatty-acid-binding protein using tryptophan mutants, *Biochem. J.* 300:827 (1994).
22. A.E.A. Thumser, J. Voysey, and D.C. Wilton, Mutations of recombinant rat liver fatty acid-binding protein at residues 102 and 122 alter its structural integrity and affinity for physiolo-

gical ligands, *Biochem. J.* 314:943 (1996).
23. L.J. Baier, J.C. Sacchettini, W.C. Knowler, J. Eads, G. Paolisso, P.A. Tataranni, H. Mochizuki, P.H. Bennett, C. Bogardus, and M. Prochazka, An amino acid substitution in the human intestinal fatty acid binding protein is associated with increased fatty acid binding, increased fat oxidation, and insulin resistance, *J. Clin. Invest.* 95:1281 (1995).
24. I.J. Ropson and C. Frieden, Dynamic NMR spectral analysis and protein folding: identification of a highly populated folding intermediate of rat intestinal fatty acid-binding protein by ^{19}F NMR, *Proc. Natl. Acad. Sci. USA* 89:7222 (1992).
25. N. Jiang and C. Frieden, Intestinal fatty acid binding protein: characterization of mutant proteins containing inserted cysteine residues, *Biochemistry* 32:11015 (1993).
26. Y. Yang, E. Spitzer, N. Kenney, W. Zschiesche, M. Li, A. Kromminga, T. Müller, F. Spener, A. Lezius, J.H. Veerkamp, G.H. Smith, D.S. Salomon, and R. Grosse, Members of the fatty acid binding protein family are differentiation factors for the mammary gland, *J. Cell Biol.* 127:1097 (1994).
27. R.A. Peeters, J.H. Veerkamp, and R.A. Demel, Are fatty acid-binding proteins involved in fatty acid transfer?, *Biochim. Biophys. Acta* 1002:8 (1989).
28. H.T. Huynh, C. Larsson, S. Narod, and M. Pollak, Tumor suppressor activity of the gene encoding mammary-derived growth inhibitor, *Cancer Res.* 55:2225 (1995).
29. B. Binas, B. Gusterson, R. Wallace, and A.J. Clark, Epithelial proliferation and differentiation in the mammary gland do not correlate with cFABP gene expression during early pregnancy, *Dev. Genet.* 17:167 (1995).
30. P.B.J. Burton, C.E. Hogben, C.L. Joannou, A.G.B. Clark, J.J. Hsuan, N.F. Totty, C. Sorensen, R.W. Evans, and M.J. Tynan, Heart fatty acid binding protein is a novel regulator of cardiac myocyte hypertrophy, *Biochem. Biophys. Res. Commun.* 205:1822 (1994).

THE ROLE OF INTRACELLULAR FATTY ACID-BINDING PROTEINS IN CELLULAR TRANSPORT OF FATTY ACIDS

Judith Storch, Fiona M. Herr, Kuo Tung Hsu, Hye Kyung Kim*, Heng Ling Liou, Ana Monroy and Elizabeth R. Smith

Rutgers University
Department of Nutritional Sciences
New Brunswick, NJ 08903

* Dept. of Food and Biotechnology, Hanseo University, Korea.

INTRODUCTION

The family of mammalian fatty acid-binding proteins (FABP) includes at least ten distinct gene products, and additional homologues continue to be reported. The FABP were first discovered in the early 1970's as abundant cytoplasmic proteins which bound long chain fatty acids (FA) *in vitro* [1-3]. In the ensuing years, a great deal of information has been obtained about their tissue distributions, ligand binding affinities and specificities, developmental regulation and, more recently, their tertiary structures [3-5] Their functions, however, remain largely undefined. Among the first members of the FABP family to be purified were those from intestine [1,6], liver [1,2] and heart [7]. In all these tissues, fatty acid flux is considerable, and thus it was proposed that the FABP are necessary for large-scale intracellular binding of fatty acids. More recently, FABP family members have been identified in other tissues that are generally considered to be lipid-active, e.g. adipose tissue [8], as well as other tissues, e.g. brain [9,10]. In addition, the FABP have been suggested as modulators not only of lipid metabolism and cellular FA flux , but also of gene expression, cell growth and differentiation, etc. [11-13] It is also of interest that a number of cell types, most notably intestinal enterocytes, express high and approximately equivalent levels of two separate FABP's [14] suggesting a functional specificity for different family members. In all of these tissues and for all of these putative functions, a role for FABP in cytoplasmic transport of FA from sites of entry and/or synthesis to sites of utilization and function, is implied.

Our laboratory has been studying how intracellular cytoplasmic binding proteins may be involved in the traffic fatty acids between subcellular membrane sites. Although it has often been hypothesized that these proteins play a role in the intracellular movement of fatty acids, their precise role and mechanism of action is not known. To address the role of FABP's in fatty acid transport to membranes, we have examined how fluorescent anthroyloxy-labeled fatty acids are transferred from fatty acid binding proteins to model membranes. These studies have begun to define the role of the FABP in cellular fatty acid traffic, and to examine potential functional differences between different FABP family members.

COMPARISON OF DIFFERENT MEMBERS OF THE FABP FAMILY

A large number of studies have examined the equilibrium binding of long chain fatty acids to the FABP family. Where direct comparisons have been made between different FABP, the fatty acid dissociation constants obtained have generally been similar, ranging from 0.5 to 3 μM [15-17]. Recently, Richieri *et al.* reported that L- and I-FABP have similar affinities for saturated fatty acids, but that L-FABP has a 5-fold higher affinity for unsaturated fatty acids [18]. The K_d's for AOFA binding to FABP are similar to those of native fatty acid and are similar for different members of the FABP family [19,20]. Thus at present, equilibrium binding studies have afforded only limited insight into potential structural or functional differences between FABP.

A comparison of the absolute transfer rates for an AOFA analog from four members of the FABP family to egg phosphatidylcholine (EPC) membranes was performed using a fluorescence energy transfer assay, as detailed previously [21,22], and the results are shown in Table 1. Notwithstanding generally similar equilibrium binding characteristics, transfer rates for the stearic acid derivative 9AS (9-AO-stearate) differ among the FABP by as much as three orders of magnitude. The relative rates of fluorescent fatty acid transfer fall into two groups, with more rapid transfer occurring from H- and A-FABP, and slower transfer from I-and L-FABP. Interestingly, these groupings coincide with the proposed evolutionary relationships among the FABP family. It has been proposed that H- and A-FABP, which share over 65% amino acid sequence identity, are closely related, having diverged approximately 240 million years ago. On the other hand, L- and I-FABP are members of a separate branch of the evolutionary tree, whose common ancestral gene diverged 690 millions of years ago [23]. It appears therefore that the rate of fatty acid transfer from FABP to membranes may be tied to their degree of evolutionary relatedness.

Table 1. Transfer of 9-anthroyloxy stearate from FABP's to membranes.

Transfer of 1.5 μM 9AS from 15 μM FABP to 300 μM SUV composed of 90 mol% egg phosphotidylcholine/10 mol% NBD-phosphotidylcholine at 25°C, pH 7.4.

	k, sec^{-1}
H-FABP	2.2 ± 0.1
A-FABP	7.4 ± 0.1
I-FABP	1.12 ± 0.02
L-FABP	0.01 ± 0.001

MECHANISMS OF FATTY ACID TRANSFER FROM FABP TO PHOSPHOLIPID MEMBRANES

Transfer of a fatty acid from FABP to an acceptor membrane can occur either by diffusion of the monomeric ligand through the aqueous phase and subsequent association with the membrane, or via collisional interaction of the FABP with the membrane, with fatty acid transfer occurring during the collision. In our studies, we sought to determine whether, in addition to the large quantitative differences observed for fatty acid transfer rates, mechanistic differences exist between different FABP family members. The approach taken was to modulate the solubility of the fatty acid in the aqueous phase and determine whether these modulations led to corresponding changes in the transfer rate from FABP. Studies with the heart and adipocyte FABP reveal that AOFA transfer from these proteins is virtually insensitive to fatty acid solubility--alterations in fatty acyl chain length and degree of unsaturation did not effect parallel changes in their rates of transfer to membranes. Furthermore, an increase in the ionic strength of the assay medium, which decreases the solubility of hydrophobic fatty acids, did not decrease the rate of AOFA transfer from these two FABP [21,24]. These results demonstrate, therefore, that fatty acid transfer from A-and H-FABP does not occur via diffusion of the ligand through the aqueous phase. In contrast, AOFA transfer from liver FABP was entirely consistent with an aqueous diffusion-controlled mechanism [19].

To examine the alternative possibility of collisional transfer of FA, the theoretical number of collisions between membranes and FABP was varied by increasing the concentration of acceptor membranes. If transfer involved diffusion through an aqueous medium, the rate of fatty acid transfer from FABP would remain constant, since in that case desorption from the FABP would be the limiting step. But in transfer by collision, the transfer rate would increase in proportion to the frequency of collision. It was found that for both adipocyte and heart FABP, the rate of AOFA transfer to membranes

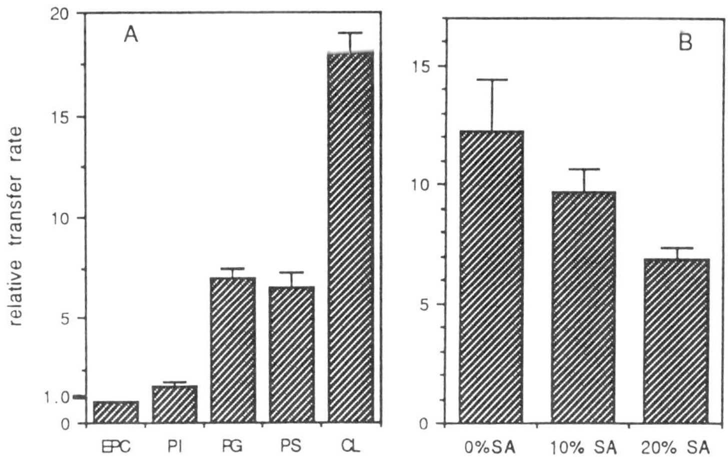

Figure 1. 2-AO-palmitate transfer from A-FABP to vesicles with a net charge. Transfer of 1.5 μM 2AP from 15 μM A-FABP to 150 μM egg phosphatidylcholine vesicles containing A) 25 mol % PI, PG, PS or CL, expressed relative to the transfer rate to zwitterionic EPC vesicles, or B) 0, 10, or 20 mol % positively charged stearylamine. Reprinted with permission from *Journal of Biological Chemistry* 269:10517-10523, 1994, American Society for Biochemistry and Molecular Biology.

increased linearly with acceptor vesicle number. This indicates that fatty acid was indeed transferred from these FABP when the protein collided with the acceptor membrane [21,24]. The rate of AOFA transfer from liver FABP to membranes, on the other hand, was independent of acceptor vesicle concentration, further indicating an aqueous diffusional FA transfer process for L-FABP [19].

Further evidence for the collisional nature of the FA transfer process from A- and H-FABP, and for the specificity of collisional transfer, comes from studies which examined the effects of acceptor membrane composition on the rate of AOFA transfer to membranes [25]. Inclusion of 25 mol% acidic phospholipids in the acceptor membranes increased the rate of transfer by as much as 20-fold. In contrast, for membranes with a net positive charge, the rate was decreased (Figure 1). These results support the hypothesis of FABP-membrane collisional interactions as the mechanism of transfer, and further suggest that positively charged amino acid residues on the surface of these FABP are important for the formation of a collisional complex with the membrane [25].

ELECTROSTATIC INTERACTIONS BETWEEN FABP AND MEMBRANES MODULATE THE FATTY ACID TRANSFER RATE

To further explore the role of positively charged surface residues on FABP in the formation of effective collisional complexes with membranes, surface lysines of A-FABP were acetylated to neutralize these positively charged residues. Circular dichroic spectra and chemical denaturant data indicated that acetylation did not alter the conformational integrity of the protein. In addition, the binding affinities of native fatty acid and AOFA to unmodified and modified protein were virtually identical [26]. The transfer of the palmitate derivative 2AP from acetylated A-FABP to EPC vesicles was ≥30-fold slower than from native protein (Figure 2). Moreover, after neutralization of positively charged lysine residues on A-FABP, 2AP transfer rate was independent of the acceptor membrane concentration and became dependent on ligand solubility. In addition, acidic phospholipids failed to stimulate the transfer rate. Collectively, the results strongly suggest that effective transfer by collision of fatty acid from A-FABP to membranes requires the presence of positively charged surface lysine residues [26]. Not surprisingly,

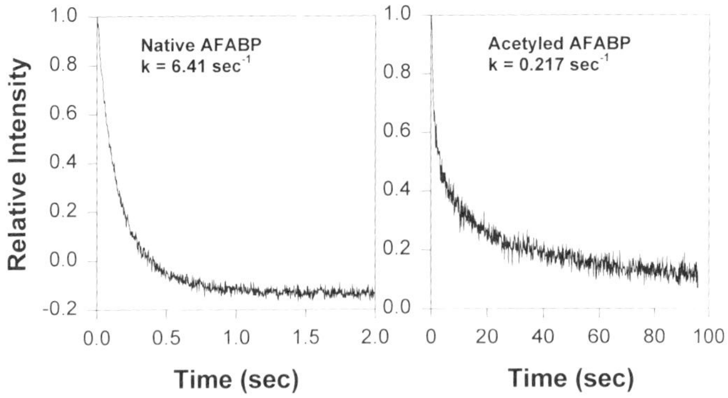

Figure 2. Transfer of 2-anthroyloxy-palmitate from native and acetylated A-FABP to egg phosphatidylcholine vesicles. 1 μM 2AP was incubated with 10 μM native or acetylated protein, and transfer to 100 μM vesicles containing 10 mol% NBD-PE was monitored at 25°C, as described [26].

similar results were found for H-FABP [22]. Thus, in the absence of collisional transfer, fatty acids will move from FABP's to membranes via aqueous diffusion.

INTERACTIONS BETWEEN FABP AND PHOSPHOLIPID MEMBRANES

Although the FABP are typically referred to as cytosolic proteins, the above-described kinetic evidence strongly suggests that specific collisional interactions between FABP's and membranes could occur. Cytochrome c is a well-studied example of a small molecular weight protein that binds to anionic phospholipid bilayers via electrostatic interactions [27]. We therefore examined whether preincubation of membranes with FABP's could prevent subsequent binding of cytochrome c. Vesicles were prepared to contain 25 mol% cardiolipin, 74 mol% egg phosphatidylcholine and 1% dansyl-phosphatidylethanolamine. Cardiolipin was chosen because of the high rate at which AOFA is transferred by collision to CL- containing membranes [25,28]. Moreover, CL is known to form associations with proteins [27]. Dansyl PE was used to monitor the binding of cytochrome c, since it quenches the dansyl fluorescence when the heme moiety of cytochrome c is bound to the membrane. Preliminary results show that preincubation with A-FABP, but not acetylated A-FABP or L-FABP, greatly reduced the subsequent binding of cytochrome c to CL-containing membranes (not shown). These results thus demonstrate the interaction of intracellular fatty acid-binding proteins with membranes.

IDENTIFICATION OF MEMBRANE-INTERACTIVE DOMAINS ON FABP

Since complete neutralization of A- and H-FABP lysine residues altered not only the rate, but also the mechanism of AOFA transfer, site-directed mutagenesis is being used to elucidate the specific residue(s) that are involved in fatty acid transfer by collision. Lysine 22 is located on the α-I helix region of H-FABP, which is hypothesized to be part of the fatty acid portal domain of the protein. Modification of lys 22 to isoleucine or glutamic acid, the latter leading to a charge reversal, resulted in a 2 to 3-fold decrease in 2AP transfer rate (Figure 3). Lysine 59 is also in the putative portal region of the protein, located in the β-2 turn that links β-strands C and D. When the neutral residue isoleucine was substituted for lysine 59, the transfer rate doubled [22]. A mutant incorporating both K22 and K59 alterations resulted in a 2AP transfer rate that was indistinguishable from that of the wild type protein. The additive nature of these results suggests that at least two separate regions on H-FABP, the α-I helix and β-2 turn, together determine the absolute rate of fatty acid transfer to membranes. AOFA transfer rates from H-FABP mutated at position 22 were virtually unaffected by the charge characteristics of the acceptor membranes; this is in distinct contrast to the wild type protein and other mutant proteins examined [22].

The mutagenesis studies thus suggest that the helical "cap" region of H-FABP plays an important role in regulating ionic interactions between the binding protein and membranes. A parallel site-directed mutagenesis study using A-FABP has further demonstrated that the helix-turn-helix "cap" region of these binding proteins plays an important role in fatty acid transfer to anionic phospholipid membranes. Neutralization of A-FABP lysine 22 on the α-I helix, or lysine 32 on the α- II helix, results in a 3-fold decrease in the rate of AOFA transfer to anionic membranes. Importantly, the α-I helix of H- and A-FABP is amphipathic. Thus, given the well-known electrostatic interactions of amphipathic helixes with membranes [29], this domain is likely to be critical for the charge-charge interactions which govern FABP-membrane interactions. Future studies

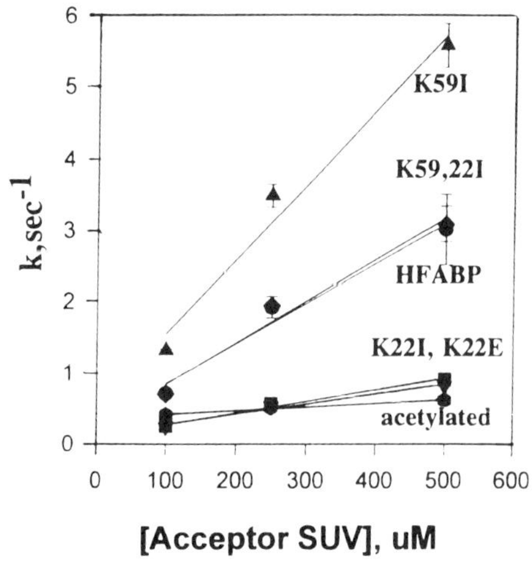

Figure 3. Effect of acceptor membrane concentration on the rate of 2AP transfer from native and mutant HFABP's. Transfer of 1 μM 2AP from 10 μM protein was monitored as a function of SUV phospholipid concentration, at 25°C and pH7.4. Reprinted with permission from *Biochemistry* 35:1296-1303, 1996, American Chemical Society.

will further define those domains of the FABP structure that are required for ligand transfer by collision and efficient interaction with membranes.

SUMMARY

The fatty acid binding proteins have long been hypothesized to play an important role in the transport and metabolism of intracellular fatty acids. Using an *in vitro* fluorescence resonance energy transfer assay, we have found large differences in the absolute transfer rates of fluorescent fatty acid from different members of the FABP family, to model phospholipid vesicles. With the exception of the liver FABP, transfer of fatty acid from all other FABP's examined (H-, A-, I-FABP) to membranes is hypothesized to involve an effective collisional interaction between the protein and the bilayer which results in ligand transfer. Direct interaction of FABP's with membranes has been demonstrated. Specific surface lysine residues on the FABP are likely to participate in the formation of effective FABP-membrane collisional complexes. Transfer of fatty acid from FABP to membranes by collision implies the targeting of fatty acid by FABP to specific intracellular sites. Ionic interactions between A-, H- or I-FABP and membrane phospholipids or acidic peptides may therefore play an important role in the intracellular trafficking of fatty acids.

REFERENCES

1. R.K. Ockner, J.A. Manning, R.B. Poppenhausen and W.K.L. Ho, A binding protein for fatty acids in cytosol of intestinal mucosa, liver, myocardium, and other tissues, *Science.* 177:56 (1972).
2. S. Mishkin, L. Stein, Z. Gatmaitan and I.M. Arias, The binding of fatty acids to cytoplasmic proteins: Binding to Z-protein in liver and other tissues of the rat, *Biochem Biophys Res Commun.* 47:997 (1972).
3. N.M. Bass, The cellular fatty acid binding proteins: Aspects of structure, regulation, and function, *Intl Review Cytol.* 3:143 (1988).
4. R.M. Kaikus, N.M. Bass and R.K. Ockner, Functions of fatty acid binding proteins, *Experientia.* 46:617 (1990).

5. L. Banaszak, N. Winter, Z. Xu, D.A. Bernlohr, S. Cowan and T.A. Jones, Lipid -binding proteins: A family of fatty acid and retinoid transport proteins, *Adv Protein Chem.* 45:89-151 (1994).
6. R.K. Ockner and J.A. Manning, Fatty acid-binding protein in small intestine: Identification, isolation, and evidence for its role in cellular fatty acid transport, *J Clin Invest.* 54:326 (1974).
7. N.C. Fournier, M. Geoffrey and J. Deshusses, Purification and characterization of a long-chain fatty acid binding protein supplying the mitochondrial β-oxidation system in the heart, *Biochim Biophys Acta.* 533:457 (1978).
8. V. Matarese and D.A. Bernlohr, Purification of murine adipocyte lipid-binding protein: Characterization as a fatty acid-and retinoic acid-binding protein, *J Biol Chem.* 263:14544 (1988).
9. A. Kurtz, A. Zimmer, F. Schnutgen, G. Bruning, F. Spener and T. Muller, The expression pattern of a novel gene encoding brain-fatty acid binding protein correlates with neuronal and glial cell development, *Development.* 120:2637 (1994).
10. L. Feng, M.E. Hatten and N. Heintz, Brain lipid-binding protein (BLBP): A novel signaling system in the developing mammalian CNS, *Neuron.* 12:895 (1994).
11. J.F.C. Glatz, T. Börchers, F. Spener and G.J. Van Der Vusse, Fatty acids in cell signaling: Modulation by lipid binding proteins, *Prostaglandins Leukotrienes and Essential Fatty Acids.* 52:121 (1995).
12. N.M. Bass, Cellular binding proteins for fatty acids and retinoids: Similar or specialized functions? *Mol Cell Biochem.* 123:191 (1993).
13. S. Sorof, Modulation of mitogenesis by liver fatty acid binding protein, *Cancer and Metastasis Reviews.* 13:317 (1994).
14. H.M. Shields, M.L. Bates, N.M. Bass, C.J. Best, D.H. Alpers and R.K. Ockner, Light microscopic immunocytochemical localization of hepatic and intestinal types of fatty acid-binding proteins in rat small intestine, *J Lipid Res.* 27:549 (1986).
15. J.B. Lowe, J.C. Sacchettini, M. Laposata, J.J. McQuillan and J.I. Gordon, Expression of rat intestinal fatty acid-binding protein in *Escherichia coli:* Purification and comparison of ligand binding characteristics with that of *Escherichia coli*-derived rat liver fatty acid-binding protein, *J Biol Chem.* 262:5931 (1987).
16. R.J.A. Paulussen, G.P.M. Jansen and J.H. Veerkamp, Fatty acid-binding capacity of cytosolic proteins of various rat tissues: Effect of postnatal development, starvation, sex, clofibrate feeding, and light cycle, *Biochim Biophys Acta.* 877:342 (1986).
17. R.G.H.J. Maatman, H.T.B. van Moerkerk, I.M.A. Nooren, E.J.J. van Zoelen and J.H. Veerkamp, Expression of human liver fatty acid-binding protein in *Escherichia coli* and comparative analysis of its binding characteristics with muscle fatty acid-binding protein, *Biochim Biophys Acta.* 1214:1 (1994).
18. G.V. Richieri, R.T. Ogata and A.M. Kleinfeld, Equlibrium constants for the binding of fatty acids with fatty acid-binding proteins from adipocyte, intestine, heart, and liver measured with the flourescent probe ADIFAB, *J Biol Chem.* 269:23918 (1994).
19. H.K. Kim and J. Storch, Free fatty acid transfer from rat liver fatty acid-binding protein to phospholipid vesicles: Effect of ligand and solution properties, *J Biol Chem.* 267:77 (1992).
20. M.G. Wootan, N.M. Bass, D.A. Bernlohr and J. Storch, Fatty acid binding sites of rodent adipocyte and heart fatty acid binding proteins: Characterization using fluorescent fatty acids, *Biochemistry.* 29:9305 (1990).
21. M.G. Wootan, D.A. Bernlohr and J. Storch, Mechanism of fluorescent fatty acid transfer from adipocyte fatty acid binding protein to membrane, *Biochemistry.* 32:8622 (1993).
22. F.M. Herr, J. Aronson and J. Storch, Role of portal region lysine residues in electrostatic interactions between heart fatty acid binding protein and phospholipid membranes, *Biochemistry.* 35:1296 (1996).
23. V. Matarese, R.L. Stone, D.W. Waggoner and D.A. Bernlohr, Intracellular fatty acid trafficking and the role of cytosolic lipid binding proteins, *Prog Lipid Res.* 28:245 (1989).
24. H.K. Kim and J. Storch, Mechanism of free fatty acid transfer from rat heart fatty acid binding protein to phospholipid membranes: Evidence for a collisional process, *J Biol Chem.* 267:20051 (1992).
25. M.G. Wootan and J. Storch, Regulation of fluorescent fatty acid transfer from adipocyte and heart fatty acid binding proteins by acceptor membrane lipid composition and structure, *J Biol Chem.* 269:10517 (1994).
26. F.M. Herr, V. Matarese, D.A. Bernlohr and J. Storch, Surface lysine residues modulate the collisional transfer of fatty acid from adipocyte fatty acid binding protein to membranes, *Biochemistry.* 34:11840 (1995).
27. F.L. Hoch, Cardiolipins and biomembrane function, *Biochim Biophys Acta.* 1113:71 (1992).
28. K.T. Hsu and J. Storch, Fatty acid transfer from liver and intestinal fatty acid binding-proteins to membranes occurs by different mechanisms, *J Biol Chem.* [In Press] (1996).
29. G.M. Anantharamaiah, M.K. Jones and J.P. Segrest, *The Amphipathic Helix,* CRC Press, Boca Raton (1993).

THE MRP8/14 COMPLEX AND NOT THE EPIDERMAL FATTY ACID-BINDING PROTEIN IS RESPONSIBLE FOR THE TRANSLOCATION OF ARACHIDONIC ACID IN HUMAN KERATINOCYTES

Gerry Hagens, Karen Roulin, Raymonde Hotz, Jean-Hilaire Saurat and Georges Siegenthaler

Department of Dermatology
University Hospital
CH-1211 Geneva 14, Switzerland

INTRODUCTION

Free fatty acids (FA) play important roles in epidermis as structural membrane components as well as bioactive agents involved in the regulation of keratinocyte proliferation and differentiation. Thus, skin is one of the most active organes of the body involved in lipid synthesis. Indeed, FA are implicated in:

lipid barrier function,
sebum secretion,
desquamation process,
membrane formation,
cell growth (eicosanoids),
gene expression (activating PPARs), and
energy production.

However, FA show high hydrophobicity, instability and form poorly soluble salts (soap) with intracellular Ca^{2+}, which is necessary for keratinocyte differentiation. FA also show potent surfactant properties which might alter cell functions. Therefore, FA are transported, solubilized and protected by FA-binding proteins (FABP).

FABPs are widely distributed and were shown to be essentially cytoplasmic proteins of 14 - 15 kDa with a high affinity for long-chain FA. The presence of different FABPs with various ligand specificity and affinity in the same tissue suggests that these proteins might have very specific and distinct functions (reviewed in Veerkamp, 1995). Since human keratinocytes contain only one member of the FABP-family (E-FABP), which cannot fulfil all the trafficking of the various FA present in these cells, one might suggest that another class of FA-binding protein must also be present to coordinate the metabolism of intracellular FAs.

Frontiers in Bioactive Lipids
Edited by Vanderhoek, Plenum Press, New York, 1996

In this article, we review observations that were made in the study of the epidermal FABP. Furthermore, we present preliminary results on the finding of a new class of FA-binding proteins, which is specific for arachidonic acid and consists of a heteromeric complex formed by the non-covalent association of the two Ca^{2+}-binding proteins MRP8 (also referred to as cystic fibrosis antigen, L1 light chain, calgranulin A, p8) and MRP14 (L1 heavy chain, calgranulin B, p14).

THE EPIDERMAL FATTY ACID-BINDING PROTEIN (E-FABP)

The PAGE-Autoradioblotting technique allows the direct visualization of labelled ligand specifically bound to proteins. Using this method several human tissue-FABPs with distinct electrophoretic mobility could be detected (Siegenthaler et al., 1993). In addition, this method allowed the detection of a new member of this family, which was subsequently purified and characterized: the E-FABP (Siegenthaler et al., 1994). An "E-FABP" of identical electrophoretic mobility was detected by SDS-PAGE and PAGE-immunoblotting in other human tissues than keratinocytes, albeit at minor amounts. E-FABP from epidermis is an essentially cytoplasmic protein, which is not detectable in membranes (Table 1) and that displays similar characteristics to FABPs of other tissues with respect to molecular mass (15 kDa), K_d (0.46 μM) and binding properties for FAs. Using the gel filtration technique in which the tritiated ligand-release is lowered compared to the PAGE-radiobinding assay, higher amounts of E-FABP could be measured, indicating that the protein is not as scarcely expressed in these cells as previously described, Table1 (Siegenthaler et al., 1994).

Table 1. Levels of E-FABP in human keratinocytes[1]

	E-FABP
Differentiating keratinocytes	
membrane	n.m.
cytosol	1661
Non-differentiated keratinocytes	
membrane	n.m.
cytosol	286
Psoriatic scales	
membrane	n.m.
cytosol	2200

[1] in pmol of bound [^{3}H]oleic acid/mg protein ; n.m., not measurable

E-FABP shows a high affinity for C_{18}-FAs. A decrease in the number of carbons or the introduction of double bonds into the fatty acid chain, decreases the affinity for the ligand (Table 2). Interestingly, E-FABP does not significantly bind arachidonic acid. Sequencing of proteolytic peptides indicated that E-FABP might be identical to the psoriasis-associated (PA)-FABP (Madsen et al., 1992).

Immunohistochemical analysis of normal and pathological epidermis and oral mucosa showed that only differentiating keratinocytes express E-FABP, whereas the germinative compartment (basal layer) was E-FABP-negative. This observation was confirmed by studying malignant epidermal tumors; indeed, E-FABP was expressed only in differentiated keratinocytes of squamous cell carcinomas but not in any cells of basal carcinomas where keratinocytes only slightly differentiate (Masouyé et al., 1996). Psoriasis is a non-malignant skin disease showing abnormal keratinocyte differentiation and proliferation (Wright and Camplejohn, 1985). In this disease we observed an overexpression of E-FABP and an alteration of its gradient expression compared to normal epidermis, supporting the concept of an alteration of fatty acid metabolism in this disease.

Table 2. Relative binding capacities of E-FABP (%)

Ligands (number of carbons : double bonds)		Binding capacity in %
Stearic acid	(18 : 0)	100
Linoleic acid	(18 : 2)	84.3
Oleic acid	(18 : 1)	61.8
Linolenic acid	(18 : 3)	18.1
Palmitic acid	(16 : 0)	3.0
Arachidonic acid	(20 : 4)	1.0
Squalene		0
Cholesterol		0
12-HETE		0
all-trans Retinol		0
all-trans-, 9-cis-13-cis-Retinoic acid		0

according to Siegenthaler et al. (1994)

We have now strong evidences that the E-FABP-like molecule previously detected in human organs other than epidermis and oral mucosa is in fact also localized in endothelial cells. However, the localization of E-FABP is restricted to endothelial cells from small vessels situated in the dermis, heart, striated muscle, small intestine, placenta and renal medulla (Siegenthaler et al., 1996). By reverse transcription polymerase chain reaction (RT-PCR) using messenger RNA isolated from human endothelial cells, Southern blotting and sequencing of the cloned RT-PCR products, we proofed that E-FABP from epidermis is identical to the E-FABP from endothelial cells.

E-FABP might be considered as an important component of the machinery that is responsible for the synthesis of a lipid barrier in skin. However, the fact that E-FABP is also found in non-keratinized cells such as oral mucosa epithelium, squamous cell carcinomas or in endothelial cells indicates that the function of E-FABP might not be restricted to the establishment of the skin lipid barrier.

Overexpression and the immunohistochemical data might indicate that E-FABP plays a role in differentiation. Another hypothesis would be that E-FABP is implicated in cellular uptake of extracellular FA and could function as a reservoir in differentiating keratinocytes, which need high amounts of FAs. The presence of E-FABP in endothelial cells of blood capillaries from tissues with a high FA-metabolism (skin, adipose tissue. heart, skeletal muscle, placenta), supports the hypothesis that E-FABP may facilitate the transport of FA from the blood compartment to the target cells. Since blood capillaries are involved in the uptake and release of various metabolites, one can speculate that endothelial cells must undergo a FA-mediated activation to perform their functions.

THE MRP8/MRP14 COMPLEX (p34)

Differentiation of keratinocytes is triggered by an increase of the intracellular calcium concentration. In epidermis, keratinocytes proliferate in the basal germinative layer and upon migration towards the uppermost horny cell layer they continue to differentiate until they become corneocytes. This differentiation process parallels with the calcium gradient measured in the epidermis (Menon et al., 1991).

Although skin is one of the more active lipid-synthesizing tissues in mammals, essential FA such as linoleic and arachidonic acids must be acquired from circulation, the latter because the epidermis lacks the Δ5- and Δ6- desaturases (Chapkin et al., 1986). Arachidonic acid and derivatives, modulate gene expression by activating the peroxisome proliferator-activated nuclear receptors (PPARs; Göttlicher et al., 1992). As linolenic acid has a very low and arachidonic acid and 12-HETE almost no affinity for E-FABP (Table 2), we reasoned that there might exist another carrier for these FA. Because of the intense intracellular traffick of these agents in cultured keratinocyte that were induced to differentiate by calcium, and in cells of inflammed tissues (i.e. lesional psoriatic skin), such a carrier should be expressed in high amounts. Thus, protein extracts of cultured keratinocytes and psoriatic scales were analyzed by gel filtration using ^{3}H-oleic acid as a tracer. We detected a radioactive peak eluting at 34 kDa (refered to as p34), which abolished by the addition of an excess of unlabelled oleic acid. The finding that the radioactive peak did not disappear upon the addition of an excess of retinoic acid, suggests that this peak binds FA in a specific and reversible manner, which could be shown by several ligand binding studies Since 20-times more p34 were found in psoriatic scales, compared to differentiating keratinocytes, scales were used to purify p34. Using gel filtration, affinity chromatography with a FA graphted Sepharose column and ion exchange chromatography we purified the p34 protein to homogeneity (as confirmed by gel filtration). Amino acid sequencing and SDS-PAGE immunoblotting, revealed that the FA-binding-protein complex of p34 is composed of two well known non-covalently associated molecules MRP8 and MRP14. Further ligand studies demonstrated that the p34 complex bound labelled oleic acid (Figure 1), and, with a lower affinity, arachidonic acid and its precursors linoleic and linolenic acid in a saturable and specific manner. On the other hand, saturated FA like palmitic and stearic acid showed a poor affinity for p34. Interestingly, the individual components of p34 did not bind the FA indicating that the binding site of p34 is formed by the quaternary structure of the complex.

Recently, the migration inhibitory factor-related proteins (MRP8 and MRP14) have been isolated and cloned molecularly (Lagasse and Clerc, 1988). Both proteins belong to the S100 family of proteins, which are characterized by their ability to bind Ca^{2+} through a helix-loop-helix tertiary structure called EF-hand (reviewed in Kligman and Hilt, 1988). *In vivo,* MRP8 and MRP14 are found predominantly as a non-covalently associated hetero- or homodimer (Edgeworth et al., 1991). In addition, heterotrimers and tetramers were detected in a Ca^{2+}-dependent manner, suggesting that binding of Ca^{2+} to MRP8 or MRP14 results in conformational changes inducing the appearance of higher molecular weight forms of the complex. Both molecules are abundantly expressed in infiltrating monocytes and granulocytes during inflammatory disorders or in myelomonocytic cells during their differentiation (reviewed in Hessian et al., 1993). The expression of MRP8 and MRP14 in

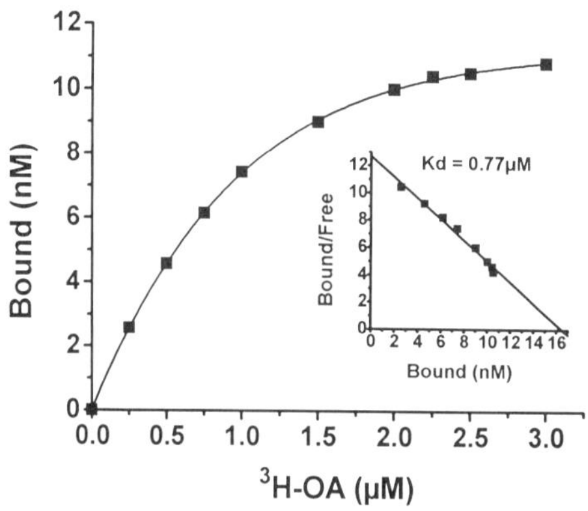

Figure 1. Binding kinetics of p34 with [^{3}H]oleic acid

epithelial cells has recently been reported (Saintigny et al., 1992). In a more detailed analysis we recently described the expression pattern of MRP8, MRP14 and p34 in non-differentiated keratinocytes, differentiating keratinocytes and in cells from abnormally differentiated psoriatic scales (manuscript in preparation). By immunoprecipitation and SDS-PAGE/immunoblotting it was possible to show that the individual components of p34 are expressed in above mentioned cell types, albeit at different levels. Furthermore, we demonstrated that the p34 complex can only be found in differentiating cells and in psoriatic scales, suggesting that the MRP-molecules must first mature during keratinocyte differentiation to become capable of forming the p34 complex. Whether post-translational modifications are implicated in complex formation is currently under investigation.

In myeloid and epithelial cells, the MRP8/MRP14 complex was shown to translocate in a Ca^{2+}-dependent way (Roth et al., 1993; Goebeler et al., 1995). In our studies we find that in keratinocytes that were induced to differentiate by Ca^{2+}, MRP8, MRP14 and p34 expression increase and that a portion of the p34 complex translocates to the membrane, suggesting that mechanisms involved in differentiation of keratinocytes regulate the Ca^{2+}-dependent expression and translocation of p34 or its components. A similar result was obtained by investigating p34 in keratinocytes located in scales of the inflammatory disease psoriasis. In contrast to the cytosolic protein fraction, the membrane protein fraction of psoriatic scales contains increased amounts of p34, supporting the idea that p34 translocation is regulated by mechanisms mediating differentiation and inflammation.

The biological function of the MRP8/MRP14 complex *in vivo* is not yet understood and observations suggest that the MRP8/14 complex might play a role in differentiation and inflammation. The ability of MRP8/14 to inhibit casein kinase I and II and the finding that MRP14 becomes phosphorylated upon certain stimuli further suggests that this complex might be involved in signal transduction (Murao et al., 1989; Edgeworth et al., 1989). Since both molecules bind Ca^{2+} it is conceivable to speculate that they may transduce Ca^{2+}-derived signals.

Since S100 proteins contain two hydrophobic and two ionic domains, homo- and heteromers are formed by the association of complementary affinity domains. The FA-binding site of p34 might be created by the arrangement of the hydrophobic domains, which form a hydrophobic cavity for FAs.

Our preliminary studies demonstrated that the p34 complex from keratinocytes represents a novel class of FA-binding proteins. As arachidonic acid represents a mediator of inflammation, we believe that its biological activities are mediated by p34. Recently, Murao *et al.* (1989) described the presence of MRP8/14 in spleen nuclei. Furthermore, arachidonic acid and its derivatives have been shown to be efficient activators of peroxisome proliferator-activated nuclear receptors (PPARs). Thus, it might be possible that above mentioned FA affect nuclear signaling pathways in a manner similar to retinoic acid, vitamin D or steorid hormones.

The p34 complex might be the specific carrier for the calcium-regulated translocation of arachidonic acid in human keratinocytes.

ACKNOWLEDGEMENTS
This investigation was supported, in part, by the Swiss National Science Foundation (grant No 32-31338.91).

References

Chapkin, R.S., Ziboh, V.A., and Marcelo, C.L., 1986, Metabolism of essential fatty acids by human epidermal enzyme preparations: evidence of chain elongation, *J. Lipid Res.* 27: 945

Edgeworth, J., Freemont, P., and Hogg, N. (1989), Ionomycin-regulated phosphorylation of the myeloid calcium-binding protein p14, *Nature* 342: 189

Edgeworth, J., Gorman, M., Bennett, R., Freemont, P., and Hogg, N., 1991, Identification of p8, 14 as a highly abundant heterodimeric calcium binding protein complex of myeloid cells, *J. Biol. Chem.* 266: 7706

Goebeler, M., Roth, J., van den Bos, C., Ader, G., and Sorg, C., 1995, Increase of calcium levels in epithelial cells induces translocation of calcium-binding proteins migration inhibitory factor-related protein 8 (MRP8) and MRP14 to keratin intermediate filaments, *Biochem. J.* 309: 419

Göttlicher, M., Widmark, E., Li, Q., and Gustafsson, J.A., 1992, Fatty acids activate a chimera of the clofibric acid-activated receptor and the glucocorticoid receptor, *Proc. Natl. Acad. Sci. USA* 89: 4653

Hessian, P.A., Edgeworth, J., and Hogg. N., 1993, MRP8 and MRP14, two abundant Ca^{2+}-binding proteins of neutrophils and monocytes, *J. Leukocyte Biol.* 53: 197

Kligman, D., and Hilt, G.C., 1988, The S100 protein family, *Trends Biochem. Sci.* 13 : 437

Kragballe, K. and Voorhees, J.J., 1987, Eicosanoids in psoriasis-15-HETE on the stage *Dermatologica* 174: 209

Lagasse, E., and Clerc, R.G., 1988, Cloning and expression of two human genes encoding calcium-binding proteins that are regulated during myeloid differentiation, *Mol. Cell. Biol.* 8: 2402

Madsen, P. Rasmussen, H.H., Leffers, H., Honoré, B., and Celis, J.E., 1992, Molecular cloning and expression of a novel keratinocyte protein (psoriasis-associated fatty acid-binding protein [PA-FABP] that is highly up-regulated in psoriatic skin and that shares similarity to fatty acid-binding proteins, *J. Invest. Dermat.* 99: 299

Masouyé, I., Saurat, J.H., and Siegenthaler, G., 1996, Epidermal fatty acid-binding protein in psoriasis, basal and squamous cell carcinomas: an immunohistological study, *Dermatology* 192: 208

Menon, G.K., and Elias, P.M., 1991, Ultrastructural localization of calcium in psoriatic and normal human epidermis, *Arch Dermatol.*,127: 57

Murao, S., Collard, F.R., and Huberman, E., 1989, A protein containing the cystic fibrosis antigen is an inhibitor of protein kinases, *J. Biol. Chem.* 264: 8356

Roth, J., Burwinkel, F., van den Bos, C., Goebeler, M., Vollmer, E., and Sorg, C., 1993, MRP8 and MRP14, S-100 like proteins associated with myeloid differentiation, are translocated to plasma membrane and intermediate filaments in a calcium-dependent manner, *Blood* 82: 1875

Saintigny, G., Schmidt, R., Shroot, B., Juhlin, L., Reichert, U., and Michel, S., 1992, Differential expression of calgranulin A and B in various epithelial cell lines and reconstructed epidermis. *J. Invest. Dermatol.* 99 : 639

Siegenthaler, G., Hotz, R., Chatellard-gruaz, D., Jaconi, S., and Saurat, J.-H., 1993, Characterization and expression of a novel human fatty acid-binding protein: the epidermal type (E-FABP), *Biochem. Biophys. Res. Commun.* 190: 482

Siegenthaler, G., Hotz, R., Chatellard-Gruaz, D., Didierjean, L., Hellman, U., and Saurat, J.-H., 1994, Purification and characterization of the human epidermal fatty acid-binding protein : localization during epidermal differentiation in vivo and in vitro, *Biochem. J.* 302: 363

Siegenthaler, G., Roulin, K., Hagens, G., Chatellard-Gruaz, D., Saurat, J.-H., and Hellman, U., 1996, A novel calcium-dependent mechanism for arachidonic acid translocation in human keratinocytes, Frontiers in bioactive lipids '96, Absract volume of the XVIth international spring symposium Washington D.C., ed. Vanderhoek, J.Y., p54.

Veerkamp, J.H, 1995, Fatty acid transport and fatty acid-binding proteins, *Proc. Nutr.Scociety* 54: 23

Wright, N.A., and Camplejohn, R.S., 1985, Psoriasis: Cell Proliferation, Churchill Livingston, Edinbrugh

INTERACTION OF FATTY ACID-BINDING PROTEINS WITH THE PEROXISOME PROLIFERATOR-ACTIVATED RECEPTOR ALPHA.

EVIDENCE FOR FABP MODULATION OF THE GENE RESPONSE TO FATTY ACID OVERLOAD.

Nathan M. Bass

Department of Medicine
University of California, San Francisco
San Francisco, CA, 94143, USA.

INTRODUCTION

The recently described peroxisome proliferator-activated receptors (PPARs)[1] are members of the nuclear receptor superfamily of transactivating factors that mediate the pleiotropic response to peroxisome proliferators.[1-4] PPARs are implicated in several important physiological processes, including adipocyte differentiation and the adaptation of hepatocytes to abnormally high fluxes of long-chain fatty acids (fatty acid overload). Long-chain fatty acids and their metabolites are important activators of the transactivating properties of PPARs, which are effected through the binding of these receptors as heterodimers with the 9 *cis* retinoic acid receptor (RXR) to peroxisome-proliferator response elements (PPREs) in the upstream promoter region of PPAR-responsive genes.[1-3] The latter include key enzymes of the extramitochondrial pathways of fatty acid oxidation in peroxisomes and endoplasmic reticulum and also the abundant cytosolic liver fatty acid-binding protein (L-FABP). L-FABP is induced both *in vivo*[5] and in cell culture[6,7] by peroxisome proliferators and also by long-chain fatty acids under conditions of impaired mitochondrial β-oxidation.[8] L-FABP contains a consensus PPRE in its 5' promoter region (Figure 1), which appears to be functional.[9] Although extensively characterized in terms of structure, binding properties and regulation,[10-13] there is less definite information on the function of the FABP family of proteins. Evidence has been presented supporting a variety of roles for these proteins in the intracellular transport/diffusion and metabolic compartmentalization of long-chain fatty acids,[14-17] in the protection of cells from the adverse effects of high concentrations of unbound fatty acids

---

[1] Abbreviations used: BSA, bovine serum albumin; FABP, fatty acid-binding protein; FACO, peroxisomal fatty acyl CoA oxidase; H-FABP, heart fatty acid-binding protein; L-FABP, liver fatty acid-binding protein; PPAR, peroxisome proliferator-activated receptor; PPRE, peroxisome-proliferator response element.

Frontiers in Bioactive Lipids
Edited by Vanderhoek, Plenum Press, New York, 1996

ACTTCTGCCTTGCCCATTCTGATTTTTATCGTTGACCATTGCTCTCAGGAGTTA

-67 -55

ATGTTTGATCCTGGCCATAAAGAAATCGACAATCAC**TGACCTATGGCCT**ATA

TTTGAGGAGGAAGAAGCCCCTTATAAAATAGCCAACAGTGGGTGGCCTGGCAGA

CAGAGCTGTTGTGGTCAGCTGTGGAAAGGAAACCTCATTGCCACC**ATG**...........

Figure 1. 5' Promoter region of the rat L-FABP gene showing the position of PPRE (bold, underlined), TATA box (underlined), transcription initiation site (arrow) and translation start codon (bold)

and also in the modulation of cell growth.[18-20] In spite of their ability to activate PPARs, long-chain fatty acids, show little binding affinity for these receptors, but are bound along with a broad array of other PPAR-activating compounds by L-FABP.[6,21,22] Thus, the possibility of a biologically important interaction between L-FABP and PPAR activation has been suggested.[9,23] This review summarizes recent work from our laboratory investigating the interaction between FABP, most notably L-FABP, and the activation of PPARα, the isotype of this family of nuclear receptors which is predominantly expressed in liver, kidney and heart muscle tissue.[24,25]

REGIONAL EXPRESSION OF L-FABP AND PPAR-RESPONSIVE GENES IN LIVER

Immunohistochemical and digitonin perfusion studies have shown that L-FABP is expressed in an acinar zone 1-2 (portal-to-central) gradient pattern in the liver.[14,26] A similar pattern of expression for L-FABP mRNA is evident from *in-situ* hybridization studies on rat liver,[27] which also show, by contrast, other key PPARα responsive genes, including cytochrome P450 4A and peroxisomal fatty acyl CoA oxidase (FACO) to be expressed in a zone 3-2 (central-to-portal) pattern. PPARα mRNA signals in normal rat liver were too weak to detect, but also showed a central-to-portal pattern in the liver of rats following treatment with clofibrate.[27] The presence of a centrizonal or acinar zone 3 predominance of PPARα[24] and its responsive gene products[28] has been observed in rodent livers by others as well. These findings indicate that while the expression of several PPARα-responsive genes follows the regional pattern of PPARα expression in the liver, L-FABP is an exception, and the expression pattern of this protein cannot be explained soley on the basis of the acinar distribution of PPARα. The data are also compatible with the possibility that L-FABP may inhibit the induction of other PPARα -responsive genes in acinar zones 1-2.[23] This possibility has been further explored using transient transactivation assays for PPARα.

INTERACTION OF ALBUMIN AND INTRACELLULAR FABPs WITH PPARα IN TRANSACTIVATION ASSAYS

We have employed transient transactivation assays of PPARα to further investigate the potential interaction between the activation of PPARα and FABPs, including albumin.[29,30] In these experiments, rat HTC hepatoma cells co-transfected with the expression vector pCMX-mPPARα[31] and a luciferase reporter gene driven by a tandem repeat PPRE from the FACO gene enhancer (PPRE-*tk*-LUC), revealed a high level of

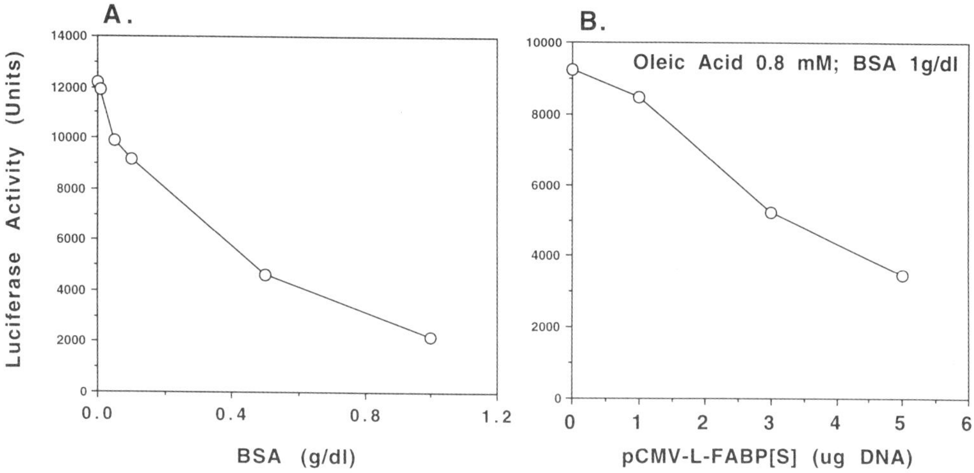

Figure 2. Effect of albumin and L-FABP on PPARα activation in HTC cells. A. Effect of increasing BSA in medium on basal activation of PPARα in transfected cells. B. Effect of increasing amount of cotransfected pCMV-L-FABP[S] on the activation of PPARα by oleic acid in the medium. Total DNA transfected was kept constant with parent plasmid pCMV.

endogenous or basal activation of the receptor in the absence of any added ligand (6-fold increase in luciferase activity compared with cells transfected with parent plasmid, pCMX, and reporter gene). Bovine serum albumin (BSA) added to the medium up to a concentration of 1 g/dl (~150 μm) produced a dose-dependent reduction in the level of basal PPARα activation (Figure 2). In addition, activation of PPARα by 0.4 mM oleic acid bound to varying amounts of albumin (0.5 - 4.0 g/dl) in the medium showed a negative exponential correlation with BSA concentration. We have also noted that BSA attenuates the activation of PPARα by clofibrate in transfection assays. Pertinent to these finding are earlier observations in primary cultured hepatocytes, in which the induction of peroxisomal enzymes of fatty acid oxidation was consistently observed to be attenuated in the presence of albumin in the medium, even at relatively low concentrations.[7,32] These data show that the attenuation by albumin of enzyme induction by peroxisome proliferators in cell culture is probably explained by the attenuation of PPARα activation. The mechanism whereby albumin exerts this effect is not established, but it seems most likely that it results from the binding of fatty acids and xenobiotic peroxisome proliferators or even more proximate activators/ligands of PPARα to albumin, which thus effectively partitions potential activators away from the intracellular compartment in which PPARα is activated.

An effect of L-FABP on PPARα activation is difficult to predict on the basis of existing information. As a facilitating factor in the intracellular diffusion of long-chain fatty acids and other small molecules,[15-17] L-FABP might enhance PPARα activation via the facilitation of ligand diffusion towards the compartment (cytosolic or nuclear) in which PPARα undergoes activation. On the other hand, L-FABP may compete with PPARα intracellularly — similar to albumin extracellularly — via an as yet poorly defined compartmentation function or via the facilitation of diffusion towards competing pathways, e.g., metabolic utilization. We addressed the question of how L-FABP interacts with PPARα by constructing and cotransfecting an expression vector for L-FABP driven by the CMV promoter.[29] Vectors tested contained the full open reading frame and Kozack sequence of rat liver L-FABP cDNA[33] in either sense (pCMV-L-FABP[S]) or antisense (pCMV-L-FABP[AS]) orientation. In experiments in which total transfected DNA was kept constant using parent plasmid, we determined that co-transfection with pCMV-L-FABP[S] attenuated both basal and oleic acid-stimulated PPARα-driven luciferase expression (Figure 2B). By contrast, co-transfection with pCMV-L-FABP[AS] had no effect on PPARα activation. Co-transfection of pCMV-L-FABP[S] also caused a shift in the dose-response curve of oleic acid activation of PPARα to the right. pCMV-L-FABP[S] cotransfection also attenuated the activation of PPARα by the peroxisome proliferating hypolipidemic drugs, clofibrate and bezafibrate. Although the bezafibrate binds to L-FABP with a Kd of ~9μM, clofibrate is much more weakly bound to this protein.[6] However, attenuation by pCMV-L-FABP[S] cotransfection of the activation of PPARα by these drugs was not proportional to their affinities for L-FABP. This is consistent with the argument that fibrate drugs do not activate PPAR directly, but rather act via a primary disruption in fatty acid metabolism, with fatty acyl ligands of L-FABP being the ultimate inducers.

L-FABP belongs to a family of at least 7 related FABP genes that differ both structurally and functionally as well. A notable difference, for example, between L-FABP and heart-type FABP (H-FABP) is in ligand specificity — H-FABP binding appears more specific for fatty acids compared to the broad affinity of L-FABP for small molecules,[22] while these proteins differ also in their mechanisms of intermembrane transfer of fluorescent fatty acids[34-36] and *in vitro* facilitated fatty acid diffusion (R.A. Weisiger, N.M. Bass, unpublished data). Using PCR, we have recently constructed an expression vector for rat H-FABP (pCR3-H-FABP) and conducted comparative studies on the effect of co-transfection of this vector with PPARα on the transcription of its reporter gene. In distinct contrast to L-FABP, co-transfection of pCR3-H-FABP resulted in a dose-related enhancement of PPARα-dependent luciferase expression. While these data are preliminary, they point to the existence of a striking divergence in structure-dependent function of tissue-specific members of the FABP family in relation to modulation of PPARα-dependent gene transcription.

SUMMARY AND CONCLUSIONS

In summary, our data support a role for albumin in the extracellular compartment and L-FABP in the intracellular compartment in dampening PPARα activation in the liver by fatty acids and xenobiotics. Although the biological significance of the attenuation of PPAR activation by binding proteins requires further study, it may serve the important function of preventing low-level, chronic induction of the peroxisomal β-oxidation pathway and hence the generation of potentially toxic reactive oxygen species which may produce tissue damage or neoplasia in the long term.[37,38] In contrast to our findings with L-FABP, cotransfection with an H-FABP expression vector promotes PPARα activation by fatty acids. These findings thus provide new evidence for structure-dependent, functional specialization of the different, tissue-specific FABP, notably in relation to the modulation of PPARα activation by fatty acids. They also suggest that there exist marked differences in the mechanisms of adaptation of different tissues to fatty acid overload.

ACKNOWLEDGMENTS

The author acknowledges the assistance of Renaissance Appel and Sandra Huling in the performance of unpublished work discussed in this manuscript. The vectors pCMX and pCMX-mPPARα used in these studies were kindly provided by Drs Barry M. Forman and Ronald M. Evans. This work is supported by grant DK 32926 from the National Institutes of Health.

REFERENCES

1. H. Keller, and W. Wahli. Peroxisome proliferator-activated receptors - a link between endocrinology and nutrition. *Trends.Endocrinol.Metab.* 4:291 (1993).

2. K. Motojima. Peroxisome proliferator-activated receptor (Ppar) - structure, mechanisms of activation and diverse functions. *Cell Struct.Funct.* 18:267 (1993).

3. W. Wahli, O. Braissant, and B. Desvergne. Peroxisome proliferator activated receptors: Transcriptional regulators of adipogenesis, lipid metabolism and more. *Chem. Biol.* 2:261 (1995).

4. B.M. Forman, P. Tontonoz, J. Chen, R.P. Brun, B.M. Spiegelman, and R.M. Evans. 15-Deoxy-Delta(12,14)-prostaglandin J(2) is a ligand for the adipocyte determination factor PPAR gamma. *Cell* 83:803 (1995).

5. N.M. Bass, R.K. Ockner, J.A. Manning, J.I. Gordon, S. Seetharam, and D.H. Alpers. Regulation of the biosynthesis of two distinct fatty acid binding proteins in rat liver and intestine: Influences of sex differences and of clofibrate. *J. Biol. Chem.* 260:1432 (1985).

6. R. Brandes, R. Kaikaus, R.K. Ockner, N. Lysenko, and N.M. Bass. Induction of fatty acid binding protein by peroxisome proliferators in primary hepatocyte cultures: relationship to the induction of peroxisomal β-oxidation. *Biochim. Biophys .Acta.* 1034:53 (1990).

7. R.M. Kaikaus, Z. Sui, N. Lysenko, N.Y. Wu, P. Ortiz de Montellano, R.K. Ockner, N.M. Bass. Regulation of pathways of extramitochondrial fatty acid oxidation and liver fatty acid binding protein by long-chain monocarboxylic fatty acids in hepatocytes. Effect of inhibition of carnitine palmitoyltransferase I. *J. Biol. Chem.* 268:26866 (1993).

8. R.M. Kaikaus, W. Chan, N. Lysenko, R. Ray, P. Ortiz de Montellano, and N.M. Bass NM. Induction of peroxisomal fatty acid β–oxidation and liver fatty acid bining protein by peroxisome proliferators: mediation via the cytochrome P-450 IVA1 ω-hydroxylase pathway. *J. Biol. Chem.* 268:9593 (1993).

9. I. Issemann, R. Prince, J. Tugwood, and S.A. Green. Role for fatty acids and liver fatty acid binding protein in peroxisome proliferation. *Biochem.Soc.Trans.* 20:824 (1992).

10. J.F.C. Glatz, and G.J. Van der Vusse. Cellular fatty acid-binding proteins: current concepts and future directions. *Mol.Cell Biochem.* 98:237 (1990).

11. J.F.C. Glatz, M.M. Vork, D.P. Cistola, and G.J. van der Vusse. Cytoplasmic fatty acid binding protein - significance for intracellular transport of fatty acids and putative role on signal transduction pathways. *Prostagland.Leuk.Essent.Fatty Acids.* 48:33 (1993).

12. T. Borchers, and F. Spener. Fatty acid binding proteins. *Cell. Lipids* 40:261 (1994).

13. J.H. Veerkamp, and R.G.H.J. Maatman. Cytoplasmic fatty acid-binding proteins: Their structure and genes. *Prog.Lipid.Res.* 34:17 (1995).

14. N.M. Bass. Fatty acid binding protein expression in the liver: its regulation and relationship to the zonation of fatty acid metabolism. *Mol. Cell. Biochem.* 98:167 (1990).

15. R.M. Kaikaus, N.M. Bass, and R.K. Ockner. Functions of fatty acid binding proteins. *Experientia* 46:617 (1990).

16. J.M. Stewart, W.R. Driedzic, and J.A. Berkelaar. Fatty-acid-binding protein facilitates the diffusion of oleate in a model cytosol system. *Biochem.J.* 275:569 (1991).

17. M.M. Vork, J.F.C. Glatz, and G.J. van der Vusse. On the mechanism of long chain fatty acid transport in cardiomyocytes as facilitated by cytoplasmic fatty acid-binding protein. *J Theor.Biol.* 160:207 (1993).

18. S. Sorof. Modulation of mitogenesis by liver fatty acid binding protein. *Cancer Metastasis.Rev.* 13:317 (1994).

19. Kurtz, A., Zimmer, A., Schnutgen, F., Bruning, G., Spener, F. and Muller, T. The expression pattern of a novel gene encoding brain fatty acid binding protein correlates with neuronal and glial cell development. *Development* 120:2637-2649, 1994.

20. Y.M. Yang, E. Spitzer, N. Kenney, W. Zschiesche, M.L. Li, A. Kromminga, T. Muller F. Spener, A. Lezius, J.H. Veerkamp, G.H. Smith, D.S. Salomon, and R. Grosse. Members of the fatty acid binding protein family are differentiation factors for the mammary gland. *J. Cell. Biol.* 127:1097 (1994).

21. J.R. Cannon, and P.I. Eacho. Interaction of LY171883 and other peroxisome proliferators with fatty-acid-binding protein isolated from rat liver. *Biochem.J.* 280:387 (1991).

22. R.G.H.J. Maatman, H.T.B. van Moerkerk, I.M.A. Nooren, E.J.J. van Zoelen, and J.H. Veerkamp. Expression of human liver fatty acid-binding protein in Escherichia coli and comparative analysis of its binding characteristics with muscle fatty acid-binding protein. *Bba-Lipid.Lipid.Metab.* 1214:1 (1994).

23. N.M. Bass. Cellular binding proteins for fatty acids and retinoids - similar or specialized functions? *Mol. Cell. Biochem.* 123:191 (1993).

24. Q. Huang, A.V. Yeldandi, K. Alvares, H. Ide, J.K. Reddy, and M.S. Rao. Localization of peroxisome proliferator-activated receptor in mouse and rat tissues and demonstration of its nuclear translocation in transfected CV-1 cells. *Int.J Oncol.* 6:307 (1995).

25. O. Braissant, F. Foufelle, C. Scotto, M. Dauca, and W. Wahli. Differential expression of peroxisome proliferator- activated receptors (PPARs): Tissue distribution of PPAR- alpha, -beta, and -gamma in the adult rat. *Endocrinology* 137:354 (1996).

26. N.M. Bass, M.E. Barker, J.A. Manning, A.L. Jones, and R.K. Ockner. Acinar heterogeneity of fatty acid binding protein expression in the livers of male, female and clofibrate-treated rats. *Hepatology* 9:12 (1989).

27. N.M. Bass, S. Huling, R.A. Appel, N.Y. Wu, and R.M. Kaikaus. Hepatic acinar expression of the peroxisome-proliferator-activated receptor (PPARα) and PPARα-responsive genes by *in situ* hybridization. *Hepatology* 22:303A (1995).

28. D.R. Bell, R.G. Bars, G.G. Gibson, and C.R. Elcombe. Localization and differential induction of cytochrome P450IVA and acyl-CoA oxidase in rat liver. *Biochem.J* 275:247 (1991).

29. N.M. Bass, Y. Sadovsky, R.A. Appel, and P.J. Kushner. Dicarboxylic fatty acids activate the peroxisome-proliferator-activated receptor (PPAR) in rat HTC hepatoma cells. *Hepatology* 18:125A (1993).

30. N.M. Bass, R.A. Appel, Y. Sadovsky, and P.J. Kushner. Transient transfection of liver fatty acid binding protein (L-FABP) inhibits activation of the peroxisome proliferator-activated receptor (PPAR). *Hepatology* 20:324A (1994).

31. S.A. Kliewer, B.M. Forman, B. Blumberg, E.S. Ong, U. Borgmeyer, D.J. Mangelsdorf, K. Umesono, and R.M. Evans. Differential expression and activation of a family of murine peroxisome proliferator-activated receptors. *Proc.Natl.Acad.Sci.USA*. 91:7355 (1994).

32. R. Hertz, J. Bar-Tana, M. Sujatta, J. Pill, F.H. Schmidt, and H.D. Fahimi. The induction of liver peroxisomal proliferation by β,β'-methyl-substituted hexadecanoic acid (Medica 16). *Biochem. Pharmacol.* 37:3571 (1988).

33. J.I. Gordon, D.H. Alpers, R.K. Ockner, and A.W. Strauss. The nucleotide sequence of rat liver fatty acid binding protein mRNA. *J. Biol. Chem.* 258:3356 (1982).

34. H.K. Kim, and J. Storch. Free fatty acid transfer from rat liver fatty acid-binding protein to phospholipid vesicles - effect of ligand and solution properties. *J. Biol. Chem.* 267:77 (1992).

35. H.K. Kim, and J. Storch. Mechanism of Free Fatty Acid Transfer from Rat Heart Fatty Acid-Binding Protein to Phospholipid Membranes - Evidence for a Collisional Process. *J. Biol. Chem.* 267:20051 (1992).

36. J. Storch, and N.M. Bass. Transfer of fluorescent fatty acids from liver and heart fatty acid-binding proteins to model membranes. *J. Biol. Chem.* 265:7827 (1990).

37. R.K. Ockner, R.M. Kaikaus, andN.M. Bass. Fatty acid metabolism and the pathogenesis of hepatocellular carcinoma. An hypothesis. *Hepatology* 18:669 (1993).

38. S. Chu, Q. Huang, K. Alvares, A.V. Yeldandi, M.S. Rao, and J.K. Reddy. Transformation of mammalian cells by overexpressing H2O2- generating peroxisomal fatty acyl-CoA oxidase. *Proc.Natl.Acad.Sci.USA*. 92:7080 (1995).

THE ROLE OF ACYLCOA BINDING PROTEIN (ACBP) AND LONG-CHAIN ACYLCOA ESTERS IN CELL REGULATION AND FUNCTION

Jens Knudsen

Institute of Biochemistry
Odense University
Campusvej 55
DK-5230 Odense M
Denmark

INTRODUCTION

ACBP was discovered independently by five different groups. First it was isolated from rat brain as a modulator of the $GABA_A$ receptor complex (Guidotti et al., 1983) and given the name Diazepam Binding Inhibitor (DBI). It was then identified as a bovine brain factor which affected cell growth (Marquardt et al., 1986; Shoyab et al., 1986). The third group identified the ACBP on its ability to modify the chain length of fatty acids synthesized by goat fatty acid synthetase (Mogensen et al., 1987) and named the protein AcylCoA Binding Protein (ACBP), and it was not until 1989 it became clear that ACBP and DBI were identical proteins (Knudsen et al., 1989). ACBP/DBI was also identified as an intestinal factor which acted as an inhibitor of glucose-induced insulin secretion from the pancreas (Chen et al., 1988). Finally, it was identified as an adrenal factor which stimulated adrenal mitochondria pregnenolone synthesis (Yanagibashi et al., 1988; Besman et al., 1989).

To what extent ACBP/DBI is a true multi-functional protein with extracellular regulatory functions is not clear. It is not synthesized with a secretory signal peptide and is therefore not likely to be secreted through the normal protein secretory pathway (Mandrup et al., 1992). Furthermore, ACBP/DBI is widely distributed, it is found in yeast (Rose et al., 1992; Knudsen et al., 1994) and in most tissues of all eukaryotic and plant species tested (Costa and Guidotti, 1991; Kolmer et al., 1994; Kragelund et al., 1996). It is therefore unlikely that it should be able to function as an extracellular signal molecule in a tissue-specific manner. The fact that nM concentrations of ACBP/DBI and its proteolytical degradation product ODN ($ACBP/DBI_{33\text{-}50}$) are able to affect glucose-induced insulin secretion from pancreas (Borboni et al., 1991; Östenson et al., 1994) and polyphosphoinositide metabolism in astrocytes (Patte et al., 1995) indicates that ACBP/DBI recognizes and interacts with a surface receptor. However, it still needs to be shown that ACBP/DBI is secreted under in vivo conditions in a tissue-specific manner in order to evaluate the physiological significance of these observations. The observed effect of ACBP/DBI on steroidogenesis is interesting and fully compatible with the ability of

ACBP/DBI to function as an intracellular acylCoA pool former and transporter. Finally, it has been reported that ACPB in brain and other tissues can be processed to smaller peptides with biological activity in cell cultures and to affect animal behaviour when injected intracerebroventricularly (see Costa and Guidotti, 1991, for review).

It is not the intention to give a fully comprehensive review on ACBP/DBI. In the present chapter I will focus on the role of ACBP/DBI as an intracellular acylCoA pool former and transporter. The presented data will show that ACBP/DBI binds long-chain acylCoA esters with high specificity and affinity and strongly indicate that ACBP/DBI indeed is able to act as an intracellular acylCoA pool former and transporter under in vivo conditions. Therefore, in the following the name ACBP will be used for ACBP/DBI.

ACBP STRUCTURE AND LIGAND BINDING

ACBP from mammals, yeast, birds and fruit fly consists of 86 amino acids, and moth larvae's ACBP contains 90 amino acids, whereas plant ACBP contains 92 residues. A comparison of 16 known sequences shows that they are closely related (identity > 25%) and most of them have been demonstrated to bind long-chain acylCoA esters (Kragelund et al., 1996). The conserved residues are found to stabilize helix-helix interaction and serve specific functional roles for ligand binding. ACBP folds very fast in a one-step folding mechanism, and it has been speculated that the conserved residues also provide a scaffold for the folding mechanism (Kragelund et al., 1996).

ACBP-like sequences have also been found as domains of larger proteins. A related DNA, encoding a transmembrane protein of 533 amino acids with a homologous domain (46% identify) with rat ACBP, has been cloned from bovine brain (Webb et al., 1987). Furthermore, the amino terminal part of enoylCoA hydratase from *C. elegans* contains a domain with 46% sequence identity to rat ACBP (Wilson et al., 1994).

The three-dimensional structure of bovine ACBP has been determined by use of multidimensional hetero-nuclear NMR. The structure has been determined to a four-helix bundle motif where the four α-helices are arranged in a left-handed up-down-down-up orientation with an overhand connection between A_2 and A_3 (Andersen and Poulsen, 1992, 1993).

The four α-helices are A_1 (Glu4 to Leu15), A_2 (Asp21 to Val36), A_3 (Gly51 to Lys62) and A_4 (Ser65 to Tyr84), and they interact mainly through hydrophobic interactions (Andersen and Poulsen, 1992, 1993). There is one binding site where long-chain acylCoA esters bind with high affinity (Rosendal et al., 1993; Rasmussen et al., 1994; Knudsen et al., 1994). The K_D for palmitoylCoA binding to bovine and yeast ACBP has, by the use of titration microcalorimetry, been determined to 0.45 10^{-13} M and 0.55 10^{-10} M, respectively (Rasmussen et al., 1994; Knudsen et al., 1994). The binding specificity shows preference for C14 to C22 acylCoA esters, and introduction of double bonds in the acyl chain does not affect binding affinity (Mikkelsen and Knudsen, 1987; Rosendal et al., 1993). The binding site for the palmitoyl chain stretches between the N-terminal of A_2 where the carboxyl part binds to the N-terminal of A_3 where the ω-end of the palmitoyl part binds. The adenosine-3'-phosphate is bound near residues of each of the four helices in an arrangement where it can form salt bridges and/or hydrogen bonds to either backbone or side-chain atoms of Ala9, Tyr28, Lys32, Lys54 and Tyr73. The polar parts of the pantetheine and the pyrophosphate are structured in the bound ligand to form an interphase with the solvent. Also the ligand forms a set of non-polar intramolecular interactions where the adenine, the pantetheine, and the palmitoyl chains are associated, so overall the structure of the bound ligand seems to be organized to protect the lipophilic palmitoyl part from the polar solvent (Kragelund et al., 1993).

GENE ORGANIZATION AND REGULATION OF EXPRESSION

The ACBP gene family in the rat comprises four processed pseudo genes and one expressed gene (Mandrup et al., 1992, 1993b). In humans two ACBP genes have been identified, one expressed gene located on chromosome two and a pseudo gene located on chromosome six (Gersuk et al., 1995). Yeast contains one ACBP gene without introns (Rose et al., 1992). The expressed rat ACBP gene is organized in four exons and three introns. There is a remarkable correspondence between the structural modules of ACBP and the exon/intron architecture of the ACBP gene. In both brain and liver transcription was initiated from two major and multiple minor initiation sites. No evidence for alternative transcript processing was obtained. The promotor region of the rat ACBP gene is located in a CpG island and lacks a conical TATA box, thus the ACBP gene expresses all the hallmarks for a typical housekeeping gene (Mandrup et al., 1992). Analysis of the promotor region of the rat ACBP gene revealed potential recognition sequences of NF-1/CTF, Sp1, AP-1 C/EBP and HNF3. In addition, specific binding to a DR-1 type element was observed. This DR-1 element was capable of binding PPARα/RXRα, PPARδ/RXRα and PPARγ2/RXRα hetero dimers. Addition of the peroxisome proliferators to H4IIC3 rat hepatoma cells led to an increase in ACBP mRNA levels, indicating that the DR-1 element is functional in the rat ACBP gene promotor (Elholm et al., 1996). The existence of a functional DR-1 element is further supported by the finding that feeding rats diets containing the peroxisome proliferators 3-thia-fatty acids, perfluoro fatty acids and clofibrates induce an increase in liver ACBP levels (Heuvel et al., 1993; Sterchele et al., 1994; Skorve et al., 1995).

Fasting for 24 hours significantly reduced tissue level of ACBP in rat liver whereas feeding high-fat diet for 48 hours caused liver ACBP level to increase. Hepatic level continued to increase with prolonged high-fat diet feeding for 28 days. Similar but smaller changes in ACBP level were observed in rat kidney whereas heart ACBP did not respond acutely to high-fat diet, but increased after prolonged exposure (28 days) (Bhuiyan and Pande, 1994). Androgens have been shown to regulate the level of ACBP mRNA in the prostatic adenocarcinoma cell line LNCaP, whereas cortisone and dexamethasone had no effect on ACBP mRNA levels (Swinnen et al., 1994). Ethanol exposure has been shown to increase ACBP mRNA expression in cultured neurons (Katsura et al., 1995a). Chronical ethanol exposure increased mice's brain ACBP mRNA level (130%), and alcohol withdrawal resulted in a temporal increase in messenger level (220%). A single ethanol dose did not affect the messenger level, and the effect of prolonged treatment could be prevented by administration of flunitrazepam (Katsura et al., 1995b). Finally, it has been shown that the ACBP level in 3T3-L1 fibroblasts increases several folds during adipocyte differentiation (Hansen et al., 1991; Baldursson et al., 1995). Although ACBP was originally thought to be a short-lived protein regulated by hormones and cAMP (Yanagibashi et al., 1988) it was later demonstrated that ACBP is a long-lived protein not acutely regulated by hormones (Brown et al., 1992; Buus et al., 1994).

TISSUE DISTRIBUTION AND INTRACELLULAR LOCALIZATION

In the rat and human brain the highest concentrations of ACBP immunoreactivity and mRNA levels are found in cerebellum, hypothalamus and reticular thalamic nucleus. The concentration of ACBP is much higher in hippocampus than in basal ganglia. In peripheral tissue of the rat the highest concentration is found in liver, testis, duodenum and adrenal cortex. In liver ACBP is evenly distributed in all cells. This is in contrast to the adrenal cortex, testis, kidney, intestine and brain where ACBP is highly concentrated in specialized cells (see Costa and Guidotti, 1991, for review).

In the adrenal cortex ACBP is found in extremely high concentrations in zona glomerulose cells. In kidney ACBP is concentrated in the epithelial cells of the convoluted tubules and the ascending limp of the loop of Henle, and in the testis ACBP is highly concentrated in Leydig cells (Bovolin et al., 1990). These cells are all characterized by having high-energy metabolism and high secretory or ion transport activity, indicating that ACBP function is closely related to energy metabolism or secretory activity. A recent investigation on expression patterns of ACBP in *Drosophila melanogaster* by Kolmer et al. (1994) strongly supports this view and also shows that ACBP is highly expressed in organs with a high fatty acid turnover. In the adult insects high concentrations of ACBP were found in the fat body and the developing gamets. The highest concentration of ACBP was found in cardia and malpighian tubules and in the testis. The distal part of the anterior malpighian tubulus, where the ACBP concentration was highest, is the main site of primary urine secretion. This is also the site of ion gradient formation because primary urine in insects is formed by generating an osmotic pressure by active K^+ transport. In rat kidney the active ion transport is occurring in the ascending limp of the loop of Henle which also has high expression of ACBP (Bovolin et al., 1990). Therefore both in mammals and *D. melanogaster* ACBP expression is high at the site of ion gradient formation, a process requiring energy. The connection between ACBP and high-energy metabolisms is further substantiated by the fact that ACBP and mitochondria colocalize in epithelial cells of the malpighian tubules (Kolmer et al., 1994). Similar distribution of ACBP and mitochondria is seen in rat testis Leydig cells (Schultz et al., 1992). In *D. melanogaster* high concentrations of ACBP were also found in outer vacuolate epitheliums of the stomodial valve of cardia. These epithelial cells are involved in the synthesis of the peritrophic membrane, a porous sleeve that lines the intestine (Kolmer et al., 1994). In human skin ACBP expression was high in the clear cells of eccrine sweat glands, the secretory cell of epidermis, whereas it was undetectable in other types of skin cells (Alho et al., 1993).

INTRACELLULAR FUNCTION OF ACBP

Studies on the intracellular function of ACBP have focused on the effect on acylCoA metabolism and steroidogenesis. Long-chain acylCoA esters serve as obligatory intermediates in lipid metabolism. In addion, there is growing evidence that long-chain acylCoA esters also play an important role as regulator of various physiological processes.

The processes affected by long-chain acylCoAs include a number of enzymes and transporters. PalmitoylCoA and 16 to 20 carbon acylCoA esters have been shown to be strong inhibitors of the rate-limiting enzyme acetylCoA carboxylase with a K_i of 5.5 nM (Ogiwara et al., 1978; Nikawa et al., 1979). Furthermore, the AMP-kinase, which phosphorylates and inactivates acetylCoA carboxylase itself, is phosphorylated in a regulatory manner by an acylCoA-regulated kinase-kinase (Carling et al., 1987; Weekes et al., 1994). The mitochondrial adenine nucleotide translocase (ANT), which is considered to be the rate-limiting step in ATP production (Lemaster and Sowers, 1979), is reversibly inhibited by long-chain acylCoA esters with $K_i < 1\ \mu M$ (Woldegiorgis et al., 1981, 1982; Shrago et al., 1995). Two key enzymes in acylCoA metabolism, the long-chain acylCoA synthetase (Pande, 1973) and the carnitinepalmitoyl-transferase (CPT) (Murthy and Pande, 1987), have both been shown to be inhibited by long-chain acylCoA esters.

Long-chain acylCoA esters are also potential messenger molecules in signal transduction. Physiological concentration of palmitoylCoA has been shown to induce rapid Ca^+ release from the terminal cisterna of sarcoplasmatic reticulum (Fulceri et al., 1994). The acylCoA ester appeared to act through the ryanodine receptor since the ryanodine receptor channel blocker ruthenium red prevented the palmitoylCoA-induced Ca^+ release. Long-chain acylCoA esters have been shown to bind to and competitively inhibit T3-

binding to the nuclear thyroid hormone receptor (K_i 0.45 μM) (Inoue et al., 1989; Li et al., 1990; Li et al., 1993). The activity of rat brain protein kinase C (PKC) in the presence of diacylglycerol, phosphatidylcholin (PS) and Ca^+ has been found to be greatly increased by μM amounts of long-chain acylCoA esters (Bronfman et al., 1988, 1989; Nesher and Boneh, 1994). The potential role of acylCoA esters as signal molecules is supported by the observation that acute stimulation of neutrophil cells with the chemotactic peptide fMet-Leu-Phe, within 5 sec raised cellular acylCoA concentration by a factor two for about 120 sec (Majumdar et al., 1991).

The recent discovery that long-chain acylCoA esters prevent binding of the *E. coli* transcription factor FadR to DNA with a K_i of 5 nM strongly indicates that long-chain acylCoA esters regulate gene transcription activity (DiRusso et al., 1992, 1993). The fact that repression of acetylCoA carboxylase and the Δ-9-desaturase by exogenous fatty acids in yeast is blocked if one or more of the fatty acid-activating genes are mutated or deleted (Kamiryo et al., 1977, 1979; Choi et al., 1996) strongly indicates that long-chain acylCoA esters are also involved in gene regulation in yeast.

The concentration of long-chain acylCoA esters in mammalian tissue is reported to be from 5 μM to 160 μM (Baekdal et al., 1996). These high concentrations of intracellular acylCoA esters together with the above described potential regulatory functions of long-chain acylCoA esters strongly emphasize the need of an intracellular acylCoA transporter and pool former. This transporter and pool former should be able to relieve product inhibition of the acylCoA synthease and form a tightly controlled pool of long-chain acylcoA esters and make them available to selected acylCoA-consuming and -regulated processes only. A further requirement to this pool former is that it can prevent binding of these very amphiphilic compounds to membranes and protect them against the very high thioesterase activity found in both the cytosol and intracellular membranes (Aas, 1971; Kurooka et al., 1972).

There is ample evidence from both in vitro and in vivo experiments that ACBP has all the characteristics to fulfil this function. First of all, ACBP binds acylCoA esters with very high specificity and affinity. Furthermore, ACBP has been shown to protect inhibition of acetylCoA carboxylase against inhibition by long-chain acylCoA esters under conditions where the molar ratio of acylCoA/ACBP is < 0.9 (Rasmussen et al., 1993). The same authors showed that ACBP in contrast to liver fatty acid binding protein (L-FABP) could relieve product inhibition of the mitochondrial long-chain acylCoA synthetase by desorbing the newly synthesized acylCoAs from the synthetase. ACBP also completely prevented inhibition of the ANT by long-chain acylCoA esters. Definitive proof that ACBP can mediate intermembrane transport of acylCoA esters in vitro was obtained by Rasmussen et al. (1994). In a test system, where acylCoA esters were dissolved in phosphotidylcholine liposomes immobilized on nitrocellulose sheets, ACBP was able to mediate transport of the esters to mitochondria, donate them to CPT1 and initiate β-oxidation. This transport ability was further substantiated by demonstrating that ACBP in a similar test system was able to mediate transport of acylCoA esters and donate them to the glycerolipid synthesizing enzymes in suspended microsomes or microsomes immobilized on a nitrocellulose membrane. Taking into account that the K_D for binding of palmitoylCoA to bovine ACBP is as low as 10^{-13} M and that 12.5-fold excess of ACBP over that of acylCoA was used in the above β-oxidation experiments, the calculated free acylCoA will be in the low femtomolar range. This indicates that it might be the ACBP/acylCoA complex and not the free acylCoA which is recognized by the mitochondrial CPT1. Bhuiyan and Pande (1994) demonstrated that ACBP both affected CPT_1 activity and its regulation with malonylCoA. At subinhibitory concentrations (10 μM) of palmitoylCoA equimolar concentrations of ACBP lowered CPT_1 activity without affecting malonylCoA regulation, increasing ACBP to 40 μM further reduced CPT_1 activities, but markedly increased the sensitivity to malonyl-CoA inhibition. High palmitoylCoA concentrations (40 μM) strongly inhibits

CPT_1. Addition of 10 μM ACBP in the presence of 40 μM palmitoylCoA markedly increased CPT_1 activity, but only slightly affected malonylCoA inhibition of the enzyme. However, increasing the ACBP concentration to 40 μM further increased CPT_1 activity and its sensitivity to malonylCoA inhibition. ACBP has also been demonstrated to be able to donate arachidonoylCoA to red cell lyso-lecithin acyltransferase (Fyrst et al., 1995). Finally, it has been shown that ACBP could prevent binding of acylCoA esters to microsomal membranes and protect them against the hydrolysis by membrane-bound acylCoA hydrolases (Rasmussen et al., 1990).

Evidence that ACBP is also able to donate acylCoA esters for regulatory processes was obtained by investigating the effect of palmitoylCoA on caffeine-induced Ca^{2+} release by the ryanodine receptor from rabbit muscle terminal cisterna. Pretreatment of terminal cisterna membranes with pM concentrations of palmitoylCoA presented in equilibrium with acylCoA complexed by ACBP (6.0 μM, molar ratio 0.9) strongly potentiated caffeine-induced Ca^{2+} release. Pretreatment of terminal cisterna membranes with similar concentrations of palmitoylCoA in the absence of ACBP did not affect caffeine-induced Ca^{2+} release (R. Fulceri, J. Knudsen and A. Benedetti, unpublished results). This result strongly indicates that ACBP, by preventing random partitioning of acylCoA into membranes, creates a soluble pool of acylCoA which it can make available for regulation of the ryanodine receptor.

The above experimental evidence strongly indicates that the ACBP in vitro can transport and donate acylCoA esters to consuming enzymes and protect acylCoA-regulated enzymes against inhibition by acylCoAs, respectively. Heterologous expression of bovine ACBP in yeast clearly demonstrated that ACBP also in vivo binds acylCoA esters and forms an intracellular pool of these esters (Mandrup et al., 1993a). Direct evidence that ACBP also in vivo is involved in acylCoA transport was obtained with a yeast ACBP knockout strain. The growth rate of the knockout strain was not affected in glucose medium and only slightly reduced in ethanol medium. However, in mixed culture the knockout strain was unable to compete with wild-type yeast during exponential growth (Schjerling et al., 1995). The knockout strain was expected to have a lowered acylCoA pool size due to the lack of ACBP. However, to our surprise we found that the acylCoA pool in the knockout strain was increased 2 to 2.5 fold. Further analysis showed that this increase in acylCoA level was caused by an increase in stearoylCoA only. De novo fatty acid synthesis in yeast is terminated by a transacylase reaction and the product released as long-chain acylCoA esters. The synthetase is strongly product-inhibited, and it was shown that this inhibition can be relieved by ACBP. We therefore suggest that the ACBP knockout yeast has a block in de novo fatty acid synthesis, caused by a reduced acylCoA transport capacity, due to the lack of ACBP.

Steroidogenesis starts with metabolism of cholesterol to pregnenolone by the inner mitochondrial membrane P450 side-chain cleavage ($P450_{scc}$) enzyme. The rate of steroid formation depends on (1) the rate of cholesterol transport to the innter mitochondrial membrane and (2) the loading rate of the $P450_{scc}$ with cholesterol. Experimental evidence indicates that the peripheral benzodiazepine receptors (PBR) play a key role in the regulation of cholesterol uptake and that ACBP (or DBI in their terminology) is vital for steroidogenesis (see Papadopoulos and Brown, 1995, for review). The experimental evidence, which includes cross-linking studies, shows that ACBP might interact with the PBR and thereby stimulate mitochondrial cholesterol uptake. ACBP has also been shown to directly stimulate the loading of $P450_{scc}$ with cholesterol. Reduction of the ACBP level in steroid-producing cell lines using oligonucleotide antisense technology completely blocked steroidogenesis. The current model for the role of ACBP in steroidogenesis is that ACBP through its interaction with PBR functions as the signal transducer of hormonal and constitutive steroidogenesis at the mitochondrial level (Papadopoulos and Brown, 1995). No link between the ability of ACBP to bind long-chain acylCoA and its function in steroidogenesis has been reported. A possible link could be the increased breakdown of cholesterol esters by the hormone-sensitive lipase during the first phase of the hormone-stimulated steroidogenesis. The released fatty acids will quickly be activated to acylCoA esters which potentially could change the saturation of ACBP with acylCoA and thereby modify its interaction with PBR.

CONCLUSION

The existing experimental evidence strongly supports that ACBP is able to function as an intracellular acylCoA binding protein. ACBP is a highly conserved housekeeping protein found in all eukaryotic species from reptiles to man and from plants to yeast. It binds acylCoA esters with high affinity and specificity. It is able to function as an acylCoA transporter and pool former both in vitro and in vivo. A possible role of this protein in steroidogenesis is in full accordance with its function as acylCoA transporter and pool former. The fact that ACBP is widely distributed in most tissue, that the concentration is high in secretory and ion-transporting cells which do not synthesize steroids, and that it is found in both plants and yeast makes it difficult to understand that ACBP can act as a specific hormone transducer in regulating steroidogenesis. However, it is very likely that ACBP plays a permissive role in steroidogenesis, related to its ability to bind acylCoA. There is no doubt that ACBP and some of its processing products show biological activity when added to cell cultures or organ perfusion media in vitro. However, it is unclear whether ACBP is secreted from specific cells in a controlled manner in vivo.

REFERENCES

Aas, M., 1971, Organ and subcellular distribution of fatty acid activating enzymes in the rat, *Biochim. Biophys. Acta* 231:32-47.

Alho, H., Vaalasti, A., Podkletnova, I., and Rechardt, L., 1993, Expression of diazepam-binding inhibitor peptide in human skin: an immunohistochemical and ultrastructural study, *J. Investig. Dermat.* 101:800-3.

Andersen, K.V., and Poulsen, F.M., 1992, Three-dimensional structure in solution of acyl-coenzyme A binding protein from bovine liver, *J. Mol. Biol.* 226:1131-41.

Andersen, K.V., and Poulsen, F.M., 1993, The three-dimensional structure of acyl-coenzyme A binding protein from bovine liver: structural refinement using heteronuclear multidimensional NMR spectroscopy, *J. Biomol. NMR* 3:271-84.

Baekdal, T., Schjerling, C.K., Hansen, J.K., and Knudsen, J., 1996, Analysis of long-chain acyl coenzyme A esters, in: *Advances in Lipid Methodology - Three*, W.W. Christie, ed., The Oily Press Ltd., Dundee, UK.

Baldursson, T., Gram, C., Knudsen, J., Kristiansen, K., and Mandrup, S., 1995, Adipocyte differentiation is dependent on the induction of the acyl-CoA binding protein, in: *Signalling Mechanisms - from Transcription Factors to Oxidative Stres*, L. Packer and K. Wirtz, Vol. H 92, Springer-Verlag.

Besman, M.J., Yanagibashi, K., Lee, T.D., Kawamura, M., Hall, P.F., and Shively, J.E., 1989, Identification of des-(Gly-Ile)-endozepine as an effector of corticotropin-dependent peripheral benzodiazepine receptor, *Proc. Natl. Acad. Sci. USA* 86:4897-901.

Bhuiyan, A.K., and Pande, S.V., 1994, Carnitine palmitoyltransferase activities: effects of serum albumin, acyl-CoA binding protein and fatty acid binding protein, *Mol. Cell. Biochem.* 139:109-16.

Borboni, P., Condorelli, L., De Stefanis, P., Sesti, G., and Lauro, R., 1991, Modulation of insulin secretion by diazepam binding inhibitor and its processing products, *Neuropharmacol.* 30:1399-1403.

Bovolin, P., Schlichting, J., Miyata, M., Ferrarese, C., Guidotti, A., Alho, H., 1990, Distribution and characterization of diazepam binding inhibitor (DBI) in peripheral tissues of rat, *Regulatory Peptides* 29:267-81.

Bronfman, M., Morales, M.N., and Orellana, A., 1988, Diacylglycerol activation of protein kinase C is

modulated by long chain acyl-CoA, *Biochem. Biophys. Res. Commun.* 152:987-92.
Bronfman, M., Orellana, A., Morales, M.N., Bieri, F., Waechter, F., Staubli, W., and Bentley, P., 1989, Potentiation of diacylglycerol-activated protein kinase C by acyl-coenzyme A thioesters of hypolipidaemic drugs, *Biochem. Biophys. Res. Commun.* 159:1026-31.
Brown, A.S., Hall, P.F., Shoyab, M. and Papadopoulos, V., 1992, Endozepine/diazepam binding inhibitor in adrenocortical and Leydig cell lines: absence of hormonal regulation, *Molec. Cell. Endocrinology* 83:1-9.
Buus, C.L., Kristiansen, K., and Knudsen, J., 1994, Turnover of acyl-CoA-binding protein (ACBP) in four different cell lines measured by using two-dimensional polyacrylamide gel electrophoresis, *Biochem. J.* 297:555-60.
Carling, D., Zammit, V.A., and Hardie, D.G., 1987, A common bicyclic protein kinase cascade inactivates the regulatory enzymes of fatty acid and cholesterol biosynthesis, *FEBS Lett.* 223:217-22.
Chen, Z.W., Agerberth, B., Gell, K., Andersson, M., Mutt, V., Östenson, C.G., Efendic, S., Barros-Söderling, J., Persson, B., and Jornvall, H., 1988, Isolation and characterization of porcine diazepam-binding inhibitor, a polypeptide not only of cerebral occurrence but also common in intestinal tissues and with effects on regulation of insulin release, *Eur. J. Biochem.* 174:239-45.
Choi, J.Y., Stukey, J., Hwang, S.Y., and Martin, C.E., 1996, Regulatory elements that control transcription activation and unsaturated fatty acid-mediated repression of the Saccharomyces cerevisiae OLE1 gene source, *J. Biol. Chem.* 271:3581-89.
Costa, E., and Guidotti, A., 1991, Diazepam binding inhibitor (DBI): a peptide with multiple biological actions, *Life Sciences* 49:325-44.
DiRusso, C.C., Heimert, T.L., and Metzger, A.K., 1992, Characterization of FadR, a global transcriptional regulator of fatty acid metabolism in Escherichia coli. Interaction with the fadB promoter is prevented by long chain fatty acyl coenzyme A, *J. Biol. Chem.* 267:8685-91.
DiRusso, C.C., Metzger, A.K., and Heimert, T.L., 1993, Regulation of transcription of genes required for fatty acid transport and unsaturated fatty acid biosynthesis in Escherichia coli by FadR, *Mol. Microbiol.* 7:311-22.
Elholm, M., Bjerking, G., Knudsen, J., Kristiansen, K., and Mandrup, S., 1996, Regulatory elements in the promoter region of the rat gene encoding the acyl-CoA-binding protein, Submitted.
Fulceri, R., Nori, A., Gamberucci, A., Volpe, P., Giunti, R., and Benedetti, A., 1994, Fatty acyl-CoA esters induce calcium release from terminal cisternae of skeletal muscle, *Cell. Calcium* 15:109-16.
Fyrst, H., Knudsen, J., Schott, M.A., Lubin, B.H., and Kuypers, F.A., 1995, Detection of acyl-CoA-binding protein in human red blood cells and investigation of its role in membrane phospholipid renewal, *Biochem. J.* 306:793-799.
Gersuk, V.H., Rose, T.M., and Todaro, G.J., 1995, Molecular cloning and chromosomal localization of a pseudogene related to the human acyl-CoA binding protein/diazepam binding inhibitor, *Genomics* 25:469-76.
Guidotti, A., Forchetti, C.M., Corda, M.G., Konkel, D., Bennet, C.D., and Costa, E., 1983, Isolation, charaterization and purification to homogeneity of an endogenous polypeptide with agonistic action on benzodiazepine receptors, *Proc. Natl. Acad. Sci. USA* 80:3531-5.
Hansen, H.O., Andreasen, P.H., Mandrup, S., Kristiansen, K., and Knudsen, J., 1991, Induction of acyl-CoA-binding protein (ACBP) and its mRNA in 3T3-L1 cells by insulin during preadipocyte to adipocyte differentiation, *Biochem. J.* 277:341-44.
Heuvel, J.P., Sterchele, P.F., Nesbit, D.J., and Peterson, R.E., 1993, Coordinate induction of acyl-CoA binding protein fatty acid binding protein and peroxisomal β-oxidation by peroxisome proliferators, *Biochim. Biophys. Acta* 1177:183-90.
Inoue, A., Yamamoto, N., Morisawa, Y., Uchimoto, T., Yukioka, M., and Morisawa, S., 1989, Unesterified long-chain fatty acids inhibit thyroid hormone binding to the nuclear receptor. Solubilized receptor and the receptor in cultured cells, *Eur. J. Biochem.* 183:565-72.
Kamiryo, T., Mishina, M., Tashiro, S.I., and Numa, S., 1977, Candida lipolytica mutants defective in an acyl-coenzyme A synthetase: isolation and fatty acid metabolism, *Proc. Natl. Acad. Sci. USA* 74:4947-50.
Kamiryo, T., Nishikawa, Y., Mishina, M., Terao, M., and Numa, S., 1979, Involvement of long-chain acyl coenzyme A for lipid synthesis in repression of acetyl-coenzyme A carboxylase in Candida lipolytica, *Proc. Natl. Acad. Sci. USA* 76:4390-4.
Katsura, M., Ohkuma, S., Jun, X., Tsujimura, A., and Kuriyama, K., 1995a, Ethanol stimulates diazepam binding inhibitor (DBI) mRNA expression in primary cultured neurons, *Mol. Brain Res.* 34:355-9.
Katsura, M., Ohkuma, S., Tsujimura, A., and Kuriyama, K., 1995b, Increase of diazepam binding inhibitor mRNA levels in the brains of chronically ethanol-treated and -withdrawn mice, *J.*

Pharmacol. Experimental Therapeutics 273:1529-33.

Knudsen, J., Højrup, P., Hansen, H.O., Hansen, H.F., and Roepstorff, P., 1989, Acyl-CoA-binding protein in rat: purification, binding characteristics, tissue concentrations and amino acid sequence, *Biochem. J.* 262:513-19.

Knudsen, J., Færgeman, N.J., Skøtt, H., Hummel, R., Børsting, C., Rose, T.M., Andersen, J.S., Højrup, P., Roepstorff, P., and Kristiansen, K., 1994, Yeast acyl-CoA-binding protein: acyl-CoA-binding affinity and effect on intracellular acyl-CoA pool size, *Biochem. J.* 302:479-85.

Kolmer, M., Roos, C., Tirronen, M., Nyöhänen, S., and Alho, H., 1994, Tissue-specific expression of the diazepam-binding inhibitor in Drosophila melanogaster: cloning, structure and localization of the gene, *Mol. Cell. Biol.* 14: 6983-95.

Kragelund, B.B., Andersen, K.V., Madsen, J.C., Knudsen, J., and Poulsen, F.M., 1993, Three-dimensional structure of the complex between acyl-coenzyme A binding protein and palmitoyl-coenzyme A, *J. Mol. Biol.* 230:1260-77.

Kragelund, B.B., Højrup, P., Jensen, M.S., Schjerling, C.K., Juul, E., Knudsen, J., and Poulsen, F.M., 1996, Fast and one-step folding of closely and distantly related homologous proteins of a four helix bundle family, *J. Mol. Biol.* 256:187-200.

Kurooka, S., Hosoki, K., and Yoshimura, Y., 1972, Some properties of long fatty acyl-coenzyme A thioesterase in rat organs, *J. Biochem. Tokyo* 71:625-34.

Lemasters, J.J., and Sowers, A.E., 1979, Phosphate dependence and atractyloside inhibition of mitochondrial oxidative phosphorylation. The ADP-ATP carrier is rate-limiting, *J. Biol. Chem.* 254: 1248-51.

Li, Q.L., Yamamoto, N., Inoue, A., and Morisawa, S., 1990, Fatty acyl-CoAs are potent inhibitors of the nuclear thyroid hormone receptor in vitro, *J. Biochem.* 107:699-702.

Li, Q.L., Yamamoto, N., Morisawa, S., and Inoue, A., 1993, Fatty acyl-CoA binding activity of the nuclear thyroid hormone receptor, *J. Cell. Biochem.* 51:458-64.

Majumdar, S., Rossi, M.W., Fujiki, T., Phillips, W.A., Disa, S., Queen, C.F., Johnston Jr., R.B., Rosen, O.M., Corkey, B.E., and Korchak, H.M., 1991, Protein kinase C isotypes and signaling in neutrophils, *J. Biol. Chem.* 266:9285-94.

Mandrup, S., Hummel, R., Ravn, S., Jensen, G., Andreasen, P.H., Gregersen, N., Knudsen, J., and Kristiansen, K., 1992, Acyl-CoA-binding protein/diazepam-binding inhibitor. Gene and pseudogenes. A typical housekeeping gene family, *J. Mol. Biol.* 228:1011-22.

Mandrup, S., Jepsen, R., Skøtt, H., Rosendal, J., Højrup, P., Kristiansen, K., and Knudsen, J., 1993a, Effect of heterologous expression of acyl-CoA binding protein (ACBP)/diazepam binding inhibitor (DBI) on the acyl-CoA level and composition in yeast, *Biochem. J.* 290:369-74.

Mandrup, S., Andreasen, P.H., Knudsen, J., and Kristiansen, K., 1993b, Genome organization and expression of the rat ACBP gene family, *Mol. Cell. Biochem.* 123:55-61.

Marquardt, H., Todaro, G.J., and Shoyab, M., 1986, Complete amino acid sequences of bovine and human endozepines. Homology with rat diazepam binding inhibitor, *J. Biol. Chem.* 261: 9727-31

Mikkelsen, J., and Knudsen, J., 1987, Acyl-CoA-binding protein from cow. Binding characteristics and cellular and tissue distribution, *Biochem. J.* 248:709-14.

Mogensen, I.B., Schulenberg, H., Hansen, H.O., Spener, F., and Knudsen, J., 1987, A novel acyl-CoA binding protein from bovine liver. Effect on fatty acid synthesis, *Biochem. J.* 241:189-92.

Murthy, M.S., and Pande, S.V., 1987, Some differences in the properties of carnitine palmitoyltransferase activities of the mitochondrial outer and inner membranes, *Biochem. J.* 248: 727-33.

Nesher, M., and Boneh, A., 1994, Effect of fatty acids and their acyl-CoA esters on protein kinase C activity in fibroblasts: possible implications in fatty acid oxidation defects, *Biochim. Biophys. Acta* 1221: 66-72.

Nikawa, J., Tanabe, T., Ogiware, H., Shiba, T., and Numa, S., 1979, Inhibitory effects of long-chain acyl coenzyme A analogues on rat liver acetyl coenzyme A carboxylase, *FEBS Lett.* 102:223-6.

Ogiwara, H., Tanabe, T., Nikawa, J., and Numa, S., 1978, Inhibition of rat-liver acetyl-coenzyme-A carboxylase by palmitoyl-coenzyme A. Formation of equimolar enzyme-inhibitor complex, *Eur. J. Biochem.* 102:223-6.

Östenson, G.-G., Ahrén, B., Karlsson, S., Knudsen, J., and Efendic, S., 1994, Inhibition by rat diazepam-binding inhibitor/acyl-CoA-binding protein of glucose-induced insulin secretion in rat, *Eur. J. Endocrinol.* 131:201-4.

Pande, S.V., 1973, Reversal by CoA of palmityl-CoA inhibition of long chain acyl-CoA synthetase activity, *Biochim. Biophys. Acta* 306:15-20.

Papadoupoulos, V., and Brown, A.S., 1995, Role of peripheral-type benzodiazepine receptor and the polypeptide diazepam binding inhibitor in steroidogenesis, *J. Steroid Biochem. Molec. Biol.* 53:103-

10.
Patte, C., Vaudry, H., Desrues, L., Gandolfo, P., Strijdveen, I., Lamacz, M., and Tonon, M.C., 1995, The endozepine ODN stimulates polyphosphoinositide metabolism in rat astrocytes, *FEBS Lett.* 362:106-10.
Rasmussen, J.T., Börchers, T., and Knudsen, J., 1990, Comparison of the binding affinities of acyl-CoA-binding protein (ACBP) and fatty acid-binding protein (FABP) for long-chain acyl-CoA esters, *Biochem. J.* 265:849-55.
Rasmussen, J.T., Rosendal, J., and Knudsen, J., 1993, Interaction of acyl-CoA binding protein (ACBP) on processes for which acyl-CoA is a substrate, product or inhibitor, *Biochem. J.* 292:907-13.
Rasmussen, J.T., Færgeman, N.J., Kristiansen, K., and Knudsen, J., 1994, Acyl-CoA-binding protein (ACBP) can mediate intermembrane acyl-CoA transport and donate acyl-CoA for β-oxidation and glycerolipid synthesis, *Biochem. J.* 299:165-70.
Rose, T.M., Schults, E.R., and Todaro, G.J., 1992, Molecular cloning of the gene for the yeast homolog (ACB) of diazepam binding inhibitor/endozepine/acyl-CoA-binding protein, *Proc. Natl. Acad. Sci. USA* 89:11287-91.
Rosendal, J., Ertbjerg, P., and Knudsen, J., 1993, Characterization of ligand binding to acyl-CoA binding protein (ACBP)/diazepam binding inhibitor (DBI), *Biochem. J.* 290:321-26.
Schjerling, C.K., Børsting, C., Hummel, R., Kristiansen, K., Hansen, J.K., and Knudsen, J., 1996, Effect of deleting the acyl-CoA binding protein (ACB1 protein) gene on lipid and acyl-CoA metabolism in yeast, *J. Biol. Chem.*. In press.
Schultz, R., Pelto-Huikko, M., and Alho, H., 1992, Expression of diazepam binding inhibitor-like immunoreactivity in rat testis is dependent on pituitary hormones, *Endocrinology* 130:3200-06.
Shoyab, M., Gentry, L.E., Marquardt, H., and Todaro, G., 1986, Isolation and characterization of a putative endogenous benzodiazepinoid (Endozepine) from bovine and human brain, *J. Biol. Chem.* 261:11968-73.
Shrago, E., Woldegiorgis, G., Ruoho, A.E., and DiRusso, C.C., 1995, Fatty acyl-CoA esters as regulators of cell metabolism, *Prostaglandins Leukot Essent Fatty Acids* 52:163-6.
Skorve, J., Rosendal, J., Vaagenes, H., Knudsen, J., Lillehaug, J.R., and Berge, R.K., 1995, Fatty acyl-CoA hydrolase activity and acyl-CoA binding protein in liver of rat treated with peroxisome proliferation 3-thia fatty acids, *Xenobiotica* 25:1181-94.
Sterchele, P.F., Vanden Heuvel, J.P., Nesbit, D.J., Davis, J.W., Shrago, E., Knudsen, J., and Peterson, R.E., 1994, Induction of hepatic acyl-CoA-binding protein and fatty acid-binding protein by perfluorodecanoic acid: examination of a possible correlation to hepatic long-chain acyl-coenzyme A levels, *Biochem. Pharmacol.* 48:955-66.
Swinnen, J.V., Esquenet, M., Heyns, W., Rombauts, W., and Verhoeven, G., 1994, Androgen regulation of the messenger RNA encoding diazepam-binding inhibitor/acyl-CoA-binding protein in the human prostatic adenocarcinoma cell line LNCaP, *Mol. Cell. Endocrinol.* 104:153-62.
Webb, N.R., Rose, T.M., Malik, N., Marquardt, H., Shoyab, M., Todaro, G.J., and Lee, D.C., 1987, Bovine and human cDNA sequences encoding a putative benzodiazepine receptor ligand, *DNA* 6:71-9.
Weekes, J., Hawley, S.A., Corton, J., Shugar, D., and Hardie, D.G., 1994, Activation of rat liver AMP-activated protein kinase by kinase in a purified, reconstituted system, *Eur. J. Biochem.* 219:751-7.
Wilson, R., Ainscouht, R., Anderson, K., Baynes, C., Berks, M., Bonfiels, J., Burton, J., Connell, M., Copsey, T., Cooper, J., Coulson, A., Craxton, M., Dear, S., Du, Z., Durbin, R., Favello, A., Fraser, A., Fulton, L., Gardner, A., Green, P., Hawkins, T., Hillier, L., Jier, M., Johnston, L., Jones, M., Kershaw, J., Kirsten, J., Laisster, N., Latreille, P., Lightning, J., Lloyd, C., Mortimore, B., O'Callaghan, M., Parsons, J., Percy, C., Rifken, L., Roopra, A., Saunders, D., Shownkeen, R., Sims, M., Smaldon, N., Smith, A., Smith, M., Sonnhammer, E., Staden, R., Sulston, J., Thierry-Mieg, J., Thomas, K., Vaudin, M., Vaughan, K., Waterston, R., Watson, A., Weinstock, L., Wilkinson-Sproat, J., and Wohldman, P., 1994, 2.2 Mb of contiguous nucleotide sequence from chromosome III of C. elegans, *Nature* 368:32-8.
Woldegiorgis, G., Shrago, E., Gipp, J., and Yatvin, M., 1981, Fatty acyl coenzyme A-sensitive adenine nucleotide transport in a reconstituted liposome system, *J. Biol. Chem.* 256:12297-300.
Woldegiorgis, G., Yousufzai, S.Y., and Shrago, E., 1982, Studies on the interaction of palmitoyl coenzyme A with the adenine nucleotide translocase, *J. Biol. Chem.* 257:14783-7.
Yanagibashi, K., Ohno, Y., Kawamura, M., and Hall, P.F., 1988, The regulation of intracellular transport of cholesterol in bovine adrenal cells: purification of a novel protein, *Endocrinology* 123:2075-82.

PHOSPHATIDYLINOSITOL TRANSFER PROTEIN FUNCTION IN THE YEAST *SACCHAROMYCES CEREVISIAE*

Melissa A. Kearns,[1] Min Fang,[1] Marcos Rivas,[1] Brian G. Kearns,[1] Satoshi Kagiwada,[1,2] and Vytas A. Bankaitis[1]

[1]Department of Cell Biology
University of Alabama at Birmingham
Birmingham, AL 35294-0005

[2]The Institute of Physical and Chemical Research (RIKEN)
Wako, Saitama, Japan

INTRODUCTION

All eukaryotic cells examined to date exhibit an array of cytosolic polypeptides that have the ability to catalyze the energy-independent transfer of lipids and/or phospholipids between membrane bilayers in vitro (Wirtz, 1991; Bankaitis et al., 1996). These phospholipid transfer proteins (PLTPs) are classified as a function of their lipid headgroup specificities in the in vitro transfer reaction, and phosphatidylinositol transfer proteins (PITPs) represent a class of oligospecific PLTPs. PITPs effect efficient transfer of PI or the zwitterionic phospholipid phosphatidylcholine (PC), and a distinct preference is exhibited towards PI as transfer substrate. While the physiological functions of PITPs, and PLTPs in general, have presented a long-standing mystery for reasons reviewed elsewhere (Wirtz, 1991; Bankaitis et al., 1996), much progress has been made in the past several years on this problem. The discovery that the *Saccharomyces cerevisiae SEC14* gene product (Sec14p) is the yeast PITP has provided the first opportunity to dissect the in vivo function of a PITP (Bankaitis et al., 1990). Immediate insights were forthcoming on this issue as genetic and biochemical analyses had identified Sec14p as a cytosolic factor that is essential for viability of the yeast cell because it is required for the biogenesis of secretory vesicles dedicated to the transport of secretory glycoproteins from a late Golgi compartment (Novick et al., 1980; Bankaitis et al., 1989; Cleves et al., 1991). The outstanding questions that remain include the following: (i) what is the mechanism of Sec14p function in the stimulation of protein export from the yeast Golgi complex?, and (ii) how does Sec14p harness its PI and PC transfer activities to biological function? In this manuscript, we will discuss the recent progress we have made on these issues.

GENETIC ANALYSIS OF SEC14p FUNCTION

The promiscuity exhibited by PITPs with regard to the sources of natural or artificial membranes that can serve as donors or acceptors in PI and PC-transfer assays when coupled with: (i) the energy-independence of the transfer reaction, and (ii) the fact that PITPs execute efficient phospholipid exchange reactions, but do not execute vectorial transfer of phospholipid, has precluded the formulation of directed attempts to reconstitute functional in vitro assays through which PITP-dependent reactions of clear physiological relevance could be studied. Nevertheless, mammalian PITP has recently been identified as a stimulatory factor in several unbiased resolution and reconstitution assays involving transmembrane signaling reactions (Thomas et al., 1993), the priming of dense core granules for regulated fusion to the plasma membrane of nueroendocrine cells (Hay and Martin, 1993; Hay et al., 1995), and the biogenesis of both constitutive secretory vesicles and dense core secretory granules from the Trans-Golgi Network of neuroendocrine cells (Ohashi et al., 1995). These reconstitutions, when coupled with the ability of mammalian PITP to effect a partial substitution for Sec14p function in yeast in a manner that requires the PI-transfer activity of PITP (Skinner et al., 1993; Alb et al., 1995), suggest the possibility that a late Golgi execution point for PITP function is conserved in mammals and yeast. Moreover, the in vitro data suggest that mammalian PITP synergizes with PI-kinases to generate dedicated metabolic pools of PI-4,5-bisphosphate (PIP_2) (Hay et al., 1995; Cunningham et al., 1995). That PITP may represent a genuine physiological stimulator of PIP_2 synthesis is suggested by evidence that PITP may associate physically with components of a growth factor signaling machine (Kauffmann-Zeh et al., 1995). By the same token, Sec14p rather effectively substitutes for PITP in these various assays (Hay and Martin, 1993; Ohashi et al., 1995; Cunningham et al., 1996), even though the yeast and mammalian PITPs share no primary sequence homology between them (Bankaitis et al., 1989; Dickeson et al., 1989). These data suggest the alternative possibility that PITP might stimulate PI-kinase activity indirectly in a rather irrelevant way (Alb et al., 1996). Thus, the authentic physiological function of PITP in mammalian cells remains a matter of some uncertainty.

A detailed analysis of PITP function in lower eukaryotic cells has been greatly facilitated by the isolation of mutations in at least seven distinct genes that render yeast completely independent of any Sec14p requirement for cell viability or Golgi secretory function (Cleves et al., 1989; Cleves et al., 1991; Bankaitis et al., 1996). Such "bypass suppressors" of normally lethal *sec14* null mutations represent both recessive (i.e. loss-of-function) and dominant (gain-of-function) mutations, and the immediate conclusion from such genetic data is that Sec14p functions to regulate Golgi secretory function and that Sec14p is not a structural component of the secretory vesicle biogenesis machinery. From the perspective of the geneticist, Sec14p is also concluded to function upstream of the gene products identified by the "bypass suppressor" screen in a regulatory pathway (that we term the Sec14p pathway) that ultimately stimulates Golgi-derived secretory vesicle biogenesis. Formally, dominant "bypass Sec14p" mutations likely identify positive effectors of this regulatory pathway, whereas recessive "bypass Sec14p" mutations identify gene products that either serve as negative effectors of the Sec14p pathway or that otherwise act to interfere with the activity of this pathway. The challenge is to determine how these various suppressor mutations effect the "bypass Sec14p" condition. The clear expectation is that analysis of such "bypass Sec14p" mutants will lend direct insights into the both the form of the Sec14p-dependent regulatory pathway that controls secretory protein exit from the yeast Golgi complex and, ultimately, the biochemistry upon which the activity of the "Sec14p pathway" depends. A tabulation of the genes identified to date in the "bypass Sec14p" suppressor mutant collection is presented in Table 1.

Below, we describe how we presently interpret the clues that each class of "bypass Sec14p" mutation has presented us, and we propose a general synthesis of these various clues into the description of an altogether novel Sec14p-dependent regulatory pathway for the assembly of Golgi-derived secretory vesicles.

THE SENSOR MODEL FOR SEC14p FUNCTION

The *cki1*, *cct1* and *cpt1* "bypass Sec14p" mutations provided immediate information concerning mechanism by which the yeast cell could become independent of the normally essential requirement for Golgi secretory function and cell viability; namely, that inactivation of the CDP-choline pathway for PC biosynthesis (see Figure 1) is sufficient to effect the "bypass Sec14p" phenotype (Cleves et al., 1991). This demonstration not only links Sec14p to the biosynthesis of one of its in vitro transfer ligands, but it also reveals a remarkable specificity with regard to PC biosynthesis and the requirement for Sec14p function. Yeast

Table 1. Listing of genes identified by "bypass Sec14p" mutations.

Bypass Sec14p mutation	Dominant/recessive character of mutation	Gene product
cki1	Recessive	Choline kinase
cct1	Recessive	Cholinephosphate cytidylyltransferase
cpt1	Recessive	Choline phosphotransferase
sac1	Recessive	Integral membrane protein of Golgi and endoplasmic reticulum
kes1	Recessive	Homologous to oxysterol binding domain of human oxysterol binding protein
BSD1-124	Dominant	Unknown
BSD2-1	Dominant	Unknown

can synthesize PC by yet another route, by the methylation of phosphatidylethanolamine (PE) (see Figure 1). Yet, inactivation of the PE-methylation pathway for PC biosynthesis affords no relief from the essential Sec14p requirement for Golgi secretory function and cell viability; even under conditions where the PE-methylation pathway is the sole route for net PC biosynthesis in the yeast cell (Cleves et al., 1991). Finally, these data formally reveal CDP-choline pathway activity to be toxic to yeast Golgi membranes (i.e. CDP-choline pathway activity is fundamentally incompatible with Golgi secretory function), and that Sec14p somehow prevents (or otherwise neutralizes) this toxicity. Whereas phospholipid

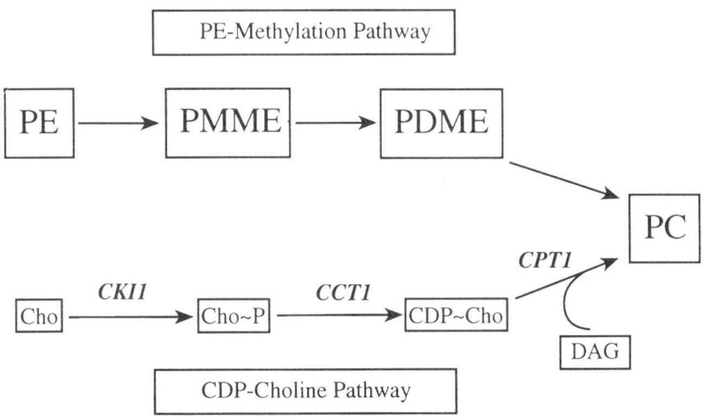

Figure 1. Phosphatidylcholine biosynthetic pathways in yeast. The CDP-choline and PE-methylation pathways for PC biosynthesis are shown. Choline, choline-phosphate, CDP-choline, phosphatidylmono- and di-methylethanolamine are indicated as Cho, Cho~P, CDP~Cho, PMME, and PDME respectively. Choline kinase, choline-phosphate cytidylyl-transferase, and cholinephosphotransferase are abbreviated as CKIase, CCTase and CPTase, respectively.

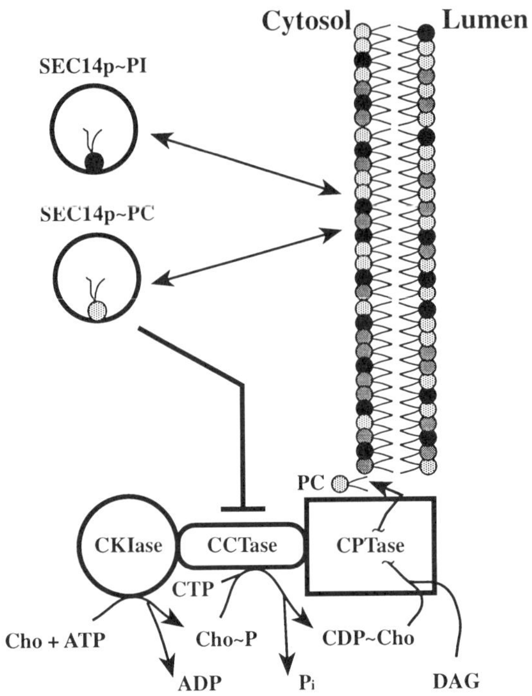

Figure 2. Sec14p as an inhibitor of CCTase activity in yeast Golgi membranes. Sec14p is concentrated on the cytosolic surface of yeast Golgi membranes (Cleves et al., 1991) where it is proposed to execute a constitutive phospholipid (PL) exchange reaction that generates a pool of PL-bound Sec14p (Sec14p~PL) comprised of both Sec14p~PC and Sec14p~PI. Sec14p~PC is proposed to serve as the negative effector of CCTase activity in yeast Golgi membranes, and the extent of CCTase inhibition is proportional to Sec14p~PC. Experimental details that provide the basis for this model have been described (McGee et al., 1994; Skinner et al., 1995).

biosynthetic pathways are not normally considered to represent lethal activities, in this case, such a conclusion is clearly warranted. Any models for Sec14p function in vivo must account not only for the specific functional interplay between Sec14p and the CDP-choline pathway, but also for the specific toxicity of the CDP-choline pathway (and not the PE-methylation pathway for PC biosynthesis) to Golgi secretory function.

The collective data from several recent reports have provided the basis for the proposal of a sensor model for the mechanism of Sec14p function in yeast (McGee et al., 1994; Skinner et al., 1995). The basic tenet of that model is that Sec14p represses the activity of the CDP-choline pathway in yeast Golgi membranes by inhibiting the activity of CCTase, the rate-determining enzyme of this pathway (Figure 2), and that the phospholipid exchange activities of Sec14p represent an intrinsic regulatory mechanism by which Sec14p controls its CCTase-directed inhibition. The Golgi execution point for Sec14p is indicated both by the phenotypes of Sec14p-deficient strains, and by biochemical and immunostaining data that indicate Sec14p is a peripheral Golgi protein (Cleves et al., 1991).

A primary prediction of the sensor model is that Sec14p does not act as a bona fide phospholipid transfer protein in vivo (Figure 2; McGee et al., 1994; Skinner et al., 1995). The evidence to this effect is several fold. First, McGee et al. (1994) have demonstrated that the primary consequence of Sec14p dysfunction in yeast is a specific increase in Golgi membrane PC content (Golgi PI content does not fluctuate), and that overproduction of Sec14p in yeast results in a marked, and specific, depression in bulk metabolic flux through the CDP-choline pathway. Second, Skinner et al. (1995) have shown that CCTase activity is very sensitive to challenge with purified Sec14p in vitro, and the data further suggest that the depression in bulk CDP-choline pathway activity observed in cells that overproduce Sec14p manifests the signature of reduced CCTase activity. The bulk effect seen upon Sec14p overproduction is interpreted to reflect the expansion of a Golgi-specific regulation of CCTase by Sec14p to a global downregulatory effect. Thus, the sensor model provides a general accounting for why specific inactivation of the CDP-choline pathway for PC biosynthesis effects bypass of the normally essential Sec14p requirement; namely, because Sec14p inhibits an antagonist of the Sec14p pathway for secretory vesicle biogenesis. What the sensor model fails to illuminate is why CDP-choline pathway activity is so detrimental to Golgi secretory function.

EVIDENCE FOR DEPLETION OF A CRITICAL GOLGI LIPID PRECURSOR POOL BY THE CDP-CHOLINE PATHWAY

One can imagine two potential mechanisms for the toxicity of CDP-choline pathway activity to Golgi membrane function: (i) that elevated PC levels in Golgi membranes might exert some detrimental structural effect, as initially proposed (Cleves et al., 1991), or (ii) that unattenuated CDP-choline pathway activity inappropriately consumes a critical Golgi diacylglycerol (DAG) pool whose integrity is required for sufficient Sec14p pathway function to drive secretory vesicle biogenesis, an idea first proposed by McGee et al. (1994). The former possibility suffers from the irrelevance of PC synthesis via the PE-methylation pathway to the Sec14p requirement, although issues of differential compartmentalization of these PC biosynthetic pathways cannot be excluded at this time. The latter possibility is more attractive as the CDP-choline pathway directly consumes DAG in the final enzymatic step for PC biosynthesis, whereas the methylation pathway involves modification of the headgroup of a preformed phospholipid (see Figure 1). If preservation of a critical Golgi lipid precursor (e.g. DAG) pool does indeed define the key target of Sec14p activity, then one would predict

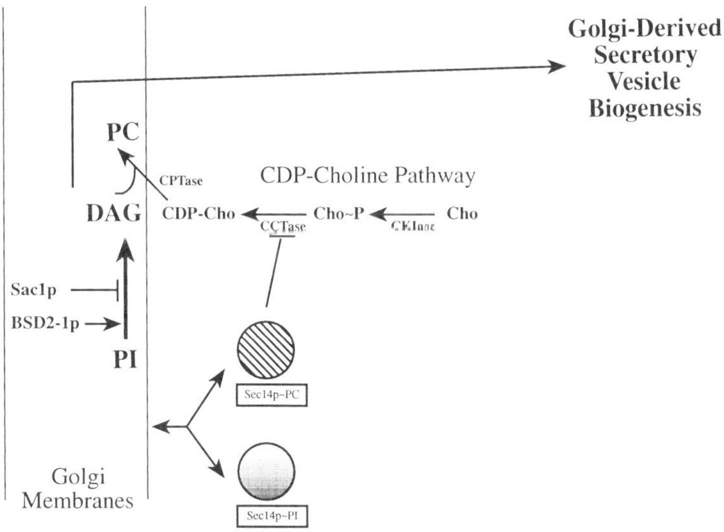

Figure 3. Preservation of a critical pool of diacylglycerol in yeast Golgi membranes may lie at the heart of the essential Sec14p requirement and the toxicity of CDP-choline pathway activity to Golgi membrane secretory function. Bypass Sec14p can be effected by sparing this DAG pool from consumption by inactivation of the CDP-choline pathway for PC biosynthesis (i.e. normally a function provided by Sec14p according to the tenets of the sensor model) or by expansion of the Golgi DAG pool to the extent that unattenuated CDP-choline pathway activity cannot deplete it to the point that the Sec14p pathway for secretory vesicle biogenesis is rendered inoperative. Of course, we cannot exclude the possibility that the critical lipid precursor is not DAG but is some other lipid precursor whose production is executed through the Golgi DAG pool.

that expansion of such a Golgi DAG pool beyond the point at which it can be consumed by the CDP-choline pathway should render cells independent of the normally essential Sec14p requirement; regardless of whether the CDP-choline pathway is active, or not. In this section, we describe evidence to support such a concept.

McGee et al. (1994) had previously reported that *sac1* and *BSD2-1* "bypass Sec14p" mutations exert their effects in the face of unattenuated CDP-choline pathway activity, and it had been noted that *sac1* and *BSD2-1* mutants exhibit phenotypes consistent with altered metabolism of inositol phospholipids. These phenotypes include allele-specific suppression of yeast actin defects (Cleves et al., 1989; Novick et al., 1989), and inositol auxotrophy (Whitters et al., 1993; Kagiwada et al., 1996). New data now indicate that the mechanism by which the *BSD2-1* mutation effects its "bypass Sec14p" phenotype is by an acceleration of bulk PI turnover. Whereas wild-type strains exhibit a bulk PI half-life of some 4.3 ± 0.3 hours, *BSD2-1* strains exhibited a bulk PI half-life of only 2.5 ± 0.6 hours. As *BSD2-1* is a dominant mutation, the data indicate that the corresponding gene product either directly, or indirectly, stimulates PI turnover. As *sac1* mutations are recessive in nature (Cleves et al., 1989; Novick et al., 1989), the *SAC1* gene product (Sac1p) is inferred to represent an inhibitor of PI-turnover. Significantly, a 15-fold overproduction of the *SAC1* gene product (Sac1p) restores wild-type rates of PI-turnover to *BSD2-1* strains and concomitantly abolishes the ability of *BSD2-1* to effect "bypass Sec14p" (Kagiwada et al., 1996). The demonstration that accelerated PI-turnover effects a "bypass Sec14p" phenotype strongly suggests that some product of PI turnover represents a critical stimulator of the "Sec14p pathway". This concept is now receiving additional support from the analysis of *sac1* mutants. Our data indicate that *sac1* mutants experience a highly elevated flux through the yeast inositol sphingolipid pathway, a particularly effective route for generating high levels of DAG in the Golgi, and that downmodulation of this metabolic flux (i.e. leading to reduced rates of DAG generation in Golgi membranes) results in the inability of *sac1* to effect "bypass Sec14p" (B.G. Kearns, T.P. McGee, and V.A. Bankaitis; manuscript in preparation). A model describing this concept is summarized in Figure 3.

DOWNSTREAM EVENTS IN THE SEC14p PATHWAY FOR GOLGI-DERIVED SECRETORY VESICLE BIOGENESIS.

The toxicity of CDP-choline pathway to Golgi membrane secretory function must b realized through inappropriate regulation of downstream elements of the Sec14p pathway Intriguing clues as to the nature of such potential elements are now being obtained. Molecula cloning experiments have now revealed the *KES1* gene to represent the wild-type allele of recessive class of bypass Sec14p mutations (M. Fang and V.A, Bankaitis, manuscript submitted) Interestingly, the *KES1* gene product (Kes1p) represents a member of a growing family of gene whose products exhibit homology to the mammalian oxysterol binding protein (OSBP); mammalian Golgi-localized protein that had initially been thought to downregulate metabolic flu through the choleterol biosynthetic pathway in response to challenge of cells with oxysterol (Dawson et al., 1989; Goldstein and Brown, 1990; Ridgway et al., 1992; Ridgway et al., 1995) As expected, bypass Sec14p is effected by loss of Kes1p function, and this effect is specific a neither individual nor combinatorial inactivation of several other yeast OSBP homologs is capabl of generating a bypass Sec14p condition. This specificity is emphasized by the finding tha inactivation of yeast Hes1p, an OSBP homolog that shares 70% primary sequence identity wit Kes1p (Jiang et al., 1994), has no effect on the cellular Sec14p requirement for Golgi secretor function and cell growth (M. Fang and V.A. Bankaitis, manuscript submitted).

KES1 exhibits striking genetic interactions with both CDP-choline pathway mutations an *BSD1-124*; the latter representing a dominant allele that effects bypass of Sec14p in the face o wild-type CDP-choline pathway activity (see Table 1). A mere doubling of *KES1* gene dosag abolishes the bypass Sec14p phenotypes associated with inactivation of the CDP-choline pathwa and the dominant *BSD1-124* mutation (M. Fang and V.A. Bankaitis, manuscript in preparation) As overexpression of downstream components in a pathway often suppresses defects in upstrean components, these data suggest that Kes1p acts downstream of the CDP-choline pathway in th Sec14p pathway for Golgi secretory vesicle biogenesis. Moreover, such data reveal that Kes1 and the *BSD1-124* gene product (Bsd1p*) share a functionally antagonistic relationship; thereb providing the first insights into how the *BSD1-124* mutation might effect bypass Sec14p.

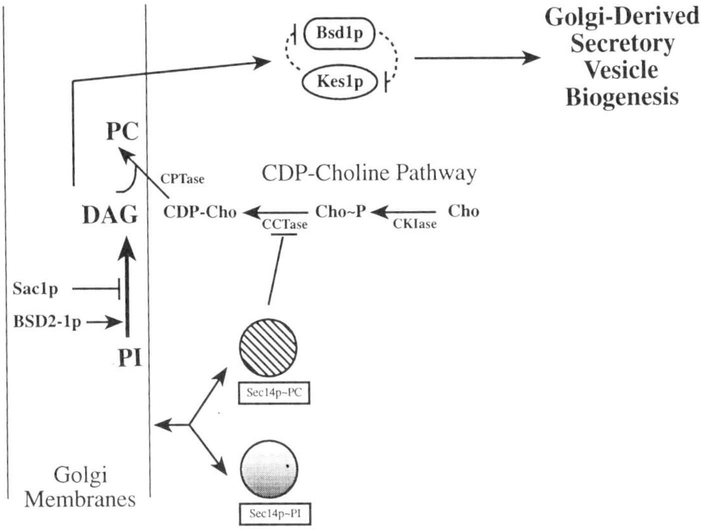

Figure 4. The Sec14p pathway for biogenesis of yeast Golgi-derived secretory vesicles. Sec14p senses the PC content of yeast Golgi membranes and the PC-bound form of Sec14p (Sec14p~PC) effects a negative feedback regulation on CDP-choline pathway activity. In the absence of Sec14p-mediated attenuation of CDP-choline pathway activity, the negative regulatory effects of Kes1p action could be potentiated either by synthesis of toxic levels of Golgi PC by the CDP-choline pathway, or by depletion of a critical Golgi pool of lipid precursor (e.g. diacylglycerol; DAG) by the CDP-choline pathway. Kes1p function is opposed by the action of the *BSD1-124* gene product (Bsd1p*) either with Kes1p as a target of inhibition by Bsd1p or as Kes1p as a Bsd1p inhibitor.

The collective data suggest that toxicity of CDP-choline pathway activity to secretory vesicle biogenesis from Golgi membranes is mediated directly or indirectly through the inappropriate activation of Kes1p, and that the net inhibitory effect of Kes1p function on the activity of the Sec14p pathway is either directed at, or responsive to, Bsd1p* function. The functional interactions between Sec14p, Bsd1p*, Kes1p and CDP-choline pathway with respect to Golgi secretory activity in yeast provide the first, and most outstanding, opportunity for detailed mechanistic studies directed at the question of how a member of the OSBP family operates in vivo. Thus, the Sec14p pathway is an altogether unique regulatory pathway that incorporates the action of a number of unique components in an unexpected integration of phospholipid metabolism with Golgi membrane dynamics. A model describing our present view of the Sec14p pathway is summarized in Figure 4.

ACKNOWLEDGEMENTS

This research was supported by a grant from the National Institutes of Health (GM44530) to V.A.B. B.G.K. and S.K. were supported by a National Science Foundation predoctoral traineeship (NSF-BIR-9256853) and a Human Frontier Science Program postdoctoral fellowship (LT-396-93), respectively.

REFERENCES

Alb, J.G., Jr., Kearns, M.A., and Bankaitis, V.A., 1996, Phospholipid metabolism and membrane dynamics. *Curr. Op. Cell Biol.* 8: In Press.

Bankaitis, V.A., Aitken, J.R., Cleves, A.E. and Dowhan, W., 1990, An essential role for a phospholipid transfer protein in yeast Golgi function. *Nature* 347: 561.

Bankaitis, V.A., Fry, M.R., Cartee, R.T., and Kagiwada, S., 1996, Phospholipid Transfer Proteins: Emerging Roles In Vesicle Trafficking, Signal Transduction, and Metabolic Regulation, R.G. Landes Co., Austin.

Bankaitis, V.A., Malehorn, D.E., Emr, S.D. and Greene, R., 1989, The *Saccharomyces cerevisiae SEC14* gene encodes a cytosolic factor that is required for transport of secretory proteins from the yeast Golgi complex. *J. Cell Biol.* 10: 1271.

Cleves, A.E., McGee, T.P., Whitters, E.A., Champion, K.M., Aitken, J.R., Dowhan, W., Goebl, M. and Bankaitis, V.A., 1991, Mutations in the CDP-choline pathway for phospholipid biosynthesis bypass the requirement for an essential phospholipid transfer protein. *Cell* 64: 789.

Cleves, A.E., Novick, P.J. and Bankaitis, V.A., 1989, Mutations in the *SAC1* gene suppress defects in yeast Golgi and yeast actin function. *J. Cell Biol.* 109: 2939.

Cunningham, E., Thomas, G.M.H., Ball, A., Hiles, I., and Cockroft, S., 1995, Phosphatidylinositol transfer protein dictates the rate of inositol trisphosphate production by promoting synthesis of PIP_2. *Current Biology* 5:775.

Dawson, P.A., Van Der Westhuyzen, D.R., Goldstein, J.L., and Brown, M. S., 1989, Purification of oxysterol binding protein from hamster liver cytosol. *J. Biol. Chem.* 264: 9046.

Dickeson, S.K., Lim, C.N., Schuyler, G.T., Dalton, T.P., Helmkamp, G.M., Jr., and Yarbrough, L.R., 1989, Isolation and sequence of cDNA clones encoding rat phosphatidylinositol transfer protein. *J. Biol. Chem.* 264:16557.

Goldstein, J.L. and Brown, M.S., 1990, Regulation of the mevalonate pathway. *Nature* 343: 425.

Hay, J.C. and Martin, T.F.J., 1993, Phosphatidylinositol transfer protein is required for ATP-dependent priming of Ca^{2+}-activated secretion. *Nature* 366: 572.

Hay, J.C., Fisette, P.L., Jenkins, G.H., Fukami, K., Takenawa, T., Anderson, R.A., and Martin, T.F.J., 1995, ATP-Dependent inositide phosphorylation required for Ca^{2+}-activated secretion. *Nature* 374: 173.

Jiang, B., Brown, J.L., Sheraton, J., Fortin, N. and Bussey, H., 1994, A new family of yeast genes implicated in ergosterol synthesis is related to the human oxysterol binding protein. *Yeast* 10: 341.

Kagiwada, S., Kearns, B.G., McGee, T.P., Fang, M., Hosaka, K,, and Bankaitis, V.A., 1996, The yeast *BSD2-1* mutation influences both the cellular requirement for phosphatidylinositol transfer protein function and the activity of the pathway for derepression of phospholipid biosynthetic gene expression. *Genetics*: In Press.

Kauffmann-Zeh, A., Thomas, G.M.H., Ball, A., Prosser, S., Cockroft, S., and Hsuan, J.J., 1995, Requirement of phosphatidylinositol transfer protein in epidermal growth factor signaling. *Science* 268:1188.

McGee, T.P., Skinner, H.B., Whitters, E.A., Henry, S.A. and Bankaitis, V.A., 1994, A phosphatidylinositol transfer protein controls the phosphatidylcholine content of yeast Golgi membranes. *J. Cell Biol.* 124: 273-287.

Novick, P., Field, C., and Schekman, R., 1980, Identification of 23 complementation groups required for post-translational events in the yeast secretory pathway. *Cell* 21: 205.

Novick, P., Osmond, B.C., and Botstein, D., 1989, Suppressors of yeast actin mutations. *Genetics* 121: 659.

Ohashi, M., de Vries, K.J., Frank, R., Snoek, G., Bankaitis, V., Wirtz, K., and Huttner, W.B., 1995, A role for phosphatidylinositol transfer protein in secretory vesicle formation. *Nature* 377: 544.

Ridgway, N.D., Dawson, P.A., Ho, Y.K., Brown, M.S. and Goldstein, J.L., 1992, Translocation of oxysterol binding protein to Golgi apparatus triggered by ligand binding. *J. Cell Biol.* 116: 307.

Ridgway, N.D. and Lagace, T.A., 1995, Brefeldin A renders chinese hamster ovary cells insensitive to transcriptional suppression by 25-hydroxycholesterol. *J. Biol. Chem.* 270: 8023.

Skinner, H.B., Alb, J.G. Jr., Whitters, E.A., Helmkamp, G.M., Jr., and Bankaitis, V.A., 1993, Phospholipid transfer activity is relevant to but not sufficient for the essential function of the yeast *SEC14* gene product. *EMBO J.* 12:4775.

Skinner, H.B., McGee, T.P., McMaster, C., Fry, M.R., Bell, R.M. and Bankaitis, V.A., 1995, Phosphatidylinositol transfer protein stimulates yeast Golgi secretory function by inhibiting choline-phosphate cytidylyltransferase activity. *Proc. Natl. Acad. Sci.* 92: 112.

Thomas, G.M.H., Cunningham, E., Fensome, A., Ball, A., Totty, N.F., Truong, O., Hsuan, J.J., and Cockroft, S, 1993, An essential role for phosphatidylinositol transfer protein in phospholipase C-mediated inositol lipid s ignaling. *Cell* 74:919.

Whitters, E.A., Cleves, A.E., McGee, T.P., Skinner, H.B. and Bankaitis, V.A., 1993, SAC1p is an integral membrane protein that influences the cellular requirement for phospholipid transfer protein function and inositol in yeast. *J. Cell Biol.* 122: 79.

Wirtz, K.W.A., 1991, Phospholipid transfer proteins. *Annu. Rev. Biochem.* 60: 73.

COMPARTMENTATION OF PROSTAGLANDIN BIOSYNTHETIC ENZYMES

William L. Smith, David L. DeWitt, Toshiya Arakawa,
Andrew G. Spencer, and Inseok Song

Department of Biochemistry
Michigan State University
East Lansing, MI 48824

INTRODUCTION

Prostaglandin endoperoxide H synthase (PGHS) catalyzes the committed step in the formation of prostaglandins and thromboxanes. There are two PGHS isozymes called PGHS-1 and PGHS-2. Both enzymes catalyze formation of PGH_2 from arachidonate. The enzymes are similar structurally and have similar kinetic properties. Moreover, both isoenzymes are inhibited by nonsteroidal anti-inflammatory drugs (NSAIDs). PGHS-1 and PGHS-2 are encoded by different genes. In general, PGHS-1 can be viewed as a constitutive enzyme whose expression appears to be regulated developmentally. PGHS-2 is an inducible enzyme which, with a few exceptions, is expressed transiently in response to growth factors, tumor promoters, or cytokines. Here we develop the concept that PGHS-1 and PGHS-2 function as part of two independent prostanoid biosynthetic systems.

EXPRESSION OF PGHS-1 AND PGHS-2

The fact that PGHS-1 is expressed constitutively suggests that this enzyme is part of an acute signaling system. Because expression of PGHS-2 involves a lag phase required for induction of the enzyme, prostanoids formed by the action of PGHS-2 are likely to function in secondary events. Prostanoid production by PGHS-1 and PGHS-2 in cultured cells appears to be initiated through distinct signaling pathways that may rely on the activation of different phospholipases.[1-3] In addition, products of PGHS-2, but not PGHS-1, can apparently function in the nucleus,[4] which suggests that PGHS-2 may independently signal directly to the nucleus. Thus, these two pathways of prostaglandin synthesis are both temporally and spatially separated. As a consequence, the same prostaglandin products could play at least two different roles in signal transduction within a single cell. Transgenic mice in which either the PGHS-1 or PGHS-2 genes have been inactivated display uniquely different phenotypes. These latter observations support the concept that each enzyme has separate physiological functions.[5-6]

Frontiers in Bioactive Lipids
Edited by Vanderhoek, Plenum Press, New York, 1996

Taken together these various studies have suggested to us (a) that prostanoids are formed in the endoplasmic reticulum through the action of PGHS-1, and to a lesser extent PGHS-2, exit cells and function through cell surface receptors to mediate so-called "housekeeping" functions such as regulation of renal H_2O and Na^+ metabolism, stomach acid secretion, and hemostasis and (b) that prostanoids are formed on the inner membrane of the nuclear envelope through the action of PGHS-2 and function at the level of the nucleus during cell replication or differentiation (Fig. 1).

THE PGHS-1 SYSTEM

Cells that express only PGHS-1 produce prostanoids rapidly in response to various hormonal stimuli, and the newly formed prostanoids act extracellularly as intercellular mediators. Human platelets[7] and rabbit renal collecting epithelia;[8] D.L. DeWitt and W.L.

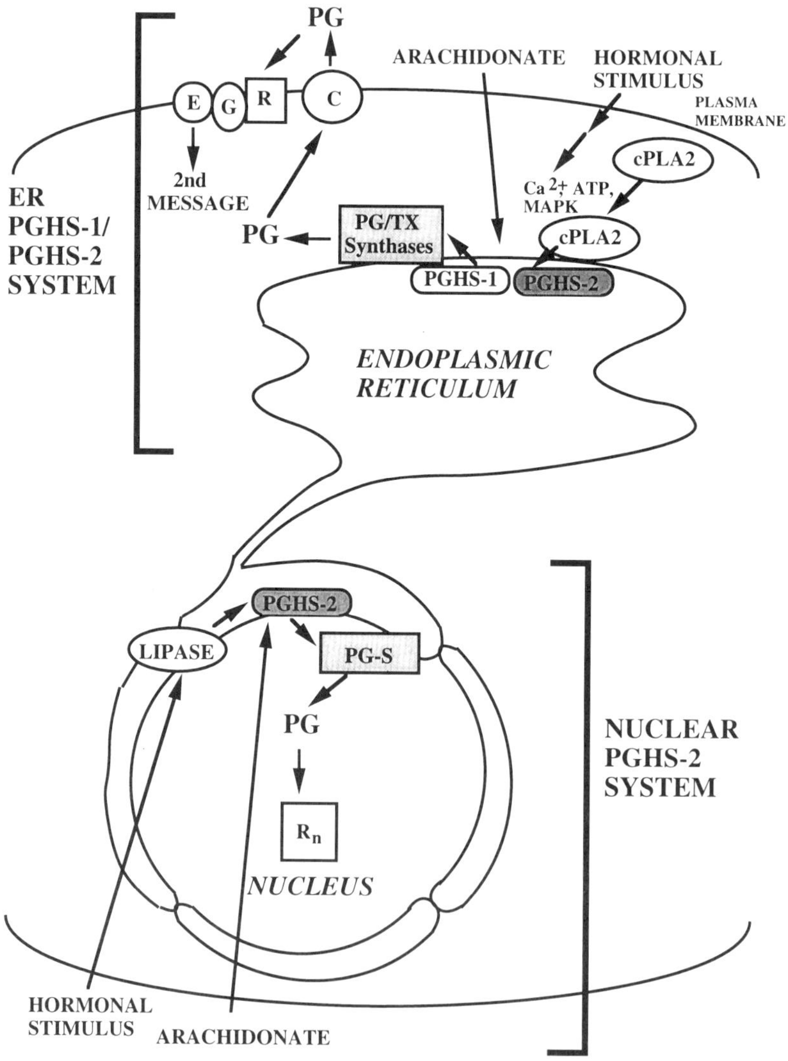

Fig. 1. Model of Prostaglandin Biosynthetic Pathways Involving PGHS-1 and PGHS-2.

Smith, unpublished results) are examples of cells which express PGHS-1 but not PGHS-2. Platelets and collecting tubule epithelia form TxA_2 and PGE_2, respectively, as their major prostanoid products, when incubated with either exogenous arachidonate or relevant hormonal stimuli (e.g., thrombin or bradykinin, respectively) that mobilize endogenous arachidonate from lipid precursors.[9-10]

Two Ca^{2+}-dependent phospholipase A_2s--an 85 kDa cytosolic (c) PLA_2 and a nonpancreatic, Type II, secretory (s) PLA_2--have been implicated in various experimental settings in prostanoid formation.[11-12] Fig. 1 shows $cPLA_2$-mediated arachidonate release coupled to the PGHS-1 synthetic system. $cPLA_2$ does function in conjunction with PGHS-1 during platelet TxA_2 formation,[13-14] probably during endothelial cell PGI_2 synthesis,[1] and in serum-starved murine 3T3 cells (Matsumoto and Smith, unpublished results). Moreover, $cPLA_2$ binds specifically to the cytoplasmic surfaces of both the ER and nuclear envelope in response to Ca^{2+} mobilization[15-16] thus becoming associated topologically with PGHS-1 which is located on the lumenal surface of the ER and NE.[4] $sPLA_2$ may also be involved in releasing arachidonate for the PGHS-1 system in some cells;[17] in those cases, $sPLA_2$ is secreted and cleaves arachidonate from phospholipids located on the cell surface, and the arachidonate can then enter the cell and subsequently the ER by simple diffusion.[18] PGHS-1, which is embedded in the lumenal surfaces of both the ER and the outer membrane of the nuclear envelope[4,19-21] converts the newly-released arachidonate to PGH_2.

Once PGH_2 is formed by the action of PGHS-1 on the lumenal surface of the ER, PGH_2 traverses the ER (to which it is freely permeable[9]) and is converted to a biological active prostanoid by enzymes such as TxA synthase[22] or PGE synthase[23-24] located on the cytoplasmic face of the ER. Newly formed prostanoids (e.g. TxA_2 or PGE_2) must exit the cell, but cell membranes are impermeant to prostanoids other than PGH_2.[9,25,26] A prostaglandin transporter, such as that recently cloned by Schuster and coworkers,[27] is most likely involved in prostanoid export.

Both TxA_2 formed by platelets and PGE_2 synthesized by collecting tubules are found extracellularly within 15-60 sec following a hormonal stimulus.[9-10] When TxA_2 is added exogenously to platelets it causes aggregation.[28] Similarly, PGE_2 added to collecting tubule epithelia inhibits vasopressin-induced tubular water reabsorption.[29] Thus, TxA_2 and PGE_2 elicit responses associated with their known physiological actions when added to the outside of cells. These responses are mediated by G protein-linked receptors which are probably present on the plasma membrane.[30-31]

Overall one can conclude from studies of the synthesis and actions of prostanoids formed by cells such as platelets and renal collecting tubules which express only PGHS-1 that prostanoids are formed rapidly via PGHS-1 systems within cells exposed to the appropriate hormonal stimulus, that the newly formed prostanoids exit the cells probably via a carrier(s), and that these prostanoids then act via specific G protein-linked receptors located on the plasma membrane (Fig. 1).

THE PGHS-2 SYSTEM

Induction of PGHS-2 in fibroblasts and monocytes expressing PGHS-1 constitutively causes relatively small 1.5-2 fold increases in the cellular PGH_2 biosynthetic activity;[32-37] thus, in those cells which already express PGHS-1, there is no dramatic increase in the total PGH_2 biosynthetic capacity following the induction of PGHS-2. PGHS-1 and PGHS-2 have very similar kinetic properties. Therefore, it is difficult to argue that in PGHS-2 is induced in fibroblasts or monocytes simply to augment the prostanoid biosynthetic capacities of these cells. Moreover, the two PGHS isozymes utilize different lipid stores of arachidonate for PGH_2 synthesis and perhaps different phospholipase systems to mobilize this arachidonate.[2-3] PGHS-1 and PGHS-2 are both found in the ER and the NE, but PGHS-2 is uniquely concentrated in the NE.[4] Taken together these results suggest that at the cellular level PGHS-1 and PGHS-2 play independent roles.

The data which have accumulated thus far suggest that there is no general rule regarding which lipase system is coupled to PGHS-2 (or PGHS-1).[1,11,24,38-40] In some cases $cPLA_2$ expression is controlled by the same factors which regulate PGHS-2[41-47] suggesting that $cPLA_2$ can couple to the PGHS-2 biosynthetic system. However, there is also evidence suggesting that $sPLA_2$ is coupled to PGHS-2 in Madin Darby canine kidney cells.[48]

Immunocytochemical and histochemical studies indicate that PGHS-1 and PGHS-2 are both present on the lumenal surface of the ER and the outer membrane of the NE but that PGHS-2 is uniquely more concentrated than PGHS-1 in the NE.[4] One explanation for these results is that a subset of PGHS-2 molecules is present on the inner membrane of the NE. Recent EM studies suggest that indeed PGHS-2 is located on both the inner and outer membrane of the NE (A.K. Spencer, J. Woods, I. Singer, W.L. Smith, unpublished results).

TxA synthase is located on the cytoplasmic surface of the ER[22] and on the NE of retinoic acid-treated HL60 cells (K. Matsumoto, I. Morita, and S. Murota, unpublished results). PGI_2 synthase has been localized to the ER and NE of vascular smooth muscle and endothelial cells.[49] TxA synthase in platelets and PGI_2 synthase in quiescent endothelial cells must be coupled to the PGHS-1 system. However, a nuclear TxA synthase activity is induced by treatment of HL60 cells with retinoic acid (K. Matsumoto, I. Morita, and S. Murota, unpublished results). Induction of PGHS-2, part of which is on the NE, is also associated with increases in PGI_2 formation by human umbilical cord endothelial cells.[50] Thus, under certain conditions both TxA synthase and PGI_2 synthase appear to be coupled to PGHS-2 and could be on the nucleoplasmic surface of the NE.

Prostanoid binding activities have been described in nuclear fractions[51] (I. Morita, unpublished results), and it is conceivable that some of the known splice variants of G protein-linked prostanoid receptors[52] are targeted to the NE. Recently, several PGJ derivatives have been shown to activate peroxisomal proliferator-activated receptor γ (PPARγ).[53,54] Thus, there may be a class of nuclear prostanoid receptors distinct from cell surface G protein-linked prostanoid receptors.

TARGETING OF PGHS-1 AND PGHS-2

The N-terminus of purified ovine PGHS-1 has the sequence ^{25}ADPGAPAPVNPCCYYP^{40} (55). ^{18}ANPCCSNPCQNRGE^{32} is the N-terminal sequence of isolated rat PGHS-2.[56] Because both PGHS-1 and PGHS-2 are membrane proteins and because the sequences near the N-termini of the unprocessed isozymes have sequences characteristic of signal peptides, it can be assumed that these peptides represent signal sequences. The signal peptides of PGHSs-1 have 24-26 amino acids, and those of PGHS-2 have 17 amino acids. The signal recognition particle of the ER is apparently capable of recognizing both peptides so there is no *a priori* reason to think that the differences in the structures of the PGHS-1 and PGHS-2 signal peptides are particularly significant in terms of targeting or processing.

All of the known PGHSs have a PTEL, STEL, or SAEL sequence at their C-termini. This sequence is similar to the KDEL sequence which functions in the retention of some resident ER proteins in the lumen of the ER.[57-59] Because PGHS-1 and PGHS-2 are found on the lumen side of the ER,[21] it was anticipated that the P/STEL sequences would serve as ER retention signals for these enzymes. The first studies of this topic indicated that modification or deletion of this sequence yielded somewhat less active mutant enzymes, but surprisingly, that there was no effect on subcellular targeting.[60,61] These latter experiments were performed with transfected *cos* cells in which ovine PGHS-1 and C-terminal mutants of PGHS-1 were overexpressed for 40 hrs. More recent studies, in which shorter expression times (e.g. 18 hrs) were used, indicate that ovine PGHS-1 from which the -PTEL region has been deleted becomes concentrated in the Golgi (I. Song and W.L. Smith, unpublished results); similar results have been obtained using constructs of PGHS-2 in which the C-terminal leucine has been converted to an asparagine. These results would be expected if the C-terminal XXEL sequences serve as ER targeting signals for PGHS-1 and PGHS-2.[62]

The mechanism(s) by which PGHS-2 is uniquely targeted to the NE is unknown. An unusual feature of PGHS-2 is the eighteen amino acid cassette located near its C-terminus. Although this was expected to contain targeting information, deletion of this region of the molecule produces an active enzyme which when stably expressed in murine 3T3 cells appears to localize in the same way as native PGHS-2 (D.L. DeWitt, unpublished results). Alternatively, targeting signals may reside in the membrane binding domain of PGHS-2. This possibility has not yet been examined.

ACKNOWLEDGMENTS

This work was supported in part by USPHS NIH Grant DK22042.

REFERENCES

1. Murakami, M., Kudo, I., and Inoue, K. (1993) *J. Biol. Chem.* **268**, 839-844
2. Murakami, M., Matsumoto, R., Austen, K. F., and Arm, J. P. (1994) *J. Biol. Chem.* **269**, 22269-22275
3. Reddy, S. T., and Herschman, H. R. (1994) *J. Biol. Chem.* **269**, 15473-15480
4. Morita, I., Schindler, M., Regier, M. K., Otto, J. C., Hori, T., DeWitt, D. L., and Smith, W. L. (1995) *J. Biol. Chem.* **270**, 10902-10908
5. Langenbach, R., Morhan, S. G., Tiano, H. F., Loftin, C. D., Ghanayem, B. I., Chulada, P. C., Mahler, J. F., Goulding, E. H., Klluckman, K. D., Kim, H. S., and Smithies, O. (1995) *Cell* **in press**
6. Morham, S. G., Langenbach, R., Loftin, C. D., Tiano, H. F., Vouloumanos, N., Jennette, J. C., Mahler, J. F., Kluckman, K. D., Ledford, A., Lee, C. A., and Smithies, O. (1995) *Cell* **in press**
7. Patrignani, P., Panara, M. R., Greco, A., Fusco, O., Natoli, C., Iacobelli, S., Cipollone, F., Ganci, A., Creminon, C., Maclouf, J., et al. (1994) *J. Pharmacol. Exp. Ther.* **271**, 1705-1712
8. Smith, W., and Bell, T. G. (1978) *Amer. J Physiol.* **235**, F451-F457
9. Marcus, A. J., Weksler, B. B., Jaffe, E. A., and Broekman, J. M. (1980) *J. Clin. Invest.* **66**, 979-986
10. Grenier, F. C., Rollins, T. E., and Smith, W. L. (1981) *Amer. J Physiol.* **241**, F94-F101
11. Bonventre, J. A. (1992) *J. Am. Soc. Nephrol.* **3**, 128-143
12. Dennis, E. A. (1994) *J. Biol. Chem.* **269**, 1057-1060
13. Riendeau, D., Guay, J., Weech, P. K., Laliberte, F., and Yergey, J. (1994) *J. Biol. Chem.* **269**, 15619-15624
14. Bartoli, F., Lin, H.-K., Ghomashchi, F., Gelb, M. H., Jain, M. K., and Apitz-Castro, R. (1994) *J. Biol. Chem.* **269**, 15625-15630
15. Glover, S., Bayburt, T., Mechthild, J., Chi, E., and Gelb, M. H. (1995) *J. Biol. Chem.* **270**, 15359-15367
16. Schievella, A. R., Regier, M. K., Smith, W. L., and Lin, L.-L. (1995) *J. Biol. Chem.* **270**, 30749-30754
17. Herschman, H. R. (1994) *Can. Metastasis Rev.* **13**, 241-256
18. Kamp, F., and Hamilton, J. A. (1993) *Biochemistry* **32**, 11074-11085

19. Rollins, T. E., and Smith, W. L. (1980) *J. Biol. Chem.* **255**, 4872-4875
20. Regier, M. K., DeWitt, D. L., Schindler, M. S., and Smith, W. L. (1993) *Arch. Biochem. Biophys.* **301**, 439-444
21. Otto, J. C., and Smith, W. L. (1994) *J. Biol. Chem.* **269**, 19868-19875
22. Ruan, K.-H., Wang, L.-H., Wu, K. K., and Kulmacz, R. J. (1993) *J. Biol. Chem.* **268**, 19483-19490
23. Tanaka, Y., Ward, S. L., and Smith, W. L. (1987) *J. Biol. Chem.* **262**, 1374-1381
24. Smith, W. L. (1992) *Am. J. Physiol.* **263**(2 Pt 2), F181-F191
25. Bito, L. Z., and Baroody, R. A. (1974) *Amer. J Physiol.* **229**, 1580-1586
26. Garcia-Perez, A., and Smith, W. L. (1984) *J. Clin. Invest.* **744**, 63-74
27. Kanai, N., Lu, R., Satriano, J. A., Bao, Y., Wolkoff, A. W., and Schuster, V. L. (1995) *Science* **268**, 866-869
28. Gorman, R. R., Fitzpatrick, F. A., and Miller, O. V. (1978) in *Advances in Cyclic Nucleotide Research*, Vol. 9 (George, W. W., and Ignarro, L. J., eds), pp. 597-609, Raven Press, NY
29. Grantham, J. J., and Orloff, J. (1968) *J. Clin. Invest.* **47**, 1154-1161
30. Coleman, R. A., Smith, W. L., and Narumiya, S. (1994) *Pharm. Rev.* **46**, 205-229
31. Ushikubi, F., Hirata, M., and Narumiya, S. (1995) *J. Lipid. Mediat.* **12**, 343-359
32. Habenicht, A. J. R., Goerig, M., Grulich, J., Rothe, D., Gronwald, R., Loth, U., Schettler, G., Kommerell, B., and Ross, R. (1985) *J. Clin. Invest.* **75**, 1381-1387
33. Raz, A., Wyche, A., Siegel, N., and Needleman, P. (1988) *J. Biol. Chem.* **263**, 3022-3028
34. Lin, A. H., Bienkowski, M. J., and Gorman, R. R. (1989) *J. Biol. Chem.* **264**, 17379-17383
35. Evett, G. E., Xie, W., Chipman, J. G., Robertson, D. L., and Simmons, D. L. (1993) *Arch. Biochem. Biophys.* **306**, 169-177
36. DeWitt, D. L., and Meade, E. A. (1993) *Arch. Biochem. Biophys.* **306**, 94-102
37. Lee, S. H., Soyoola, E., Chanmugam, P., Hart, S., Sun, W., Zhong, H., Liou, S., Simmons, D., and Hwang, D. (1992) *J. Biol. Chem.* **267**, 25934-25938
38. Nalefski, E. A., Sultzman, L. A., Martin, D. M., Kriz, R. W., Towler, P. S., Knopf, J. L., and Clark, J. D. (1994) *J. Biol. Chem.* **269**, 18239-18249
39. Balsinde, J., Barbour, S. E., Bianco, I. D., and Dennis, E. A. (1994) *Proc. Natl. Acad. Sci. U. S. A.* **91**, 11060-11064
40. Barbour, S. E., and Dennis, E. A. (1993) *J. Biol. Chem.* **268**, 21875-21882
41. Hoeck, W. G., Ramesha, C. S., Chang, D. J., Fan, N., and Heller, R. A. (1993) *Proc. Natl. Acad. Sci. U. S. A.* **90**, 4475-4479

42. Hulkower, K. I., Wertheimer, S. J., Levin, W., Coffey, J. W., Anderson, C. M., Chen, T., DeWitt, D. L., Crowl, R. M., Hope, W. C., and Morgan, D. W. (1994) *Arthritis Rheum.* **37**, 653-661
43. Roshak, A., Sathe, G., and Marshall, L. A. (1994) *J. Biol. Chem.* **269**, 25999-26005
44. Chepenik, K. P., Diaz, A., and Jimenez, S. A. (1994) *J. Biol. Chem.* **269**, 21786-21792
45. Doerfler, M. E., Weiss, J., Clark, J. D., and Elsbach, P. (1994) *J. Clin. Invest.* **93**, 1583-1591
46. Marshall, L. A., Bolognese, B., and Roshak, A. (1994) *J. Lipid Med.* **10**, 295-313
47. Angel, J., Berenbaum, F., Le-Denmat, C., Nevalainen, T., Masliah, J., and Fournier, C. (1994) *Eur. J. Biochem.* **226**, 125-131
48. Schaefers, J. H., Haselmann, J., and Goppelt-Struebe, M. (1995) *Biochim. Biophys. Acta* **in press**
49. Smith, W. L., DeWitt, D. L., and Allen, M. L. (1983) *J. Biol. Chem.* **258**, 5922-5926
50. Jones, D. A., Carlton, D. P., McIntyre, T. M., Zimmerman, G. A., and Prescott, S. M. (1993) *J. Biol. Chem.* **268**, 9049-9054

51. Rao, C. V., and Mitra, S. B. (1982) *Methods Enzymol.* **86**, 192-202
52. Namba, T., Sugimoto, Y., Negishi, M., Irie, A., Ushikubi, F., Kakizuka, A., Ito, S., Ichikawa, A., and Narumiya, S. (1993) *Nature* **365**, 166-170
53. Forman, B. M., Tontonoz, P., Chen, J., Brun, R., Spiegelman, B. M., and Evans, R. M. (1995) *Cell* **83**, 803-812
54. Tontonoz, P., Hu, E., Graves, R. A., Budavari, A. I., and Spiegelman, B. M. (1994) *Genes Dev.* **8**, 1224-1234
55. DeWitt, D. L., and Smith, W. L. (1988) *Proc. Natl. Acad. Sci. USA* **85**, 1212-1416
56. Sirois, J., and Richards, J. S. (1992) *J. Biol. Chem.* **267**, 6382-6388
57. Nothwehr, S. F., and Stevens, T. H. (1994) *J. Biol. Chem.* **269**, 10185-10188
58. Pelham, H. R. B. (1990) *TIBS* **15**, 483-485
59. Tang, B. L., Wong, S. H., Low, S. H., and Hong, W. (1992) *J. Biol. Chem.* **267**, 7072-7076
60. Regier, M. K., Otto, J. C., DeWitt, D. L., and Smith, W. L. (1995) *Arch. Biochem. Biophys.* **317**, 457-463
61. Ren, Y., Loose-Mitchell, D. S., and Kulmacz, R. J. (1995) *Arch. Biochem. Biophys.* **316**, 751-757
62. Munro, S., and Pelham, H. R. B. (1987) *Cell* **48**, 899-907

DIFFERENTIAL CONTROL OF CYCLOOXYGENASE CATALYSIS IN PGH SYNTHASE ISOFORMS: ROLE OF HYDROPEROXIDE INITIATOR

Richard J. Kulmacz, Wei Chen, and Lee-Ho Wang

Division of Hematology, Deptartment of Internal Medicine
University of Texas Health Science Center at Houston
Houston, TX 77030

INTRODUCTION

Prostaglandin H synthase (PGHS) catalyzes the conversion of arachidonate to PGG_2, a key control step in prostanoid biosynthesis. Two isoforms of PGHS are known (Herschman, 1996). PGHS-1 is present at more or less constitutive levels in many cells, and has been assigned a housekeeping role. PGHS-2 is strongly induced by cytokines and mitogens in some cells, and is believed to play an important role in inflammatory processes. Induction of PGHS-2 in fibroblasts and macrophages is accompanied by large increases in arachidonate release and prostanoid synthesis (Reddy and Herschman, 1994). Inhibition of PGHS-2 expression blocked the mitogen-dependent increase in prostaglandin synthesis, but not the arachidonate release; the constitutive PGHS-1 remained in a latent form, unable to utilize the released arachidonate (Reddy and Herschman, 1994). Thus, cyclooxygenase catalysis by PGHS-1 and PGHS-2 is differentially controlled in these cells which have both isoforms.

Cyclooxygenase catalysis by PGHS-1 has long been known to require initiation by hydroperoxide; cyclooxygenase activity was found to be suppressed by hydroperoxide scavengers even in the presence of abundant fatty acid substrate (Smith and Lands, 1972). Before the discovery of the second PGHS isoform, hydroperoxide scavengers, such as glutathione peroxidase (GSP), were proposed to play a major role in controlling cellular prostaglandin synthesis (Marshall et al., 1987). The observation of differential catalytic control in the two PGHS isoforms (Reddy and Herschman, 1994) prompted us to compare their hydroperoxide activator requirements.

EVALUATION OF HYDROPEROXIDE INITIATOR REQUIREMENTS

Our approach was to "titrate" purified preparations of PGHS-1 and PGHS-2 with GSP. In this approach, a fixed amount of PGHS is reacted with excess arachidonate in the presence of a variable amount of GSP (Kulmacz and Lands, 1983). The cyclooxygenase velocity decreases approximately linearly with increasing GSP, and extrapolation affords an estimate of how many GSP units are needed to suppress cyclooxygenase activity completely (Kulmacz and Lands, 1983). For ovine PGHS-1, cyclooxygenase suppression required about 75 GSP units/cyclooxygenase unit, whereas suppression of human PGHS-2 activity required about 670 GSP units/cyclooxygenase unit (Kulmacz and Wang, 1995). These results were confirmed in reactions using arachidonate metabolites instead of oxygen consumption to monitor the cyclooxygenase activity (Kulmacz and Wang, 1995). The cyclooxygenase activity of human PGHS-2 thus was considerably more resistant than the ovine PGHS-1 activity to inhibition by hydroperoxide scavenger, indicating that the activation process was more efficient in PGHS-2. Similar results were found for ovine PGHS-2 by ourselves and others (Capdevila et al., 1995).

A mathematical model of the GSP competition experiment was used to derive quantitative estimates of the hydroperoxide activator requirements for the PGHS isoforms (Kulmacz and Wang, 1995). In this model, changes in the hydroperoxide (PGG_2) level are calculated from the rates of PGG_2 generation (via cyclooxygenase activity) and PGG_2 removal (via PGHS peroxidase and GSP activities), each responding to the existing PGG_2 level in a saturable fashion:

$$\begin{aligned} \Delta PGG_2/\Delta t \quad = \quad & + (\text{PGHS Cycloox units})/(1+K_p/[PGG_2]) \\ & - (\text{PGHS Perox units})/(1+K_{mPGHS}/[PGG_2]) \\ & - (\text{GSP units})/(1+K_{mGSP}/[PGG_2]) \end{aligned}$$

The units of PGHS cyclooxygenase and peroxidase, and of GSP are known from direct measurements, as are the K_m values for PGHS peroxidase and GSP. An arbitrary K_p value, the PGG_2 level giving half-maximal cyclooxygenase activation, is then assumed, and the cyclooxygenase reaction kinetics are simulated by numerical integration to predict the peak cyclooxygenase velocity. For each assumed K_p value, the peak cyclooxygenase velocity is predicted for a series of GSP levels, generating a line reflecting the susceptibility to inhibition by hydroperoxide scavenger at that K_p (Figure 1). Extrapolation to the x-axis predicts the GSP/cyclooxygenase ratio needed to completely suppress cyclooxygenase activity ("suppression ratio") at a particular K_p value. Plotting the predicted suppression ratio as a function of the assumed K_p value (Figure 2) shows the expected inverse relationship between the two variables. In the final step, the relation in Figure 2 is used as a standard curve to convert the experimentally observed suppression ratios to estimates of the K_p values for purified PGHS-1 (21 nM) and PGHS-2 (2.3 nM). This result makes it clear that efficient activation of the cyclooxygenase requires about an order of magnitude lower hydroperoxide in PGHS-2 than in PGHS-1.

HYDROPEROXIDE REQUIREMENT IN MIXTURES OF PGHS ISOFORMS

Despite the large difference in hydroperoxide activator requirements of the purified PGHS isoforms, initiation of PGHS-2 cyclooxygenase could conceivably lead to activation of any PGHS-1 in the same cellular compartment, leading to a lower effective K_p for the PGHS-1. To clarify this issue, a mixture of 80% PGHS-1 and 20% PGHS-2 was examined for its sensitivity to inhibition by GSP (Kulmacz and Wang, 1995). It was found that the PGHS-1 initiation efficiency in the mixture was much the same as in isolation, indicating that the presence of active PGHS-2 does not automatically activate nearby PGHS-1 cyclooxygenase. The two isoforms are thus able to respond independently to the ambient hydroperoxide level set by cellular peroxidases.

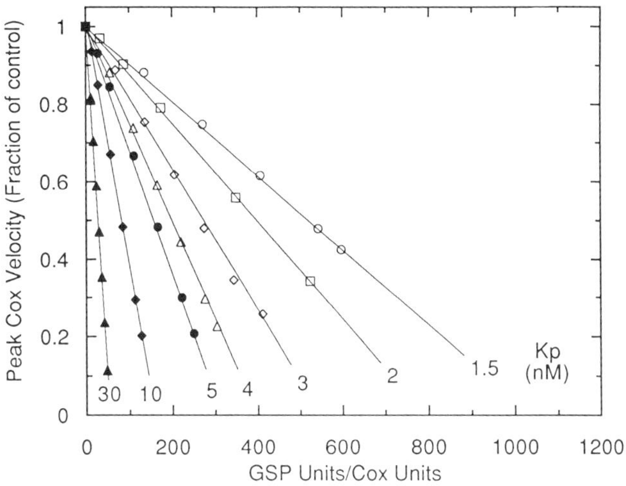

Figure 1. Predicted inhibition of cyclooxygenase by added GSP. Cyclooxygenase reaction kinetics for mixtures of a fixed amount of PGHS and a variable amount of GSP were simulated by numerical integration at the indicated assumed values of Kp.

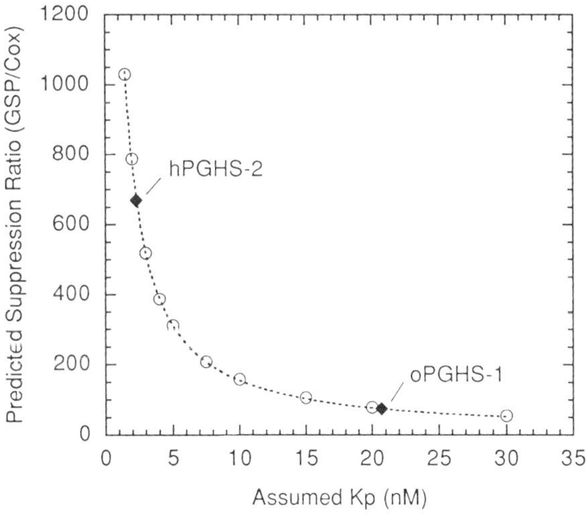

Figure 2. Effect of Kp value on predicted suppression ratio. Suppression ratios were obtained from Figure 1.

DIFFERENTIAL CONTROL OF CATALYSIS IN THE PGHS ISOFORMS

As a result of its lower hydroperoxide requirement, much of the PGHS-2 cyclooxygenase will be catalytically active at cellular hydroperoxide levels where almost all the PGHS-1 cyclooxygenase in the same compartment remains latent (Figure 3). Such differential control by hydroperoxide level might help explain the observed discrimination against PGHS-1 in fibroblasts and macrophages (Reddy and Herschman, 1994). Cellular hydroperoxide levels are difficult to measure directly, but 10^{-9} M levels would be

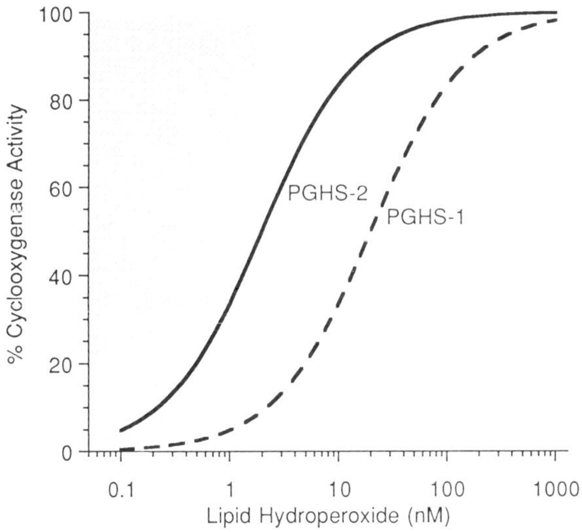

Figure 3. Predicted effects of hydroperoxide level on cylooxygenase activation for PGHS-1 (Kp = 20 nM) and PGHS-2 (Kp = 2 nM). Selective PGHS-2 activity is expected at hydroperoxide levels in the shaded range.

consistent with the large excess of GSP activity in many tissues (Marshall et al., 1987). Under these circumstances, the K_p of PGHS-1 would lead to only a slow trickle of prostanoid synthesis, whereas the K_p of PGHS-2 would permit the conversion of any available arachidonate to prostaglandin without raising the hydroperoxide concentration to levels which might affect other redox-sensitive cellular components.

Acknowledgements This work was supported in part by NIH grant GM 52170.

REFERENCES

Capdevila, J.H., Morrow, J.D., Belosludtsev, Y.Y., Beauchamp, D.R., DuBois, R.N., and Falck, J.R., 1995, The catalytic outcomes of the constitutive and the mitogen inducible isoforms of prostaglandin H2 synthase are markedly affected by glutathione and glutathione peroxidase(s), *Biochemistry* 34:3325-3337.

Herschman, H.R., 1996, Prostaglandin synthase 2, *Biochim. Biophys. Acta* 1299:125-140.

Kulmacz, R.J., and Lands, W.E.M., 1983, Requirements for hydroperoxide by the cyclooxygenase and peroxidase activities of prostaglandin H synthase, *Prostaglandins* 25:531-540.

Kulmacz, R.J., and Wang, L.-H., 1995, Comparison of the hydroperoxide initiator requirements for the cyclooxygenase activities of prostaglandin H synthase-1 and -2, *J. Biol. Chem.* 270:24019-24023.

Marshall, P.J., Kulmacz, R.J., and Lands, W.E.M., 1987, Constraints on prostaglandin synthesis in tissues, *J. Biol. Chem.* 262:3510-3517.

Reddy, S.T., and Herschman, H.R., 1994, Ligand-induced prostaglandin synthesis requires expression of the TIS10/PGS-2 prostaglandin synthase gene in murine fibroblasts and macrophages, *J. Biol. Chem.* 269:15473-15480.

Smith, W.L., and Lands, W.E.M., 1972, Oxygenation of polyunsaturated fatty acids during prostaglandin biosynthesis by sheep vesicular gland, *Biochemistry* 11:3276-3285 .

REGULATION OF EXPRESSION AND THE FUNCTIONAL ROLE OF CYCLOOXYGENASE-2

Kirsi Narko, Ari Ristimäki, Pazit Ben-Av and Timothy Hla

Department of Molecular Biology, Holland Laboratory, American Red Cross, Rockville, MD 20855, USA

THE CYCLOOXYGENASE ENZYME

Cyclooxygenase (Cox), also known as Prostaglandin H Synthase (E.C.1.14.99.1) is a rate limiting enzyme in a synthesis of prostanoids. Once the cellular phospholipase A_2 has released arachidonic acid from membrane phospholipids, the Cox enzyme oxidizes arachidonic acid to PGG_2 by the cyclooxygenase activity and then reduces PGG_2 to PGH_2 by the peroxidase activity. The PGH_2 is converted to various bioactive prostanoids by tissue specific isomerases (1). Cox enzyme is an integral membrane protein located in endoplasmic reticulum and nuclear membrane (2). It is irreversibly self-inactivated after a limited amount of catalysis and then rapidly degraded by proteases (3).

GENE STRUCTURE OF COX -2

Two cyclooxygenase genes Cox-1 and Cox-2 have been identified and they share approximately 60% identity at amino acid level (4). While the Cox-1 gene is constitutively expressed in many cell types, Cox-2 is induced by a wide variety of extracellular agents including cytokines, growth factors and tumor promoters. Rosen et al. (6), using low-stringency hybridization with Cox-1 probe was the first to predict the existence of the Cox-2 gene. Since then the Cox-2 cDNA has been isolated as an immediately-early gene from e.g.*v-src* transformed chick embryo fibroblasts (7), murine fibroblasts (8), rat granulosa cells (9) and human vascular endothelial cells (4). The human Cox-2 gene (8.5 kb) consists of 10 exons. While the ORF of the Cox-2 cDNA is 80% identical to the Cox-1 ORF, the 3'- untranslated region (3'-UTR) is highly divergent. The largest exon (exon 10) in the Cox-2 gene encodes 410 bp of the coding region and the entire 2.5 kb 3'-UTR. 3'-UTR of Cox-2 gene contains 22 AUUUA mRNA instability elements (10). AUUUA sequence is highly conserved and it is often repeated multiple times in the 3'-UTR regions of genes encoding short lived cytokines, proto-oncogenes and nuclear transcription factors (11). These AU-rich (ARE) motifs have been shown to function as RNA destabilizing elements e.g. in GM-CSF, *c-fos and c-myc* mRNAs (12, 13, 14). These elements may also be involved in rapid turnover of the Cox-2 mRNA.

REGULATION OF EXPRESSION BY IL-1 AND DEXAMETHASONE

The IL-1 family of cytokines are important regulators of inflammation (15). IL-1 induces Cox-2 and the induction seems to involve activation of transcription as well as post-transcriptional stabilization of the mRNA. We have demonstrated that IL-1 stimulates Cox-2 expression in a sustained manner in human umbilical vein endothelial cells (HUVEC), ECV304 cells, IMR-90 lung fibroblasts and rheumatoid synovial fibroblasts cells (16, 17). After IL-1 treatment, elevation of Cox-2 mRNA can be observed as early as after 30 min. However, maximal levels are achieved at 24 h and the stimulation remains evident after incubation for 96 h. In spite of the higher steady state levels of Cox-2, the activation of transcription by IL-1 is only transient as determined by nuclear run-on analysis (16). In the presence of transcription inhibitors DRB or actinomycin D, the effect of IL-1 is potentiated, suggesting that IL-1 induces a burst of transcription of Cox-2 gene followed by a stabilization of the pre-existing mRNA. Cycloheximide enhances the induction of IL-1 resulting in "superinduction", which is presumably related to the ability of cycloheximide to inhibit the degradation of the Cox-2 mRNA by preventing synthesis of proteins involved in degradation process.

Anti-inflammatory glucocorticoid dexamethasone (dex) is known to suppresses Cox-2 expression induced by growth factors, serum and cytokines (reviewed in Reference 5). Dexamethasone is known to regulate several genes by both transcriptional and post-transcriptional level. We have investigated the mechanisms involved in the dexamethasone down-regulation of IL-1 induced Cox-2 mRNA expression in IMR-90 cells and synovial fibroblasts (18). Dexamethasone treatment resulted suppression of IL-1 induced Cox-2 mRNA levels whether it was added concurrently with IL-1 or was added 16 h post-IL-1 treatment. Our data suggests that dexamethasone destabilizes IL-1 induced Cox-2 mRNA. Dexamethasone was unable to suppress Cox-2 mRNA levels when transcription was inhibited with DRB or actinomycin D suggesting that the ability of dexamethasone to down-regulate Cox-2 mRNA requires the transcriptional induction of mRNA instability factors.

REGULATORY ROLE OF THE 3'-UTR OF COX-2

The Cox-2 transcripts exhibit heterogeneity. In many cell types a predominant 4.1-4.6 kb isoform and a less abundant 2.8 kb isoform are observed. Three canonical polyadenylation signals (AAUAAA) are present in the 3-'UTR of the Cox-2 gene and utilization of these sites would yield transcripts of 4.6, 4.4 and 4.1 kb respectively. We have demonstrated that the 2.8 kb Cox-2 transcript is derived by alternative polyadenylation signal (AUUAAA) in exon 10. Interestingly, the glucocorticoid dexamethasone down regulates the 2.8 kb isoform with a slower kinetics than the large transcript isoforms (18). This may be due to the existence of only one evolutionally conserved AU-rich element in the 2.8 kb transcript while there is three elements present in long isoforms.

To investigate if the Cox-2 3'-UTR is involved in post-transcriptional regulation, we prepared the following chimeric cDNA constructs; Cox-1 ORF+Cox-2 3'-UTR and Cox-1 ORF with Cox-2 3'-UTR in which the first 6 AUUUA containing ARE1 motif was deleted. Constructs were subcloned into expression vector pcDNAI/Neo and transiently transfected into transformed human embryonic kidney (293) cells. Expression of constructs was not modulated by IL-1. Surprisingly dexamethasone induced both constructs two fold (unpublished data). Cotransfection with human glucocorticoid receptor (hGR) potentiated the induction. In addition, the Cox-2 ORF has been found to be less stable than Cox-1 in trasfected COS cells suggesting that mRNA instability motifs may exist in both the Cox-2 ORF and the 3'-UTR (4).

COX-2 IN RHEUMATOID ARTHRITIS

Prostaglandins are important mediators of inflammation in rheumatoid arthritis (RA) (19, 20). The nonsteroidal anti-inflammatory drugs (NSAID), which are inhibitors of Cox enzyme are extensively used in the treatment of RA. IL-1 is a major proinflammatory cytokine present in synovial fluid and peripheral blood of patients with RA, and IL-1 levels have been found to correlate with disease activity (21). Injection of IL-1 into joints has been reported to induce inflammatory synovitis (22). The inflammation of RA is sensitive to corticosteroids. One effect of corticosteroids is to reduce the production of prostaglandins by synovial tissues. *In vitro* data suggest that the inhibition of both Cox and cytosolic phospholipase A_2 expression may be important in the corticosteroid inhibition of prostaglandin production. (23, 24). When the *in vivo* expression of Cox polypeptide was studied in synovia from patients with RA and osteoarthritis (OA), significantly more staining was observed in RA than in synovia from OA and normal human synovia (25). Treatment with dexamethasone have been shown to decrease inflammation and reduce Cox immunostaining in rats with streptococcal cell wall or adjuvant induced arthritis (25). Cox-1 and Cox-2 polypeptides and mRNA have been found to been expressed in rheumatoid synovial explants and cultured rheumatoid synoviocytes under basal culture conditions. However, Cox-2 appeared to be the major regulated form and expression of Cox-2 was stimulated with IL-1 and PMA, and induction was suppressed by dexamethasone (17). Modulation of Cox-2 expression in rheumatoid synovial tissues may be an important mechanism regulating inflammation in RA.

Angiogenesis is a process by which new blood vessels grow by sprouting from established ones and it is an important component of chronic synovitis and tumor development. PGE_2 is the major product of the cyclooxygenase pathway in rheumatoid synovial tissues and it is a potent inducer of angiogenesis but the mechanism of its action is not known (26). We have examined the effect of $PGF_{2\alpha}$, PGE_1 and PGE_2 on the expression of a potent and specific angiogenic factor VEGF (27). PGE_1 and PGE_2 but not $PGF_{2\alpha}$ induced VEGF expression in RA synovial fibroblasts. Furthermore, IL-1 induced sustained up-regulation of VEGF. Thus in response to an inflammatory cytokine and prostanoid action, the RA synovial fibroblast is capable of expressing an angiogenic signal by up-regulation of VEGF expression. PGE_1 and PGE_2 mediate their action via interaction with subtypes of prostaglandin receptors. The EP1 and EP2 but not EP3 and EP4 subtypes of PGE receptors were found to be expressed in synovial fibroblasts. Activation of the EP1 receptor has been associated with a rise of intracellular calcium, possibly via activation of phospholipase C, and activation of EP2 receptor is assosiated with increase in cellular cAMP (28, 29). Cholera toxin and forskolin, which induce intracellular cAMP levels, also induced VEGF mRNA. However, down-regulation of protein kinase C did not influence the PGE_2 effect (27). These data suggest that EP2/Gs/cAMP/protein kinase A pathway may be involved in the VEGF induction by PGE_2 in RA synovial fibroblasts.

ROLE OF COX IN DEVELOPMENT OF TUMORS

Several epidemiological studies indicate that NSAID such as aspirin and sulindac reduce the risk of colon cancer (30, 31, 32, 33). The mechanisms by which NSAIDs inhibit colon carcinogenesis is not known. Prostaglandins and Cox enzyme may be involved in the initiation and the promotion of carcinogenesis because a major action of NSAIDs is the inhibition of Cox. Prostaglandin levels have been reported to be higher in human colon tissues than in surrounding normal tissues (34) and Cox-2 mRNA and protein levels has been shown to be markedly increased in the colon tumors but not in normal colon tissues from same patients (35, 36, 37). To test if the over-expression of the Cox enzyme is contributing to the tumorigenesis, immortalized ECV cells were stably transfected with the expression vector containing Cox-1 ORF. ECV-Cox-1 cells expressed high level of enzyme constitutively whereas no expression was detected in control cells. In-vitro the growth rate, apoptosis and adhesive properties were not altered by Cox-1 over-expression. However when cells were transplanted in nude mice, result was exaggerated tumor development (unpublished data). This suggests involvement of products of the cyclooxygenase pathway in initiation or maintenance of the tumor development.

CONCLUSIONS

Our studies as well as data from other laboratories point out the crucial role of Cox in inflammatory diseases and cancer. Post-transcriptional regulation of the mRNA stability is a widely used mechanism to regulate gene expression. Because enhanced Cox-2 expression is correlated with proliferative and chronic inflammatory condition, it would be of interest to determine the contribution of the Cox-2 mRNA stability mechanisms and dysregulation in various pathological conditions. Induction of the angiogenic response may be a potential mechanism via which prostanoids regulate proliferative diseases. For example the induction of VEGF expression by prostaglandins may be an important mechanism in inflammatory angiogenesis. Elucidation of the specific steps of the regulation of the Cox gene expression, and the function of prostanoids may provide additional avenues for the development of novel anti-inflammatory and chemotherapeutic agents.

REFERENCES

1. Needleman, P., Turk, J., Jakschik, B.A., Morrison, A.R. and Lefkowith, J.B. (1986) Ann. Rev. Biochem. 55, 69-102.
2. Smith, W.L. and Marnett, L.J. (1991) Biochim. Biophys. Acta. 1083, 1-17.
3. Chen, Y.P., Bienkowski, M.J., and Marnett, L.J. (1987) J. Biol. Chem. 262, 16892-16899.

4. Hla, T. and Neilson, K. (1992) Proc. Natl. Acad. Sci. USA. 89, 7384-7388.
5. Herschman, H.R. (1994) Cancer Metastasis Rev. 13, 241-256.
6. Rosen, G.D., Birgenmeier, T.M., Raz, A. and Holzman, M.J. (1989) Biochem. Biophys. Res. Commun. 164, 1358-1365.
7. Xie, W., Chipman, J.G., Ropertson, D.L., Erikson, R.L. and Simmons, D.L. (1991) Proc. Natl. Acad. Sci. USA. 88, 2692-2696.
8. Kujubu, D.A., Fletcher, B.S., Varnum, B.C., Lim, R.W. and Herschman, H.R. (1991) J. Biol. Chem. 266, 12866-12872.
9. Sirois, J., Levy, I.O., Simmons, D.I. and Richards, J.S. (1993) J. Biol. Chem. 268, 12199-12206.
10. Appleby, S.B., Ristimäki, A., Neilson, K., Narko, K. and Hla, T. (1994) Biochem. J. 302, 723-727.
11. Caput, D., Beutler, B., Thayer, R., Brown-Smith, S. and Cerami, A. (1986) Proc. Natl. Acad. Sci. USA. 83, 1670-1674.
12. Shaw, G. and Kamen, R. (1986) Cell 46, 659-667.
13. Shyu, A-B., Belasco, J.G. and Greenberg, M.E. (1991) Genes Dev. 5, 221-231.
14. Jones, T.R. and Cole, M.D. (1987) Mol. Cell. Biol. 7, 4513-4521.
15. Dinarello, C.W. (1988) FASEB J. 2, 108-115.
16. Ristimäki, A., Carfinkel, S., Wassendorf, J. and Hla, T. (1994) J. Biol. Chem. 296, 11769-11775.
17. Crofford, L.J., Wilder, R.L., Ristimäki, A.P., Sano, H., Remmers, E.F., Epps, H.R. and Hla, T. (1994) J. Clin. Invest. 93, 1095-1101.
18. Ristimäki, A., Narko, K. and Hla, T. (1996) Biochem. J. In Press.
19. Salmon, J.A., Higgs, G.A., Vane, J.R., Bitensky, L., Chayen, J., Henderson, B. and Cashman, B. (1983) Ann. Rheum. Dis. 42, 36-39.
20. Davies , P., Bailey, P.J., Goldenberg, M.M. and Ford-Hutchinson, A.W. (1984) Annu. Rev. Immunol. 2, 335-357.
21. Estgate, J.A., Symons, J.A., Wood, N.C., Grinlinton, F.M., diGiovine, S.F. and Duff, G.W. (1988) Lancet ii 706-709.
22. Andreis, M., Stansty, P. and Ziff, M. (1974) Arthritis Rheum. 17, 537-551.
23. Fu, J.-I., Masferrer, J.L., Seivert, K., Raz, A. and Needleman, P. (1990) J. Biol. Chem. 265, 16737-16740.
24. Koehler, L., Hass, R., DeWitt, D.L., Resch, K. and Goppelt-Strube, M. (1990) Biochem. Pharmacol. 40, 1307-1316.
25. Sano, H., Hla, T., Maier, J.A.M., Crofford, L.J., Case, J.P., Maciag, T. and Wilder, R.L. (1992) J. Clin. Invest. 89, 97-108.
26. Ziche, M., Jones, J. and Gullino, P.M. (1982) J. Natl. Canc. Inst. 69,475-482.
27. Ben-Av, P., Crofford, L.J. Wilder, R.L. and Hla, T. (1995) FEBS Lett. 372, 83-87.
28. Funk, C.D., Furci, L., Fitzgerald, G.A., Grygorczyk, R., Rochette, C., Bayne, M.A., Abramovitz, M. and Metters, K.M. (1993) J.Bol. Chem. 268, 26767-26772.
29. Regan, J.W., Bailey, T.J., Pepperl, D.J., Pierce, K.L., Bogardous, A.M., Donello, J.E., Fairbairn, C.E., Kedzie, K.M., Woodward, D.F. and Gill, D.W. (1994) Mol. Pharmacol. 46, 213-220.
30. Thun, M.J., Namboodiri, M.M. and Heath, C.W. (1991) N. Eng. J. Med. 325, 1593-1596.
31. Rosenberg, L., Palmer, J.R., Zauner, A.G., Warshauer, M.E., Stolley, P.D. and Shapiro, S.A. (1991) J. Natl. Cancer. Inst. 83, 355-358.
32. Suh, O., Mettlin, C. and Petrelli, N.J. (1993) Cancer (Phila.) 72, 1171-1177.
33. Greenberg, E.R., Baron, J.A., Freeman, D.H, jr., Mandel, J.S. and Haile, R. (1993) J. Natl. Cancer. Inst. 85, 912-916.
34. Bennet, A., Civier, A., Hensby, C.N., Melhuishin, P.B. and Stamford, I.F. (1987) Gut. 28, 315-318.
35. Eberhart, C.E., Coffey, R. J., Radhika, A., Giardiello, F. M., Ferrenbach, S. and Dubois, R.N. (1994) Gastroenterology 107, 1183-1188.
36. Kargman, S.L., O'Neill, G.P., Vickers, P.J., Evans, J.F., Mancini, J.A. and Jothy, S. (1995) Cancer Res. 55, 2556-2559.
37. Sano, H., Kawahito, Y., Wilder, R.L., Hashiramoto, A., Mukai, S., Asai, K., Kimura, S., Kato, H., Kondo, M. and Hla, T. (1995) Cancer Res. 55, 3785-3789.

THE REGULATION OF PROSTAGLANDIN E_2 SYNTHESIS BY INTERLEUKIN-1ß AND CERAMIDE IN HUMAN FIBROBLASTS: EFFECTS ON CYCLOOXYGENASE-1, CYCLOOXYGENASE-2, AND PHOSPHOLIPASE A_2 GENE EXPRESSION

Leslie R. Ballou,[1,2,3] Kanyawim Kirtikara,[2] and Stanley J. F. Laulederkind[2]

[1]The Department of Veterans Affairs Medical Center and The Departments of [2]Medicine and [3]Biochemistry, The University of Tennessee, Memphis, TN, 38104.

INTRODUCTION

One of the most recently identified membrane lipids capable of giving rise to metabolically active signal transduction molecules is sphingomyelin (SM). Ceramide, which is the immediate product of SM hydrolysis, and other SM metabolites (e.g., sphingosine and sphingosine-1-phosphate) have been shown to function in a variety of signaling and physiological processes [1-9].

We have previously shown that both sphingosine and ceramide (added as the C_2 analog) enhanced interleukin-1ß (IL-1)-mediated prostaglandin E_2 (PG) production in human dermal fibroblasts[10-12]. We also showed that IL-1 induced the hydrolysis of SM to yield ceramide suggesting that endogenous SM metabolites, generated in response to IL-1, may play an important role in IL-1 signal transduction [12]. In an effort to elucidate the mechanism(s) by which IL-1 and ceramide interact to modulate PG production, we have examined the effects of IL-1 and ceramide on the expression of genes encoding the two isoforms of cyclooxygenase (COX-1 and COX-2) and phospholipase A_2 (primarily via studies on the "cytosolic" enzyme, $cPLA_2$) each of which represents a rate-limiting step in PG synthesis.

Our most recent results indicate that ceramide increases COX expression but not PLA_2 expression which accounts for its inability to stimulate PG production in the absence of IL-1. In the presence of IL-1 (which induces both COX and PLA_2), the effects of ceramide and IL-1 together on COX expression are approximately additive, accounting for increased PG production. Interestingly, even though COX-2 epxression is significanltly enhanced within 2 hours of IL-1 treatment, no PG is produced for at least another 6 hours. The time at which PG synthesis begins correlates precisely with the time at which COX-1 and $cPLA_2$ expression increase. Together, our results suggest that the induction of "constitutive" COX-1 may play a significant role in IL-1-induced PG production. The availability of arachidonic acid also plays a critical role with respect to regulating PG production in these cells given the high levels of inducible COX-2 expression at short time points, but the absence of PG production until much later when $cPLA_2$ is expressed. Studies are underway using COX-1 and COX-2 "knockouts" to elucidate the precise role of the COX isoforms in mediating IL-1 and ceramide-induced PG production.

Frontiers in Bioactive Lipids
Edited by Vanderhoek, Plenum Press, New York, 1996

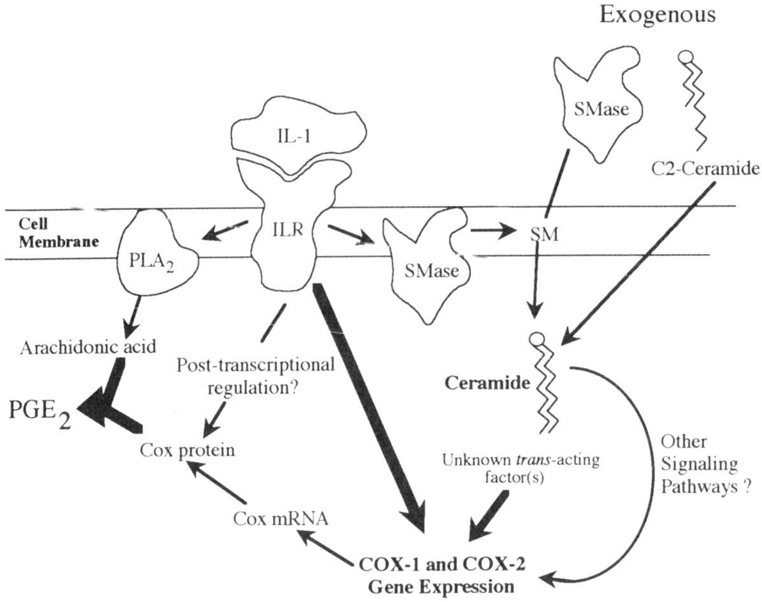

Figure 1. The relationship between IL-1-induced PG production and SM metabolism.[12] IL-1 induces PG synthesis by enhancing PLA_2 and COX expression/activity. IL-1 also induces SM turnover. Treatment of cells with sphingomyelinase or the addition of exogenous ceramide enhances IL-1-mediated PG production and ceramide induces COX gene expression suggesting that ceramide may modulate IL-1-induced PG production at the level of Cox gene expression. At present, the biochemical steps linking SM turnover and induction of COX expression are unknown. IL-1R, interleukin-1 receptor; SM, sphingomyelin; SMase, sphingomyelinase; COX, cyclooxygenase; PLA_2, phospholipase A_2; Adapted from Ballou, et al., J. Biol. Chem. 267, 20044-20050 (1992).

MATERIALS AND METHODS

Materials-Anti-PG antibody was obtained from Upstate Biotechnology (Lake Placid, NY), [^{3}H]PG, [^{35}S]methionine (translational grade) and α-[^{32}P]dCTP were obtained from Du Pont-New England Nuclear (Boston, MA). Human recombinant IL-1ß was purchased from R and D Systems (Minneapolis, MN). D-*erythro*-C_2-ceramide was kindly provided by Dr. Alijca Bielawska, Duke University Medical Center.

Cell culture-Human dermal fibroblasts were cultured as previously described[12]. Where indicated, cells were treated with IL-1ß (0.05 ng/ml) and C_2-ceramide as indicated. In some cases IL-1 and ceramide were added together, along with the appropriate vehicle controls. At these concentrations of C_2-ceramide did not affect cell morphology or viability . In [^{35}S]methionine cell labeling experiments, cells were treated as above except they were seeded in 6-well culture plates (2 ml/well). Following a 72-h incubation, the cells were washed three times with DMEM (-methionine) containing 1% fetal calf serum followed by the addition of DMEM (-methionine) containing 1% fetal calf serum and [^{35}S]methionine (100 μCi/ml). In experiments requiring RNA extraction, cells were seeded in 10 cm petri dishes (10 ml/dish) as described above.

Prostaglandin E_2 measurement - At the times indicated, PG in the media was measured by radioimmunoassay (RIA) as previously described [10,12].

RNA extraction and Northern blot hybridization- Total RNA was extracted by the RNAzol method (Tel-Test, Inc., Friendswood, TX). For Northern blot analysis, RNA was electrophoresed on denaturing formaldehyde gel, blotted onto nylon membranes, and probed. Radiolabeled cDNA probes (COX-1, COX-2, $cPLA_2$ and GAPDH) were hybridized to filters at 65°C in a high stringency hybridization buffer containing 0.5 M $NaPO_4$, pH 7.2, 7% SDS and 1 mM EDTA for 16 h. Each filter was washed twice at room temperature in 2X SSC for 15 min followed by one wash in 2X SSC, 2% SDS at 65°C for

40 min and finally washed for 10 min in 0.1X SSC at room temperature. The blots were then autoradiographed on the Kodak X-Omat AR films (Eastman Kodak, Rochester, NY) with intensifying screens at -80°C for the appropriate amount of time.

Immunoprecipitation and SDS-PAGE-The medium from human dermal fibroblasts cultured in 6-well plates was aspirated, and the plates were cooled on ice. Cell monolayers were lysed in buffer containing 50 mM Tris, pH 7.5, 150 mM NaCl, protease inhibitors (1 mM PMSF, 25 mM Benzamidine, 10 μM leupeptin and 10 μM pepstatin), 5 mM DTT, 0.1 mM Na_2VO_4, 0.1 mM Na_2MoO_4, 30 mM p-nitrophenylphosphate, 750 μM ATP, and 1% NP-40. Lysates were clarified by centrifugation at 14,000 rpm for 30 min and incubated with either rabbit polyclonal anti-COX-1, anti-COX-2 (Oxford Biomedical Research) or anti-$cPLA_2$ antiserum (a gift from Genetics Institute, Boston, MA). After 1 h of incubation at 4°C, the antigen-antibody complex was bound to Protein-G Plus agarose (Oncogene Science, Cambridge, MA). The agarose was washed three times with 10 mM Tris, pH 7.5, 500 mM NaCl, 2 mM EDTA, 0.2% NP-40 and 1 mM PMSF, twice with 10 mM Tris, pH 7.5, 150 mM NaCl, 2 mM EDTA, 0.2% NP-40 and 1 mM PMSF, and once with 10 mM Tris, pH 7.5, and 1 mM PMSF before solubilizing the agarose bound protein in Laemli's sample buffer. The samples were run on SDS-PAGE (7.5%). The gels were fixed, treated with Enhance (DuPont NEN), dried, and exposed to the Kodak X-Omat AR film.

RESULTS

Ceramide enhances IL-1-induced PG Production

Figure 2 shows that treatment of human foreskin fibroblasts with IL-1 induces PG production in a time-dependent manner but only after 8 hours of incubation. Treatment of cells with C_2-ceramide alone does not induce PG production but the addition of C_2-ceramide along with IL- significantly increases PG production.

COX-2 is Rapidly Expressed but no PG is Produced

As shown in Fig. 3 untreated cells expressed extremely low levels of COX-2 mRNA however, IL-1 or ceramide rapidly induced COX-2 mRNA 20 and 5-fold over control values after 40 and 80 minutes, respectively. Even though IL-1-induced COX-2 mRNA levels began to decrease after 4 hours, mRNA expression remained ≈10-fold higher than control levels for 24 hours; C_2-ceramide-induced COX-2 mRNA expression was more transient, returning to basal levels by 4 hours. Immunoprecipitation of COX-2 proteins from $[^{35}S]$-methionine labeled cells confirmed that respective mRNAs were proportionally

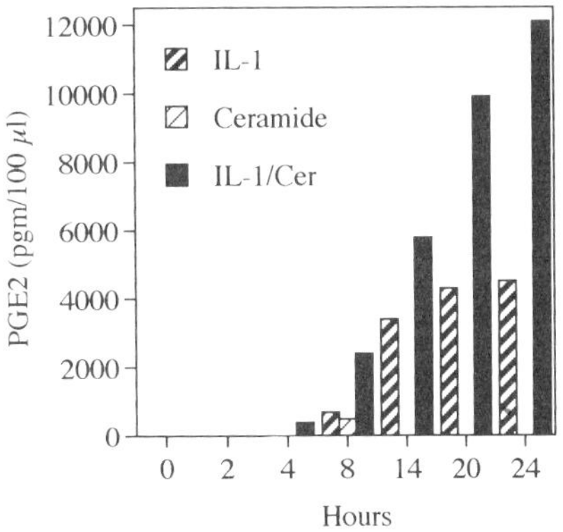

Figure 2. The Effect of IL-1 and ceramide on PG production.
Cells were treated with IL-1, ceramide, or a combination of both and at the times indicated and PG present in the media was measured by RIA.

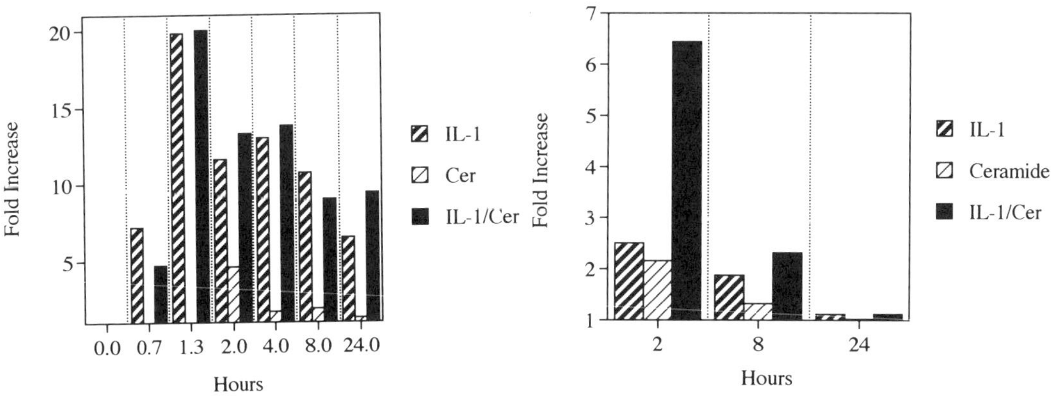

Figures 3 and 4. The effect of IL-1 and ceramide on COX-2 mRNA and protein levels. Following exposure of cells to IL-1 or ceramide for the times indicated, RNA was extracted and analyzed by Northern blot using a COX-2 cDNA probe or proteins labeled with ^{35}[S]methionine and immunoprecipitated with COX-2 antibody.

translated resulting in the formation of COX-2 protein in accordance with the induction of message (Fig. 4). When COX-2 mRNA and protein were maximally induced (e.g., after 80 minutes exposure to IL-1) no PG could be detected by RIA. As shown in Fig.2, PG accumulation did not actually begin until after 8 hours exposure to IL-1 by which time COX-2 mRNA and protein levels were actually declining. The induction of COX-2 mRNA and protein by ceramide was far less pronounced than IL-1 and no PG could be detected even after 24 hours (Fig. 2). When cells were treated with IL-1 and ceramide together, COX-2 mRNA induction was essentially identical to levels induced IL-1 by itself however, COX-2 protein levels after 2 hours were significantly higher than protein induced by IL-1 or ceramide alone.

COX-1 Expression Correlates with PG Production.

Unlike COX-2 expression, IL-1 did not enhance COX-1 mRNA levels at early time points, but only after 8 hours. At 24 hours IL-1 treatment increased COX-1 mRNA levels more than 3 fold over zero time controls. IL-1-induced COX-1 protein levels also increased

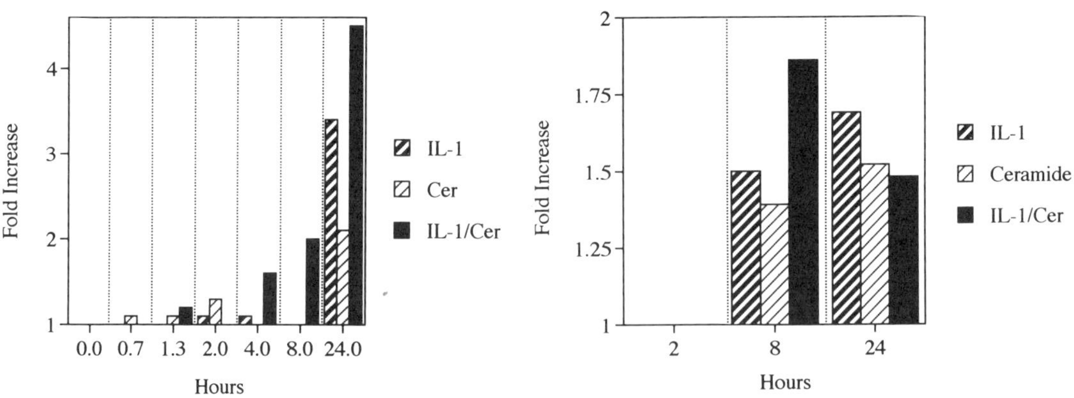

Figures 5 and 6. The effect of IL-1 and ceramide on COX-1 mRNA and protein. Following exposure of cells to IL-1 or ceramide for the times indicated, RNA was extracted and analyzed by Northern blot using a COX-1 cDNA probe or proteins labeled with ^{35}[S]methionine and immunoprecipitated with COX-1 antibody.

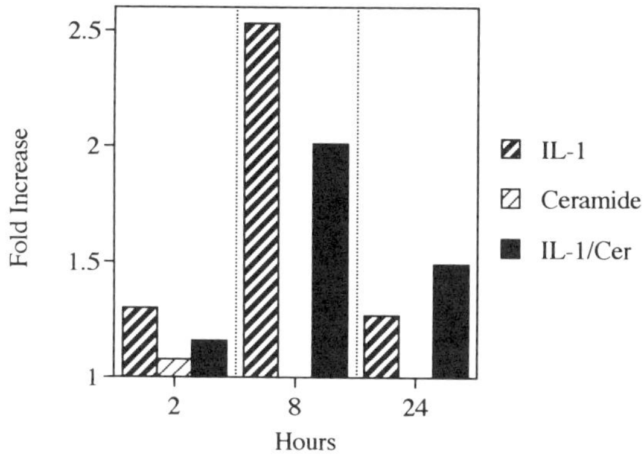

Figure 7. The effect of IL-1 and ceramide on $cPLA_2$ expression.
Following exposure of cells to IL-1 and/or ceramide for the times indicated, RNA was extracted and analyzed by Northern blot using a $cPLA_2$ cDNA probe or proteins labeled with ^{35}[S]methionine and immunoprecipitated with $cPLA_2$ antibody.

in accordance with the mRNA levels: -fold over control levels at 24 hours (Fig. 6). Most notably, the time course of IL-1 induced COX-1 expression and further increases in COX-1 expression when C2-ceramide is added, directly correlated with the time of induction and amount of PG produced (see Fig. 2). Interestingly, while ceramide alone induced a significant increase in COX-1 mRNA and protein by 24 hours no PG was produced indicating that some factor other than COX expression must account for the inability of ceramide to support PG production.

IL-1 Induces $cPLA_2$ Expression: Correlation with PG Production.

Because ceramide induced both COX-1 and COX-2 (albeit with different time-courses) without increasing PG production, we examined the effect of ceramide (and IL-1) on $cPLA_2$ mRNA and protein expression to determine whether arachidonic acid availability (regulated by $cPLA_2$ expression) may account for the lack of PG production even though

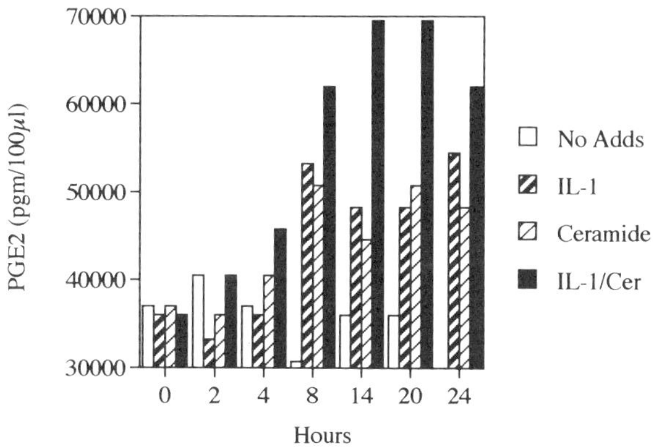

Figure 8. Excess arachidonic acid and PG production.
To address the question of substrate (arachidonic acid) availability versus COX expression as the rate-limiting step in IL-1/ceramide-induced PG production, we added excess arachidonic acid in fresh media to cells which had been treated for various times with either IL-1, C2-ceramide, or a mixture of both. After a 15 minute incubation with excess arachidonic acid, PG production was measured.

COX levels are significantly increased. No changes in $cPLA_2$ expression were detected during the first 6 hours of exposure to IL-1, C2-ceramide, or a combination of both but levels did increase by 8 hours of incubation in response to IL-1 (Fig. 7). C2-ceramide alone had no effect on either $cPLA_2$ mRNA or protein levels at any time point tested. These results suggest that the inability of ceramide alone to induce PG production, despite its ability to induce COX enzymes, may be due to a lack of arachidonic acid resulting from the inability of ceramide to stimulate $cPLA_2$ expression and/or enzyme activity.

Effect of Exogenous Arachidonic Acid on IL-1 and Ceramide Treated Cells.

As shown in Fig. 8 the addition of arachidonic acid to ceramide treated cells resulted in PG production to the same levels as that of IL-1 while having no effect on untreated cells supporting the notion that lack of substrate likely accounts for the inability of ceramide to induce PG production even though COX is expressed. Interestingly, the addition of substrate to cells expressing high levels of COX-2 ($\approx$ 2 hour exposure to IL-1) did not result in any increase in PG production but did result in higher PG levels at later time points when COX-1 expression was increased. No changes in $cPLA_2$ expression were detected during the first 6 hours of exposure to IL-1, C2-ceramide, or a combination of both but expression did increase by 8 hours of incubation in response to IL-1. C2-ceramide alone had no effect on either $cPLA_2$ mRNA or protein levels at any time point tested. These results further support the notion that PG production is more directly correlated with COX-1 than COX-2.

SUMMARY AND CONCLUSIONS

In untreated fibroblasts COX-1 mRNA and protein were easily detectable while COX-2 mRNA and protein was minimally expressed. In response to IL-1, COX-2 mRNA increased dramatically (over 20-fold) within 2 hours and remained $\approx$10-fold higher than controls over 24 hours. Ceramide induced a transient, 5-fold increase in COX-2 mRNA at 2 hours which decreased significantly by 4 hours and ceramide added along with IL-1 did not alter the pattern of COX-2 mRNA induction. Interestingly, even though COX-2 mRNA was dramatically increased by 2 hours, no PG was produced after 2, 4 or even 8 hours suggesting that COX-2 may not be rate-limiting. In sharp contrast to the pattern of COX-2 mRNA expression, COX-1 mRNA was essentially unaffected by IL-1 for at least 8 hours but by 24 hours, mRNA increased over 3-fold; ceramide alone induced a 2-fold increase in COX-1 mRNA while the combination of IL-1 and ceramide resulted in more than a 4-fold increase. Unlike the increase in COX-2 mRNA which did not result in PG production, increases in COX-1 mRNA in response to IL-1 and IL-1/ceramide correlated well with the induction of PG production. The addition of exogenous arachidonic acid, even when COX-2 was highly expressed, did not result in enhanced PG production, but did so at later time points when COX-1 expression was increased. Together, our results suggest that the induction of COX-1 may play a significant role in IL-1-induced PG production and that ceramide enhances COX expression via a different mechanism than IL-1 since the effects of IL-1 and ceramide are $\approx$ additive with respect to COX induction. The availability of arachidonic acid may also be limiting with respect to PG production in these cells given the high levels of COX-2 expression at short time points, but the absence of PG production until much later. Studies are underway using COX-1 and COX-2 "knockouts" to elucidate the precise role of COX isoforms in mediating IL-1/ceramide-induced PG production.

REFERENCES

1. Y. A. Hannun, The sphingomyelin cycle and the second messenger function of ceramide, J. Biol. Chem. 269:3125 (1994).
2. L. R. Ballou, Sphingolipids and cell function, Immunolology Today, 13:339 (1992).
3. L.R. Ballou, S.J.F. Laulederkind, E.F. Rosloniec, and R. Raghow, Ceramide signalling and the immune response, Biochimica et Biophysica Acta (in press) (1996).

4. R. M. Bell, Y. A. Hannun, and A. H. Merril, Jr., eds. Advances in Lipid Research, Sphingolipids Part A: Functions and Breakdown Products, vol. 25, Academic Press, San Diego, Ca., 1993.
5. R. M. Bell, Y. A. Hannun, and A. H. Merrill, Jr., eds. Advances in Lipid Research, Sphingolipids Part B: Regulation and Function of Metabolism, vol. 26, Academic Press, San Diego, Ca., 1993.
6. Y. A. Hannun and L. M. Obeid, Ceramide: an intracellular signal for apoptosis, TIBS 20:73 (1995).
7. R. Kolesnick, Signal transduction through the sphingomyelin pathway, Mol. Chem. Neuropathol. 21:287 (1994).
8. R. Kolesnick and D. W. Golde, The sphingomyelin pathway in tumor necrosis factor and interleukin-1 signaling, Cell 77:325 (1994).
9. R. Kolesnick and Z. Fuks, Ceramide: A signal for apoptosis or Mitogenesis?, J. Exp. Med. 181:1949 (1995).
10. L. R. Ballou, S. C. Barker, A. E. Postlethwaite, and A. H. Kang, Sphingosine potentiates IL-1-mediated prostaglandin E_2 production in human fibroblasts, J. Immunol. 145:4245 (1990).
11. M. Candela, S. C. Barker, and L. R. Ballou, Sphingosine synergistically stimulates tumor necrosis factor α-induced prostaglandin E_2 production in human fibroblasts, J. Exptl. Med. 174:1363 (1991).
12. L. R. Ballou, C. P. Chao, M. A. Holness, S. C. Barker, and R. Raghow, Interleukin-1-mediated PGE_2 production and sphingomyelin metabolism. Evidence for the regulation of cyclooxygenase gene expression by sphingosine and ceramide, J. Biol. Chem. 267:20044 (1992).

REGULATION OF VASCULAR SMOOTH MUSCLE CELL 12-LIPOXYGENASE BY GROWTH FACTORS AND INFLAMMATORY CYTOKINES

Rama Natarajan and Wei Bai

Department of Diabetes, Endocrinology and Metabolism
City of Hope Medical Center
Duarte, CA 91010

INTRODUCTION

The process of atherosclerosis involves a complex interaction of several growth factors, lipids and cytokines[1]. Each of the major cell types involved, namely, platelets, endothelial cells, monocytes and vascular smooth muscle cells can produce and also respond to these agents thus setting up a network of autocrine/paracrine effects responsible for the development and progression of the fatty streak. Excessive proliferation and migration of vascular smooth muscle cells (VSMC) has been implicated as critical events in atherosclerosis and restenosis after balloon angioplasty. However, the cellular and molecular mechanisms involved in these processes are not very clear. Further, very few studies have addressed the role of free lipids in these events.

The interaction of growth factors or cytokines with their specific cell surface receptors can lead to the stimulation of a cascade of biochemical events including the activation of several phospholipases which in turn can result in the release of several second messengers such as inositol trisphosphate, which leads to calcium mobilization from intracellular stores, as well as diacylglycerol (DAG) which in turn can activate protein kinase C[2-5]. DAG is also a rich source of arachidonic acid which can be released by the action of the enzyme DAG lipase[6]. Arachidonic acid can also be directly released from membrane phospholipids by the activation of phospholipase A2 (PLA2)[2] and PLA2 activation has been implicated in the action of several cytokines. It has been suggested that the release of arachidonic acid from the *sn*-2 position of membrane phospholipids may be one of the signals leading to cell proliferation[7]. Arachidonic acid can then be further metabolized by three major pathways, the cyclooxygenase pathway which leads to the formation of prostaglandins and thromboxane, the lipoxygenase (LO) pathway which forms hydroxyeicosatetraenoic acids (HETEs) and leukotrienes and thirdly, the cytochrome P-450 monooxygenase pathway which leads to the formation of epoxides such as epoxyeicosatrienoic acids as well as HETEs[8]. The production of LO products such as 12- and 15-HETE has been shown in several vascular tissues including cultured VSMC, endothelial cells, aorta and coronory arteries.

Studies show that there are at least two major isoforms of 12-lipoxygenases (12-LO)[9-11]. The platelet-type has been cloned from human platelets and the megakaryocytic cell line,

HEL[12,13]. The other 12-LO, namely the leukocyte-type, has been detected in porcine leukocytes[14], VSMC[15] and also in human adrenal glomerulosa cells[16], human monocytes, endothelial and VSMC[17]. The cDNA for porcine leukocyte 12-LO[14], and murine leukocyte 12-LO[18] have been cloned. The porcine leukocyte 12-LO is only 65% homologous to the human platelet 12-LO[12-14] whereas it is 87% homologous to human 15-LO[14,19].

Increasing evidence has implicated the LO enzymes and their products namely HETEs (from arachidonic acid) and hydroxyoctadecadienoic acids (HODEs) (from linoleic acid) in the pathogenesis of atherosclerosis. The LO enzyme has been suggested to mediate the oxidative modification of LDL[20,21]. Furthermore, Yla-Herttuala and coworkers have shown that 15-LO mRNA and protein are expressed in macrophage-rich areas of atherosclerotic lesions[22]. Atherosclerotic lesions also show higher levels of 15-LO enzymatic activity than corresponding normal arteries[23] and increased levels of HETEs have been found in the aortas of atherosclerotic rabbits[24]. Increased amounts of LO products have been shown to be present in early atherosclerotic lesions[25]. Analysis of the lipid oxidation products in human atherosclerotic lesions revealed that the oxidation of polyunsaturated fatty acids therein was mainly mediated by the LO enzyme[26]. We also have new immunohistochemical evidence supporting the presence of a leukocyte-type of 12-LO in human and porcine atherosclerotic lesions (unpublished observations). LO products have direct mitogenic effects in endothelial cells[27]. They also have potent chemotactic effects in VSMC, inducing VSMC migration at concentrations as low as 1 pM[28]. The LO pathway has also been implicated in the vasopressor and renin inhibitory effects of angiotensin II[29,30].

We have recently shown that angiotensin II, a potent VSMC growth factor, can increase 12-HETE formation and also intracellular 12-LO activity in porcine VSMC[15]. Thus Table 1 shows the dose-dependent stimulation of 12-HETE release by AII in these VSMC. AII could also increase the expression of the leukocyte-type 12-LO mRNA as well as protein in these VSMC[15]. Furthermore, in other studies, we showed that the hypertrophic effects of AII were attenuated by a specific LO blocker, baicalein, but not by a specific cyclooxygenase blocker, ibuprofen (Fig. 1)[31]. In addition, we noted that the 12-LO product, 12-HETE, had direct hypertophic effects in VSMC as determined by its ability to increase total cell protein content (Fig. 2)[31]. It is seen in the Fig. that 12-HETE was as potent as AII while 15-HETE, at the same dose, was not effective under these conditions. 12-HETE was also able to increase the formation of the key matrix protein, fibronectin, similar to the effect of AII[31]. We have also recently shown that platelet-derived growth factor (PDGF), a potent VSMC mitogen and chemoattractant, also caused a marked increase in 12-LO activity and expression in VSMC[32]. In view of the key role played by cytokines in atherosclerosis, we have examined whether inflammatory cytokines such as interleukin-1β (IL-1), IL-4 and IL-8 have the ability to induce the 12-LO enzyme in porcine VSMC.

IL-1β and IL-8 have documented mitogenic effects in VSMC[33,34]. Further, IL-8 is a VSMC chemoattractant and can induce VSMC to migrate[34]. In VSMC, IL-4 can increase the expression of the cell adhesion molecule VCAM-1 which has been shown to play a major role in the genesis of atherosclerosis[35]. The present studies show that some of their vascular effects may be mediated by upregulation of the 12-LO pathway.

MATERIALS AND METHODS

Cell culture. Primary cultures of porcine VSMC were obtained as described[15]. Cells were maintained in Dulbecco's modified Eagle's medium and 10% fetal calf serum (FCS).

Measurement of immunoreactive 12-HETE and intracellular 12-LO enzyme activity. Cell associated 12-HETE was extracted from cell pellets and quantitated by a specific radioimmunoassay as described earlier[15,36]. 12-LO enzyme activity was determined as

Table 1. Effect of Angiotensin II on released 12-HETE levels in Porcine VSMC

	12-HETE ($pg/10^6$ cells)
Basal	108 ± 19
AII (10^{-10} M)	116 ± 23
AII (10^{-9} M)	168 ± 28*
AII (10^{-8} M)	215 ± 16**
AII (10^{-7} M)	223 ± 30**

12-HETE levels were measured by a specific RIA of extracted cell supernatants after a 10 min treatment with or without AII. *, $p < 0.05$ vs. basal; **, $p < 0.01$ vs. basal.

described[15] by examining the conversion of substrate arachidonic acid to 12-HETE by sonicates of cells that had been pretreated with the cytokines. 12-HETE was quantitated by our validated reverse phase HPLC technique[15,36].

12-LO mRNA. Quiescent cells in DME medium containing 0.4% FCS and 0.2% BSA were treated with the cytokines. For 12-LO mRNA expression, total cellular RNA was subjected a specific reverse transcriptase PCR method[15,16] to amplify leukocyte-type 12-LO mRNA. GAPDH mRNA amplification was used an internal control[15].

DNA synthesis. Serum-starved cells in 24 well dishes were treated with the cytokines for 24 hr in DME medium containing 0.4% FCS and 0.2% BSA as well as indomethacin (2.5 μM). [^{3}H]thymidine (1μCi/ml) was added during the last 6 hr. Washed cells were treated with trichloroacetic acid and solubilized with 0.3N NaOH. Radioactivity in the resulting solutions was quantititated by liquid scintillation counting.

RESULTS

We initially examined whether these cytokines could affect the formation of 12-HETE in porcine VSMC. Table 2 shows that a 6 hr treatment with IL-1, IL-4 as well as IL-8 led to significant increases in the levels of immunoreactive cell-associated 12-HETE formed in these cells. The stimulatory effects were seen by 2 hr, were maximum at 6 hr and still evident at 24 hr. Furthermore, treatment for 6 hr also led to an increase in intracellular 12-LO enzyme activity as measured by the conversion of arachidonic acid to 12-HETE by sonicates of treated cells. This is also depicted in Table 2.

In order to examine whether these effects are transcriptionally regulated, we examined the effects of these cytokines on leukocyte-type 12-LO mRNA expression in the porcine VSMC. 12-LO mRNA expression was studied by a specific RT-PCR method. Fig. 3 shows that all these three cytokines had potent stimulatory effects on 12-LO mRNA expression (333 bp PCR product). 12-LO cDNA amplification, the positive control for PCR, is shown in the far right. After correction for the internal control for PCR, namely GAPDH mRNA amplification (284 bp PCR product shown in the lower panel) using densitometry, it is seen that all three interleukins led to a potent 3-6 fold increase in the level of 12-LO mRNA.

With a view to investigating the functional significance of 12-LO activation by these agents, we initially evaluated whether these three cytokines have growth- promoting effects

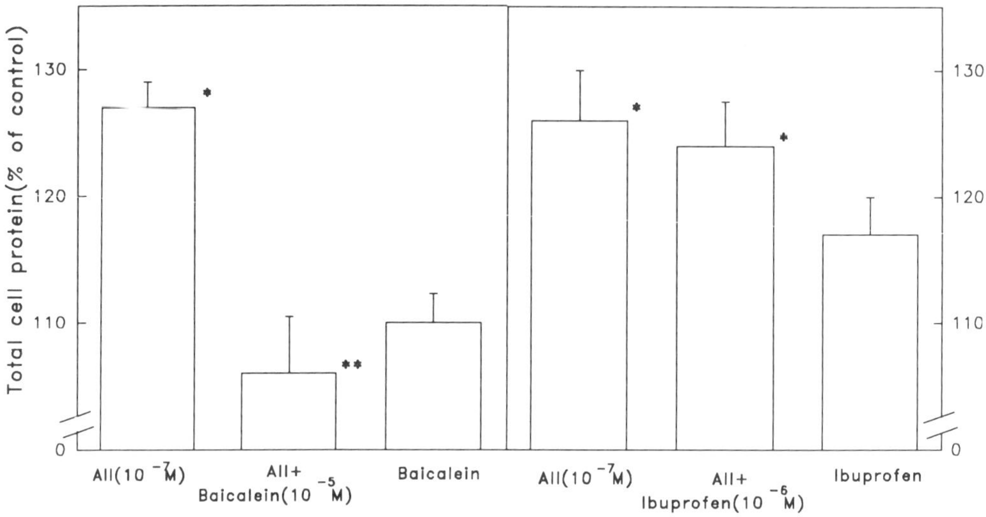

Fig. 1. The effect of a specific lipoxygenase inhibitor (baicalein) and a cyclooxygenase inhibitor (ibuprofen) on angiotensin II (AII)-induced protein synthesis in porcine VSMC. *, $p<0.01$ vs basal; **, $p<0.03$ vs AII. Reproduced from Ref. 31 with permission.

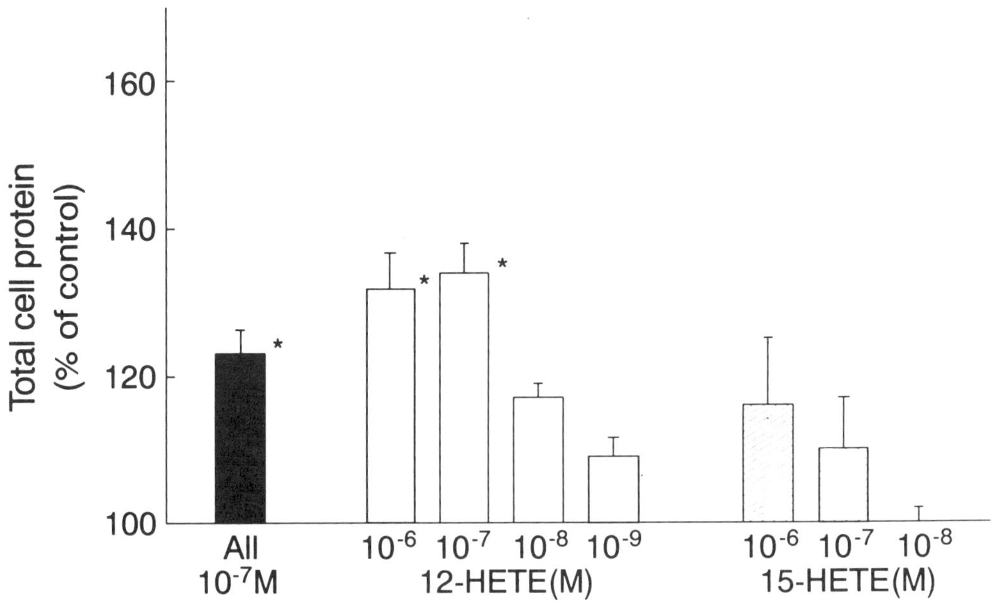

Fig. 2. Direct effects of 12- and 15-HETE on total cell protein content in porcine VSMC. *, $p<0.02$ vs basal. Effect of AII is shown for comparison. Reproduced from Ref. 31 with permission.

Table 2. Effect of interleukins on cell associated 12-HETE and 12-LO enzyme activity.

	12-HETE (pg/10^6 cells)	12-LO activity: 12-HETE peak height (fold over control)
Control	861 ± 55	---
IL-1β (5ng/ml)	1130 ± 60*	1.51 ± 0.11**
IL-4 (5ng/ml)	1595 ± 118**	1.59 ± 0.08**
IL-8 (5ng/ml)	1409 ± 90**	1.33 ± 0.04**

Cell associated 12-HETE after 6 hr incubations with porcine VSMC was quantitated by RIA. 12-LO activity was studied by HPLC of extracted cell sonicates. *, $p < 0.05$ vs. control; **, $p < 0.01$ vs. control.

in porcine VSMC. Since earlier studies have indicated that the mitogenic effects of IL-1 can be unmasked only in the presence of a cyclooxygenase blocker, we examined the effects of IL-1, -4 and -8 on VSMC DNA synthesis as measured by [^{3}H]thymidine incorporation. The results indicated that they do indeed increase DNA synthesis under these conditions, with IL-4 being the most potent of the three (IL-1, 1.35 ± 0.13 fold over control; IL-4, 1.72 ± 0.1 fold; IL-8, 1.4 ± 0.16 fold). Furthermore, in preliminary experiments, we have shown that the effects of all the three interleukins were attenuated by pretreatment with a LO blocker baicalein (1μM), thus suggesting that their growth promoting effects may be mediated at least in part by activation of the LO pathway.

DISCUSSION

The signal transduction mechanisms of action of cytokines have recently elicited much interest owing to their key roles in the processes of inflammation, immune reactions and atherosclerosis. While several lines of evidence have documented their agonist roles on the

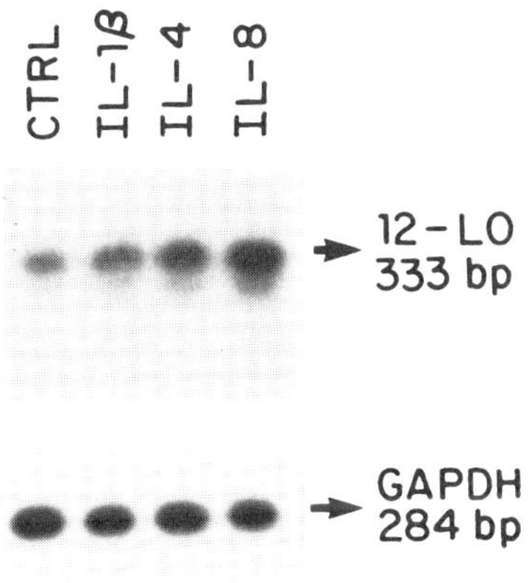

Fig. 3. Regulation of leukocyte-type 12-LO mRNA expression by IL-1β, IL-4 and IL-8 in porcine VSMC treated for 24 hr with these cytokines. RT-PCR was run as described under Methods. The upper panel shows results obtained by hybridization with a 12-LO oligonucleotide probe and the lower panel is the control for PCR and obtained by hybridization with a GAPDH probe.

cyclooxygenase pathway, few studies have examined their effects on the LO pathway. Thus recently it was shown that treatment of monocytes with IL-4 or IL-13 could stimulate the 15-LO pathway by inducing 15-LO mRNA and protein[37,38]. Studies such as those showing that 15-LO mRNA and protein are colocalized with epitopes of oxidized LDL in the macrophage-rich areas of the atherosclerotic plaque, have implicated the LO pathway in the atherogenic process. The results of our present studies showing that inflammatory cytokines such as IL-1, -4, and -8 are potent inducers of 12-LO activity and 12-LO mRNA expression in VSMC lend further support to this concept.

Earlier studies have had difficulty in observing direct mitogenic effects of these interleukins in vitro with VSMC cultures and this has been attributed to the potent stimulatory effects of these cytokines on the cyclooxygenase pathway which leads to the formation of growth inhibitory prostanoids. Thus, IL-1β and IL-8 have been demonstrated to have mitogenic effects on VSMC in the presence of the cyclooxygenase inhibitor, indomethacin[33,34]. The present studies have confirmed these observations in porcine VSMC. Furthermore, we have shown for the first time that that IL-4 has mitogenic effects in VSMC and in fact was the most potent of the three. In preliminary experiments, we have also verified that the observed increases in DNA synthesis are also accompanied by increases in cell number, thus indicating a true mitogenic response. In addition ,the observation that baicalein could attenuate their effects, suggests that the LO pathway may, at least in part, mediate their growth effects. The inflammatory effects of IL-8 has been attributed not only to its mitogenic effects but also to its chemotactic effects[34]. Since 12-LO products such as 12-HETE are potent inducers of VSMC migration, it is also likely that they mediate the chemotactic effects of these cytokines. Thus, it is possible to envision an imbalance between activation of the LO and cyclooxygenase pathway in pathological situations leading to an increased accummulation of inflammatory, chemotactic and growth-promoting LO products.

The mechanisms by which LO products lead to increased VSMC growth and migration have not been evaluated in these present studies, but evidence indicate that they can initiate several growth-related signalling events such as activation of oncogenes, protein kinase C and mitogen activated protein kinase[39]. In summary, LO enzymes activated by growth factors or cytokines in vascular and inflammatory cells can form products that have several physiological and pathological actions. These include potent growth and chemotactic effects which therefore implicate them in the pathogenesis of diseases such as atherosclerosis, hypertension and diabetes. We have recently developed a specific molecular technique to block the 12-LO mRNA by using a specific ribozyme or catalytic RNA which efficiently cleaves porcine leukocyte-type 12-LO[40]. Hence, pharmacological approaches as well as molecular strategies using ribozyme or antisense technology to efficiently block these pathways may serve as new therapeutic modalities to combat cardiovascular disease.

Acknowledgments. The authors would like to thank Dr. Jerry Nadler for helpful discussions. This work was supported by grants from the National Institutes of Health to RN (PO1 HL55798 and R29 HL48920).

REFERENCES

1. R. Ross, The pathogenesis of atherosclerosis, *Nature* 37:801-809(1993).
2. M. Liscovitch, Crosstalk among multiple signal-activated phospholipases. *Trends Biochem. Sci.* 17:393-399(1992).
3. M.J. Berridge, Inositol trisphosphate and diacylgycerol as second messengers. *Biochem. J.* 220:345-360 (1984).
4. Y. Nishizuka, Role of protein kinase C in cell surface signal transduction and tumor promotion, *Nature* 308:693-698(1984).

5. K.K. Griendling, S.E. Rittenhouse, T.A. Brock, L.S. Ekstein, and M.A. Gimbrone, Sustained diacylglycerol formation from inositol phospholipids in angiotensin II-stimulated vascular smooth muscle cells. *J. Biol. Chem.* 261:5901-5906(1986).
6. R.L. Bell, D.A. Kennerly, N. Stanford, and P.W. Majerus, Diglyceride lipase: A pathway for arachidonate release from human platelets, *Proc. Natl. Acad. Sci. USA.* 76:3238-3241(1979).
7. J.B.A Millar, and E. Rozengurt, Arachidonic acid release by bombesin. A novel postreceptor target for heterologous mitogenic desensitization, *J. Biol. Chem.* 265:19973-19979(1990).
8. W.L. Smith, The eicosanoids and their biochemical mechanisms of action, *Biochem. J.* 259:315-324(1989).
9. S. Yamamoto, Mammalian lipoxygenases: molecular structures and functions, *Biochim. Biophys. Acta.* 1128:117-131(1992).
10. C.D. Funk, Molecular Biology in the eicosanoid field in "*Progress in Nucleic acid Research and Molecular Biology*" , W.E. Cohn, K. Moldave, eds., Academic Press, Inc., 1993, 45:67-98(1993).
11. E. Sigal, The molecular Biology of mammalian arachidonic acid metabolism, *Am. J. Physiol.* 260 (Lung Cell. Mol. Physiol.):L13-L28 (1991).
12. C.D. Funk, L. Furci, and G.A. FitzGerald, Molecular cloning, primary structure and expression of the human platelet/erythroleukemia cell 12-lipoxygenase, *Proc. Natl. Acad. Sci. USA* 87:5638-5642(1990).
13. T. Izumi, S. Hoshiko, O. Radmark, and B. Samuelsson, Cloning of the cDNA for human 12-lipoxygenase, *Proc. Natl. Acad. Sci. USA* 87:7477-7481(1990).
14. T. Yoshimoto, H. Suzuki, S. Yamamoto, T. Takai, C. Yokoyama, and T. Tanabe, Cloning and sequence analysis of the cDNA for arachidonate 12-lipoxygenase of porcine leukocytes, *Proc. Natl. Acad. Sci. USA* 87:2142-2146(1990).
15. R. Natarajan, J.L. Gu, J. Rossi, N. Gonzales, L. Lanting, L. Xu, and J. Nadler, Elevated glucose and angiotensin II increase 12-lipoxygenase activity and expression in porcine aortic smooth muscle cells, *Proc. Natl. Acad. Sci. USA* 90:4947-4951(1993).
16. J.L Gu, R. Natarajan, J. Ben-Ezra, G. Valente, S. Scott, T. Yoshimoto, S. Yamamoto, J.J. Rossi, and J. Nadler, Evidence that a leukocyte-type of 12-lipoxygenase is expressed and regulated by angiotensin II in human adrenal glomerulosa cells, *Endocrinology* 134:70-77(1994).
17. J.A. Kim, J.L. Gu, R. Natarajan, J. Esteban, J.A. Berliner, and J. Nadler, A leukocyte-type of 12-lipoxygenase is expressed in human vascular and mononuclear cells: evidence for upregulation by angiotensin II. *Arterioscler. Throm. Vasc. Biol.* 15:942-948(1995).
18. X. Chen, U. Kurre, N.A. Jenkins, N.G. Copeland, and C.D. Funk, cDNA cloning, expression, mutagenesis c-terminal iso-leucine, genomic structure, and chromosomal localizations of murine 12-lipoxygenases, *J. Biol. Chem.* 269:13979-13987(1994).
19. E. Sigal, C.S. Craik, E. Highland, D. Grunberger, L.L. Costelo, R.A.F. Dixon, and J.A. Nadel, Molecular cloning and primary structure of human 15-lipoxygenase, *Biochem. Biophys. Res. Commun.* 157:457-464(1988).
20. S. Parthasarathy, E. Wieland, and D. Steinberg, A role for endothelial cell lipoxygenase in the oxidative modification of low density lipoprotein, *Proc. Natl. Acad. Sci. USA.* 86:1046-1050(1989).
21. C.P. Sparrow, S. Parthasarathy, and D. Steinberg, Enzymatic modification of low density lipoprotein by purified lipoxygenase plus phospholiase A2 mimics cell-mediated oxidative modification, *J. Lipid Res.* 29:745-753 (1988).
22. S. Yla-Hertualla, M. Rosenfeld, S. Parthasarathy, C. Glass, E. Sigal, J. Witztum, D. Steinberg, Colocalization of 15-LO mRNA and protein with epitopes of oxidized low density lipoprotein in macrophage-rich areas of atherosclerotic lesions, *Proc. Natl. Acad. Sci. USA.* 87:6959-6963(1990).
23. P. Henricksson, M. Hamberg, and U. Diczfalusy, Formation of 15-HETE as a major hydroxyeicosatetraenoic acid in the atherosclerotic vessel wall, *Biochim. Biophys. Acta.* 834:272-274(1985).
24. T. Wang, and W. Powell, Increased levels of monohydroxy metabolites of arachidonic and linoleic acids in LDL and aorta from atherosclerotic rabbits, *Biochem. Biophys. Acta.* 1084:129-138(1991).
25. H. Kuhn, J. Belkner, S. Zaiss, T. Fahrenklemper, and S. Wohlfeil, Involvement of 15-lipoxygenase in early stages of atherogenesis, *J. Exp. Med.* 179:1903-1911(1994).
26. V.A. Folcik, R.A. Nivar-Aristy, L.P. Krajewski, and M.A. Cathcart, Lipoxygenase contributes to the oxidation of lipids in human atherosclerotic plaques, *J. Clin. Invest.* 96:504-510(1995).
27. B.N.Y. Setty, J.E. Graeber, and M.J. Stuart, The mitogenic effects of 15- and 12-hydroxyeicosatetraenoic acids on endothelial cells may be mediated via diacylglycerol kinase inhibition. *J. Biol. Chem.* 262:17613-17622(1987).
28. J. Nakao, T. Ooyama, H. Ito, W.C. Chang, and S. Murota, Comparative effects of lipoxygenase products of arachidonic acid on rat aortic smooth muscle cell migration, *Atherosclerosis* 44:339-342(1982).

29. I. Antonipillai, R. Horton, R. Natarajan, and J. Nadler, A 12-lipoxygenase product of arachidonate metabolism is involved in angiotensin action on renin release, *Endocrinology* 125:2028-2034(1989).
30. N. Stern, M. Golub., K. Nozawa, M. Berger, E. Knoll, N. Yanagawa, R. Natarajan, M.L. Tuck, and J. Nadler, Selective inhibition of angiotensin II-mediated vasoconstriction by lipoxygenase blockade, *Am. J. Physiol.* 257:H434-H443(1989)
31. R. Natarajan, N. Gonzales, L. Lanting, and J. Nadler, Role of the lipoxygenase pathway in angiotensin II-induced vascular smooth muscle cell hypertrophy, *Hypertension* 23[suppl I]:I-142-I-147(1994).
32. R. Natarajan, W. Bai, J.L. Gu, V. Rangarajan, and J. Nadler, Regulation of 12-lipoxygenase by platelet-derived growth factor in vascular smooth muscle cells, in: *"Advances in Prostaglandin, Thromboxane and Leukotriene Research.* B. Samuelsson, R. Paoletti, and P.W. Ramwell, eds, Raven Press, New York, pp 435-437(1995).
33. P. Libby, J. Stephen, C. Warner, Interleukin 1: a mitogen for human vascular smooth muscle cells that induces the release of growth-inhibitory prostanoids, *J. Clin. Invest.* 81:487-498, 1988.
34. T. Yue, X. Wang, C. Sung, B. Olson, P. Mckenna, J. Gu, and G.Z. Feuerstein, Interleukin-8. A mitogen and chemoattractant for vascular smooth muscle cells, *Circ. Res.* 75:1-7, 1994.
35. H. Li, M.I. Cybulsky, M.A. Gimbrone, and P. Libby, Inducible expression of vascular cell adhesion molecule by vascular smooth muscle cells in vitro and within rabbit atheroma, *Am. J. Pathol.* 143:1551-1559(1993)
36. J. Nadler, R. Natarajan, and N. Stern, Specific action of the lipoxygenase pathway in mediating angiotensin II-induced aldosterone synthesis in isolated adrenal glomerulosa cells, *J. Clin. Invest.* 80:1763-1769 (1987).
37. D.J. Conrad, H. Kuhn, M. Mulkins, E. Highland, and E. Sigal, Specific inflammatory cytokines regulate the expression of human monocyte 15-lipoxygenase, *Proc. Natl. Acad. Sci. USA* 89:217-222(1992).
38. G.M. Nasser, J.D. Morrow, L.J. Robert II, F.J. Lakkis, and K.F. Badr, Induction of 15-lipoxygenase by IL-13 in human blood monocytes, *J. Biol. Chem.* 269:27631-27634(1994).
39. R. Natarajan, N. Stern, and J. Nadler, Arachidonic acid metabolites on renin and vascular smooth muscle cell growth, in: Contemporary Endocrinology: *Endocrinology of the Vasculature,* J.R. Sowers, ed., Humana Press, Totowa, NJ(1996)(In Press).
40. J.L. Gu, D. Veerapanane, J.J. Rossi, R. Natarajan, L. Thomas, and J. Nadler, Ribozyme-mediated inhibition of expression of leukocyte-type 12-lipoxygenase in porcine aortic vascular smooth muscle cells, *Circ. Res.* 77:14-20(1995).

MANIPULATIONS OF THE ARACHIDONIC ACID CASCADE WITH LIPOXYGENASE GENE-INACTIVATED MICE

Colin D. Funk and Duxin Sun

Center for Experimental Therapeutics
University of Pennsylvania
Philadelphia, PA 19104-6100

INTRODUCTION

The lipoxygenases are a family of enzymes that stereospecifically insert molecular oxygen onto defined positions of arachidonic acid giving rise to various eicosanoids including leukotrienes, lipoxins and hydro(pero)xy-eicosatetraenoic (H(P)ETE) acids (1). There are at least four lipoxygenase (LO) forms in mice, each encoded by a separate gene. The precise roles of the various metabolites are poorly understood in many cases. In our laboratory, we have carried out the cloning and inactivation of three murine LO genes (5-LO, leukocyte-type 12-LO and platelet-type 12-LO) to examine potential biological functions of eicosanoids.

EXPERIMENTAL PROCEDURES

Generation of 5-LO Deficient Mice. Mice unable to synthesize leukotrienes were generated by targeted disruption of the 5-lipoxygenase gene as previously described (2).

Preparation of Leukocyte-type 12-LO Gene Targeting Vector. The complete mouse L-12LO gene was cloned previously (3) from a strain 129 Sv genomic library. A 6.9 kb XbaI fragment was excised and cloned into pGEM3zf(-) (Promega). A blunt ended neo cassette under control of the polII promoter and 3' bovine growth hormone polyadenylation sequence was inserted into the StuI site within exon 3. A thymidine kinase gene was cleaved from pMC1-TK with XhoI and inserted into the SalI site in the polylinker. The plasmid was linearized with a SalI site within the polylinker of pMC1-TK.

Generation of L-12LO Deficient Mice. DNA (25 μg) was electroporated into D3H mouse embryonic stem (ES) cells as previously described (2). After 8 days selection in G418 (300 μg/ml) and ganciclovir (2 μM; gift of Syntex, Inc., Palo Alto, CA) colonies were picked, placed in 96-well plates and processed for cell freezing and DNA Southern blot analysis. The

Frontiers in Bioactive Lipids
Edited by Vanderhoek, Plenum Press, New York, 1996

DNA was digested with StuI/BamHI, electrophoresed in a 1 % agarose/TAE gel, blotted to nylon membrane and hybridized with a random prime ^{32}P-labeled 240-bp fragment external to the targeting vector. Hybridization to a 10 kb band, in addition to the wild type 5.4 kb band, was indicative of targeted events. Targeted ES cells (10-15) were injected into 3.5 day C57BL/6 blastocysts and the embryos were transferred into ICR pseudopregnant recipients. Male chimera (evidence of agouti coat color) offspring were mated with C57BL/6 females to obtain germline transmitting progeny. All mice used in these studies were a mixed C57BL/6 x 129 Sv genetic background.

Preparation of Macrophages. Resident peritoneal macrophages (PM) were collected after i.p. injection of 5 ml Dulbecco's modified Eagle's medium (DMEM) with 10 % fetal bovine serum (complete medium), washed once with phosphate-buffered saline (PBS), resuspended in complete medium, and plated onto culture dishes. After 2 h at 37 °C in a 95 % air/5 % CO_2 incubator, the macrophages were washed once with DMEM to remove non-adherent cells.

Immunoblot Analysis. Macrophages were sonicated in PBS containing 1 mM phenylmethylsulfonyl fluoride and 60 µg/ml soybean trypsin inhibitor and homogenate protein was subjected to electrophoresis in a 10 % SDS-polyacrylamide mini-gel with stacking gel. The protein was electoblotted to Hybond EC nitrocellulose and was probed with an anti-mouse L-12LO antiserum (3) at 1:100 dilution. An enhanced chemiluminescence detection system was used (Amersham).

Enzyme Activity Assay and RP-HPLC Analysis. Macrophages were incubated with arachidonic acid (100 µM) plus A23187 (2 µM) for 15 min at 37 °C. For direct analysis of HETEs, without prior extraction, the samples were loaded onto an Altex Ultrasphere 5-µm ODS-silica column (Beckman, 25 x 0.46 cm) using a solvent system of MeCN/MeOH/H_2O/HAc (350:150:250:1) and flow rate of 1.5 ml/min.

Peritoneal Macrophage Antitumor Activity. PM were plated onto 96-well plates (2.5 x 10^5/well) in complete medium, and stimulated with LPS (100 ng/ml) + IFN-γ (500 U/ml) or A23187 (2 µM) at 37 °C for 24 h. After stimulation, PM were washed three times with DMEM. P815 murine mastocytoma cells and ^3H-thymidine (0.5 µCi/ml) were added to PM (PM/P815 ratio 2:1 or 10:1). After 20 h incubation, cells were collected on glass fiber filters (Inotech Biosystems Int.) with a cell harvester and the incorporated radioactivity was quantitated by liquid scintillation counting. ^3H-thymidine incorporation into PM or P815 cells alone were used as controls.

RESULTS AND DISCUSSION

5-Lipoxygenase Deficient Mice

5-LO deficient mice generated by targeted gene disruption are unable to synthesize leukotrienes (2). When these mice are subjected to inflammatory or injurious challenges (e.g. bacterial and parasite infections, PAF-induced shock, allergic airway inflammation, etc.) they respond differently than wild-type control mice. The mice react adversely in some models (e.g. infection with *Listeria monocytogenes*) while usually the phenotype of these mice is less severe (e.g. PAF-induced shock) or unaffected in other models (2). Recent studies indicate that 5-LO products are important for the development of airway hyperresponsiveness and are partially involved in eosinophil recruitment in allergic airway inflammation in studies performed with the 5-LO deficient mice and wild-type controls (4).

Leukocyte-Type 12-Lipoxygenase Deficient Mice

Manipulation of the leukocyte-type 12LO (L-12LO) gene by targeted gene disruption enabled the generation of homozygous deficient mice (Fig.1).

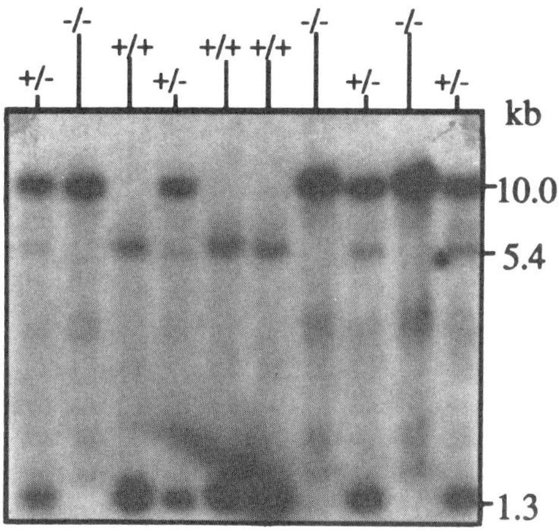

Figure 1. Targeted disruption of the murine L-12LO gene. Southern blot analysis to genotype offspring of F_1 intercrosses was performed. Tail DNA was subjected to StuI/BamHI digestion, electrophoresed and blotted to nylon membrane. An internal probe within the targeting vector (exons 2-6 plus intervening introns) was used for hybridization. The expected 10 kb targeted allele is seen only in homozygous (-/-) and heterozygous (+/-) samples but not wildtype (+/+) which hybridizes to two bands of 5.4 and 1.3 kb.

Protein immunoblotting revealed a 75 kDa band in macrophages from wild-type mice and a similar band of approximately 50 % intensity was detected in samples from heterozygous mice. The immunoreactive L-12LO band was absent in macrophages from homozygous mice (Fig.2).

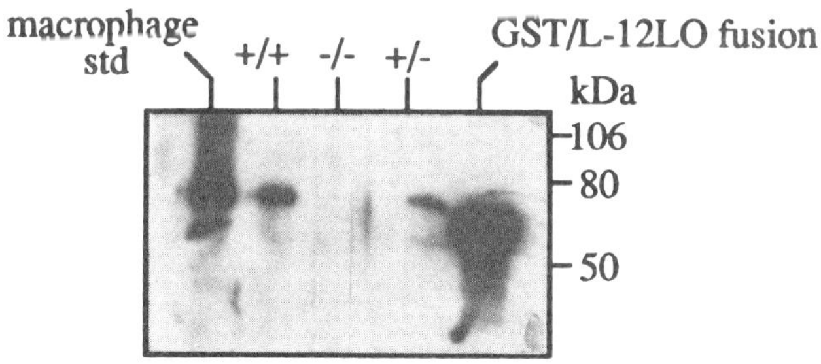

Figure 2. Protein immunoblot analysis to verify absence of L-12LO protein in homozygous L-12LO$^{-/-}$ mice. Proteins (5 μg) were electrophoresed, blotted to nitrocellulose and probed with a mouse L-12LO antiserum (1:100 dilution) with detection by enhanced chemiluminescence. The 61 kDa immunoreactive standard (5 ng) is a GST/N-terminal L-12LO fusion protein. The positive L-12LO immunoreactive band at 75 kDa is seen in control wild-type (+/+) macrophages and heterozygous (+/-) samples but not in homozygous (-/-) samples.

Macrophages stimulated with calcium ionophore A23187 from control L-12LO$^{+/+}$ mice were capable of synthesizing products that co-migrated with 12-H(P)ETE standards on RP-HPLC using arachidonic acid substrate (Fig.3). In contrast, macrophages from L-12LO$^{-/-}$ mice made virtually undetectable amounts of 12-HETE but interestingly there was apparent substrate diversion into the 5-LO pathway (increase in 5-HETE production) when exogenous arachidonic acid was added to the cells. Macrophages from wild-type mice do not normally produce significant levels of 5-HETE under the given incubation conditions. The enhanced 5-HETE production in the deficient mice was not due to a compensatory increase in 5-lipoxygenase gene expression. To determine that only the L-12LO gene was disrupted, and not the P-12LO gene, which is closely linked on chromosome 11, platelets were obtained from both L-12-LO$^{-/-}$ mice and littermate controls and lipoxygenase activity was measured with exogenous arachidonic acid. Both groups of platelets synthesized equivalent amounts of 12-HETE.

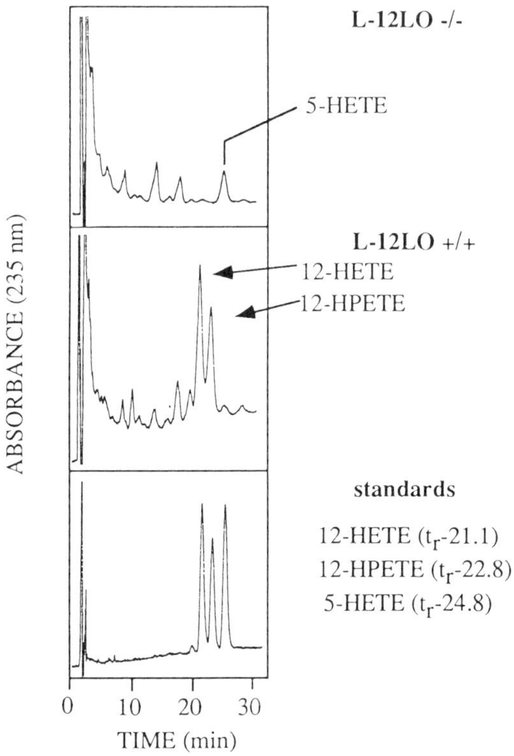

Figure 3. RP-HPLC chromatograms of products formed from arachidonic acid by resident peritoneal macrophages of L-12LO$^{+/+}$ and L-12LO$^{-/-}$ mice. Macrophages from L-12LO$^{+/+}$ or L-12LO$^{-/-}$ mice were incubated with A23187 (2 μM) and arachidonic acid (100 μM) for 15 min and subjected to RP-HPLC as described in Experimental Procedures.

L-12LO$^{-/-}$ mice grow normally and are fertile. Gross macroscopic examination of external and internal organs revealed no evidence for abnormalities. Sections through pituitary gland, adrenals, pancreas and spleen (tissues known to contain L-12LO in some species) showed no histologic evidence for defects nor did bone marrow preparations reveal abnormal precursor cells. Macrophages from L-12LO$^{-/-}$ mice appeared morphologically normal.

Total blood cell counts were normal and there was no change in leukocyte differentials. The reticulocyte counts in L-12LO$^{-/-}$ mice were not significantly different from heterozygous and wild-type littermates. Hemoglobin and hematocrit values were not affected. L-12LO is the apparent species equivalent of human and rabbit 15-LO. Disruption of this pathway in mice does not appear to lead to any obvious changes in the reticulocyte maturation pathway, although 15LO-mediated oxygenation of membrane lipids has been proposed to be a major player in this process (5,6). Further evaluation of these reticulocytes at the electron microscopy level could reveal differences, however.

Macrophage Function in L-12LO$^{-/-}$ Mice

Since the peritoneal macrophage is by far the most abundant site of L-12LO gene expression in mice as assessed by RNAse protection assays (data not shown), we have carried out several different functional tests in these cells to define a potential role of the 12/15-LO pathway. We hypothesized that the ability of L-12LO to directly oxygenate membrane lipids may affect membrane function and thus alter phagocytic activity or bactericidal function, for example. When IgG-sensitized sheep red blood cells were incubated with resident peritoneal macrophages, rosette formation was significantly diminished in L-12LO$^{-/-}$ mice. Fc receptor-mediated phagocytosis was not significantly altered, however. Although the functional significance of this finding is unclear, it could relate to FcRI occupancy to enhance phospholipase A_2 activity and subsequent metabolism via the arachidonic acid cascade (7), possibly through L-12LO. The ability of macrophages to phagocytose Texas Red-conjugated yeast particles and mannose receptor activity (indices of opsonin-independent phagocytic pathway) were not altered. Antitumor cell macrophage activity, measured as the ability of macrophages to suppress [^{3}H]thymidine incorporation into P815 murine mastocytoma cells, was not affected in the deficient mice (Fig.4), nor was the production of nitric oxide by cells stimulated with various agents.

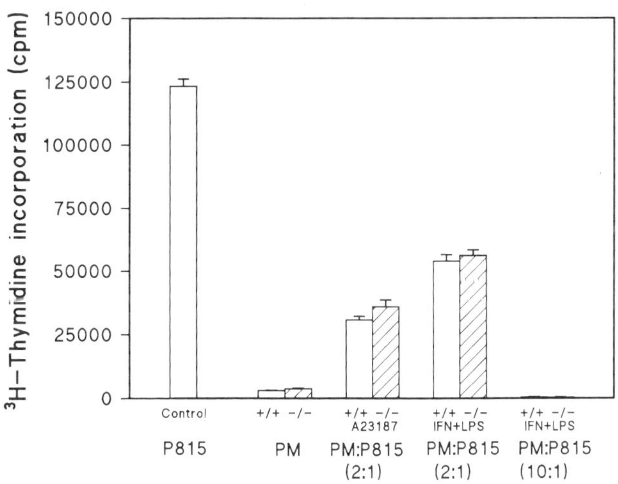

Fig. 4 Antitumor cell macrophage functional assay to test importance of L-12LO pathway. The suppression of ^{3}H-thymidine uptake in P815 murine mastocytoma cells by peritoneal macrophages was measured. See Experimental Procedures section for details. +/+, wild-type; -/-, L-12LO homozygous deficient. Each value is the mean±SEM performed in triplicate from 2 to 4 experiments.

Peritoneal macrophages can undergo apoptosis *in vitro* when activated by certain bacterial products, either by nitric oxide dependent or independent mechanisms. We tested the involvement of the L-12LO pathway on programmed cell death in LPS/IFN-γ activated cells by examining DNA fragmentation with gel electrophoresis and strand breaks labeled *in situ* with fluorescein dUTP by terminal transferase. Unstimulated macrophages from both L-12LO homozygous mice and wild-type controls showed no evidence of DNA fragmentation. In contrast, LPS/IFN-γ-challenged cells displayed the DNA fragmentation ladder characteristic of cells undergoing apoptosis. There were variations in the extent of fragmentation between experiments but no drastic differences were seen with L-12LO$^{-/-}$ macrophages. Likewise, the *in situ* cell death detection values (L-12LO$^{+/+}$, 9.0 % vs L-12LO$^{-/-}$, 12.8 %, low dose LPS/IFN-γ; 12.1 % vs 8.0 %, high dose LPS/IFN-γ) were not statistically significant.

The ability of L-12LO$^{-/-}$ mice to counter infection with *Listeria monocytogenes*, a gram-positive, facultative intracellular bacterium, was not disrupted as approximately equal numbers of viable bacteria were recovered in spleens and livers of infected mice. The lethal dose was not altered by disruption of the L-12LO gene.

Thus, the biological importance of L-12LO remains an enigma and it is hoped that future studies will reveal its function not only in macrophages but in other tissues known to express the enzyme. LO-deficient mice should be useful for elucidation of eicosanoid functions in health and disease.

ACKNOWLEDGMENTS

Supported by NIH grants HL53558 and RCDA HL02710.

REFERENCES

1. C.D. Funk. Molecular biology in the eicosanoid field. *Prog. Nucl. Acids Res. and Mol. Biol.* 45: 67 (1993).
2. X.-S. Chen, J.R. Sheller, E.N. Johnson and C.D. Funk. Role of leukotrienes revealed by targeted disruption of the 5-lipoxygenase gene. *Nature* 372:179 (1994).
3. X.-S. Chen, U. Kurre, N.A. Jenkins, N.G. Copeland, and C.D. Funk. cDNA cloning, expression, mutagenesis, intracellular localization, and gene chromosomal assignment of mouse 5-lipoxygenase. *J. Biol. Chem.* 269:13979 (1994).
4. C.G. Irvin, Y.-P. Tu, J.R. Sheller and C.D. Funk. 5-Lipoxygenase products are necessary for ovalbumin-induced airway responsiveness in mice. manuscript submitted (1996).
5. S. Rapaport and T. Schewe. The maturational breakdown of mitochondria in reticulocytes. *Biochim. Biophys. Acta* 864:471 (1986).
6. H. Kühn and A.R. Brash. Occurrence of lipoxygenase products in membranes of rabbit reticulocytes. *J. Biol. Chem.* 265:1454 (1990).
7. D.O. Adams and T.A. Hamilton. The cell biology of macrophage activation. *Ann. Rev. Immunol.* 2:283 (1984).

DUAL REGULATORY ROLE OF CYCLOOXYGENASE AND LIPOXYGENASE AND THEIR PRODUCTS IN CELL SURVIVAL AND APOPTOSIS

Dean G. Tang,[1] John D. Taylor,[2] Arthur T. Porter,[1] and Kenneth V. Honn[1]

[1]Department of Radiation Oncology
[2]Department of Biological Sciences
Wayne State University
Detroit, MI 48202

Aoptosis, or programmed cell death, is a genetically encoded cell suicide program defined by characteristic morphologic, biochemical, and molecular changes resulting in nonpathologic cell loss. A large number of distinct cellular phenotypes set apoptosis apart from another cell death process, i.e., necrosis.[1,2] Apoptosis plays a key role in physiological processes such as embryonic development, maturation of the host immune system, and in maintaining tissue and organ homeostasis. Apoptosis has also been implicated in a variety of pathological conditions exemplified by cardiac infarction, atherosclerosis, Alzeimer's disease and other neurodegenic diseases, HIV, tumorigenesis and tumor progression. A multitude of factors have been implicated in regulating/modulating apoptosis; these include: (i) oncogenes/tumor suppressor genes exemplified by p53, bcl-2 family (bcl-2, bcl-XL, bcl-Xb, bcl-XS, bax, BAG-1, bad, bak, A1, Mcl-1), myc, ras, abl, raf, Rb-1, and Waf 1; (ii) growth factor/growth factor receptors represented by NGF/NGF receptor, TNF-α/Fas, TGF-β/TGF receptor, IGF-1/IGF receptor, and PDGF/PDGF receptor; (iii) intracellular signal transducers such as protein kinase C, tyrosine kinases and protein phosphatases, lipid signaling molecules such as ceramide, and Ca^{2+}; (iv) cell cycle regulators exemplified by cdc-2 and E2F; (v) reactive oxygen species; (vi) extracellular matrix regulators/signal transducers (extracellular matrix proteins such as fibronectin and transmembrane integrin receptors); (vii) specific endonucleases such as Ca^{2+}- and Mg^{2+}-dependent DNase; and (viii) cytoplasmic proteases typified by ICE (interleukin 1-converting enzyme) family.[1-10] The major impact of apoptosis on cancer research is manifested primarily in three areas: oncogenesis, tumor homeostasis, and the mechanism of action of cytotoxic antitumor drugs.[11] Most anti-tumor agents such as radiation (by generating oxygen radicals) and chemotherapeutic drugs kill tumor cells by inducing apoptosis. Likewise, development by tumor cells of resistance to these treatments is mostly a result of loss of response to apoptosis induction.[1,11]

AA, an essential component of the cell membrane phospholipids is primarily released through the action of phospholipase A_2. Released AA is metabolized via three major biochemical pathways: (i) the cyclooxygenase (COX) pathway leading to the generation of prostaglandins, prostacyclin, and thromboxane; (ii) the lipoxygenase (LOX) pathway giving rise to various hydroperoxy (HPETEs) and hydroxy (HETEs) fatty acids as well as leukotrienes; and (iii) the P450-dependent epoxygenase pathway generating EETs. Various AA metabolites have been implicated in a wide variety of growth-related signaling pathways involving ras,[12] interferon-α,[13] EGF,[14] cAMP,[15] protein kinase C,[16] mitogen-activated kinases[17] and fos.[18] Numerous studies have demonstrated a strong correlation between growth factor promoted cell proliferation and generation of various cyclooxygenase products, primarily prostaglandins[19]. Similarly, eicosanoids derived from lipoxygenase pathways as well as epoxygenase pathways of AA metabolism also play an essential role in mediating the growth factor-stimulated cell growth. Examples include 15-HETE as a mitogenic regulator of T-lymphocytes,[20] 12-HETE and leukotriene B_4 as growth stimulators of epidermal cells,[21] 12 HETE stimulation of keratinocyte DNA synthesis,[22] 12- and 15-HETEs as mediators of insulin and EGF-stimulated mammary epithelial cell proliferation[23] and as synergistic effectors of bFGF- and PDGF-regulated growth of vascular endothelial cells and smooth muscle cells.[24] Our work has also demonstrated that 12(S)-HETE is a positive growth regulator for some tumor cells and vascular endothelial cells and that the mRNA levels for 12-lipoxygenase as well as the ability to biosynthesize 12(S)-HETE correlate positively with the capacity of tumor cells to metastasize.[25-27]

Mammalian cells express 5-, 12-, and 15-lipoxygenases. Our work has documented that the 12-lipoxygenase expression/function are related to tumor growth and metastasis.[25-27] The 12-lipoxygenase enzymes include three subtypes. The first is human platelet-type 12-lipoxygenase which metabolizes only AA (but not C-18 fatty acids such as linoleic acid) to form exclusively 12(S)-HETE.[28] The second is porcine leukocyte-type 12-lipoxygenase which metabolizes both AA and linoleic acid thus generating 12(S)-HETE as well as small amounts of 15(S)-HETE.[28] The third type of 12-LOX (sometimes termed epithelial 12-LOX) shares more homology with 15-LOX and leukocyte-type 12-LOX than with platelet-type 12-LOX and catalyzes the formation of both 12(S)-HETE and 15(S)-HETE.[28-29] Most tumor cells appear to express the platelet-type 12-lipoxygenase.[25-27]

Both cyclooxygenase (COX) and lipoxygenase (LOX) products of AA metabolism, in addition to positively regulating cell growth, may also be involved in modulating cell survival and apoptosis. Many prostaglandins (Table 1) exogenously administered have been shown to induce apoptosis. The administration of a synthetic analog of PGE_2 in mice induces thymocyte apoptosis.[30] Similarly, exogenous PGE_2 induces apoptosis of freshly isolated lymphocytes[31] and immature normal as well as malignant B lymphocytes[32] and a positive correlation has been observed between induction of lymphocyte apoptosis and PGE_2 production by macrophages infected with HIV.[33] Also, apoptosis of ovarian surface epithelial cells is inhibited by indomethacin whose effect could be reversed by exogenous

Table 1. COX Metabolites and Apoptosis

Metabolite	Cell type	Apoptosis	Reference
PGE_2	Immature normal & lymphoma cells	Induction	Brown et al. Clin. Immunol. & Pharmacol., 1992
$PGE_2/PGF_{2\alpha}$	Ovine follicle cells	Induction	Ackerman and Murdoch Prostaglandins, 1993
PGA_2/PGJ_2	Murine L1210 leukemia cells	Induction	Kim et al., FEBS, 1993
TXA_2	Immature thymocytes	Induction	Ushikubi et al., JEM, 1993
PGE_2	Human lymphocytes	Induction	Pica et al., Adv PG TX, LT Res., 1995
PGE_2	Activated human immature thymocytes	Suppression	Goetzl et al., J. Immunol., 1995
PGA_2	MCF7 breast carcinoma cells	Induction	Gorospe & Holbrrok Cancer Res., 1996

PGE_2 or $PGF2\alpha$.[34] The PGE_2-induced apoptosis may be mediated by elevating intracellular cAMP levels since agents which elevate cAMP concentrations have been shown to induce thymocyte apoptosis.[35] Other effects such as increased c-Myc protein expression may also be involved.[31] Cyclopentenone prostaglandins such as PGA_2 and Δ^{12}-PGJ_2 have also been reported to induce tumor cell apoptosis possibly through arresting cells in the G2/M phase and synthesizing novel death-related genes.[36] TxA_2 and the thromboxane Ai2 receptor have been implicated in DNA fragmentation and apoptosis.[37] Intriguingly, PGE_2 and LTB_4 also have been reported to protect $CD4^+CD8^+$ thymocytes from apoptosis induced by both T cell receptor-dependent and -independent mechanisms involving elevation of intracellular cAMP.[38,39] Thymocyte apoptosis initiated by TxA_2 can also be blocked by exogenous PGE_2.[37] The reason for this apparent discrepancy is unclear, but this may be related to different experimental conditions, different cell lines, and different doses of PGE_2 used. In contrast to the conflicting results observed with exogenous prostaglandins, cyclooxygenase inhibitors (NSAID) have been consistently demonstrated to trigger apoptosis of cultured cells (Table 2).[40-42] Thus, sulindac sulfide and sulfone inhibit colon carcinoma cell (HT-29) growth by inducing apoptosis.[41,42] Likewise, multiple NSAIDs including diflunisal, indomethacin, acemethacin, diclofenac, mefenamic acid, flufenamic acid, niflumic acid, ibuprofen, and carprofen cause apoptosis in chicken embryo fibroblasts.[40] The mechanisms of apoptosis induction by these NSAIDs are very complicated and in some cases they may not involve prostaglandin production. Induction of apoptosis in HT-29 cells by sulindac sulfide involves inhibition of cell proliferation and cell cycle arrest caused by inhibition of the activity of cyclin-dependent kinases.[42] In contrast, fibroblast apoptosis induced by indomethacin and other NSAIDs mentioned above could not be inhibited by PGE_2. Interestingly, the chronic treatment of the fibroblasts with these NSAIDs resulted in increased mRNA levels of both COX-1 and COX-2 and increased protein levels of COX-2. Based on the fact that all these NSAIDs inhibit the COX activity, the authors concluded that COXs and their products may play a key role in preventing apoptosis.[40] This conclusion recently gained support from the report that overexpression of COX-2 confers on rat intestinal epithelial cells resistance to apoptosis induction.

Similarly, LOXs and their products have also been implicated in regulating cell survival and apoptosis (Table 3). Exogenous lipid hydroperoxides such as 15-HPETE induces HIV-infected human T cells.[44] The high sensitivity of these cells is due to their inability to convert 15-HPETE to 15-HETE owing to a marked reduction in glutathione peroxidase activity.[44] Lipoxygenase inhibitors such as ETYA and NDGA have been shown to inhibit $TNF\alpha$-induced apoptosis of murine fibrosarcoma cells (L929).[45] Interestingly, in this study the lipoxygenase product 12(S)-HETE also provided a protective effect on

Table 2. COX Inhibitors, COX Overexpression, and Apoptosis

Treatment	Cell type	Apoptosis	Reference
Sulindac sulfide & sulfone	HT-29 colon carcinoma	Induction	Piazza et al., Cancer Res., 1995
Sulindac sulfide	HT-29 colon carcinoma	Induction	Shiff et al., JCI, 1995
26 NSAIDs	v-src-transformed CEF	Induction	Lu et al., PNAS, 1995
NSAIDs	HT-29 colon carcinoma	Induction	Shiff et al., Exp. Cell Res., 1996
COX2 over-expression	Rat intestinal epithelial cells	Resistance	Tsuji & DuBois Cell, 1996

Table 3. LOX/LOX Products and Apoptosis

Treatment	Cell type	Apoptosis	Reference
5-LOX inhibitors (A63162/SC41661)	Myelogenous leukemia	Induction	Anderson et al., PG LT Essent. Fat Acids 1992
15-HPETE	HIV-infected human T cells	Induction	Sandstrom et al., JBC, 1994
TNF-α	L929 cells	LOX-mediated	O'Donnell et al., Biochem. J., 1995
Blocking 5-LOX	Human lung cancer cells	Induction	Avis, et al., JCI 1996
Blocking 12-LOX	Multiple cancer cells	Induction	Tang, et al., PNAS, 1996

TNFα-induced cell death.[45] On the other hand, 5-lipoxygenase inhibitors can cause apoptosis of human leukemia blast cells[46] and blocking the 5-LOX-mediated growth factor signaling inhibited the growth of human lung cancer cells by inducing apoptosis[47], thus suggesting that in some cells 5-LOX may function as a survival factor.

Our most recent work[48] demonstrated that 12-LOX and probably, 15-LOX, but not 5-LOX, plays an essential role in apoptosis. Rat W256 cells of the monocytoid origin express platelet-type 12-LOX and synthesize 12(S)- and 15(S)-HETEs as their major AA lipoxygenase metabolites.[49,50] Down-regulation of 12-lipoxygenase gene expression by antisense oligonucleotides triggered time- and dose-dependent apoptosis. An antisense oligo targeted to 12-LOX coding sequence time-dependently down-regulated the 12-LOX protein[48] and induced dramatic W256 cell apoptosis (Fig. 1). The W256 cell death triggered by this antisense oligo could be partially blocked by exogenous 12(S)- or 15(S)-

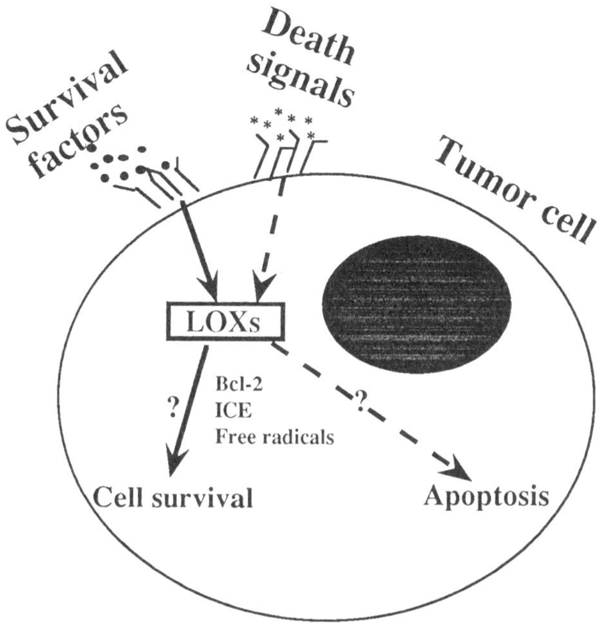

Figure 1. A model illustrating the dual role of LOXs in regulating cell survival as well as apoptosis. See text for details.

HETE but not by 5(S)-HETE. The antisense oligo treatment also resulted in a significant decerase in the ratio of bcl-2/bax, a critical determinant of apoptosis. As expected, overexpression of bcl-2 protein provided partial death-sparing effect. General inhibitors of the lipoxygenase activity as well as 12-lipoxygenase-selective inhibitors also caused significant W256 cell death.[48]

W256 cells are not the only cell type that responds to the above treatment by undergoing apoptosis. Other tumor cells such as RBL-1 (rat basophilic leukemia cells), MTLn-3 (rat mammary adenocarcinoma cells), and HEL (human erythroleukemia cell line) cells all undergo apoptosis following treatment with NDGA, although the doses required vary with different cell lines.[51] In contrast, normal cells such as rat aortic endothelial cells (RAEC) do not undergo apoptosis upon treatment with NDGA or other inhibitors, suggesting that tumor cells differ significantly from normal cells with respect their usage of LOXs in regulating cell survival and apoptosis.

Based upon the above discussion, it is possible to present a model for the dual role of LOXs in regulating tumor cell survival and apoptosis. For some tumor cells or under certain circumstances, LOXs/LOX prodcuts are the actual mediators of the growth effect initiated by survival factors (growth factors, hormones, cytokines, etc). In this case, the LOX system is critical for the cell survival. Blocking the gene expression of LOXs or their enzymatic activities with inhibitors will trigger cell death through apoptosis. The prototypical example is W256 cells presented above. For other tumor cells or under different circumstances, LOXs/LOX products are the actual mediators of apoptosis triggered by death signals. The prototypical example for this case is the apoptosis induced by TNF-α.[45] Whether LOXs/LOX products play a pro-survival or pro-apoptotic role will depend on the cell type and the immediate environmental milieu cells are in contact with (i.e., survival or death factors). This dual-function model also applies to the COX/COX products which may induce cell death or protect cells from apoptosis. Future work needs to focus on the intracellular signaling mechanisms implicated in COX/LOX-regulated cell survival and apoptosis.

REFERENCES

1. G. Majno, and I. Joris. Apoptosis, oncosis, and necrosis. An overview of cell death. Am. J. Pathol. 146:3 (1995).
2. Z.N. Oltvai, and S.J. Korsmeyer. Checkpoints of dueling dimers foil death wishes. Cell 79:189 (1994).
3. B.F. Trump, and I.K. Berezesky. Calcium-mediated cell injury and cell death. FASEB J. 9:219 (1995).
4. S.J. Korsmeyer, X.M. Yin, Z.N. Oltvai, D.J. Veis-Novack, and G.P. Linette. Reactive oxygen species and the regulation of cell death by the bcl-2 gene family. Biochim. Biophys. Acta 1271:63 (1995).
5. S.J. Martin, and D.R. Green. Protease activation during apoptosis: Death by a thousand cuts? Cell 82:349 (1995).
6. D.M. Hockenbery. Bcl-2, a novel regulator of cell death. Bioessays 17:631 (1995).
7. G. Kroemer, P. Petit, N. Zamzami, J. Vayssiere, and B. Mignotte. The biochemistry of programmed cell death. FASEB J. 9:1277 (1995).
8. S. Cory. Regulation og lymphocyte survival by the bcl-2 gene family. Annu. Rev. Immunol. 13:513 (1995).
9. R. Rubin, and R. Baserga. Insulin-like growth factor-I receptor. Its role in cell proliferation, apoptosis, and tumorigenicity. Lab. Invest. 73:311 (1995).
10. E. Ruoslahti, and J. Reed. Anchorage dependence, integrins, and apoptosis. Cell 77:477 (1994).
11. B. W. Stewart. Mechanisms of apoptosis: Integration of genetic, biochemical, and cellular indicators. J. Natl. Cancer Inst. 86:1286 (1994).
12. J-W. Han, F. McCormick, and I.G. Macara. Regulation of Ras-GAP and neurofibromatosis-1 gene product by eicosanoids. Science 252:576 (1991).
13. G.E. Hannigan, and B.R. Williams. Signal transduction by interferon-α through arachidonic acid metabolism. Science 251:204 (1991).

14. M. P. Peppelenbosch, L. G.J. Tertoolen, W.J. Hage, and S.W. de Laat. Epidermal growth factor-induced actin remodeling is regulated by 5-lipoxygenase and cyclooxygenase products. Cell 74:565 (1993).
15. D.G. Tang, I.M. Grossi, K.Q. Tang, C.A. Diglio, and K.V. Honn. Inhibition of TPA and 12(S)-HETE-stimulated tumor cell adhesion by prostacyclin and its stable analogs: Rationale for their antimetastatic effects. Int. J. Cancer 60:418 (1995).
16. D.G. Tang, Y.Q. Chen, C.A. Diglio, and K.V. Honn. PKC-dependent effects of 12(S)-HETE on endothelial cell vitronectin and fibronectin receptor. J. Cell Biol. 121:689 (1993).
17. G.N. Rao, A.S. Baas, W.C. Glasgow, T.E. Eling, M.S. Runge, and W.R. Alexander. Activation of mitogen-activated protein kinases by arachidonic acid and its metabolites in vascular smooth muscle cells. J. Biol. Chem. 269, 32586-32591, (1994).
18. U. Danesch, P.C. Weber, and A. Sellmayer. Arachidnoic acid increases c-fos and Egr-1 mRNA in 3T3 fibroblasts by formation of prostaglandin E2 and activation of protein kinase C. J. Biol. Chem. 269:27258 (1994).
19. R.D. Nolan, R.M. Danilowicz, and T.E. Eling. Role of arachidonic acid metabolism in the mitogenic response of BALB/c 3T3 fibroblasts to epidermal growth factor. Mol. Pharmacol. 33:650 (1988).
20. J.M. Bailey, R.W. Bryant, C.E. Low, M.B. Papillo, and J.Y. Vanderhoeck. Regulation of T-lymphocyte mitogenesis by the leukocyte product 15-hydroxy-eicosatetraenoic acid (15-HETE). Cell Immunol. 67:112 (1982).
21. C. Chan, E. Duhamel, and A. Ford-Hutchinson, A. Leukotriene B4 and 12-hydroxyeicosatetraenoic acid stimulates epidermal proliferation in vivo in the guinea pig. J. Invest. Dermatol. 85:333 (1985).
22. K. Kragballe, and J.D. Fallon. Increased aggregation and arachidonic acid transformation by psoriatic platelets: evidence that platelet-derived 12-hydoxy-eicosatetraenoic acid increases keratinocyte DNA synthesis in vitro. Arch. Dermatol. Res. 278:449 (1986).
23. G.K. Bandyopadhyay, W. Imagawa, D.R. Wallace, and S.J. Nandi. Proliferative effectors of insulin and epidermal growth factor on mouse mammary epithelial cells in primary culture. Enhancement by hydroxyeicosatetraenoic acids and synergism with prostaglandin E2. J. Biol. Chem. 263:7567 (1988).
24. B.N. Yamaja Setty, J.E. Graeber, and M.J. Stuart. The mitogenic effect of 15- and 12-hydroxyeicosatetraenoic acid on the endothelial cells may be mediated via diacylglycerol kinase inhibition. J. Biol. Chem. 262:17613 (1987).
25. D.G. Tang, C. Renaud, S. Stojakovic, C.A. Diglio, A. Porter, and K.V. Honn. 12(S)-HETE is a mitogenic factor for microvascular endothelial cells: Its potential role in angiogenesis. Biochem. Biophys. Res. Commun. 211:462 (1995).
26. K.V. Honn, D.G. Tang, X. Gao, I.A. Butovich, B. Liu, J. Timar, and W. Hagmann. 12-Lipoxygenases and 12(S)-HETE: role in cancer metastasis. Cancer Metastasis Rev. 13:365 (1994).
27. D.G. Tang, and K.V. Honn. 12-Lipoxygenase, 12(S)-HETE, and cancer metastasis. Annals New York Acad Sci. 744:199 (1994).
28. C.D. Funk. Molecular biology in the eicosanoid field.Nucleic Acid Res. Mol. Biol. 45:67 (1993).
29. J. Freire-Moar, A. Alavi-Nassab, M. Ng, M. Mulkins, and E. Sigal. Cloning and characterization of a murine macrophage lipoxygenase. Biochim. Biophys. Acta 1254:112 (1995).
30. A. Mastino, M. Piacentini, and S. Grelli. Induction of apoptosis in thymocytes by prostaglandin E_2 in vivo. Develop Immunol. 2:263 (1992).
31. F. Pica, O. Franzese, C. D'Onofrio, L. Paganini, C. Favalli, E. Bonmassar, and E. Garaci. Effect of PGE_2 on c-myc and Bcl-2 production and programmed cell death in human lymphocytes. Adv. Prostaglandin Thromboxane Leukotriene Res. 23:457 (1995).
32. D.M. Brown, G.L. Warner, J.E. Ales-Martinez, D.W. Scott, and R.P. Phipps. Prostaglandin E_2 induces apoptosis in immature normal and malignant B lymphocytes. Clin. Immuno. Immunopathol. 63:221 (1992).
33. A. Mastino, S. Grelli, M. Piacentini, S. Oliverio, C. Favalli, C.F. Perno, and E. Garci. Correlation between induction of lymphocyte apoptosis and prostaglandin E_2 production by macrophages infected with HIV. Cell Immunol. 152:120 (1993).

34. R.C. Ackerman, and W.J. Murdoch. Prostaglandin-induced apoptosis of ovarian surface epithelial cells. Prostaglandins 45:475 (1993).
35. D.J. McConkey, S. Orrenius, and M. Jondal. Agents that elevate cAMP stimulate DNA fragmentation in thymocytes. J. Immunol. 145: 1227 (1990).
36. I. Kim, J.W. Lee, H-W. Sohn, H-S. Kim, and S-H. Kim. Prostaglandin A_2 and Δ^{12}-prostaglandin J_2 induce apoptosis in L1210 cells. FEBS Lett. 321:209 (1993).
37. F. Ushikubi, Y. Aiba, K. Nakamura, T. Namba, M. Hirata, O. Mazda, Y. Katsura, and S. Narumiya. Thromboxane Ai2 receptor is highly expressed in mouse immature thymocytes and mediates DNA fragmentation and apoptosis. J. Exp. Med. 178:1825 (1993).
38. E.J. Goetzl, S. An, and L. Zheng. Specific suppression by prostaglandin E_2 of activation-induced apoptosis of human $CD4^+8^+$ T lymphoblasts. J. Immunol. 154:1041 (1995).
39. E.J. Goetzel, S. An, and W.J. Smith. Specificity of expression and effects of eicosanoid mediators in normal physiology and human diseases. FASEB J. 9:1051 (1995).
40. X. Lu, W. Xie, D. Reed, W.S. Bradshaw, and D.L. Simmons. Nonsteroidal antiinflammatory drugs cause apoptosis and induce cyclooxygenases in chicken embryo fibroblasts. Proc. Natl. Acad. Scid. USA 92:7961 (1995).
41. G.A. Piazza, A.L.K. Rahm, M. Krutzsch, G. Sperl, N.S. Paranka, P.H. Gross, K. Brendel, R.W. Burt, D.S. Alberts, R. Pamukcu, and D.J. Ahnen. Antineoplastic drugs sulindac sulfide and sulphone inhibit cell growth by inducing apoptosis. Cancer Res. 55:3110 (1995).
42. S.J. Shiff, L. Qiao, L-L. Tsai, and B. Rigas. Sulindac sulfide, an aspirin-like compound, inhibits proliferation, causes cell cycle quiescence, and induces apoptosis in HT-29 colon adenocarcinoma cells. J. Clin. Invest. 96:491 (1995).
43. M. Tsujii, and R. DuBois. Alterations in cellular adhesion and apoptosis in epithelial cells overexpressing prostaglandin endopeorxide synthase 2. Cell 83:493 (1995).
44. P.A. Sandtrom, P.W. Tebbey, S. Van Cleave, and T.M. Buttke. Lipid hydroperoxides induce apoptosis in T cells displaying a HIV-associated glutathione peroxidase deficiency. J. Biol. Chem. 269:798 (1994).
45. V.B. O'Donnell, S. Spycher, and A. Azzi. Involvement of oxidants and oxidant-generating enzyme(s) in tumor-necrosis-factor-a-mediated apoptosis: role for lipoxygenase pathway but not mitochondrial respiratory chain. Biochem. J. 310:133 (1995).
46. K.M. Anderson, J. Levin, A. Jejah, T. Seed, and J.E. Harris. Induction of apoptosis in blood cells from a patient with acute myelogenous leukemia by SC41661, a selective inhibitor of 5-lipoxygenase. Prostaglandin Leukotriene Essen. Fat. Acid 48, 323-326 (1993).
47. I.M. Avis, M. Jett, T. Boyle, M.D. Vos, T. Moody, A.M. Treston, A. Martinez, and J.L. Mulshine. Growth control of lung cancer by interruption of 5-lipoxygenase-mediated growth factor signaling. J. Clin. Invest. 97:806 (1996).
48. D.G. Tang, Y.Q. Chen, and K.V. Honn. Arachidonate lipoxygenases as essential regulators of cell survival and apoptosis. Proc. Natl. Acad. Sci. USA. 1996, in press.
49. B. Liu, J. Timar, J. Howlett, C.A. Diglio, and K.V. Honn. Lipoxygenase metabolite of arachidonic and linoleic acids modulate the adhesion of tumor cells to endothelium via regulation of protein kinase C. Cell Regul. 2:1045 (1991).
50. Y.Q. Chen, Z.M. Duniec, B. Liu, W. Hagmann, X. Gao, K. Shimoji, L.J. Marnett, C.R. Johnson, and K.V. Honn. Endogenous 12(S)-HETE production by tumor cells and its role in metastasis. Cancer Res. 54:1574 (1994).
51. D.G. Tang, and K.V. Honn Tumor cell apoptosis induced by NDGA, a lipoxygenase inhibitor, involves lipid peroxidation and depletion of GSH. Sumitted.

ASPIRIN SWITCHES BIOSYNTHETIC CIRCUITS TRIGGERING NOVEL EICOSANOIDS DURING CELL-CELL INTERACTIONS THAT ARE POTENT INHIBITORS OF NEUTROPHIL MIGRATION AND TUMOR CELL PROLIFERATION

Charles N. Serhan* and Joan Clària

Center for Experimental Therapeutics and Reperfusion Injury
Brigham and Women's Hospital and Harvard Medical School
Boston, MA 02115

INTRODUCTION

Aspirin [acetylsalicylic acid (ASA)] is the lead nonsteroidal drug (NSAID) and is widely used to relieve inflammation.[1, 2] Recent results indicate that its use in low doses is also associated with new, previously unappreciated beneficial effects, including reduced risk of heart disease[3] and decreased incidence of lung, colon and breast cancer,[4] and *in vitro* it can inhibit nuclear factor–κB transcription, which may be relevant in treatment of patients with human immunodeficiency virus.[5] ASA's target is the cyclooxygenase activity of prostaglandin G/H synthase (PGHS). It does not inhibit the lipoxygenases (LOs).[6-8] Although blockage of PGs and TXs can account for many of ASA's pharmacologic properties, the mechanism of action of this important drug in several clinical settings including lowering the risk of human cancer is still a subject of interest and debate.[1, 2]

Eicosanoid biosynthesis by transcellular and cell–cell interactions is now recognized as an important means of both amplifying and generating new mediators, particularly those produced by LOs.[9] In humans, lipoxin (LX) biosynthesis is an example of LO–LO interactions via transcellular routes (reviewed in Serhan[10]). LXs show counterregulatory actions in several tissues. They inhibit polymorphonuclear leukocyte (PMN) transmigration,[11] modulate adhesion to endothelial cells (EC)[12] and inhibit chemotaxis of PMN and eosinophils.[13,14] In the microcirculation, LXs are vasodilatory and inhibit leukotriene B_4 (LTB_4)–induced inflammatory responses.[15] EC–leukocyte interactions (i.e., rolling, adhesion, etc.) are central to host defense,[16] and transcellular metabolism between PMN and EC[17] can promote generation of LTs by EC.[18] Earlier results showed that ASA inhibits prostacyclin production by EC, which is an important mechanism in understanding its vascular impact.[6-8] Recently, we observed that ASA therapy before angioplasty elevates

*Address correspondence to C.N. Serhan, Director, Center for Experimental Therapeutics and Reperfusion Injury, Brigham and Women's Hospital, 75 Francis Street, Boston, MA 02115

LO–derived eicosanoids [i.e., LTs and (5*S*,6*R*,15*S*)-5,6,15-trihydroxy-7,9,13-*trans*-11-*cis*-eicosatetraenoic acid (LXA_4)] during atherosclerotic plaque rupture in patients.[19] Thus, potent vasoconstrictors [e.g., leukotriene C_4 and D_4 (LTC_4 and LTD_4)] and vasodilatory LXA_4 are still generated in vessel lumen when ASA is taken. Here, we review our recent results indicating that human EC–PMN coincubations in the presence of ASA generate a previously unreported class of eicosanoids. The compounds in turn regulate EC–PMN interactions. Moreover, 15-epi-LX are also generated by the human adenocarcinoma cell line (A549 cells) interacting with PMN, which are enhanced when ASA is present. Of particular interest, 15-epi-LXB_4 inhibits tumor cell proliferation *in vitro*.[20]

MATERIALS AND METHODS

Synthetic (5*S*,6*R*,15*R*)-5,6,15-trihydroxy-7,9,13-*trans*-11-*cis*-eicosatetraenoic acid (15–epi–LXA_4)–methyl ester was prepared by total organic synthesis[21] by Prof. N. A. Petasis (Department of Chemistry, University of Southern California). Human umbilical vein EC (HUVEC) and PMN were obtained as in Clària and Serhan[22]. Cell viabilities of HUVEC and PMN were determined by their ability to exclude trypan blue. Human type II epithelial (A549) cells from adenocarcinoma were obtained from American Type Culture Collection (Rockville, MD). PCR, HPLC and GC-MS analyses were performed as in Clària and Serhan[22] and Clària et al.[20]

RESULTS AND DISCUSSION

Endothelial Cell-PMN Interactions

Coincubations of HUVEC and PMN in the presence of ASA resulted in the formation of several previously undescribed eicosanoids. RP–HPLC analysis of the fractions obtained from cytokine [interleukin–1_β (IL–1_β)]–treated HUVEC coincubated with PMN showed four compounds, which each displayed a triplet of absorbing bands characteristic of conjugated tetraene–containing chromophores indicative of the LX basic structure (maxima at 301 nm and shoulders at 288 and 316 ± 2 nm; see Clària and Serhan[22]). When [^{3}H]–arachidonic acid was added to ASA–treated HUVEC–PMN suspensions (80×10^6 cells/ml, HUVEC/PMN cell ratio of 1:10), materials beneath the tetraene-containing peaks (Fig. 1A) carried both the tritiated label and UV chromophores. Without ASA these products were not detected in coincubations nor were they generated by ASA treatment of either PMN or HUVEC incubated separately (data not shown). Also, HUVEC treated with IL–l_β (155 ± 18 ng; *n* = 4) and incubated with ASA generated 5–6 times higher levels than those without IL–1_β treatment (27 ± 6 ng; *n* = 4). Since neither HUVEC nor PMN significantly further metabolizes exogenous LXs, these findings indicate that the novel compounds originated from transcellular routes.

To characterize parameters leading to formation of these novel eicosanoids during HUVEC/PMN coincubations, immunoreactive LX was monitored since the new compound crossreacted with the LXA_4 antisera.[22] ASA dramatically increased formation, while cells incubated for 20 min at 37°C with either indomethacin or salicylate or exposed to dexamethasone for 24 hr did not result in their appearance (Fig. 1A). Formation of the new products was also observed at lower ASA concentrations (100 and 250 μM; 145 ± 14 and 166 ± 31 ng/plate, respectively). Indomethacin added together with ASA prevented this increase. Production from endogenous sources of arachidonate was obtained when HUVEC treated with IL–1_β were exposed to ASA and coincubations were challenged with thrombin (1 unit/ml) plus 100 nM fMet-Leu-Phe or 5 μM ionophore A_{23187} or 100 nM phorbol

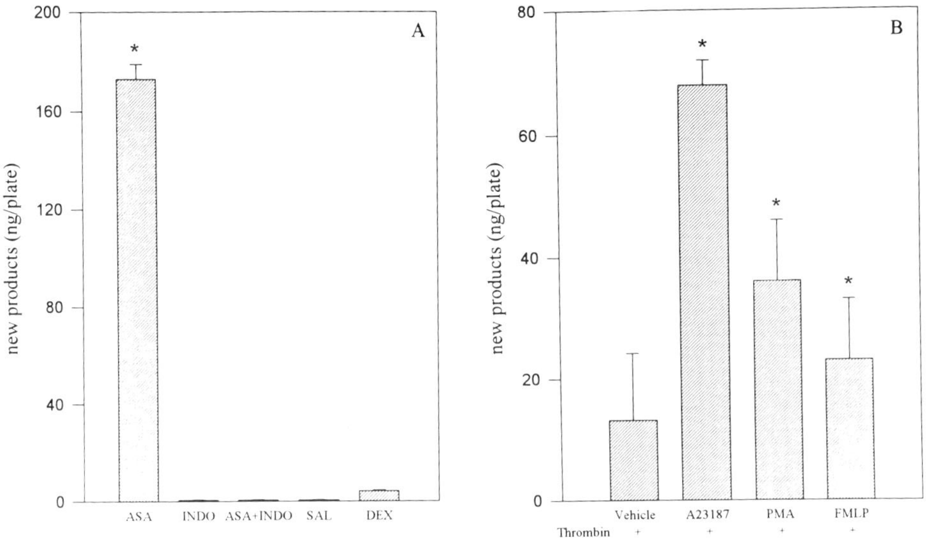

Figure 1. (*A*) Selective generation by ASA. Confluent HUVEC were exposed for 24 hr at 37°C to 1 ng of IL–1_β per ml and treated for 20 min at 37°C with 500 μM ASA, 500 μM indomethacin (INDO), 500 μM ASA plus 500 μM INDO (A+I), 500 μM salicylate (SAL), or 2 μM dexamethasone (DEX). HUVEC–PMN coincubations were for 30 min at 37°C as in Figure 1*A*. HUVEC were preincubated with 2 μM DEX for 24 hr. Quantitation of immunoreactive LX was determined by ELISA. Values are expressed in ng/plate (mean ± SEM; $n = 3$). *, $P < 0.01$, treatment with vs. without ASA. (B) Generation from endogenous sources. IL–1_β–treated HUVEC were exposed to 500 μM ASA or vehicle (0.1% EtOH) for 20 min at 37°C and to 1 unit of thrombin per ml for 2 min and coincubated with PMN, followed by addition of 5 μM A_{23187} or 100 nM fMet-Leu-Phe or 100 nM PMA for 30 min at 37°C. Quantitation was as in Fig. 1*A*. Values are expressed in ng/plate (mean ± SEM; $n = 2$, with six determinations). *, $P < 0.05$ vs. control values without ASA.

12-myristate 13–acetate (PMA) (Fig. 1B). Formation of the new products (15-epi-LX) from endogenous sources was also observed when cells were stimulated with either 5 μM A_{23187} (45 ± 10 ng/plate) or 100 nM PMA (28 ± 8 ng/plate) alone but was not statistically significant with either 1 unit of thrombin per ml or 100 nM fMLP added separately with these cell numbers (Fig. 1B and data not shown). These results indicate that these unique arachidonic acid–derived products were formed following receptor–ligand activation and that activation of both cell types was required for their biosynthesis.

Stable Analogs of Aspirin-Triggered 15-epi-LX and LXs

Biosynthetic studies revealed that 15*R*-HETE was released by activated endothelial cells when ASA was added to the incubations. The 15*R*-HETE from HUVEC was converted by activated PMN to 15-epi-LX via a route similar to that demonstrated for 15*S*-HETE-dependent production of LX. In contrast to the LX route initiated by the 15-lipoxygenase, this 15*R*-HETE route involves PGHS-2 that has been acetylated by ASA. The formation of 15-epi-LX by EC and PMN interactions is of particular interest because these cells are an initial site of interaction during PMN recruitment and inflammation. LXA_4 and 15-epi-LXA_4 both inhibit PMN adhesion and transmigration[22] suggesting that their production at sites of inflammation may be an important regulatory action of both ASA-triggered and LO-generated LX. Since LXA_4 is rapidly metabolized at C-15, a series of analogs was designed and tested to evaluate the role of the C-15 alcohol chirality and omega end of LXA_4. Each of the structural analogs 15(*R*/*S*)methyl-, 15-cyclohexyl- and 16-phenoxy-LXA_4 proved to be potent inhibitors of PMN transmigration and adhesion and resistant to inactivation. In contrast, 15-deoxy-LXA_4 was inactive in these bioassays.[23] These findings confirm the bioactions of 15-epi-LXA_4 and LXA_4 and suggest that LXA_4 stable analogs are useful tools to evaluate LX actions and therapeutic potentials.

Tumor Cell (Epithelial)-PMN Interactions

We next investigated the transcellular biosynthesis of these novel eicosanoids during costimulation of human alveolar type II epithelial cells (A549 cell line from adenocarcinoma) treated with ASA and human neutrophils. Interleukin-1β (IL-1β)-primed A549 cells showed elevated levels of PGHS-2 mRNA and generated 15-hydroxyeicosatetraenoic acid (15-HETE). Again, ASA markedly increased 15-HETE formation, whereas treatment with an inhibitor of cytochrome P450 resulted in ~50% reduction of 15-HETE, implicating both PGHS-2 and cytochrome P450 as two independent routes in 15-HETE biosynthesis in these cells. Maximal production of 15-HETE by A549 cells occurred at 24 hours of cytokine (IL-1β) exposure and declined thereafter. Chiral analysis revealed that ~85% of ASA-triggered epithelial-derived 15-HETE carries its carbon 15 alcohol group in the *R* configuration. Costimulation of A549 cells and PMN led to production of both LXA_4 and LXB_4, as well as 15-epi-LXA_4 and 15-epi-LXB_4 (9.5 ± 0.5 ng/10^7 A549 cells). 15-epi-LXA_4 accounted for 88% of the total amount of LXA_4 produced.[20]

Are There Endogenous Mediators of ASA Actions in Humans?

ASA's Inflammatory Action. Costimulation also produced substantial amounts (77.2 ± 8.2 ng/10^7 A549 cells) of peptidoleukotrienes (pLTs). When ASA was added to these coincubations, not only was an increase in the amounts of LXs observed, but a parallel decrease in the generation of pLTs was also noted. The ability of ASA to inhibit LT biosynthesis is consistent with a transcellular cellular mechanism where the formation of 15*R*-HETE by ASA-treated A549 cells is a substrate for the leukocyte 5-lipoxygenase-dependent generation of 15-epi-LX. Concomitant with the biosynthesis of a 15*R*-5(6)-epoxytetraene by 5-LO, LTA_4 formation from endogenous arachidonate is blocked and thus leads to a decrease in LT production by both cell types, since LTA_4 is not available for LTB_4 generation in the PMN, nor is it released for transcellular conversion to pLT by A549 cells.[20] This transcellular mechanism provides a potential link between ASA's ability to reduce inflammation *in vivo* and ASA-triggered 15*R*-HETE's ability to block pLT and enhance 15-epi-LX biosynthesis. Since 15-epi-LXA_4 and LXA_4 block PMN adhesion and migration, they may serve as local endogenous mediators of ASA's beneficial action in humans.

ASA's Antiproliferative Action: A Role for 15-epi-LX? ASA reduces the risk and incidence of cancer,[4,24] yet the mechanism of action is not known (see Marnett[25] and Weissmann[1] for review and discussion). Some other NSAIDs share this ability, yet ASA appears to possess a unique impact in inducing a protective mechanism compared to other NSAIDs. Therefore, we sought evidence for a potential link in ASA's unique ability to generate 15-epi-LX to its antiproliferative effect.

LXs are vasodilators and potent regulators of leukocyte responses, such as inhibition of chemotaxis, adhesion to endothelial cells and transmigration across epithelium (*vide supra* and reviewed in Serhan[10]). pLTs, in particular LTC_4 and LTD_4, possess both vasoconstrictor and proinflammatory actions as well as stimulate the growth of several cell types including fibroblasts, smooth-muscle and glomerular epithelial cells.[26] LXs reverse the vasoconstrictor action of LTD_4 in rat renal hemodynamics and block LTC_4-stimulated hematopoiesis[27] (reviewed in Serhan[10]). Because ASA enhances 15-epi-LX formation and

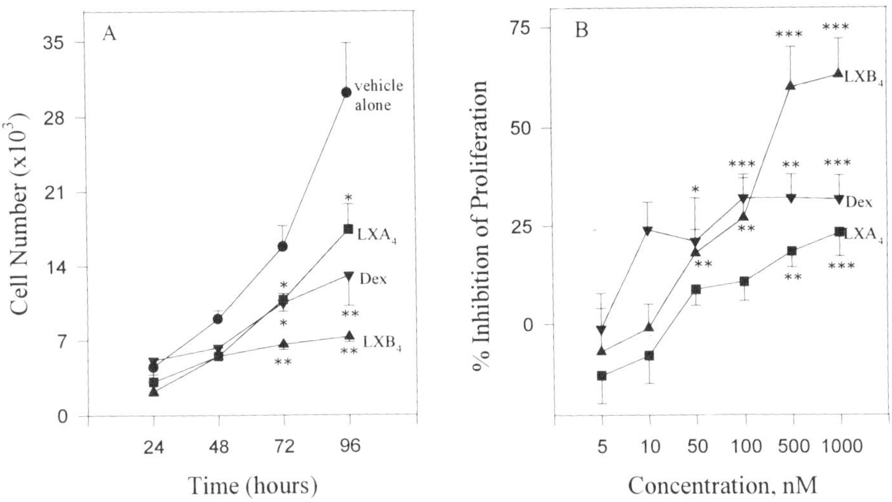

Figure 2. LX inhibits A549 cell proliferation. (*A*) Time-dependent growth arrest. A549 cells in 96-well plates were treated with either vehicle (0.15% EtOH) or equimolar concentrations (10^{-6} M) of LXA_4, LXB_4 or Dex for up to 96 hours at 37 °C. At the indicated intervals, cells were harvested for the MTT assay. Data are means ± SEM of 3-7 experiments performed in quadruplicate. *, $P < 0.05$ and **, $P < 0.005$ for compounds vs. vehicle. (*B*) Concentration dependence. A549 cells were exposed to LXA_4, LXB_4 or Dex at the indicated concentrations for 72 hours at 37 °C. Results are means ± SEM of 5-8 experiments performed in quadruplicate. Results are expressed as the percent inhibition of proliferation relative to vehicle. *, $P < 0.05$, **, $P < 0.025$ and ***, $P < 0.005$ for compounds vs. vehicle.

inhibits pLT biosynthesis,[20,22] these eicosanoids may play counterregulatory actions on cell proliferation and contribute to ASA's protective mechanisms in human cancer. To this end, we tested the effect of these LO products on epithelial cell proliferation (Fig. 2) and compared their actions to that of dexamethasone, a well-established inhibitor, employing a soluble microculture tetrazolium (MTT) assay.[28] These experiments were performed with synthetic LXA_4 and LXB_4, which were available in sufficient quantities for bioassay, rather than the biologically-derived 15-epi-LX, which are the major LX produced by these cells. As shown in Figure 2, LXA_4 and LXB_4 inhibited A549 cell proliferation in a time- (A) and dose-dependent (B) fashion. LXA_4 and LXB_4 as well as dexamethasone [1 μM] inhibited A549 cell proliferation after 72 and 96 hours of treatment (Figure 2A). After 72 hours, LXA_4 shared the anti-proliferative properties (Fig. 2) observed for dexamethasone with these cells (cf. Croxtall and Flower).[29] The half-maximum inhibition (IC_{50}) for LXA_4 was ~80 nM compared to that of dexamethasone, which was ~7 nM. LXB_4 at concentrations of 0.5 and 1 μM was ≈3 times more active than either LXA_4 or dexamethasone (Fig. 2B). Of interest, both LXA_4 and LXB_4 showed essentially equal potency for blocking A549 cell growth when each was added repeatedly (i.e., 24-hour intervals) to the cells for 3 consecutive days (data not shown, $n = 3$, d = 4), suggesting that LX may be inactivated by these epithelial cells. Results from additional experiments employing direct cell enumeration and measurement of total cellular protein content (data not shown, $n = 3$, d = 4) paralleled those obtained with MTT assay, thus confirming the anti-proliferative actions of LXs in A549 cells. Furthermore, a blockage of DNA synthesis, as determined by ^{3}H-thymidine incorporation, occurred when A549 cells were exposed for 72 hours to concentrations of 50 nM or higher of LXB_4 or dexamethasone. After incubation of A549 cells with the compounds, the viability of the cells, as determined by trypan blue exclusion assay, was found to be ~98%,

Table 1. LX and 15-epi-LX actions on cell proliferation.

Compound [10^{-7} M]	% inhibition	P value[1]
Vehicle	1.6 ± 10.3	NS
LXA_4	16.3 ± 5.2	NS
15-epi-LXA_4	20.1 ± 1.9	< 0.025
LXB_4	34.0 ± 8.5	< 0.005
15-epi-LXB_4	79.3 ± 0.3	< 0.001[2]
Dexamethasone	42.1 ± 5.4	< 0.005

Cells (2000 A549 cells/well) were grown in 96-well plates and exposed to vehicle (0.15% vol/vol in EtOH/F-12K media) or equimolar concentrations (10^{-7} M) of LXA_4, 15-epi-LXA_4, LXB_4, 15-epi-LXB_4, or dexamethasone for 72 hours at 37°C. Values represent ± SEM from 3 to 7 experiments performed in quadruplicate and are expressed as percent inhibition of cell growth.
[1]P values denote statistical differences as compared to cells alone.
[2]$P < 0.001$ for 15-epi-LXB_4 as compared to LXB_4.

indicating that these compounds were not cytotoxic within the range of concentrations used in these experiments.

The 15-hydroxy epimeric forms of LXA_4 and LXB_4 (15-epi-LXA_4 and 15-epi-LXB_4, respectively), which were the dominant forms of LX isolated from costimulation of leukocytes and the tumor cells, proved to also be potent inhibitors of epithelial cell proliferation (Table 1). At equimolar levels (100 nM), 15-epi-LXA_4 inhibited A549 cell growth to a similar extent as LXA_4. On the other hand, 15-epi-LXB_4 isolated from conversion of 15*R*-HETE by activated PMN and added back to the A549 cells gave a more potent anti-proliferative activity than LXB_4 (~80% vs. ~34% inhibition proliferation, $P < 0.001$; Table 1). This compound was characterized by UV, HPLC and GC-MS (C value: 23.3). Diagnostic ions for the OTMS methyl ester were m/z 173 (base peak), 203, 289, 379 and less prominent ions at 482 (M^+-100).[20] In contrast, other structurally related eicosanoids including 15-epi-*trans*-LXA_4, 11-*trans*-LXB_4, 8,9-acetylenic-LXB_4, peptidoleukotrienes (LTC_4 and LTD_4) and the LX precursors (15*S*- and 15*R*-HETE) each tested at 10^{-6} - 10^{-9} M were unable to significantly inhibit A549 cell proliferation (data not shown, n = 3-5, d = 4). These results indicate that LX and 15-epi-LX gave a stereoselective action in blocking cell proliferation in A549 cells. We also tested LXA_4 and LXB_4 with human skin fibroblasts to determine if they were antiproliferative for this cell type. At 100 nM, both LXA_4 and LXB_4 inhibited proliferation of fibroblasts. LXB_4 gave 38.0 ± 7.5% inhibition and LXA_4 10.7 ± 1.8% compared to dexamethasone (29.7 ± 0.4%) as positive control (n = 3).

SUMMARY

Cytokine regulation of eicosanoid-generating enzymes can enhance or "turn on" the formation of LXs at "later stages" of inflammation or tissue injury during cell-cell interactions.[10] The results summarized here establish a previously unrecognized biosynthetic circuit involving PGHS-2 interactions with the 5-LO as well as P450-derived 15*R*-HETE interactions with leukocyte 5-LO to generation of 15-epi-LX (see Fig. 3). These enzymes are likely to be in place before ASA therapy is administered. The finding that ASA triggers as well as enhances biosynthesis of a previously unrecognized class of products from cell-cell interactions *in vitro* suggests that drugs such as ASA may influence inflammation, differentiation, or proliferation by pirating endogenous biosynthetic mechanisms to produce different mediators. Thus, it is tempting to propose that eicosanoid

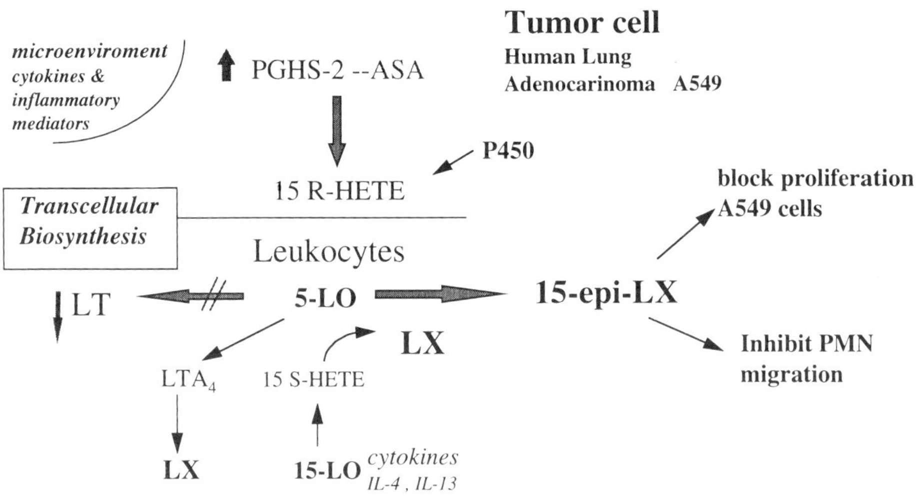

Figure 3. Aspirin switches eicosanoid circuits: proposed scheme for generating 15-epi-LXs. ASA-acetylated PGHS-2 and/or P450 activities contribute to 15*R*-HETE generation. Epithelial-derived 15*R*-HETE undergoes transcellular conversion by leukocyte 5-LO to a 15-epi-5(6)-epoxytetraene intermediate, which is common to both 15-epi-LXA_4 and 15-epi-LXB_4. The complete stereochemistry of 15-epi-LXA_4 has been determined,[22] and the stereochemical assignment of 15-epi-LXB_4 is tentative.

biosynthesis and profiles are switched by the presence of ASA and various cytokines in the local microenvironment. This does not exclude potential "shunting" of arachidonic acid to LO pathways but rather opens previously unrecognized biosynthetic routes that produce bioactive 15-epi-LX. 15-epi-LXA_4 blocks adhesion of PMN and 15-epi-LXB_4 inhibits tumor cell (A549 cell) proliferation. The ability of ASA to "switch circuits" and stimulate 15-epi-LX formation provides new avenues for investigating mechanisms underlying important actions of ASA in humans.

Acknowledgments

These studies were supported by NIH GM38765 and P01-DK50305 grants (C.N.S.). We thank Prof. N. Petasis (Department of Chemistry, University of Southern California) for providing reference samples of synthetic 15-epi-LXA_4 methyl ester. We also thank M. Halm Small for assistance in manuscript preparation and Min Lee for expert technical assistance.

REFERENCES

1. G. Weissmann, Aspirin, *Sci. Am.* 264:84 (1991).
2. R.J. Flower, Cyclooxygenase inhibitors: an overview, in: *Prostaglandins, Leukotrienes and Lipoxins: Biochemistry, Mechanism of Action, and Clinical Applications*, J.M. Bailey, ed., Plenum, New York (1985).
3. Steering Committee of the Physicians' Health Study Research Group, Final report on the aspirin component of the ongoing Physicians' Health Study, *N. Engl. J. Med.* 321:129 (1989).
4. D.M. Schreinemachers and R.B. Everson, Aspirin use and lung, colon, and breast cancer incidence in a prospective study, *Epidemiology* 5:138 (1994).
5. E. Kopp and S. Ghosh, Inhibition of NF-κB by sodium salicylate and aspirin, *Science* 265:956 (1994).
6. J.R. Vane, Inhibition of prostaglandin synthesis as a mechanism of action for aspirin-like drugs, *Nature (London) New Biol.* 231:232 (1971).

7. B. Samuelsson, From studies of biochemical mechanisms to novel biological mediators: prostaglandin endoperoxides, thromboxanes and leukotrienes, in: *Les Prix Nobel: Nobel Prizes, Presentations, Biographies and Lectures*, Almqvist & Wiksell, Stockholm (1982).
8. J.R. Vane, Adventures and excursions in bioassay: the stepping stones to prostacyclin, in: *Les Prix Nobel: Nobel Prizes, Presentations, Biographies and Lectures*, Almqvist & Wiksell, Stockholm (1982).
9. A.J. Marcus and D.P. Hajjar, Vascular transcellular signaling, *J. Lipid Res.* 34:2017 (1993).
10. C.N. Serhan, Lipoxin biosynthesis and its impact in inflammatory and vascular events, *Biochim. Biophys. Acta* 1212:1 (1994).
11. S.P. Colgan, C.N. Serhan, C.A. Parkos, C. Delp-Archer, and J.L. Madara, Lipoxin A_4 modulates transmigration of human neutrophils across intestinal epithelial monolayers, *J. Clin. Invest.* 92:75 (1993).
12. A. Papayianni, C.N. Serhan, and H.R. Brady, Lipoxin A_4 and B_4 inhibit leukotriene-stimulated interactions of human neutrophils and endothelial cells, *J. Immunol.* 156:2264 (1996).
13. T.H. Lee, C.E. Horton, U. Kyan-Aung, D. Haskard, A.E. Crea, and B.W. Spur, Lipoxin A_4 and lipoxin B_4 inhibit chemotactic responses of human neutrophils stimulated by leukotriene B_4 and N-formyl-L-methionyl-L-leucyl-L-phenylalanine, *Clin. Sci.* 77:195 (1989).
14. O. Soyombo, B.W. Spur, and T.H. Lee, Effects of lipoxin A_4 on chemotaxis and degranulation of human eosinophils stimulated by platelet-activating factor and N-formyl-L-methionyl-L-leucyl-L-phenylalanine, *Allergy* 49:230 (1994).
15. J. Raud, U. Palmertz, S.E. Dahlén, and P. Hedqvist, Lipoxins inhibit microvascular inflammatory actions of leukotriene B_4, *Adv. Exp. Med. Biol.* 314:185 (1991).
16. T.M. Carlos and J.M. Harlan, Leukocyte-endothelial adhesion molecules, *Blood* 84:2068 (1994).
17. H.R. Brady and C.N. Serhan, Adhesion promotes transcellular leukotriene biosynthesis during neutrophil-glomerular endothelial cell interactions: inhibition by antibodies against CD18 and L-selection, *Biochem. Biophys. Res. Commun.* 186:1307 (1992).
18. H.-E. Claesson and J. Haeggström, Human endothelial cells stimulate leukotriene synthesis and convert granulocyte released leukotriene A_4 into leukotrienes B_4, C_4, D_4 and E_4, *Eur. J. Biochem.* 173:93 (1988).
19. D.A. Brezinski, R.W. Nesto, and C.N. Serhan, Angioplasty triggers intracoronary leukotrienes and lipoxin A_4. Impact of aspirin therapy, *Circulation* 86:56 (1992).
20. J. Clària, M.H. Lee, and C.N. Serhan, Aspirin-triggered lipoxins (15-epi-LX) are generated by the human lung adenocarcinoma cell line (A549)-neutrophil interactions and are potent inhibitors of cell proliferation, *Mol. Med.*, submitted.
21. K.C. Nicolaou, J.Y. Ramphal, N.A. Petasis, and C.N. Serhan, Lipoxins and related eicosanoids: biosynthesis, biological properties, and chemical synthesis, *Angew. Chem. Int. Ed. Engl.* 30:1100 (1991).
22. J. Clària and C.N. Serhan, Aspirin triggers previously undescribed bioactive eicosanoids by human endothelial cell-leukocyte interactions, *Proc. Natl. Acad. Sci. USA* 92:9475 (1995).
23. C.N. Serhan, J.F. Maddox, N.A. Petasis, I. Akritopoulou-Zanze, A. Papayianni, H.R. Brady, S.P. Colgan, and J.L. Madara, Design of lipoxin A_4 stable analogs that block transmigration and adhesion of human neutrophils, *Biochemistry* 34:14609 (1995).
24. A.J. Marcus, Aspirin as prophylaxis against colorectal cancer, *N. Engl. J. Med.* 333:656 (1995).
25. L.J. Marnett, Aspirin and the potential role of prostaglandins in colon cancer, *Cancer Res.* 52:5575 (1992).
26. L. Baud, J. Sraer, J. Perez, M.-P. Nivez, and R. Ardaillou, Leukotriene C_4 binds to human glomerular epithelial cells and promotes their proliferation in vitro, *J. Clin. Invest.* 76:374 (1985).
27. L. Stenke, P. Reizenstein, and J.Å. Lindgren, Leukotrienes and lipoxins--new potential performers in the regulation of human myelopoiesis, *Leukemia Res.* 18:727 (1994).
28. M.C. Alley, D.A. Scudiero, A. Monks, M.L. Hursey, M.J. Czerwinski, D.L. Fine, B.J. Abbott, J.G. Mayo, R.H. Shoemaker, and M.R. Boyd, Feasibility of drug screening with panels of human tumor cell lines using a microculture tetrazolium assay, *Cancer Res.* 48:589 (1988).
29. J.D. Croxtall and R.J. Flower, Lipocortin 1 mediates dexamethasone-induced growth arrest of the A549 lung adenocarcinoma cell line, *Proc. Natl. Acad. Sci. USA* 89:3571 (1992).

THE TARGETING OF LEUKOCYTES BY 5-OXO-EICOSANOIDS

Joseph T. O'Flaherty[1], Mitsuyuki Kuroki[1], Larry W. Daniel[1], Robert L. Wykle[1], Andrew B. Nixon[1], and Silvano Sozzani[2]

[1]Wake Forest University, Winston-Salem, NC 27157 and
[2]Istituto di Ricerche Farmacologiche, Milan, Italy

INTRODUCTION

The identities and roles of arachidonic acid (AA) metabolites in human diseases are often elusive. For example, allergens cause tissues to secrete agents that attract eosinophils (Eo) and induce these cells to release granule enzymes, superoxide anion, and bioactive molecules that lead to the organ dysfunctions seen in allergy[1]. Zileuten, an anti-lipoxygenase drug, relaxes the airways of asthmatics, perhaps by blocking the synthesis of an Eo-targeting eicosanoid[2-4]. The drug clearly inhibits allergen-induced Eo fluxes to lung as well as bronchospasm in primates and lower mammals[5,6]. Thus, an eicosatetraenoate (ETE) may link Eo to allergy. Among the eicosanoids occupying human allergen-reactive sites, 15(S)-hydroxy-ETE (15-HETE), prostaglandins, and thromboxanes are not chemotactic for Eo; leukotriene (LT)B_4 attracts Eo but is more active on polymorphonuclear neutrophils (PMN); and peptido-LTs, lipoxins, and 5(S)-hydroxy-ETE (5-HETE) act weakly on both cell types[1,7-10]. These agents are unlikely to mediate Eo-based lesions *in vivo*. However, a newly defined 5-HETE analog, 5-oxoETE, has *in vitro* actions suggesting that it could be such a mediator.

EARLY STUDIES WITH 5-HETE IN PMN

Shortly after the discovery that cells convert AA to LTB_4 and 5-HETE, LTB_4 was found to be a potent stimulator of PMN migration, degranulation, and superoxide anion production. 5-HETE proved 1,000-fold weaker than LTB_4 as a chemotactic factor (CF) and had little impact on exocytosis or superoxide[11,12]. Nonetheless, 5-HETE possessed actions that LTB_4 lacked: it increased the ability of platelet-activating factor (PAF), but not other CFs (C5a and *N*-formyl-MET-LEU-PHE [FMLP] as well as LTB_4) to induce PMN migration, degranulation, and superoxide production[12]. It also stimulated degranulation and superoxide-generation in PMN primed by cytokines such as granulocyte-macrophage colony-stimulating factor (GM-CSF)[13,14]. Interestingly, both PAF and GM-CSF are implicated in human allergy[4,10].

Frontiers in Bioactive Lipids
Edited by Vanderhoek, Plenum Press, New York, 1996

DISCOVERY OF 5-OxoETE AND 5-Oxo-15-Hydroxy-ETE

While probing the cellular processing of ETEs, Powell *et al.*[15] found porcine leukocytes converted LTB_4 and 12-hydroxy-ETE to 12-oxo counterparts by a NAD^+-dependent 12-dehydrogenase pathway that ultimately formed eicosatrienoates. Since the trienoate and 12-oxo products of these reactions are less potent on PMN than their parent compounds[16,17], the pathway seemingly serves to inactivate 12-hydroxy-ETEs. Human PMN differed from pig cells in that they convert LTs to 5-oxo, not 12-oxo, products *via* a microsomal $NADP^+$-dependent dehydrogenase which acts on 5(S)-hydroxy-ETEs with C 6-7 *trans* (*e.g.*, 5-HETE) rather than C 6-7 *cis* (*e.g.*, LTB_4) double bonds[18,19]. Among various ETEs and LTs, the enzyme prefers 5-HETE although it readily converts 5(S),15(S)-diHETE to 5-oxo-15-hydroxy-ETE[18]. Resting human PMN, however, mostly acylate 5-HETE into storage glycerolipid, convert it to 5,20-diHETE, and reduce 5-oxoETE to 5-HETE *via* a NADPH-dependent activity[18,20]. Hence, these cells accumulate little[18] or no[19,20] 5-oxoETE if incubated with 5-HETE. PMN treated with direct protein kinase C activators, contrastingly, amass 5-oxoETE in preference to 5-HETE, perhaps because of rises in $NADP^+$/NADPH ratios: the activators promote reactions that oxidize NADPH to $NADP^+$ and thus promote $NADP^+$-dependent, while dampening NADPH-dependent, enzyme activities[19]. The impact of physiologic stimuli on 5-oxoETE accumulation is mostly untested. In any case, human Eo and monocytes (Mono) also form 5-oxoETE from 5-HETE[21,22], and, as reviewed elsewhere[17], free radicals attack AA to yield 5-oxoETEs. Thus, 5-oxoETEs form in diverse enzymatic and non-enzymatic pathways. Indeed, a non-enzymatic reaction first directed researchers to 5-oxoETEs: Morita *et al.* treated Eo with AA to generate an Eo CF, which in later work examining the products of free radical-oxidized AA, proved to be identical to 5-oxo-15-hydroxy-ETE[23,24]. This ETE provided the initial clue indicating that 5-dehydrogenation may activate, rather than deactivate, 5-hydroxy-ETEs[24].

ACTIONS OF 5-OxoETEs ON HUMAN LEUKOCYTES

Polymorphonuclear Neutrophilic Leukocytes

5-OxoETE stimulates PMN chemotaxis and has intrinsic although weak degranulating activity. While failing to trigger superoxide formation, it nonetheless enhances the potency of PAF, but not other CFs, in stimulating PMN not only to produce superoxide anion but also to degranulate[12,17]. It also degranulates PMN that are primed by GM-CSF, granulocyte colony-stimulating factor, or tumor necrosis factor-α[13,14]. 5-OxoETE stimulates PMN to translocate cytosolic secretory vesicles to the cell surface (defined by tracking vesicular alkaline phosphatase)[14] and triggers PMN aggregation[17]. Since the last two responses are associated with the expression of adhesins, the oxoETE likely promotes PMN attachment to many cellular and matrix surfaces[14]. In all these respects, 5-oxoETE has the same action as, but is ~10-fold stronger than 5-HETE and ~3-fold stronger than 5-oxo-15-hydroxy-ETE[17,25,26]. Degranulation and superoxide responses to the ETEs (±PAF or a cytokine) require PMN first be treated with cytochalasin B. The chemotactic, secretory vesicle, aggregating, and other effects of the ETEs need no such treatment[17].

Monocytic Leukocytes

5-OxoETE and, with ~2-fold and 10-fold less potency, respectively, 5-oxo-15-hydroxy-ETE and 5-HETE induce Mono chemotaxis, actin polymerization, and expression of ICAM-1 and CD-18 adhesins. The chemotactic activity of the ETEs is synergistic with those of two cytokines, monocyte chemotactic proteins (MCP)-1 and -3, but not with FMLP[27].

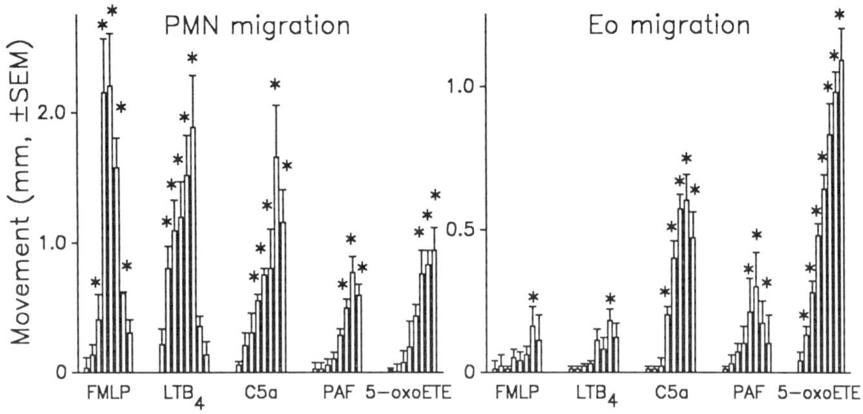

Figure 1. PMN (left panel) and Eo (right panel) responses to selected CFs. Data are given in sets of 8 histograms. The first histogram of each set gives responses to 1 pM; the next 7 histograms give response to 10-fold increasing amounts (to 10μM) of each stimulus. Results are in mm of cell movement toward stimuli (N>7, as assayed under agarose[29]). Stimulus vehicle caused Eo to move <0.1 mm, values that are included in results. Asterisks indicate responses significantly ($p<.05$, Students paired-t test) greater than those to stimulus vehicle.

Eosinophilic Leukocytes

5-OxoETE is chemotactic for Eo[21,28]. In our hands[29], it induces Eo to migrate at levels far lower than the most potent known CF, C5a, or weaker CFs like PAF, FMLP, LTB_4 (figure 1), RANTES, and MCP-1 (not shown). 5-OxoETE also causes Eo to move greater distances than do other CFs. While it has little degranulating activity, 5-oxoETE enhances the ability of PAF to degranulate Eo: in the presence of 500 nM 5-oxoETE, 10 pM PAF takes on the degranulating potency of 100 nM PAF (figure 2). 5-OxoETE likewise enhances Eo exocytosis responses to C5a, LTB_4, and, to a lesser extent, FMLP[29]. Finally, 5-oxoETE degranulates Eo primed by GM-CSF. Since GM-CSF shares a receptor system with the allergic cytokine, IL-5, 5-oxoETE also likely acts synergistically with IL-5 in Eo. 5-HETE is 1,000- to 10,000-fold and 5-oxo-15-hydroxy-ETE 30- to 300-fold weaker than 5-oxoETE in all of these assays[29].

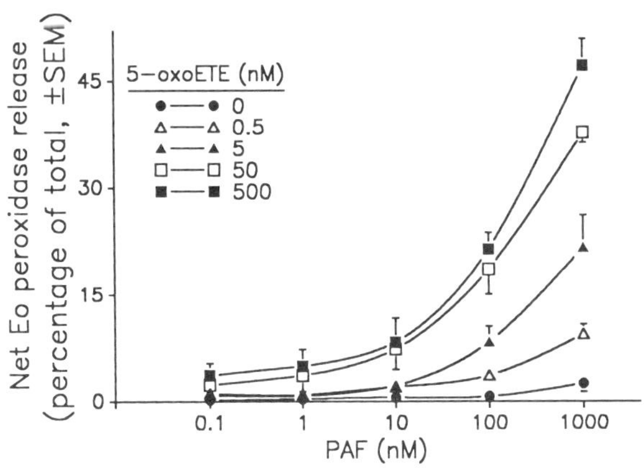

Figure 2. Effect of 5-oxoETE on Eo degranulation responses to PAF. Eo were concurrently challenged with PAF and 5-oxoETE in the presence of 5 μg/ml cytochalasin B for 15 min. Data (means of 6 experiments) are given as the percentage of total cell Eo peroxidase (a granule-bound enzyme) released minus that released by 5-oxoETE alone (this value was <9%).

MECHANISMS OF ACTION

5-Hydroxy-ETEs

Early studies indicated that 5-HETE stimulates PMN to mobilize storage Ca^{2+} and consequently raise cytosolic Ca^{2+} levels[12]. In evoking this response, it uses a down-regulatable pathway: 5-HETE desensitized PMN Ca^{2+} responses to itself but not to LTB_4 or other CFs. Moreover, analogs of 5-HETE such as 5,15-diHETE and 5,20-diHETE likewise stimulate PMN Ca^{2+} transients. These diHETEs cross-desensitized to 5-HETE while other ETEs are inactive (*e.g.* 15-HETE) or act (*e.g.* 12(R)- and 12(S)-HETE) *via* a pathway that cross-desensitizes to LTB_4 rather than 5-HETE[16,17]. Finally, 5(S)-HETE is 30-fold weaker than its unnatural isomer, 5(R)-HETE; its Ca^{2+}-mobilizing actions are pertussis toxin-sensitive; and it stimulates GTPase and GTP binding activity in PMN plasma membranes[12,13]. Further analyses, *e.g.* [^{3}H]5-HETE binding studies, are hampered by the ability of PMN to acylate 5-HETE into storage lipids at 4 or 37°C[13]. Nonetheless, the data imply 5-HETE and its analogs operate on PMN *via* a G protein-linked, stereospecific receptor system. This system does not involve receptors for LTB_4, prostaglandins E_2 and D_2, or various CFs[12,13,17,30].

5-OxoETEs

5-OxoETE, 5-oxo-15-hydroxy-ETE, 5-HETE, 5-hydroxy-15-oxoETE, 5,15-diHETE, and/or 5,20-diHETE show decreasing potency on PMN[17,26], Mono[27], and Eo[21,26,29]. These ETEs desensitize PMN, Mono, and Eo to themselves and each other but not to various CFs[26-29]. Moreover, 5-oxoETE, similar to 5-HETE, operates not only on PMN but also Mono and Eo by a pertussis toxin-sensitive but LTB_4 receptor antagonist-resistant pathway[13,14,27,28]. Finally, 15-oxoETE neither stimulates these leukocytes nor cross-desensitizes to 5-oxoETE[17,27-29]. We conclude that 5-oxoETEs are especially potent members of a 5-HETE family of (presumptive) receptor agonists.

Leukocyte Targeting

Two related points merit emphasis. First, 5-HETE-like agonists are most effective on Eo, showing less potency and a more limited range of influences on Mono and PMN[27,29]. Second, 5-oxoETE is uniquely selective for Eo: relative Eo chemotactic potencies for 5-oxoETE, 5-oxo-15-hydroxy-ETE, and 5-HETE are ~1000, 10, and 1, respectively, but ~10, 3, and 1 for PMN and Mono. Thus, among the various ETEs tested, 5-oxoETE stands apart in provoking Eo responses at low picomolar levels[29]. This may indicate that putative receptors for 5-HETE differ in agonist specificity and therefore structure among the three cell types, or, alternatively, that these leukocytes types differ in ability to metabolize and thus inactivate the ETEs.

POST-RECEPTOR SIGNAL TRANSDUCTION PATHWAYS

5-OxoETE along with its analogs stimulate cells to activate mitogen-activated protein kinases (MAPK). PMN express 44, 42, and 40 kDa proteins that react with various antibodies to MAPK plus a 41 kDa protein that reacts with two antibodies to p38 Mpk. All three MAPK antigens undergo an increase in phosphotyrosine content that is accompanied by a slowing in their electrophoretic mobility in PMN challenged by 5-oxoETE, 5-oxo-15-hydroxy-ETE, or 5-HETE (relative potencies: 10, 3, and 1, respectively). There is no comparable change in p38 Mpk antigen[14]. 5-OxoETE-treated PMN have increased activity that phosphorylates myelin basic protein, as defined by *in vitro* gel kinase assays. This activity migrates at 40-44 kDa. The effect of 5-HETE on MAPK is pertussis toxin-sensitive but independent of cell Ca^{2+} or LTB_4 receptors. 5-OxoETE, 5-oxo-15-hydroxy-ETE, and 5-HETE (relative potencies: ~10, 3, and 1, respectively) also induce PMN to phosphorylate and activate an immediate down-stream target of MAPK, cytosolic phospholipase A_2 ($cPLA_2$), as defined by $cPLA_2$ electrophoretic mobility retardation and arachadonoyl-glycerophosphocholine deacylation assays[34]. While 5-oxoETE does not by itself trigger AA release, it induces large amount of AA to be released in cells primed by GM-CSF. GM-CSF likewise primes MAPK responses to 5-oxoETE[14]. 5-OxoETE also works synergistically with CFs to promote PMN release of AA[31]. These results relate to Eo and Mono. Eo express 44 and 42 but not 40 kDa MAPK antigens. Both proteins are tyrosine phosphorylated and gel-shifted in Eo exposed to ≥ 1 pM 5-oxoETE, 1 nM 5-oxo-15-hydroxy-ETE, or 10 nM 5-HETE[29]. Moreover, 5-oxoETE, while having little intrinsic impact on AA release in Mono, greatly increases the amount of AA release in Mono stimulated by MCP-1[27]. We accordingly suggest that putative 5-HETE receptors couple to the activation of the MAPK-$cPLA_2$ axis of cell stimulation in not only PMN but also Mono and Eo. This coupling may, in turn, lead to a variety of cellular responses relevant to allergy.

CONCLUSIONS AND FUTURE DIRECTIONS

Eo-based allergic reactions in humans involve a series of mediators: CFs, principally PAF and C5a; the eosinophilatropic cytokines, IL-5 and GM-CSF; and possibly an unspecified AA metabolite whose production is initiated 5-lipoxygenase. Relevant to the last agent, work from three laboratories points to oxo analogs of 5-HETE. 5-OxoETE, in particular, is the strongest known Eo CF. It also distinguishes itself from most other CFs in its extraordinarily high selectivity for Eo over PMN and Mono. The oxoETE has potent cooperative interactions with PAF, C5a, and at least one eosinophilatrophic cytokine, GM-CSF, in causing exocytosis, MAPK activation, AA release, and, likely, other Eo responses. By mobilizing $cPLA_2$, the oxoETE may induce Eo to make PAF ($cPLA_2$ may be the initial rate-limiting enzyme in the deacylation-acetylation pathway of PAF synthesis). By mobilizing MAPK, 5-oxoETE may activate MAPK-dependent pathways that regulate the transcription and translation of various proteins including eosinophilatrophic cytokines. The data thus allow that 5-oxoETE is a central mediator in a network of signals that regulates the pro-inflammatory capacities of Eo and thereby directs the development of allergic reactions. In spite of its intriguing *in vivo* actions, however, existing studies fall short of implicating 5-oxoETEs in allergy. Future studies must define the *in vivo* abilities of these compounds to stimulate Eo and impact other events common to allergy like bronchospasm, mucus secretion (5-oxoETE induces [ED_{50} ~40 pM] K^+ and Cl^- secretion in guinea pig intestinal epithelial cells[32]), capillary leakage, and collagen formation (5-HETE stimulates [100 pM] osteoclasts[33]); seek their presence in allergic reactions; and test their ability to promote, and form in, non-allergic reactions (5-oxoETE stimulates [ED_{50} ~2 nM] PMN-vascular endothelium attachment reactions[32]). Development of antagonists for the 5-HETE agonist family would help immensely in defining the functions of 5-oxoETEs. Finally, we need molecular as well as binding analyses of putative 5-HETE receptors. Clearly, these directions offer fascinating opportunities to explore factors mediating human disease.

REFERENCES

1. M.B. Resnick and P.F. Weller, Mechanisms of eosinophil recruitment, *Am J Respir Cell Mol Biol.* 8:349 (1993).
2. K.P. Hui, I.K. Taylor, G.W. Taylor, P. Rubin, J. Kesterson, N.C. Barnes, and P.J. Barnes, Effect of a 5-lipoxygenase inhibitor on leukotriene generation and airway responses after allergen challenge in asthmatic patients, *Thorax.* 46:184 (1991).

3. E. Israel, P. Rubin, J.P. Kemp, J. Grossman, W. Pierson, S.C. Siegel, D. Tinkelman, J.J. Murray, W. Busee, A.T. Segal, J. Fish, H.B. Kaiser, D. Ledford, S. Wenzel, R. Rosenthal, J. Cohn, C. Lanni, H. Pearlman, P. Karahalios, and J.M. Drazen, The effect of inhibition of 5-lipoxygenase by zileuton in mild-to-moderate asthma, *Ann Intern Med.* 119:1059 (1993).
4. E. Israel, Moderating the inflammation of asthma: inhibiting the production or action of products of the 5-lipoxygenase pathway, *Ann Allergy.* 72:279 (1994).
5. W. A. Abraham, A. Ahmed, A. Cortes, M.W. Sielczak, W. Hinz, J. Bouska, C. Lanni, and R.L. Bell, The 5-lipoxygenase inhibitor zileuton blocks antigen-induced late airway responses, inflammation and airway hyperresponsiveness in allergic sheep, *Eur J Pharmacol.* 217:119 (1992).
6. C.D. Wegner, R.H. Gundel, W.M. Abraham, E.S. Schulman, M.J. Kontny, E.S. Lazer, C.A. Homon, A.G. Graham, C.A. Torcellini, C.C. Clarke, P. Jager, W.W. Wolyniec, L.G. Letts, and P.R. Farina, The role of 5-lipoxygenase products in preclinical models of asthma, *J Allergy Clin Immunol.* 91:917 (1993).
7. E. Morita, J-M. Schroder, and E. Christophers, Differential sensitivities of purified human eosinophils and neutrophils to defined chemotaxins, *Scand J Immunol.* 29:709 (1989).
8. C.S. Spada, A.L. Nieves, A.H-P. Krauss, and D.F. Woodward, Comparison of leukotriene B_4 and D_4 effects on human eosinophil and neutrophil motility in vitro, *J Leuk Biol.* 55:183 (1994).
9. S. O, B.W. Spur, and T.H. Lee, Effects of lipoxin A_4 on chemotaxis and degranulation of human eosinophils stimulated by platelet-activating factor and *N*-formyl-L methionyl-L-leucyl-L-phenylalanine, *Allergy.* 49:230 (1994).
10. P.L.B. Bruijnzeel, P.H.M. Kuijper, S. Rihs, S. Betz, R.A.J. Warringa, and L. Koenderman, Eosinophil migration in atopic dermatitis I: Increased *migratory* responses to *N*-formyl-methionyl-leucyl-phenylalanine, neutrophil-activated factor, platelet-activating factor, and platelet factor 4, *J Invest Derm.* 100:137 (1993).
11. L. Nagy, T.H. Lee, J. Goetzl, W.C. Pickett, and A.B. Kay, Complement receptor enhancement and chemotaxis of human neutrophils and eosinophils by leukotrienes and other lipoxygenase products, *Clin Exp Immunol.* 47:541 (1982).
12. A.G. Rossi and J.T. O'Flaherty, Bioactions of 5-hydroxyicosatetraenoate and its interaction with platelet activacting factor, *Lipids.* 26:1184 (1991).
13. J.T. O'Flaherty and A.G. Rossi, 5-Hydroxyicosatetraenoate stimulates neutrophils by a stereospecific, G protein-lined mechanism, *J Biol Chem.* 268:14708 (1993).
14. J.T. O'Flaherty, M. Kuroki, J. Wijkander, E. Yee, S.L. Lee, P.K. Smitherman, R.L. Wykle, and L.W. Daniel, 5-Oxo-eocisanoids and hematopoietic cytokines cooperate in stimulating neutrophil function and the MAP kinase pathway, *J Biol Chem.* In press.
15. S.L. Wainwright and W.S. Powell, Mechanism for the formation of dihydro metabolites of 12-hydroxy-eicosanoids, *J Biol Chem.* 266:20899 (1991).
16. W.S. Powell, M. Hashefi, J.R. Falck, K. Chauhan, J. Rokach, S.S. Wang, E. Mills, and R.J. MacLeod, Effects of oxo and dihydro metabolites of 12-hydroxy-5,8,10,14-eico*satetraeno*ic acid on chemotaxis and cytosolic calcium levels in human neutrophils, *J Leuk Biol.* 57:257 (1995).
17. J.T. O'Flaherty, J.F. Cordes, S.L. Lee, M. Samuel, and M.J. Thomas, Chemical and biological characterization of oxo-eicosatetraenoic acids, *Biochim Biophys Acta.* 1201:505 (1994).
18. W.S. Powell, F. Gravelle, and S. Gravel, Metabolism of 5(*S*)-hydroxy-6,8,11,14-eicosatetraenoic acid and other 5(*S*)-hydroxyeicosanoids by a specific dehydrogenase in human polymorphonuclear leukocytes, *J Biol Chem.* 267:19233 (1992).
19. W.S. Powell, F. Gravelle, and S. Gravel, Phorbol myristate acetate stimulates the formation of 5-oxo-6,8,11,14-eicosatetraenoic acid by human neutrophils by activating NADPH oxidase, *J Biol Chem.* 269:25373 (1994).
20. J.T. O'Flaherty, R.L. Wykle, J. Redman, M. Samuel, and M. Thomas, Metabolism of 5-hydroxy-eicosatetra-enoate by human neutrophils: Production of a novel ω-oxidized derivative, *J Immunol.* 137:3277 (1986).
21. W.S. Powell, D. Chung, and S. Gravel, 5-Oxo-6,8,11,14-eicosatetraenoic acid is a potent stimulator of human eosinophil migration, *J Immunol.* 154:4123 (1995).
22. Y. Zhang, and W.S. Powell, synthesis of 5-oxo-6,8,11,14-eicosatetraenoic acid by human monocytes, *Proc Fed Assoc Soc Exp Biol,* 8:A356 (1994).
23. E. Morita, J-M. Schroder, and E. Christophers, Identification of a novel and highly potent eosinophil chemotactic lipid in human eosinophils treated with arachidonic acid, *J Immunol.* 144:1893 (1990).
24. U. Schwenk, E. Morita, R. Engle, and J-M. Schroder, Identification of 5-oxo-15-hydroxy-6,8,11,13-eicosatetraenoic acid as a novel and potent human eosinophil chemotactic eicosanoid, *J Biol Chem.* 267:12482 (1992).
25. J.T. O'Flaherty, J. Cordes, J. Redman, and M.J. Thomas, 5-Oxo-eicosatetraenoate, a potent human neutrophil stimulus, *Biochem Biophys Res Comm.* 192:129 (1993).
26. W.S. Powell, R.J. MacLeod, S. Gravel, F. Gravelle, and A. Bhaker, Metabolism and biological effects of 5-

oxo-eicosanoids on human neutrophils, *J Immunol.* 156:336 (1996).
27. S. Sozzani, D. Zhou, M. Locati, S. Bernasconi, W. Luini, A. Mantovani, and J.T. O'Flaherty, Stimulating properties of 5-oxo-eicosanoids for human monocytes. Synergism with monocyte chemotactic protein-1 and -3, *Submitted.* (1996).
28. U. Schwenk, and J-M. Schroder, 5-Oxo-eicosanoids are potent eosinophils chemotactic factors, *J Biol Chem.* 270:15029 (1995).
29. J.T. O'Flaherty, M. Kuroki, A.B. Nixon, J. Wijkander, E. Yee, S.L. Lee, P.K. Smitherman, R.L. Wykle, and L.W. Daniel, 5-Oxo-eicosatetraenoate is a broadly active, eosinophil-selective stimulus for human granulocytes, *J Immunol.* In press.
30. A.G. Rossi and J.T. O'Flaherty, Prostaglandin binding sites in human polymorphonuclear neutrophils, *Prostaglandins.* 37:641 (1989).
31. J. Wijkander, J.T. O'Flaherty, A.B. Nixon, and R.L. Wykle, 5-Lipoxygenase products modulate the activity of the 85-kDa phospholipase A_2 in human neutrophils, *J Biol Chem.* 270:26543 (1995).
32. W.S. Powell, S. Gravel, P. Stamatiou, and R.J. Macleod, Effects of 5-oxoETE on neutrophils, eosinophils, and intestinal epithelial cells, *Adv Prostaglandin Thromboxane Leukotriene Res* 23:431 (1995).
33. W.E. Gallwitz, G.R. Mundy, C.H. Lee, M. Quiao, G.D. Roodman, M. Rafery, S.J. Gaskell and L. F. Bonewald, 5-Lipoxygenase metabolites of arachidonic acid stimulate isolated osteoclasts to resorb calcified matrices, *J Biol Chem.* 268:10087 (1993).

THE ROLE OF ANANDAMIDE AND RELATED FATTY ACID ETHANOLAMIDES AS ENDOGENOUS LIGANDS FOR THE CB1 AND CB2 CANNABINOID RECEPTORS

Christian C. Felder[1], Amie Nielsen[1], Eileen M. Briley[1], M. Palkovits[1], J. M. Richardson[2], R. M. Riggin[2], G. W. Becker[2]

[1]The Laboratory of Cell Biology, National Institute of Mental Health, Bethesda, MD 20892
[2]Lilly Research Labs, Eli Lilly and Co., Indianapolis, IN 46285

Introduction

Fatty acids are best known for their role as structural components of lipid membranes but they also can act as messenger molecules with targets located both inside and outside the cell. In particular, the polyunsaturated fatty acids, such as arachidonic acid, can modulate a variety of cellular targets such as protein kinase C, ion channels, and possibly gene regulation. Arachidonic acid can be further metabolized to over 50 eicosanoids, some for which specific receptors have been identified. A novel arachidonic acid containing molecule, anandamide, was identified during a search for an endogenous agonist for the CB1 cannabinoid receptor and its structure was found to be the condensation of ethanolamine and arachidonic acid linked through an amide bond. Anandamide was further shown to be an agonist for both CB1 and CB2 cannabinoid receptors expressed in mammalian fibroblasts and to functionally activate the inhibition of adenylate cyclase. Other potential cannabinoid receptor agonists have been identified including ethanolamides of mead acid, dihomo-γ-linolenic acid, and adrenic acid. This manuscript will serve as an overview of the cannabinoid receptors system and include discussion of potential endogenous agonists for this family of receptors. Although little is known about the physiologic significance of the anandaergic nervous system, some speculation will be provided based on the current literature.

Cannabinoid Receptors

The active principle in marijuana, Δ-9-tetrahydrocannabinol (Δ-9-THC), has been used therapeutically and recreationally for centuries, yet only recently has a receptor for this plant product been positively identified. The first Δ-9-THC receptor cloned, the CB1 cannabinoid receptor (Matsuda et al., 1990), was found widely distributed in mammalian brain (Herkenham et al., 1990) and less so in the periphery with lower amounts found in testes and ovary (Galiegue et al., 1995). The CB1 receptor is a member of the superfamily of G protein coupled receptors characterized by a single polypeptide whose hydrophobic domains span the plasma membrane seven times. Using the selective cannabinoid agonist, 3H-CP55,940, high levels of cannabinoid receptor binding were found localized to specific regions of human and rat brain with highest levels found in cerebellum, hippocampus, basal ganglia, and cortex (Herkenham et al., 1990). Binding was sparse or absent from areas of brain stem, and medulla, and thalamus. A second cannabinoid receptor, CB2, was identified in the human promyelocytic leukemic cell line, HL60, and was found to have a 44% overall sequence similarity to the CB1 receptor which rose to 68% identity in the regions thought to define the ligand binding pocket (Munro et al., 1993) The CB2 receptor has been localized almost exclusively to the immune system with highest levels found in areas of spleen and lymph nodes rich in B cells and macrophages (Munro et al., 1993; Lynn and Herkenham, 1994). A splice variant of the CB1 receptor, CB1A, was subsequently identified in human and rat and found to be expressed at lower levels than CB1, but to have similar distribution and pharmacological properties (Shire et al., 1995). The behavioral and therapeutic effects of cannabinoids such as antiemisis, hypothermia, analgesia, immunosuppression, ataxia, sedation, antinociception, and antiglaucoma are thought to be mediated by one or both of these cannabinoid receptor subtypes.

Signal Transduction of CB1 and CB2 Receptors

Cannabinoid agonists were shown to functionally inhibit adenylate cyclase activity in rat brain preparations long before the cloning of the CB1 receptor providing the first evidence of its membership in the superfamily of G protein coupled receptors (Howlett and Fleming, 1984). Transfection and stable expression of single CB1 and CB2 receptor subtype cDNA in mammalian cell lines not expressing endogenous cannabinoid receptors provided the opportunity to study the mechanism of receptor subtype specific signal transduction. It was found that application of cannabinoid agonist to CB1 receptor expressing cells attenuated adenylate cyclase through a pertussis toxin sensitive G protein, probably Gi (Felder et al., 1992). This effect was not observed in control cells, indicating that it was CB1 receptor mediated. Additionally it was shown that cannabinoid agonists stimulated arachidonic acid release and inhibited arachidonic acid uptake equally in both control and CB1 receptor expressing cells at concentrations of cannabinoid agonist several fold higher than their reported Ki's. Similar non-CB1 receptor-mediated release of cytoplasmic calcium stores was also observed for several cannabinoid agonists. Therefore, both cannabinoid receptor-dependent signaling and cannabinoid receptor-independent modulation of cellular processes are possible consequences of cannabinoid agonist application.

CB1 receptors also modulate ion channel function. Cannabinoid agonists inhibit N-type voltage-dependent calcium channels in both NG108-15 (Caufield and Brown, 1992) and N-18 neuroblastoma cells (Mackie et al., 1993) which express endogenous cannabinoid receptors. When expressed in AtT-20 cells, the CB1 receptor couples to the inhibition of Q-type voltage-dependent calcium channels (Mackie et al., 1995). In all three of these cell lines, the inhibition is pertussis toxin sensitive suggesting Gi or Go involvement. Cannabinoid agonists also activate an inwardly rectifying potassium current in CB1 transfected AtT-20 cells (Mackie et al., 1995) through a pertussis toxin sensitive mechanism.

The CB2 receptor exhibits similar inhibition of adenylate cyclase in splenocytes (Kaminski et al., 1994) where it is naturally expressed or in transfected fibroblasts (Felder et al., 1995, Baywhich et al., 1995). Unlike the CB1 receptor, however, the CB2 receptor does not couple to ion channel modulation when expressed in AtT-20 cells, but further efforts will be required to determine if this is true for endogenously expressed CB2 receptors (Felder et al., 1995).

Discovery of Anandamide and Fatty Acid Ethanolamides

The discovery of the CB1 receptor initiated a search for an endogenous agonist for this receptor. Taking advantage of the knowledge that cannabinoid receptor agonists share a common property of hydrophobicity, a molecule was isolated from porcine brain lipid extracts that was identified as arachidonylethanolamide by mass spectrometry (Devane et al., 1992). Given the common name anandamide, derived from the ancient Indian Sanskrit word for inner bliss and tranquility, this novel eicosanoid displayed specific binding and functional properties of a cannabinoid receptor agonist. Anandamide was also isolated from calf brain during a search for natural molecules that might bind to and modulate the L-type voltage-sensitive calcium channel indicating it may have several biological activities (Johnson et al., 1993). Cannabinoid agonist activity of anandamide was verified in binding and functional studies using cell lines transfected with CB1 and CB2 receptors (Felder et al., 1993; Felder et al., 1995). Anandamide had similar affinities for the CB1 and CB2 receptors and stimulated identical signaling pathways as those observed for classical cannabinoid agonists. Cannabimimetic behavioral effects, such as hypothermia, ataxia, and sedation, further supported its role as a cannabinoid receptor agonist (Fride and Mechoulam, 1993; Crawley et al., 1993). Structure activity studies of anandamide revealed several potential fatty acid ethanolamide agonists with essentially equal affinity for the CB1 (Felder et al, 1993; Priller et al., 1995; Pinto et al., 1994, Adams et al., 1995) and CB2 receptor (Felder et al., 1995), but only two, docosatetraenylethanolamide, and dihomo-γ-linolenylethanolamide have been isolated but not quantitated in mammalian brain (Hanus et al., 1993). In general, however, the pharmacological affinity and potency of anandamide and related fatty acid ethanolamides are quite low compared to most neurotransmitters and their receptors. Studies in neurons in primary culture demonstrated the release of a family of fatty acid ethanolamides through a depolarization- and calcium influx-dependent mechanism (Di Marzo et al., 1994). Therefore, a number of possible fatty acid ethanolamides may be generated, each with its own potential biological activity.

Early studies on the biological effects of fatty acid ethanolamides focused on N-palmitoylethanolamine due to its antiinflammatory properties (Coburn et al., 1954; Kuehl et al., 1957). N-palmitoylethanolamine was first isolated from mammalian tissue and its enzymatic synthesis from ethanolamine and palmitic acid shown in rat and guinea pig brain liver and muscle (Bacher et al., 1965). Extensive research on N-acyl phospholipids and their metabolism, and biological activity in ischemic tissue occurred in the 1980's (for a review see Schmid et al., 1990). Of the fatty acid ethanolamides isolated, anandamide is the only one for which a receptor and signaling has been identified. The discovery of anandamide has refocused efforts in this area of lipid research.

Biosynthesis of Fatty Acid Ethanolamides

Two alternative pathways have been proposed for the formation of anandamide and presumably other fatty acid ethanolamide species. The first requires the release of arachidonic acid and ethanolamine by the activation of phospholipase A2 and phospholipase D respectively, followed by their condensation to form anandamide, a reaction catalyzed by anandamide synthase (Kruska and Gross, 1994; Devane and Axelrod, 1995). Anandamide synthase activity was independent of addition of ATP or CoA, was optimum at an alkaline pH, had a preference for arachidonic acid and structurally closely related fatty acids, was inhibited by serine protease inhibitors such as PMSF, and was found in brain plasma membranes, synaptosomes, and microsomes. It is a bit troubling however that the Km's for ethanolamine and arachidonic acid were quite high compared to levels thought to be present in intracellular compartments. A recent study demonstrated strong similarity in the characteristics of the putative anandamide synthase with a degradative anandamide amidase activity suggesting that the synthase was actually the amidase working in reverse (Ueda et al., 1995).

The second proposed pathway requires the presence of a storage form of anandamide in which arachidonic acid is found in an amide linkage to ethanolamine in the third and polar head position of phosphatidylethanolamine (N-arachidonylphosphatidylethanolamine, NAPE). This N-acylated phospholipid belongs to a larger family of N-acylated phospholipids first discovered in wheat flour (Bornstein, 1965) and later described as products of ischemia in heart and brain (Schmid et al., 1990). However, the arachidonic acid containing NAPE, N-arachidonylPE, was never isolated in these studies. The presence of N-arachidonylPE in neurons in culture was suggested though not directly demonstrated (Di Marzo et al., 1994). This study did however, demonstrate the degradation of synthetic 3H-NarachidonylPE in neuronal membranes and the formation of anandamide following exogenous addition of phospholipase D which provides tantalizing preliminary evidence that the second pathway may exist in brain. Studies in rat testis identified the presence of N-arachidonylPE in quantities 46 times higher than levels of anandamide suggesting that sufficient mass of the substrate occurs to generate anandamide (Sugiura et al., 1996). It is not clear if sufficient N-arachidonylPE occurs naturally in brain to serve as substrate for anandamide release or if this molecule serves as an intermediate acceptor of fatty acid from other phospholipid species as previously suggested (Schmid et al., 1990).

Metabolism of Fatty Acid Ethanolamides

If one considers the role of anandamide to be as a neurotransmitter or messenger molecule, then removal of anandamide through metabolism and or transport would provide the off signal analogous to the removal of most neurotransmitters. Anandamide degradation into arachidonic acid and ethanolamine appears to be the most active route of its removal, having been demonstrated in a variety of tissues including neuroblastoma and glioma cells (Deutsch and Chin, 1993), brain regions rich in CB1 receptors and in liver (Desarnaud et al., 1995), porcine brain microsomal fractions (Ueda et al., 1995), and in bovine brain synaptosomal and microsomal fractions (Devane and Axelrod, 1994). Anandamide is taken up into neurons through a saturable mechanism where it is rapidly degraded to arachidonic acid and ethanolamine (Di Marzo et al., 1994) indicating that transport across the plasma membrane is first required before degradation occurs. Since most synthesized anandamide appears to remain within the neuron, it is not clear how it may act as an agonist to plasma membrane resident receptors. Anandamide can also be metabolized by the same enzymes that metabolize arachidonic acid to a number of potentially bioactive eicosanoids in mammalian tissue (Ueda, et al., 1995b; Hampson et al., 1996). It is not clear if ethanolamine containing eicosanoids have any physiologic significance.

Distribution of Anandamide in Brain and Peripheral Tissues

Levels of anandamide were first estimated to occur at 0.4 pmol/gm (133 pg/gm) in whole porcine brain where it was first isolated (Devane et al., 1992), and recently quantitated in porcine and bovine brain at 173 pmol/gm (60 ng/gm) and 101 pmol/gm (35 ng/gm) respectively (Schmid et al., 1995). A recent study reports levels of anandamide in rat testis to be considerably lower (6 pmol/gm, 2.1 ng/gm)(Sugiura et al., 1996). However, anandamide has never been isolated from human tissue or fluids. Furthermore, levels of anandamide have not been measured in regions of rat brain or in tissues such as spleen where CB2 receptors have been shown to be expressed at high levels. We have recently isolated and quantitated anandamide by liquid chromatography/mass spectrometry in various tissues and fluids from postmortem human and rat. Selected tissues were extracted with toluene, semipurified against authentic anandamide standard by thin layer chromatography, and analyzed by LC/MS/MS. Anandamide was found in all regions of human brain tested with highest levels (108 pmol/gm tissue wet weight) occurring in hippocampus, followed by cerebellum, striatum, and parahippocampal cortex, regions expressing high levels of CB1 receptor. Thalamus, where CB1 receptor levels are sparse, contained levels of anandamide similar to those observed in striatum. Corresponding areas of rat brain showed relatively little differences with an average value of 22 ± 4.8 pmol/gm tissue wet weight. Human and rat spleen contained anandamide (human:14 ± 4 pmol/gm, rat: 6 ± 0.8 pmol/gm tissue wet weight) suggesting that it may act as an endogenous agonist in the immune system where CB2 receptors are expressed in high levels. Only trace quantities of anandamide were detected in human serum, plasma, or cerebral spinal fluid suggesting that its role is not hormonal in nature. Small amounts of anandamide were detected in human heart and rat skin, areas expressing little if any CB1 or CB2 receptors. Levels of anandamide measured were collected from rat within minutes postmortem and within 2 hours postmortem for human specimens reducing the possibility that levels were an artifact of postmortem delay (Schmid et al., 1995).

Conclusion

Whether anandamide is a putative neurotransmitter acting on plasma membrane receptors within the same cell in which is formed, is a transcellular or transynaptic messenger diffusing readily out of cells due to its hydrophobic nature to cross the extracellular space, or is an intracellular messenger leading to such actions as mobilization of calcium, release of arachidonic acid, or regulation of gene expression remains to be determined. It may function in one or all of these capacities depending on the cell type in which it is formed. The apparent mismatch between anandamide levels and cannabinoid receptor distribution may have several possible explanations. It is possible that more cannabinoid receptor subtypes may have yet to be identified or that anandamide has additional physiological roles within the brain and periphery. A recent report demonstrated an inhibitory effect of anandamide on gap junctions in glia that was independent of cannabinoid receptor activation (Venance et al., 1995). The levels of anandamide measured do not indicate the state of activation of producing cells and additional studies will be required in cell culture models to determine regulation of formation and potential physiological significance of this novel eicosanoid.

Adams, I. B., Ryan, W., Singer, M., Thomas, B. F., Compton, D. R., Razdan, R. K., Martin, B. R., 1995, Evaluation of cannabinoid receptor binding and in vivo activities for anandamide analogs. J. Pharmacol. Exp. Ther. 273:1172-1181.

Bachur, N. R., Masek, K., Melmon, K. L., Udenfriend, S., 1965, Fatty acid amides of ethanolamine in mammalian tissues. J. Biol. Chem. 240:1019.

Bayewitch, M., Avidor-Reiss, T., Levy, R., Barg, J., Mechoulam, R., Vogel, Z., 1995, The peripheral cannabinoid receptor: adenylate cyclase inhibition and G protein coupling. FEBS Lett. 375:143-147.

Bomstein, R. A., 1965, A new class of phosphatides islolated from wheat flour. Biochem. Biophys. Res. Comm. 21: 49-54.

Caulfield, M. P., Brown, D. A., 1992, Cannabinoid receptor agonists inhibit Ca^{2+} current in NG108-15 neuroblastoma cells via a pertussis toxin-sensitive mechanism. Br. J. Pharmacol. 106:231-232.

Coburn, A. F., Graham, C. E., Haninger, J., 1954, The effect of egg yolk in diets on anaphylactic arthritis (passive arthus phenomema) in guinea pig. J. Exp. Med. 190: 425.

Crawley, J. N., Corwin, R. L., Robinson, J. K., Felder, C. C., Devane, W. A., Axelrod, J., 1993, Anandamide, an endogenous ligand of the cannabinoid receptor, induces hypomotility and hypothermia in vivo in rodents. Pharmacol. Biochem. Behav. 46:967-972.

Desarnaud, F., Cadas, H., Piomelli, D., 1995, Anandamide amidohydrolase activity in rat brain microsomes. Identification and partial characterization. J. Biol. Chem. 270:6030-6035.

Devane, W. A., Axelrod, J., 1994, Enzymatic synthesis of anandamide, an endogenous ligand for the cannabinoid receptor, by brain membranes. Proc. Natl. Acad. Sci. USA 91:6698-6701.

Devane, W. A., Hanus, L., Breuer, A., Pertwee, R. G., Stevenson, L. A., Griffin, G., Gibson, D., Mandelbaum, A., Etinger, A., Mechoulam, R., 1993, Isolation and structure of a brain constituent that binds to the cannabinoid receptor. Science 258:1946-1949.

Di Marzo, V., Fontana, A., Cadas, H., Schinelli, S., Cimino, G., Schwartz, J. C., Piomelli, D., Formation and inactivation of endogenous cannabinoid anandamide in central neurons., 1994, Nature 372:686-691.

Felder, C. C., Veluz, J. S., Williams, H. L., Briley, E. M., Matsuda, L. A., 1992, Cannabinoid agonists stimulate both receptor- and non-receptor-mediated signal transduction pathways in cells transfected with and expressing cannabinoid receptor clones. Mol. Pharmacol. 42:838-845.

Felder, C. C., Briley, E. M., Axelrod, J., Simpson, J. T., Mackie, K., Devane, W. A., 1993, Anandamide, an endogenous cannabinoid agonist, couples to receptor mediated signal transduction. Proc. Natl. Acad. Sci. USA 90:7656-7660.

Felder, C. C., Joyce, K. E., Briley, E. M., Mansouri, J., Mackie, K., Lai, Y., Ma, A. L., 1995, Comparison of the pharmacology and signal transduction of the human cannabinoid CB1 and CB2 receptors. Mol. Pharmacol. 48:443-450.

Galiegue, S., Mary, S., Marchand, J., Dussossoy, D., Carriere, D., Carayon, P., Bouaboula, M., Shire, D., Le Fur, G., Casellas, P., 1995, Expression of central and peripheral cannabinoid receptors in human immune tissues and leukocyte subpopulations. Eur. J. Biochem. 232:54-61.

Hampson, A. J., Hill, W. A., Zan-Phillips, M., Makriyannis, A., Leung, E., Eglen, R. M., Bornheim, L. M., 1995, Anandamide hydroxylation by brain lipoxygenase metabolite structures and potencies at the cannabinoid receptor. Biochim. Biophys. Acta. 1259:173-179.

Hanus, L., Gopher, A., Almog, S., Mechoulam, R., 1993, Two new unsaturated fatty acid ethanolamides in brain that bind to the cannabinoid receptor. J. Med. Chem. 36:3032-3034.

Herkenham, M., Lynn, A. B., Little, M. D., Johnson, M. R., Melvin, L. S., de Costa, B. R., Rice, K. C., 1990, Cannabinoid receptor localization in brain. Proc. Natl. Acad. Sci. USA 87:1932-1936.

Howlett, A. C., Fleming, R. M., 1984, Cannabinoid inhibition of adenylate cyclase. Pharmacology of the response in neuroblastoma cell membranes. Mol. Pharmacol. 26:532-538.

Johnson, D. E., Heald, S. L., Dally, R. D., Janis, R. A., 1993, Isolation, identification and synthesis of an endogenous arachidonic amide that inhibits calcium channel antagonist 1,4-dihydropyridine binding. Prostaglandins Leukot. Essent. Fatty Acids 48:429-437.

Kaminski, N. E., Koh, W. S., Yang, K. H., Lee, M., Kessler, F. K., 1994, Suppression of the humoral immune response by cannabinoids is partially mediated through inhibition of adenylate cyclase by a pertussis toxin-sensitive G-protein coupled mechanism. Biochem. Pharmacol. 48:1899-1908.

Kruszka, K. K., Gross, R. W., 1994, The ATP- and CoA-independent synthesis of arachidonoylethanolamide. A novel mechanism underlying the synthesis of the endogenous ligand of the cannabinoid receptor. J. Biol. Chem. 269:14345-14348.

Kuehl, F. A., Jacob, T. A., Ganley, O. H., Ormond, R. E., Meisinger, M. A. P., 1957, The identification of N-(2-hydroxyethyl)-palmitamide as a naturally occuring anti-inflammatory agent. J. Am. Chem. Soc. 79, 5577-5578.

Lynn, A. B., Herkenham, M., 1994, Localization of cannabinoid receptors and nonsaturable high-density cannabinoid binding sites in peripheral tissues of the rat: Implications for receptor-mediated immune modulation by cannabinoids J. Pharmacol. Exp. Therap. 268:1612-1623.

Mackie, K., Lai, Y., Westenbroek, R., Mitchell, R., 1995, Cannabinoids activate an inwardly rectifying potassium conductance and inhibit Q-type calcium currents in AtT20 cells transfected with rat brain cannabinoid receptor. J. Neurosci.15:6552-6561.

Mackie, K., Devane, W. A., Hille, B., 1993, Anandamide, an endogenous cannabinoid, inhibits calcium currents as a partial agonist in N18 neuroblastoma cells. Mol. Pharmacol. 44:498-503.

Matsuda, L. A., Lolait, S. J., Brownstein, M., Young, A., Bonner, T. I., 1990, Structure of a cannabinoid receptor and functional expression of the cloned cDNA. Nature 346: 561-564.

Munro, S., Thomas, K. L., Abu-Shaar, M., 1993, Molecular characterization of a peripheral receptor for cannabinoids. Nature 365:61-65.

Pinto, J. C., Potie, F., Rice, K. C., Boring, D., Johnson, M. R., Evans, D. M., Wilken, G. H., Cantrell, C. H., Howlett, A. C., 1994, Cannabinoid receptor binding and agonist activity of amides and esters of arachidonic acid. Mol. Pharmacol. 46:516-522.

Priller, J., Briley, E. M., Mansouri, J., Devane, W. A, Mackie, K., Felder, C. C., 1995, Mead ethanolamide, a novel eicosanoid, is an agonist for the central (CB1) and peripheral (CB2) cannabinoid receptors. Mol. Pharmacol. 48:288-292.

Schmid, H. H., Schmid, P. C., Natarajan, V., 1990, N-acylated glycerophospholipids and their derivatives. Prog. Lipid Res. 29:1-43.

Schmid, P. C., Krebsbach, R. J., Perry, S. R., Dettmer, T. M., Maasson, J. L., Schmid, H. H. O., 1995, Occurance and postmortem generation of anandamide and other long-chain N-acylethanolamides in mammalian brain. FEBS Lett. 375, 117-120.

Shire, D., Carillon, C., Kaghad, M., Calandra, B., Rinaldi-Carmona, M., Le Fur, G., Caput, D., Ferrara, P., 1995, An amino-terminal variant of the central cannabinoid receptor resulting from alternative splicing. J. Biol. Chem. 270:3726-3731.

Sugiura, T., Kondo, S., Sukagawa, A., Tonegawa, T., Nakane, S., Yamashita, A., Keizo, W., 1996, Enzymatic synthesis of anandamide, an endogenous cannabinoid receptor ligand, through N-acylphosphatidylethanolamine pathway in testis: Involvement of Ca^{2+}-dependent transacylase and phosphodiesterase activities. Biochem. Biophys. Res. Commun. 218:113-117.

Ueda, N., Kurahashi, Y., Yamamoto, S., Tokunaga T., 1995, Partial purification and characterization of the porcine brain enzyme hydrolyzing and synthesizing anandamide. J. Biol. Chem. 270:23823-23827.

Ueda, N., Yamamoto, K., Kurahashi, Y., Yamamoto, S., Ogawa, M., Matsuki, N., Kudo, I., Shinkai, H., Shirakawa, E., Tokunaga, T., 1995, Oxygenation of arachidonylethanolamide (anandamide) by lipoxygenases. Adv. Prostaglandin Thromboxane Leukot. Res. 23:163-165.

Venance, L., Piomelli, D., Glowinski, J., Giaume, C., 1995, Inhibition by anandamide of gap junctions and intercellular calcium signalling in striatal astrocytes. Nature 376:590-594.

EP RECEPTOR SUBTYPE-DEPENDENCE OF REGULATION OF IMMUNE CELLULAR FUNCTIONS BY PROSTAGLANDIN E2[a]

Li Zeng, Songzhu An, and Edward J. Goetzl[b]

Departments of Medicine and Microbiology-Immunology, University of California Medical Center, San Francisco, CA 94143-0711

INTRODUCTION: IMMUNE CELL PRODUCTION AND EFFECTS OF EICOSANOIDS

The recognition that some eicosanoids are potent inflammatory mediators was based on initial findings of the smooth muscle, microvascular, epithelial cell and nociceptive neural effects of prostaglandins (PGs) and peptidoleukotrienes, which mimicked the major humoral manifestations of tissue inflammation (1-3). The leukocytic component of inflammation also was shown to be reproduced by leukotriene B4 (LTB4) and several di-hydroxy-eicosatetraenoic acids (di-HETEs) from the 15-lipoxygenase (15-LO) pathway, with only modulatory influences of cyclooxygenase products, peptidoleukotrienes and lipoxins (4,5). The eicosanoids often are generated first at sites of inflammation by resident cells responding to the inciting stimulus. For example, neutrophils accumulate in the lungs of individuals lacking alpha1-antitrypsin as a result of the chemotactic activity of LTB4 produced by alveolar macrophages activated through binding of uninhibited elastase (6). Neutrophil recruitment to the sites of immune complex deposition (7) and substance P release from sensory nerves (8) similarly are responses in part to LTB4 that is produced by activated resident mast cells, as evidenced by the lesser neutrophil responses in mast cell-deficient mice. In inflammatory reactions, the effects of individual eicosanoid mediators may be manifested separately (9), but most often there are substantial interactions both among the eicosanoids and between eicosanoids and mediators from other systems (10-12).

The later recognition of immunoregulatory roles of eicosanoids was founded on data less clearly interpretable than results of studies of inflammation. Two principal problems complicating analyses of eicosanoids in immunity were the extreme phenotypic heterogeneity of normal populations of T and B cells, and the difficulties assigning general significance to some findings in cultured lines of B and T cells. The apparent complexity of some effects of eicosanoids on immunity also is partially attributable to the ability of many immune cells to generate as well as respond to a range of eicosanoid mediators. In isolation, macrophages and dendritic cells produce several types of prostaglandins and leukotrienes (13), mast cells generate PGD2 and peptidoleukotrienes (14) and activated B cells elaborate 5-HETE and LTB4 (15). The sometimes greater production of eicosanoids by cooperative cell mixtures

Frontiers in Bioactive Lipids
Edited by Vanderhoek, Plenum Press, New York, 1996

may reflect expression by one type of cell of only a single enzyme in a pathway, as for LTB4 generation by LTA4 hydrolase in T cells, that lack 5-lipoxygenase (5-LO), only after being supplied with LTA4 by mononuclear phagocytes (16). The generation of eicosanoids by immune cells also is altered qualitatively and quantitatively by cytokine-driven differentiation and activation of the cells. Recently, immune stimuli have been shown to increase the activities of 15-LO, 5-LO, LTC4 synthase and a uniquely inducible second form of cyclooxygenase (COX II), suggesting specific mechanisms for localized induction of each pathway. COX II, which contains all of the amino acid sequences required for full catalytic activity of the constitutively-expressed COX I, is induced by interleukin (IL-1), tumor necrosis factor (TNF) and many growth factors and mitogens in mononuclear phagocytes, fibroblasts, endothelial cells, and other cells by a glucocorticoid-suppressible mechanism (17). COX I and II also have different profiles of susceptibility to inhibition by a range of non-steroidal anti-inflammatory drugs (17). Monocyte 15-LO is selectively induced by IL-4 at the mRNA level, through a mechanism specifically inhibited by interferon-gamma (IFN-gamma) and cortisone (18). Immunological and biochemical activation of 5-LO for LT biosynthesis requires translocation to a nuclear membrane site, in association with an 18 kDa 5-LO-activating protein (FLAP) and another protein(s) induced by transforming growth factor-beta (TGF-beta) in cooperation with TNF or granulocyte-macrophage colony-stimulating factor (GM-CSF) (17,18). Other immune cell-derived cytokines influence the expression of 5-LO and FLAP in the course of cell-cell interactions. For example, monocyte levels of 5-LO and FLAP are augmented by a required combination of GM-CSF and IL-3 from lymphocytes (19). The recently-cloned LTC4 synthase also is activated by perinuclear association with FLAP (20). Generation of many eicosanoid mediators thus is controlled by specific immune cellular mechanisms responsive to immune cytokines.

A clearer and more distinctive pattern of immunofunctional activities of some eicosanoids emerges when consideration of responses is restricted to those evoked in defined populations of immune cells at one stage of differentiation by picomolar to nanomolar concentrations of an eicosanoid. Such stringently-selected data show potent effects of eicosanoids on specific immune contributions of macrophages, subsets of B and T cells, natural killer (NK) cells and subsets of cytotoxic lymphocytes. PGE1, PGE2 and LTB4 suppress numerous components of macrophage phagocytic activation, and concurrently promote their immune differentiation, as reflected in increased expression of specialized receptors and pathways of generation of monokines (21,22). For example, LTB4 both induces production of IL-1, IL-6 and TNF, and increases by 1,000-fold the sensitivity of macrophage cytokine generation to IL-2 by upregulating IL-2Rs. The B cell responses to eicosanoids similarly are conditioned by their state of differentiation. PGE2 induces apoptosis in immature mouse B lymphocytes (23), but non-cytotoxically inhibits several aspects of IL-4- and LPS-induced activation and IgM production of a subset of resting mature B cells, and concurrently promotes their synthesis of IgG1 and IgE (24,25). Effects also may be stimulus-dependent, as PGE2 enhances proliferation and secretion of IgM, IgG and IgA by human tonsillar B cells stimulated with anti-CD40 and IL-10, but inhibits IgG and IgE production evoked by a combination of anti-CD40 and IL-4 (26). LTB4 sequentially initiates differentiation of human resting B cells, then induces proliferation and finally enhances Ig production by the most mature subset in synergy with IL-2 and IL-4 (27).

Subsets of T cells differ in their susceptibility to eicosanoids, but the quantitatively most prominent effects are directed to cytokine secretion. PGE2 inhibits T cell proliferative responses, with preferential potency for $CD4^+$ (functional helper) Th cells, and differentially suppresses cytokine production by $CD4^+$ Th cells, through transcriptional mechanisms that are directed selectively to IL-2 and IFN-gamma from Th1 cells, and IL-3 and GM-CSF from both the Th1 and Th2 subsets, but not IL-4 and IL-5 from Th2 cells (28,29). LTB4 mediates immunosuppression by both enhancing proliferation of $CD8^+$ T cells and inhibiting proliferation of CD4$^+$ cells (30). Many of the effects of LTB4 in immunity are attributable to augmentation of IFN-gamma production by T helper cells and reduction in IFN-gamma production by T suppressor/cytotoxic cells (31). PGE2 alters the morphology and suppresses lytic activity of IL-2-activated NK cells, whereas LTB4 augments natural cytotoxic cell activities by increasing both binding to target cells and lytic efficiency (32). In some instances, immune cell-derived PGE2 has functional effects on other immune cells of sufficient magnitude to be a primary mediator of interactions determining the response. One example is the suppression of production of IFN-gamma by CD4$^+$ T cells, activated in the presence of stimulatory concentrations of IL-12, that is attained by PGE2 from monocytes, macrophages and other accessory cells (33).

The establishment of myriad potential sites of production and actions of eicosanoids in mammalian tissues has directed much of current research to elucidation of mechanisms of specificity, that might explain very discrete effects of the mediators in organ systems and selective benefits of pharmacological inhibitors of eicosanoids in some human diseases. Although selectively-inducible cellular patterns of generation and degradation of each type of eicosanoid are one determinant of their biological roles, recent evidence suggests that expression of structurally and functionally distinct sets of receptors (Rs) is the predominant mechanism for adaptive regulation of regional tissue responsiveness to eicosanoid mediators (17,34,35). All major types of eicosanoid Rs characterized structurally are of the heterotrimeric G protein- associated superfamily with seven intramembranous domains (17). Known structures to date are limited to prostanoid and thromboxane Rs, with the exception of one R of the inflammatory peptide class found to bind lipoxin A4 (36). The primary variable that defines an eicosanoid R is affinity of binding of the principal ligand compared to natural variants and synthetic analogs of related structure. However, other elements of eicosanoid receptor structure, that distinguish subtypes, determine very relevant differences in tissue distribution, signal transduction and susceptibility to ligand-induced desensitization.

The recent discovery of several distinct subtypes of some eicosanoid receptors suggests that the effects of an eicosanoid on any one type of immune cell may be conditioned by the relative levels of expression of functionally different Rs for that eicosanoid, which may vary with the source, history of stimulation and continuing interactions with other immune cells. Each such subtype of Rs is distributed distinctively in different types of cells and tissues, as in the predominance of EP3-subtype PGE2 Rs in kidneys and uterus, in contrast to the wider representation of the EP4-subtype that includes lung tissues, leukocytes and immune organs (17). Whereas a high density of Rs for an eicosanoid is the primary determinant of localization of activity, the nature and magnitude of the response elicited also depend on multiple levels of regulation of effector pathways. For example, different isoforms of EP3Rs have been identified in several tissues, that result from alternative splicing of mRNA encoding the carboxyl-terminal tail (17,37). Each murine and bovine isoform of EP3R transduces a different profile of constitutive and PGE2-evoked biochemical responses (38), that is presumed to result from association with two or more heterotrimeric G proteins (39). Interactions of Rs with different G proteins also may affect membrane localization and intracellular traffic, as in the range of susceptibility of human EP3R isoforms to PGE2-induced desensitization based on differences in cytoplasmic tail structure and sites of phosphorylation (37).

CHARACTERISTICS OF THE EP SUBTYPES OF PGE2 RECEPTORS

The structurally distinct family of G protein-associated receptors (Rs) that recognize many of the eicosanoids are classified primarily as types based principally on major ligand-binding properties, but also exhibit substantial differences in the range of subtypes and isoforms of each subtype, mechanisms of regulation of expression and signaling, and cellular and tissue distribution (17; Table 1). Recent elucidation of selected structural determinants of eicosanoid receptor functions has confirmed some similarities with other G protein-coupled Rs and exposed unique elements that relate predominantly to pathways of signaling and down-regulation.

Table 1. Subtypes of mammalian EP Rs for PGE2

Subtype	Binding Preferences		Signaling Mechanisms	Immune Cell Effects
	Natural Eicosanoids	Synthetic Agents		
EP1	PGE2>PGE1>>PGF2a>PGD2	I,SC,S	I [Ca^{++}]i	
EP2	PGE2=PGE1>>>PGF2a,PGD2	B,M/B	I [cAMP]i	Ig class-switch/B cells Thymocyte MMPs, migration
EP3	PGE2=PGE1>PGF2a,PGD2	S,M,M/B	I [Ca^{++}]i D [cAMP]i	Monocyte migration Pre-T cell MMP,migration
	<u>Isoforms</u>			
	Cytoplasmic Tail			
	EP3 I (I') 10 + 31 aa			
	EP3 II 10 + 29 other aa			
	EP3 III 10 + 6 other aa			
	EP3 IV 10 + 9 other aa + the 6 aa of III			
EP4	PGE2=PGE1>>PGF2a,PGD2	M,M/B	I [cAMP]i	Ig class-switch/B cells NK cell MMPs, migration Thymocyte MMPs, migration

I= iloprost , SC= SC-19220, S= sulprostone, M= misoprostol, M/B= M&B 28767, B=butaprost; I= increased, D= decreased, aa= amino acids.

The structurally divergent subtypes of EP Rs for PGE2 exhibit differences in ligand-binding specificity and signal transduction pathways. In contrast, the isoforms of EP3 Rs differ only in the amino acid sequence of the cytoplasmic tail after the initial 10 amino acids and are presumed to arise from alternative splicing of the precursor mRNA product of a single gene (37). The human EP3 R isoforms possess the same ligand binding affinity and specificity, and signaling pathways in CHO cell transfectants, but transduce signals of different magnitude after binding PGE2 at the same concentration (37). The rank-order of human EP3 R effectiveness for M&B 28767 suppression of increases in [cAMP]i elicited by forskolin was I > II > III >> IV. For M&B 28767-induced increases in $[Ca^{++}]i$ transduced by human EP3 Rs, the order of effectiveness was II,I > IV > III. Some members of the parallel set of EP3 R isoforms isolated from bovine adrenal gland and mouse tissues differ in signaling from any of the human isoforms in transducing increases in [cAMP]i (17). These increases in [cAMP]i have been shown to be attributable to coupling of bovine EP3-B and -C to Gs and of bovine EP3-D and mouse EP3-gamma to both Gs and Gi, as contrasted with the coupling of [cAMP]i- suppressive bovine EP3-A and mouse EP3-alpha and -beta Rs only to Gi (39).

Another functional distinction delineated so far only in human isoforms of EP3 Rs in CHO cell stable transfectants are differences in agonist-induced desensitization (37). Although a 60 min preincubation at 37°C with an optimal concentration of M&B 28767 resulted in similar 24% to 32% suppression of binding of PGE2 and of increases in $[Ca^{++}]i$, shorter and longer exposures revealed significant differences (37). EP3 R isoforms III and IV were down-regulated maximally at 15 min (mean= 48%) and 60 min (mean= 32%), respectively, with rapid recovery that was nearly complete by 3 hr and complete by 6 hr. In contrast, EP3 R isoform II was down-regulated maximally in the same studies only after 3 hr, and this was maintained for at least 12 hr. Thus the short carboxyl-terminal cytoplasmic tails of isoforms III and IV, that lack sites for phosphorylation, are associated with rapid desensitization and re-expression, whereas the longer cytoplasmic tail of isoform II with 5 potential sites of phosphorylation predisposes to slow and sustained desensitization of EP3 Rs.

COUPLING OF SUBTYPES OF EP RS TO SPECIFIC IMMUNE CELLULAR FUNCTIONS

The ability to quantify mRNA expression, ligand binding, and signal transduction specific for each subtype of EP R in cultured lines and purified native populations of mononuclear leukocytes has permitted assignment of some PGE2- elicited responses to one or two predominant subtypes. Ligand-independent induction of separate down-regulation of one subtype of EP R individually has facilitated confirmation of the coupling of one or more subsets to distinctive functional effects of PGE2 on immune cells.

Human monocytic cells of the U937 line co-express EP4 and EP3 Rs, that were detected by reverse transcription PCR analyses and Northern blots, as well as inhibition of binding of [3H]PGE2 by sulprostone (EP1/EP3-specific) and M&B 28767 (EP4/EP3/EP2-specific), but not SC-19220 (EP1-specific) or butaprost (EP2- specific) (40). Pretreatment of U937 cells for 1 hr with PGE2 coordinately downregulated EP4 and EP3 Rs, whereas the protein kinase C activator phorbol 12- myristate 13-acetate (PMA) selectively downregulated EP4 Rs. That pretreatment with PGE2, but not PMA, suppressed the subsequent stimulation by PGE2 of U937 cell chemokinesis across micropore filters in vitro, suggested that EP3 Rs mediate stimulation of migration of U937 cells.

Quiescent mouse splenic B cells have been shown recently to express mRNA encoding EP1, EP2, EP3B and EP4 Rs. However, only EP2 and EP4 R-selective agonists exert significant functional effects, including prevention of enlargement of B cells incubated with IL-4 and/or lipopolysaccharide (LPS), and suppression of expression of class II MHC and CD23 (41). The stimulation of immunoglobulin class-switching to IgE in the mouse B cells by IL-4 and LPS was enhanced directly by PGE2, solely through EP2 and EP4 Rs by a cyclic AMP- dependent mechanism (41). A similar coupling of EP4 Rs exclusively to some functions of NK cells was observed in the elicitation of chemotaxis and enhancement of secretion of matrix metalloproteinases (MMPs) by PGE2 (42). The RNK-16 line of rat large granular lymphocytes (LGLs), that exhibit NK cell cytotoxic activity similar to that of IL-2-

Table 2. Implications of PGE2 R (EP R) subtype diversity and functional selectivity

1. The subtypes of EP Rs expressed by any cell are a major determinant of responses to PGE2
2. Ligands, cytokines and other mediators regulate the representation of EP R subtypes on a cell and thereby the responses to PGE2
3. The subtypes of EP Rs expressed establish the susceptibility of PGE2-evoked responses to EP R-specific antagonists and inhibitors of transductional biochemical pathways
4. EP R subtype-selective agonists and antagonists are expected to affect physiological and pathological processes more selectively than PGE2, when the dominant responding cells express more than one subtype

differentiated LGLs of rat spleen, express EP3 and EP4 Rs for PGE2, according to the results of competitive binding and biochemical analyses. Secretion of MMP-1 and -3, that predominate in RNK-16 cells, was enhanced significantly by 1 to 100 nM PGE2 and equally active concentrations of EP4 R-preferential agonists, as assessed by Western blots and zymography (42). PGE2 and EP4 R-directed agonists stimulated chemokinesis of RNK- 16 cells across micropore filters coated with Matrigel, to simulate a basement membrane. This MMP-dependent chemokinetic response was suppressed nearly completely by a concentration of MMP-specific inhibitor that did not affect RNK-16 cell chemokinesis through a micropore filter without Matrigel (42).

The representation of subtypes of EP Rs on T cells appears to be the most complex of the immune cells studied, with differences in distribution possibly attributable to the developmental stage. The HSB.2 line of human leukemic pre-T cells, that express CD2 and CD7, but not CD3, CD4 or CD8, bear EP3, EP4 and lower levels of EP2 Rs, as demonstrated by selective agonist competitive binding studies and biochemical responses (43). The cytosolic and secreted levels of the predominant MMP-9 of HSB.2 pre-T cells were increased by PGE2 and EP3 R-preferential agonists, as assessed by Western blots and zymographic analyses. A prerequisite for EP3 R-coupled increases in $[Ca^{++}]i$, that are attributable to release of intracellular stores, was suggested by the capacity of pretreatment with thapsigargin to prevent PGE2-mediated increases in HSB.2 cell content and secretion of MMP-9. The increases in mRNA encoding MMP-9 in HSB.2 cells after PGE2 stimulation implies that the EP3 R acts in part by a transcriptional mechanism (43). In contrast to the pre-T cells of the HSB.2 line, the thymocyte-like $CD4^+8^+3^{lo}$ human lymphoblastoma cells of the Tsup-1 line bind and respond to PGE2 predominantly through EP2 and EP4 Rs. In Tsup-1 cells and a subset of human blood T cells, PGE2 and EP2/EP4 R-directed agonists increase the content, secretion and cell-surface representation of MMP-2 and -9 (44). The requirement for EP2/EP4 R-mediated enhancement of MMPs in Tsup-1 cell and blood T cell migration through Matrigel-coated micropore filters was demonstrated by the suppressive effect of MMP inhibitors, that did not reduce the level of migration through filters without Matrigel. Thus whether EP3 Rs or EP2/EP4 Rs mediate increases in T cell MMPs and MMP-dependent migration through Matrigel is predominantly a function of the type and possibly stage of differentiation of the T cells.

CONCLUSIONS

At this point in the evolution of our understanding of cellular mechanisms for specific responses to eicosanoids, one can formulate only a tentative model to synthesize possible contributions of receptors, signal transduction pathways, innate cellular capabilities and the effects of other mediators and cellular systems. Present data make clear a dominant role of receptor heterogeneity and subtype specificity in determining the responses of cells to PGE2 and presumably other eicosanoid mediators (Table 2). The predominant subtype of EPR expressed by any immune cell, and possibly other cells, is a function of the cell type, interactions with other cells, connective tissue elements and fluid-phase factors. All of the subtypes bind PGE2 with equally high affinity, but utilize different transductional signals to attain a distinctive array of functional responses. The predominance of one or two subtypes of receptors for one eicosanoid also conditions susceptibility to pharmacological agents from selective agonists to inhibitors of biochemical pathways. Thus more precise knowledge of these mechanisms will permit increasingly refined approaches to development of potent specific drugs.

NOTES

a-The studies described were supported by grant HL-31809 from the National Institutes of Health.

b-Correspondence to: Edward J. Goetzl, M.D., Immunology and Allergy-UB8B, Box 0711, University of California, 533 Parnassus, San Francisco, CA 94143; Telephone= 415-476-5339, FAX= 415-476-6915.

REFERENCES

1. B. Samuelsson, E. Granstrom, K. Green, M. Hamberg, and S. Hammarstrom. Prostaglandins. Annu. Rev. Biochem. 44:669-695 (1975).
2. R.A. Lewis, K.F. Austen, and R.J. Soberman. Leukotrienes and other products of the 5-lipoxygenase pathway. Biochemistry and relation to pathobiology in human diseases. *N. Engl. J. Med.* 323:645-655 (1990).
3. H.A. Lehr, A. Guhlmann, D. Nolte, D. Keppler, and K. Messmer. Leukotrienes as mediators in ischemia-reperfusion injury in a microcirculation model in the hamster. J. Clin. Invest. 87:2036-2041 (1991).
4. E.J. Goetzl and W.C. Pickett. The human PMN leukocyte chemotactic activity of complex hydroxy-eicosatetraenoic acids (HETEs). *J. Immunol.* 125:1789-1791 (1980).
5. A.W. Ford-Hutchinson. Leukotriene B4 in inflammation. *Crit. Rev. Immunol.* 10:1-23 (1990).
6. R.C. Hubbard, G. Fells, J. Gadek, S. Pacholok, J. Humes, and R.G. Crystal. Neutrophil accumulation in the lung in a1-antitrypsin deficiency. Spontaneous release of leukotriene B4 by alveolar macrophages. *J. Clin. Invest.* 88:891- 897 (1991).
7. B.F. Ramos, Y. Zhang, R. Qureshi, and B.A. Jakschik. Mast cells are critical for the production of leukotrienes responsible for neutrophil recruitment in immune complex-induced peritonitis in mice. *J.Immunol.* 147:1636-1641 (1991).
8. I. Iwamoto, S. Tomoe, H. Tomioka, and S. Yoshida. Leukotriene B4 mediates substance P-induced granulocyte infiltration into mouse skin. *J.Immunol.* 151:2116-2123 (1993).
9. D.E. Griswold, E.F. Webb, and L.M. Hillegass. Induction of plasma exudation and inflammatory cell infiltration by leukotriene C4 and leukotriene B4 in mouse peritonitis. *Inflammation* 15:251-258 (1991).
10. T.H. Lee, P. Lympany, A.E.G Crea, and B.W. Spur. Inhibition of leukotriene B4 -induced neutrophil migration by lipoxin A4: Structure-function relationships. *Biochem. Biophys. Res. Commun.* 180:1416-1421 (1991).

11. M.E. Wiles, R. Welbourn, G. Goldman, H.G Hechtman, and D. Shepro. Thromboxane-induced neutrophil adhesion to pulmonary microvascular and aortic endothelium is regulated by CD18. *Inflammation* 15:181-199 (1991).
12. J. Palmblad, R. Lerner, and S.H. Larsson. Signal transduction mechanisms for leukotriene B4-induced hyperadhesiveness of endothelial cells for neutrophils. *J.Immunol.* 152:262-269 (1994).
13. M.S. Balter, G.B. Toews, and M. Peters-Golden. Different patterns of arachidonate metabolism in autologous human blood monocytes and alveolar macrophages. *J. Immunol.* 142:602-613 (1989).
14. A.N. Fonteh, and F.H. Chilton. Mobilization of different arachidonate pools and their roles in the generation of leukotrienes and free arachidonic acid during immunologic activation of mast cells. *J. Immunol.* 150:563-571 (1993).
15. P.G. Schulam, and W.T. Shearer. Evidence for 5-lipoxygenase activity in human B cell lines: A possible role for arachidonic acid metabolites during B cell signal transduction. *J. Immunol.* 144:2696-2701 (1990).
16. P.-J. Jakobsson, D. Steinhilber, B. Odlander, O. Radmark, and H.-E. Claesson. On the expression and regulation of 5-lipoxygenase in human lymphocytes. *Proc. Natl. Acad. Sci.* USA 89:3521-3525 (1992).
17. E.J. Goetzl, R.A. Lewis, and M. Rola-Pleszczynski. *Cellular Generation, Transport and Effects of Eicosanoids: Biological Roles and Pharmacological Intervention.* Eds. N.Y. Academy of Sciences, N.Y., vol. 744 (1995).
18. M. Pouliot, P.P. McDonald, P. Borgeat, and S.R. McColl. Granulocyte/macrophage colony-stimulating factor stimulates the expression of the 5-lipoxygenase- activating protein (FLAP) in human neutrophils. *J. Exp. Med.* 179:1225-1232 (1994)
19. W.L. Ring, C.A. Riddick, J.R. Baker, D.A. Munafo, and T.D. Bigby. Lymphocytes stimulate expression of 5-lipoxygenase and its activating protein in monocytes in vitro via granulocyte-macrophage colony-stimulating factor and interleukin 3. *J. Clin. Invest.* 97:1293-1301 (1996).
20. B.K. Lam, J.F. Penrose, G.J. Freeman, and K.F. Austen. Expression cloning of a cDNA for human leukotriene C4 synthase, an integral membrane protein conjugating reduced glutathione to leukotriene A4. *Proc. Natl. Acad. Sci. USA* 91:7663-7667 (1994).
21. R. Keller, R. Keist, and P.H. Van der Meide. Modulation of tumoricidal activity induced in bone marrow-derived mononuclear phagocytes by interferon-gamma or Corynebacterium parvum by interferon-beta, tumor necrosis factor, PGE2 and transforming growth factor-beta. *Int. J. Cancer* 49:796-807 (1991).
22. S. Schreiber, J.S. Blum, J.C. Chappel, W.F Stenson,. P.D. Stahl, S.L. Teitelbaum, and S.L. Perkins. Prostaglandin E specifically upregulates the expression of the mannose-receptor on mouse bone marrow-derived macrophages. *Cell Regul.* 1:403-413 (1990).
23. D.M. Brown, G.L. Warner, J.E. Ales-Martinez, D.W. Scott, and R.P. Phipps. Prostaglandin E2 induces apoptosis in immature normal and malignant B lymphocytes. *Clin Immunol. Immunopathol.* 63:221-229 (1992).
24. P. Garrone, L. Galibert, F. Rousset, S.M. Fu, and J. Banchereau. Regulatory effects of prostaglandin E2 on the growth and differentiation of human B lymphocytes activated through their CD40 antigen. *J. Immunol.* 152:4282-4290 (1994).
25. R.L. Roper, and R.P. Phipps. Prostaglandin E2 and cAMP inhibit B lymphocyte activation and simultaneously promote IgE and IgG1 synthesis. *J. Immunol.* 149:2984-2991 (1992).
26. K.A. Yamaoka, H.-E Claesson, and A. Rosen, Leukotriene B4 enhances activation, proliferation, and differentiation of human B lymphocytes. *J. Immunol.* 143:1996-2000 (1989).

27. B. Dugas, N. Paul-Eugene, J. Cairns, J. Gordon, A. Calenda, J. M. Mencia- Huerta, and P. Braquet. Leukotriene B4 potentiates the expression and release of Fc epsilon RII/CD23, and proliferation and differentiation of human B lymphocytes induced by IL-4. *J. Immunol.* 145:3406-3411 (1990).
28. J.S. Goodwin, A.D. Bankhurst, and R.P. Messner. Suppression of human T-cell mitogenesis by prostaglandin. *J. Exp. Med.* 146:1719-1726 (1977).
29. M. Betz, and B. S. Fox. Prostaglandin E2 inhibits production of Th1 lymphokines but not of Th2 lymphokines. *J. Immunol.* 146:108-113 (1991).
30. D.G. Payan, A. Missirian-Bastian, and E.J. Goetzl. Human T-lymphocyte subset specificity of the regulatory effects of leukotriene B4. *Proc. Natl. Acad. Sci. U.S.A.* 81:3501-3505 (1984).
31. M. Rola-Pleszczynski, L. Bouvrette, D. Gingras, and M. Girard. Identification of interferon-gamma as the lymphokine that mediates leukotriene B4-induced immunoregulation. *J. Immunol.* 139:513-517 (1987).
32. L. Gagnon, M. Girard, A.K. Sullivan, and M. Rola-Pleszczynski. Augmentation of human natural cytotoxic cell activity by leukotriene B4 mediated by enhanced effector-target cell binding and increased lytic efficiency. *Cell Immunol.* 110:243-252 (1987).
33. C.M.U. Hilkens, A. Snijders, H. Vermeulen, P.H. van der Meide, E.A. Wierenga, and M.L. Kapsenberg. Accessory cell-derived IL-12 and prostaglandin E2 determine the IFN-gamma level of activated human $CD4^+$ T cells. *J. Immunol.* 156:1722-1727 (1996).
34. E.J. Goetzl, J. Yang, M. Xia, L. Zeng, S.W. So, and S. An. *Diverse mechanisms of specificity of human receptors for eicosanoids. In: Cellular Generation, Transport and Effects of Eicosanoids: Biological Roles and Pharmacological Intervention.* Goetzl, E.J. Lewis, R.A. and Rola-Pleszczynski, M. eds. N.Y. Acad. Sci., N.Y. vol. 744, pp.146-154 (1995).
35. E.J. Goetzl, S. An, L. Zeng, and and W.L Smith. Eicosanoid mediators of homeostasis and host defense: Specificity of generation, recognition and effects. *FASEB J.* 9:1051-1058 (1995).
36. S. Fiore, J.F. Maddox, H.D. Perez, and C.N. Serhan. Identification of a human cDNA encoding a functional high-affinity lipoxin A4 receptor. *J. Exp. Med.* 180:253-260 (1994).
37. S. An, J. Yang, L. Zeng, S.W. So, and and E.J. Goetzl. Isoforms of the EP3-subtype of human prostaglandin E2 receptor transduce both calcium and cyclic AMP signals. *Biochem.* 33:14496-14502 (1995).
38. H. Hasegawa, M. Negishi, and A Ichikawa. Two isoforms of the prostaglandin E receptor EP3 subtype different in agonist-independent constitutive activity. *J. Biol. Chem.* 271:1857-1860 (1996).
39. M. Negishi, A. Irie, Y. Sugimoto, T. Namba, and A. Ichikawa. Selective coupling of prostaglandin E receptor EP3D to Gi and Gs through interaction of alpha-carboxylic acid of agonist and arginine residue of seventh transmembrane domain. *J. Biol. Chem.* 270:16122-16127 (1995).
40. L. Zeng, S. An and E.J. Goetzl. Independent down-regulation of EP2 and EP3 subtypes of the prostaglandin E2 receptors on U937 human monocytic cells. *Immunol.* 86:620-628 (1995).
41. E.R. Fedyk, J.M. Ripper, D.M. Brown and R.P. Phipps. A molecular analysis of PGE receptor (EP) expression on normal and transformed B lymphocytes: Coexpression of EP1, EP2, EP3beta and EP4. *Molec. Immunol.* 33:33-45 (1996).
42. L. Zeng, S. An and E.J. Goetzl. Regulation of RNK-16 cell matrix metalloproteinases selectively by the EP4-subtype of prostaglandin E2 receptor. *Biochem.* in press (1996).

43. L. Zeng, S. An and E.J. Goetzl. Prostaglandin E2 (PGE2) stimulation of secretion of matrix metalloproteinase-9 (MMP-9) from HSB.2 human T cells mediated by the EP3-subtype of PGE2 receptors (Rs). *FASEB J.* in press (1996).
44. D. Leppert, S.L. Hauser, J.L. Kishiyama, S. An, L. Zeng and E.J. Goetzl. Stimulation of matrix metalloproteinase-dependent migration of T cells by eicosanoids. *FASEB J.* 9:1473-1481 (1995).

CO-EXPRESSION OF SPECIFIC G-PROTEIN α SUBUNITS ALTERS THE AFFINITY OF THROMBOXANE A_2 RECEPTORS

Perry V. Halushka*†, Claire J. Allan*†, Michael Martin*†, Thomas A. Morinelli*†, Katsuhiro Higashiura*†G. Patrick Meier‡†, and Thomas W. Gettys *.

†Department of Cell and Molecular Pharmacology, *Department of Medicine, and ‡Department of Pharmaceutical Sciences, Medical University of South Carolina, Charleston, SC29425

INTRODUCTION

Activation of thromboxane A_2 (TXA_2) receptors results in shape change and aggregation of platelets[1], and in vascular smooth muscle cells, constriction and either a mitogenic[2, 3] and/or a hypertrophic[4] response. As a result of these effects, TXA_2 has been implicated as a potential pathological factor in a variety of diseases[5]. This series of effects may be exerted through potentially different TXA_2 receptor subtypes, although their existence is still a matter of debate[6, 7].

Controversy exists regarding the existence of TXA_2 receptor subtypes in human platelets and vascular smooth muscle cells (VSMC)[6, 7]. Much of the evidence supporting the existence of subtypes of TXA_2 receptors has been derived from pharmacological and radioligand binding studies using a series of 13-azapinane TXA_2 analogs[8-10]. From these studies it was concluded that TXA_2 receptors were different between platelets and smooth muscle and that 13-azapinane TXA_2 analogs could discriminate between these subclasses. Morinelli and co-workers[11] provided further pharmacological evidence that platelet and vascular TXA_2 receptors were different using analogs of 10, 10 difluoro-TXA_2.

TXA_2 receptors on platelets exist in two affinity states. The apparent high affinity receptor which is responsible for shape change has a K_d in the pM range for the agonist ligand ^{125}I-BOP, while the low affinity receptor thought to mediate aggregation and secretion possess a K_d in the low nM range[12-15]. It is uncertain if these 2 affinity states represent distinct receptor proteins or a single receptor protein coupled to different guanine nucleotide binding proteins. In contrast, TXA_2 receptors on VSMC exist as a single class of binding sites for ^{125}I-BOP with an affinity comparable to the high affinity platelet receptor[10, 16, 17].

The TXA_2 receptor cDNA isolated originally from human placenta is 100% homologous with the megakaryocyte-like MEG-01 or HEL cells[18], K562 myelogenous leukemia cells [19] and human mesangial cells [6, 7]. A human TXA_2 receptor cDNA has been cloned from a human endothelial cell cDNA library[20] and has a unique carboxyl terminal tail due to alternative splicing. It has recently been shown that this splice variant is also present in human platelets[21], (Halushka, unpublished observations). For these cloned receptors, it is uncertain if they more closely resemble the platelet or vascular receptor subtype, since no determination of their binding characteristics using discriminating 13-azapinane TXA_2 analogs have been conducted.

It has been postulated that different affinity states of the TXA_2 receptor may represent the coupling of a single TXA_2 receptor protein to different cell specific complements of G protein α-subunits [7, 19]. Platelet TXA_2 receptors have been reported to couple to $G\alpha_q$, $G\alpha_{12}$, $G\alpha_{13}$ and $G\alpha_{i2}$[22-25]. The platelet aggregation response, which is regulated in part by increasing intracellular calcium, appears to be regulated by an interaction of the TXA_2 receptor with $G\alpha_q$. The G protein(s) transducing platelet shape change have not been clearly defined. One of the earliest events associated with platelet shape change is alkalinization of intraplatelet cytosol, which may be regulated by an interaction between the TXA_2 receptor and $G\alpha_{12}$ and/or $G\alpha_{13}$, since these Gα subunits have been reported to modulate the activity of the Na^+/H^+ exchanger [26, 27]. To date there have been no studies investigating the consequences of expressing a cloned TXA_2 receptor with different Gα subunit proteins.

This study was designed to determine the ligand binding characteristics and the effects of interactions with Gα subunits for the cloned HEL (platelet-like) TXA_2 receptor. The megakaryocyte-like HEL cell served as a model of human platelet TXA_2 receptors since its TXA_2 receptor is pharmacologically similar to the platelet aggregation type receptor [28-30].

MATERIALS AND METHODS

The following were generous gifts: a partial clone of the human placenta TXA_2 receptor cDNA from Dr. M. Abramowitz, Merck Frosst Canada Inc., Pointe Claire-Dorval, Quebec. COS-7 cells and the cDNA for $G\alpha_{i2}$ were from Dr. S. Lanier, Dept. of Pharmacology, Medical University South Carolina. Full length wild type murine $G\alpha_{12}$ and $G\alpha_{13}$ from Dr. Mel Simon, California Institute of Technology, Pasadena CA. Human $G\alpha_q$ cDNA was from Dr. Juan Codina and Dr. Lutz Birnbaumer, Baylor College of Medicine, Houston, TX. ^{125}I-BOP was synthesized as described previously [31].

Cell Culture and Transfection.

COS-7 cells were maintained in DMEM supplemented with 10% Serum Plus, 1% Penicillin, Streptomycin, Fungizone (JRH Biochemicals). Cells were split the day before transfection and used at a confluency of >80%.

Co-Transfection of the HEL Cell TXA_2 receptor with Gα subunits

The TXA_2 receptor cDNA was cloned from the HEL cell using RT-PCR as previously described[18]. TXR cDNA (10μg) was transfected using the DEAE Dextran technique and the transfected COS-7 cells were maintained in culture for a further 60 hrs. DNA for transfections was isolated using the Maxiprep technique. For the co-transfection studies, 5μg of cDNA for TXR, and either $G\alpha_{12}$, $G\alpha_{13}$, $G\alpha_q$ or $G\alpha_{i2}$ were transfected into COS-7 cells by the DEAE-Dextran technique. The total amount of DNA transfected was maintained at 15μg with pcDNA3 plasmid.

Radioligand Binding Studies

Transfected COS-7 cells were washed twice with 5ml of 25mM HEPES, 2mM EDTA, pH 6.5 supplemented with 10μM indomethacin, PMSF (1.7μl/ml of a 75mg/ml stock), benzamidine (0.01M) and detached from the plates by scraping into the same buffer and homogenized. Membranes were prepared by centrifugation at 100,000 X g for 45 min in a Beckman LC-50 ultracentrifuge, the pellets were resuspended in 1-2ml of incubation buffer and homogenized.

Radioligand binding studies with transfected COS-7 cell membranes were performed using 2.5-5μg membrane protein in a final volume of 200μl as described previously [32]. Non-specific binding was defined as the amount of radioactivity bound in the presence of the TxA_2 receptor antagonist (-)L657925 (10μM).

For the co-transfection experiments, COS-7 cell membranes were incubated in a binding buffer composed of 25mM HEPES (pH 6.5 or 7.4), 125mM NaCl, 10mM $MgCl_2$ and 5mM

KCl with 10μM indomethacin, PMSF and benzamidine (0.01M). Experiments with GTPγS (100μM) were conducted for 2 hours at 30°C using the same incubation buffer.

Western blotting of G proteins.

C-terminal antisera were raised for each specific $G\alpha$ subunit, according to previously published procedures[33]. Solubilized COS-7 cell membranes were resolved by SDS-PAGE, transferred to Immobilon membranes (Millipore Inc.) and probed with the specific $G\alpha$ protein antisera. $G\alpha_{12}$ and $G\alpha_{13}$ were detected with ^{125}I-labeled goat anti-rabbit IgG as described[33] while $G\alpha_q$ and $G\alpha_{i2}$ were detected using goat anti-rabbit IgG conjugated to horse radish peroxidase and chemiluminescence with the ECL system (Amersham).

Data Analysis

Data from equilibrium binding experiments were analyzed according to the method of Scatchard [34] using the LIGAND computer program of Munson and Rodbard to determine K_d (receptor affinity) and B_{max} (receptor density) values [35]. All radioligand binding data were subjected to both one-site and two-site model fits. The two-site model was not statistically significantly better than the one-site model so the latter was adopted. Statistical analyses were performed using an ANOVA with significance set at $p < 0.05$.

RESULTS

Since only a single human gene has been identified for the TXA_2 receptor, this raised the possibility that the high affinity state of the platelet TXA_2 receptor associated with shape change could be the result of an interaction(s) with a specific G protein(s). $G\alpha_{12}$ and/or $G\alpha_{13}$ have been associated with cytosolic alkalinization, an early event in platelet activation and may be coupled to the platelet TXA_2 receptor [24]. Coupling of receptors to their respective G proteins may alter the affinity of the receptor for its ligands. Therefore, we chose to determine if co-transfection of the receptor with $G\alpha_{12}$ and/or $G\alpha_{13}$ would increase the affinity of the TXA_2 receptor for its agonist ligand ^{125}I-BOP, mimicking the high affinity platelet TXA_2 receptor. The K_d for COS-7 cells transfected with TXA_2 receptor alone was 1.12 ±0.05 nM (Table 1). Co-transfection of the TXA_2 receptor with $G\alpha_{12}$ had no apparent effect on the K_d for ^{125}I-BOP (Table 1). Co-transfection of the TXA_2 receptor with $G\alpha_{13}$ caused a significant decrease ($p<0.05$) in the K_d compared to TXA_2 receptor alone, reflecting an increased affinity of the HEL cell TXA_2 receptor for ^{125}I-BOP. Co-transfection of TXA_2 receptor cDNA with both $G\alpha_{12}$ and $G\alpha_{13}$ also gave a further significant decrease ($p<0.05$) in the K_d. The affinity of ^{125}I-BOP to bind at human platelet TXA_2 receptors is increased at pH 6.5 versus 7.4 [15]. At pH 7.4 COS-7 cells transfected with the TXA_2 receptor alone had a K_d of 5 ± 0.3 nM which was significantly ($p<0.05$) decreased to 2.9 ± 0.2 nM (n=3) when co-transfected with both $G\alpha_{12}$ and $G\alpha_{13}$. There was no significant difference in the B_{max} values for these experiments and only a single class of binding sites was detected.

Table 1. Radioligand binding studies in COS-7 cell membranes co-transfected with TXA_2 receptor (TXR), $G\alpha_{12}$, $G\alpha_{13}$, $G\alpha_q$ or $G\alpha i_2$ either individually or in combination.

	K_d (nM)	B_{max}(pmol/mg protein)
TXR	1.12 ± 0.05	13 ± 1.3
TXR + $G\alpha_{12}$	1.08 ± 0.08	17 ± 5.2
TXR +$G\alpha_{13}$	0.64 ± 0.06*	11 ± 3
TXR +$G\alpha_{12}$ +$G\alpha_{13}$	0.38 ± 0.08*	7.6 ± 1.2
TXR‡	1.17 ± 0.11	36 ± 3.2
TXR + $G\alpha_q$‡	0.7 ± 0.07*	32 ± 3.7
TXR + $G\alpha_{i2}$‡	0.92 ± 0.12	30 ± 4

The values are the mean ± S.E.M. n=4-6. * $p<0.05$ compared to TXR alone by ANOVA. ‡Separate series of transfection experiments. Reprinted by permission of the authors[18].

The TXA_2 receptor is also coupled to $G\alpha_q$[22] and $G\alpha_{i2}$ [25]. In a second series of experiments, we determined their effects on the affinity of the TXA_2 receptor for ^{125}I-BOP.

Co-transfection of the receptor with $G\alpha_q$ significantly ($p<0.05$) decreased the K_d of ^{125}I-BOP from 1.17± 0.11nM to 0.7 ± 0.07nM (n=6) (Table 1). No alteration in receptor affinity for ^{125}I-BOP was observed when COS-7 cells were co-transfected with TXA_2 receptor and $G\alpha_{i2}$ (Table 1). The B_{max} was unaltered in any of the co-transfection studies with $G\alpha_q$ and $G\alpha_{i2}$ compared to TXA_2 receptor alone (Table 1).

In order to establish that the $G\alpha$ subunits were expressed, Western blotting for the respective proteins was conducted. Western blotting of COS-7 cell membranes showed little or no endogenous expression of $G\alpha_{12}$ and $G\alpha_{13}$ [18]. Transfection of the $G\alpha_{13}$ cDNA, resulted in a significant increase in the apparent amount of protein. Although transfection of $G\alpha_{12}$ resulted in an apparent increase in its expression, it was considerably less than that of $G\alpha_{13}$ or the other $G\alpha$ subunits. COS-7 cells showed a basal endogenous expression of $G\alpha_q$ and $G\alpha_{i2}$, which was increased markedly upon transfection of the appropriate cDNA.

Effect of GTPγS on TXA_2 Receptor Affinity for ^{125}I-BOP in the Presence of $G\alpha_{12}$ and $G\alpha_{13}$.

Stable, non-hydrolyzable analogs of GTP will uncouple agonist-occupied receptors from the receptor-G protein complex. Due to the slow rate of GDP-GTP exchange and the reportedly low affinity of $G\alpha_{12}$ and $G\alpha_{13}$ for GTP analogues [36-38], we performed the incubations with GTPγS for 2 hours. There was no effect of GTPγS on COS-7 cell membranes transfected with either TXA_2 receptor alone or co-transfected with $G\alpha_{12}$ (Table 2). In contrast, in COS-7 cell membranes co-transfected with TXA_2 receptor and $G\alpha_{13}$ there was a significant decrease ($p<0.05$) in affinity for ^{125}I-BOP in the presence of GTPγS (Table 2). In the absence of GTPγS there was a significant ($p<0.05$) increase in the affinity of the receptor for ^{125}I-BOP when co-transfected with TXA_2 receptor and $G\alpha_{13}$, consistent with the data in Table 1.

Table 2. Effect of 100μM GTPγS on ^{125}I-BOP binding in COS-7 cell membranes transfected with TXR, $G\alpha_{12}$ and $G\alpha_{13}$.

	K_d (nM)	B_{max}(pmol/mg protein)
No GTPγS		
TXR	1.24 ± 0.07	21 ± 0.9
TXR +$G\alpha_{12}$	1.2 ± 0.11	31 ± 2.1
TXR +$G\alpha_{13}$	0.78 ± 0.05*	25 ± 1.5
Plus 100μM GTPγS		
TXR	1.17 ± 0.07	18 ± 0.7
TXR +$G\alpha_{12}$	1.10 ± 0.04	32 ± 1
TXR +$G\alpha_{13}$	1.04 + 0.04	29 + 1

Results are mean ± S.E.M (n=3). * $p<0.05$ compared to TXR alone by ANOVA. Reprinted by permission of the authors [18].

DISCUSSION

We investigated the possibility that the affinity state of the HEL cell TXA_2 receptor could be increased, mimicking the vascular receptor, by co-expressing it with different $G\alpha$ subunits in COS-7 cells. This study demonstrates that the TXA_2 receptor cloned from HEL cells changes its affinity state when co-expressed with specific G-protein α subunits. The low affinity platelet TXA_2 receptor mediates aggregation in a calcium-dependent manner which is predicted to be regulated via $G\alpha_q$. No data are available for possible G proteins mediating platelet shape change via the apparent high affinity TXA_2 receptor, which appears to be the same as the vascular receptor.

$G\alpha_{12}$ and $G\alpha_{13}$ activate Na^+/H^+ exchange [26, 27] and since alkalinization of platelet cytosol is one of the earliest events accompanying shape change, the possibility exists that the interaction of the TXA_2 receptor with one or both of these G proteins would allow it to assume a high affinity state. The radioligand binding studies support the notion that $G\alpha_{13}$ can increase the affinity of the HEL cell TXA_2 receptor for ^{125}I-BOP. It has been suggested previously that both $G\alpha_{12}$ and $G\alpha_{13}$ are capable of associating with the human platelet TXA_2 receptor [24]. In light of this observation we were surprised by the lack of any effect of $G\alpha_{12}$ on receptor affinity for ^{125}I-BOP at either pH 6.5 or 7.4, or in the presence of GTPγS when transfected with the TXA_2 receptor. Western blot analyses of COS-7 cell membranes showed that $G\alpha_{12}$ was expressed but at a much lower amount compared to $G\alpha_{13}$ although all $G\alpha$ subunits were transfected in the same plasmid, pcDNA3. The reason for the lower $G\alpha_{12}$ expression is unclear. Immunological screening of several cell lines, including platelets and HEL cells, has revealed an apparent lower level of expression of $G\alpha_{12}$ compared to $G\alpha_{13}$ [39]. Expression of a GTPase -deficient mutant of $G\alpha_{12}$ and $G\alpha_{13}$ resulted in an apparent lower functional effect for $G\alpha_{12}$ compared to $G\alpha_{13}$ [40]. Interestingly, when the HEL cell TXA_2 receptor was co-expressed with both $G\alpha_{12}$ and $G\alpha_{13}$, we observed a further decrease in the K_d compared to the receptor plus $G\alpha_{13}$ alone. This was unexpected because $G\alpha_{12}$ alone had no effect on the affinity of the TXA_2 receptor. However, it is not a simple stoichiometric phenomenon since doubling the amount of $G\alpha_{13}$ cDNA transfected did not produce a further lowering of the K_d and doubling the amount of $G\alpha_{12}$ cDNA transfected did not affect receptor affinity (unpublished observation). $G\alpha_q$, the G protein thought to be involved in platelet aggregation, also increased the affinity of the cloned TXA_2 receptor for ^{125}I-BOP, although not to the same extent as $G\alpha_{13}$. $G\alpha_{i2}$, like $G\alpha_{12}$, had no effect on receptor affinity.

This study presents the first pharmacological data that at least two of the G protein α subunits thought to interact with human platelet TXA_2 receptors are capable of regulating its affinity state. It must be borne in mind, however, that the transfection system used in the current studies is artificial, with the expession levels of receptor and G protein being in excess to the normal *in vivo* cell situation. In addition, the changes in affinity induced by $G\alpha_{13}$ and $G\alpha_q$, were only modest in magnitude, but probably specific since $G\alpha_{12}$ and $G\alpha_{i2}$ did not produce any effects on the affinity of the receptor. Further experiments using reconstituted systems where the molar ratios of receptor and G protein can be accurately controlled will address more directly the stoichiometric requirement and rate controlling steps in the regulation of receptor affinity for agonist ligands by $G\alpha$ subunits.

GTPγS failed to alter the affinity of the cloned TXA_2 receptor for ^{125}I-BOP when expressed in the absence of $G\alpha$ subunits. The lack of effect of GTPγS on the expressed TXA_2 receptor alone is consistent with a previous observation on the cloned K562 cell TXA_2 receptor expressed in HEK 293 cells [41]. We were able to observe a small, but significant effect of GTPγS in COS-7 cell membranes co-transfected with TXA_2 receptor and $G\alpha_{13}$, but not with TXA_2 receptor plus $G\alpha_{12}$ (Table 2). GTPγS decreased the affinity of the receptor for ^{125}I-BOP in the presence of $G\alpha_{13}$ to a value not significantly different to the affinity of the receptor alone. This supports the notion that $G\alpha_{13}$ interacting with the TXA_2 receptor can increase its apparent affinity for ^{125}I-BOP, and the interaction between the TXA_2 receptor and $G\alpha_{13}$ was disrupted by GTPγS.

Whether a separate gene for a high affinity TXA_2 receptor for ^{125}I-BOP exists remains to be determined. The recent discovery of the splice variant of the HEL cell TXA_2 receptor in endothelial cells and[20] and also in platelets [21] raises the possibliity that this receptor interacting with specific G protein α-subunits will possess the biochemical and pharmacologic characteristics of the high affinity platelet and vascular smooth muscle cell TXA_2 receptor. The endothelial cell receptor is a potential candidate for a high affinity TXA_2 receptor; unfortunately, the TXA_2 receptor antagonist [^{3}H]SQ29548 was used for pharmacological characterization, so a comparison cannot be made with the current data [20]. Since the splice variant receptor is 100% homologous with the HEL cell TXA_2 receptor up to the carboxy terminal tail, if it can assume a high affinity state, it would most likely arise from an interaction with one or more G proteins. The observation that the EP_3 receptor subtypes have their affinity state altered by interactions with different G proteins [42] and are alternatively spliced, raises the possibility that for the super family of eicosanoid receptors, the regulation of receptor affinity state by specific G proteins is a common phenomenon.

In summary, we have shown that the affinity state of the cloned TXA_2 receptor from the megakaryocyte-like HEL cell can be altered by co-transfection with different $G\alpha$ subunits.

ACKNOWLEDGEMENTS

This work was supported in part by NIH grant HL36836; South Carolina Affiliate of the American Heart Association, the Eli Lilly Pharmaceutical Company (CJA), an Institutional Postdoctoral Fellowship (CJA) and DK42486 (TWG).

REFERENCES

1. Halushka, P.V., D.E. Mais, P.R. Mayeux and T. Morinelli, Prostaglandin, thromboxane and leukotriene receptors. Annu Rev Pharmacol Toxicol, 29:213 (1989).
2. Hanasaki, K., T. Nakano and H. Arita, Receptor-mediated mitogenic effect of thromboxane A_2 in vascular smooth muscle cells. Biochem. Pharmacol., 40:2535 (1990).
3. Morinelli, T.A., K.E. Meier, L. Zhang and W.H. Newman, Thromboxane A_2/prostaglandin H_2 stimulated mitogenesis of coronary artery smooth muscle cells involves activation of MAP kinase and S6 kinase. J Biol Chem, 269(8):5693 (1993).
4. Dorn II, G.W., M.W. Becker and M.G. Davis, Dissociation of the contractile and hypertrophic effects of vasoconstrictor prostanoids in vascular smooth muscle. J. of Biol. Chem., 267:1 (1992).
5. Davis-Bruno, K.L. and P.V. Halushka, Molecular pharmacology and therapeutic potential of thromboxane A_2 receptor antagonists. Adv. in Drug Research, 25:173 (1994).
6. Hirata, M., Y. Hayashi, F. Ushikubi, Y. Yokota, R. Kageyama, S. Nakanishi and S. Narumiya, Cloning and expression of cDNA for a human thromboxane A_2 receptor. Nature, 349:617 (1991).
7. Nüsing, R.M., M. Hirata, A. Kakizuka, T. Edi, K. Ozawa and S. Narumiya, Characterization and chromosomal mapping of the human thromboxane A_2 receptor gene. J Biol Chem, 268:25253 (1993).
8. Mais, D.E., D. Saussy, LJr., A. Chaikhouni, P.J. Kochel, D.R. Knapp, N. Hamanaka and P.V. Halushka, Pharmacologic characterization of human and canine thromboxane A_2/prostaglandin H_2 receptors in platelets and blood vessels: Evidence for different receptors. J Pharmacol Exp Ther, 233:418 (1985).
9. Mais, D.E., C. Dunlap, N. Hamanaka and P.V. Halushka, Further studies on the effect of epimers of thromboxane A_2 antagonists on platelets and veins. Eur J Pharmacol, 111:125 (1985).
10. Masuda, A., D.E. Mais, J. Oatis, EJr. and P.V. Halushka, Platelet and vascular thromboxane A_2/prostaglandin H_2 receptors: Evidence for different subclasses in the rat. Biochem Pharmacol, 42:537 (1991).
11. Morinelli, T.A., A.K. Okwu, D.E. Mais, P.V. Halushka, V. John, C.K. Chen and J. Fried, Difluorothromboxane A_2 and stereoisomers: Stable derivatives of thromboxane A_2 with differential effects on platelets and blood vessels. Proc Natl Acad Sci, 86:5600(1989).
12. Morinelli, T.A., S. Niewiarowski, J.L. Daniel and J.B. Smith, Receptor-mediated effects of a PGH_2 analogue (U46619), on human platelets. Am J Physiol, 253:H1035 (1987).
13. Dorn, G.W., Distinct platelet thromboxane A_2/prostaglandin H_2 receptor subtypes. J Clin Invest, 84:1883 (1989).
14. Dorn, I., G.W. and A. DeJesus, Human platelet aggregation and shape change are coupled to separate thromboxane A_2-prostaglandin H_2 receptors. Am. J. Physiol., 260:H327 (1991).
15. Mayeux, P.R., T.A. Morinelli, T.C. Williams, E.S. Hazard, D.E. Mais, J.E. Oatis, D.A. Baron and P.V. Halushka, Differential effect of pH on thromboxane A_2/prostaglandin H_2 receptor agonist and antagonist binding in human platelets. J Biol Chem, 266:13752 (1991).
16. Sessa, W.C., P.V. Halushka, A. Okwu and A. Nasjletti, Characterization of the vascular thromboxane A_2/prostaglandin endoperoxide receptor in rabbit aorta regulation by dexamethasone. Circ Res, 67:1562 (1990).
17. Masuda, A., R. Mathur and P.V. Halushka, Testosterone increases thromboxane A_2 receptors in cultured rat aortic smooth muscle cells. Circ Res, 69:638(1991).
18. Allan, C.J., K. Higashiura, M. Martin, T.A. Morinelli, D.T. Kurtz, O. Geoffroy, G.P. Meier, T.W. Gettys and P.V. Halushka, Characterization of the Cloned HEL Cell Thromboxane A_2 Receptor: Evidence that the Affinity State Can be Altered by $G\alpha_{13}$ and $G\alpha_q$. J. Pharmacol. Exp. Therap., In press. (1996).
19. D'Angelo, D.D., M.G. Davis, S. Ali and G.W. Dorn II, Cloning and pharmacologic characterization of a thromboxane A_2 receptor from K562 (human chronic myelogenous leukemia) cells. J. Pharm. Exp. Ther., 271:1034 (1994).

20. Raychowdhury, M.K., M. Yukawa, L.J. Collins, S.H. McGrail, K.C. Kent and J.A. Ware, Alternative splicing produces a divergent cytoplasmic tail in the human endothelial thromboxane A_2 receptor. J. Biol. Chem., 269(30):19256 (1994).
21. Hirata, T., F. Ushikubi, A. Kakizuka, M. Okuma and S. Narumiya, Two Thromboxane A_2 Receptor Isoforms in Human Platelets. Opposite Coupling to Adenylyl Cyclase with Different Sensitivity to Arg^{60} to Leu Mutation. J. Clin. Invest., 97:949 (1996).
22. Shenker, A., P. Goldsmith, C.G. Unson and A.M. Spiegel, The G protein coupled to the thromboxane A_2 receptor in human platelets is a member of the novel Gq family. J Biol Chem, 266:9309(1991).
23. Knezevic, I., C. Borg and G.C. Le Breton, Identification of Gq as one of the G-proteins which copurify with human platelet thormboxane A_2/prostaglandin H_2 receptors. J Biol Chem, 268:26011(1993).
24. Offermanns, S., K.L. Laugwitz, K. Spicher and G. Schultz, G proteins of the G_{12} family are activated via thromboxane A_2 and thrombin receptors in human platelets. Proc. Natn. Acad. Sci., 91:504 (1994).
25. Ushikubi, F., K. Nakamura and S. Narumiya, Functional Reconstitution of Platelet Thromboxane A_2 Receptors with G_q and G_{i2} in Phospholipid Vesicles. Molecular Pharmacology, 46:808 (1994).
26. Voyno-Yasenetskaya, T., B.R. Conklin, R.L. Gilbert, R. Hooley, H.R. Bourne and D.L. Barber, $G\alpha13$ Stimulates Na-H Exchange. J. Biol Chem, 269:4721 (1994).
27. Dhanasekaran, N., M.V.V.S. Vara Prasad, S.J. Wadsworth, J.M. Dermott and G. van Rossum, Protein kinase C-dependent and -independent activation of Na^+/H^+ exchanger by $G\alpha_{12}$ class of G proteins. J. Biol. Chem., 269(16):11802 (1994).
28. Nakajima, M., M. Yamamoto, F. Ushikubi, M. Okuma, M. Fujiwara and S. Narumiya, Expression of thromboxane A_2 receptor in cultured human erythroleukemia cells and its induction by 12-0-tetradecanoylphorbol-13-acetate. Biochem Biophys Res Com, 158:958 (1989).
29. Mayeux, P.R., D.E. Mais, C. Carr and P.V. Halushka, Human erythroleukemia cells express functional thromboxane A_2/prostaglandin H_2 receptors. J Pharmacol Exp Ther, 250:923 (1989).
30. Matsuda, K., R.S. Mathur, M.E. Ullian and P.V. Halushka, Sex Steroid Regulation of Thromboxane A_2 Receptors in Cultured Rat Aortic Smooth Muscle Cells. Prostaglandins, 49:183 (1995).
31. Morinelli, T.A., J.E. Oatis, A.K. Okwu, D.E. Mais, P.R. Mayeux, A. Masuda, D.R. Knapp and P.V. Halushka, Characterization of an $[^{125}I]$-labelled thromboxane A_2/prostaglandin H_2 receptor agonist. J Pharmacol Exp Ther, 251:557 (1989).
32. Allan, C.J. and P.V. Halushka, Characterization of human peripheral blood monocyte thromboxane A_2. J. Pharmacol. Exp. Therap., 270:446(1994).
33. Gettys, T.W., K. Sheriff-Carter, J. Moomaw, I.L. Taylor and J.R. Raymond, Characterization and use of crude α-subunit preparations for quantitative immunoblotting of G proteins. Anal. Biochem., 220:82 (1994).
34. Scatchard, G., The interaction of proteins with small molecules and ions. Ann N Y Acad Sci, 51:660 (1949).
35. Munson, P.J. and D. Rodbard, LIGAND: A versatile computerized approach for characterization of ligand binding systems. Anal Biochem, 107:220 (1980).
36. Kozasa, T. and A.G. Gilman, Purification of recombinant G proteins from Sf9 cells by hexahistidine tagging of associated subunits. J. Biol. Chem., 270:1734(1995).
37. Harhammer, R., B. Nurnberg, K. Spicher and G. Schultz, Purification of the G-protein G_{13} from rat brain membranes. Biochem. J., 303:135 (1994).
38. Singer, W.D., R.T. Miller and P.C. Sternweis, Purification and Characterization of the α subunit of G13. J. Biol. Chem., 269:19796 (1994).
39. Spicher, K., F. Kalkbrenner, A. Zobel, R. Harhammer, B. Nürnberg, A. Söling and G. Schultz, G12 and G13 α-subunits are immunochemically detectable in most membranes of various mammalian cells and tissues. Biochemical and Biophysical Research Communications, 198(3):906 (1994).
40. Prasad, M.V.V.S.V., J.M. Dermott, L.E. Heasley, G.L. Johnson and N. Dhanasekaran, Activation of Jun Kinase/Stress-activated Protein Kinase by GTPase-deficient Mutants of $G\alpha_{12}$ and $G\alpha_{13}$. J. Biol. Chem., 270:18655 (1995).

41. D'Angelo, D.D., M.G. Davis, S. Ali and G.W. Dorn II, Cloning and pharmacologic characterization of a thromboxane A_2 receptor from K562 (human chronic myelogenous leukemia) cells. J. Pharmacol. Exp. Therap., 271:1034 (1994).
42. Negishi, M., A. Irie, Y. Sugimoto, T. Namba and A. Ichikawa, Selective Coupling of Prostaglandin E Receptor EP3D to G_i and G_s through Interaction of α–Carboxylic Acid of Agonist and Arginine Residue of Seventh Transmembrane Domain. J. Biol. Chem., 270:16122 (1995).

THE ROLE OF CERAMIDE IN THE CELLULAR STRESS RESPONSE

Ghassan S. Dbaibo[1], Lina M. Obeid[2], and Yusuf A. Hannun[2]

[1] Department of Pediatrics
[2] Department of Medicine
Duke University Medical Center
Durham, NC 27710

INTRODUCTION

The cellular response to stressful environmental conditions is complex and only partially understood. Cells can be stressed by nutritional starvation, infection, physical or chemical conditions (e.g. heat, extreme pH), genotoxic drugs, or other conditions. During evolution, various stress responses were preserved to allow cells or organisms to withstand different forms of stress by organizing the repair of proteins and DNA that had been damaged by the particular stressful event. The specific stress response is probably determined and regulated by the nature of the insulting condition. Despite the importance of these processes, our knowledge remains limited[1-3].

Membrane sphingolipids are emerging as a reservoir of signaling molecules analogous to that provided by the glycerolipids. Long thought to function mostly as structural components of the cell membrane, as antigens, or as receptors for viral and bacterial products[4], sphingolipids and their metabolic products are now recognized as bioactive molecules with important roles in cell regulation[5-7]. The first evidence supporting a role for sphingolipids in signal transduction came from the finding that sphingosine, the lipid backbone molecule of lysosphingolipids, can reversibly inhibit protein kinase C[8]. More recently, studies from several laboratories revealed that ceramide, the structural backbone of sphingolipids, has multiple biologic activities which are predominantly growth-suppressive[9,10]. Specifically, ceramide was found to induce cell cycle arrest[11,12] and apoptosis[13,14]. These findings led us to speculate about the potential role of ceramide in modulating the cellular stress response. In this chapter we present our current understanding of the mechanisms by which ceramide exhibits its growth-suppressive activities with emphasis on recent developments in this rapidly evolving field.

Table 1. Inducers of ceramide accumulation

Inducer[1,2]	Cell Type	Biologic Effect
TNFα, γ-interferon, Vitamin D_3	HL-60 (promyelocytic leukemia)	Monocytic differentiation
NGF	T9 (glioma)	Glial differentiation
TNFα, Fas	Jurkat (T cell leukemia), SKW6.4 (B-lymphoblastoid) MCF-7 (breast carcinoma)	Apoptosis Clonal deletion
REAPER	Schneider cells (*Drosophila*)	Apoptosis
Anti-IgM, Steroids, Complement, CD 28	WEHI & JY25 (B lymphoma), primary T cells, Jurkat	Immune regulation, Clonal deletion, Apoptosis
Interleukin-1	Fibroblasts, EL-4 (leukemia)	cox gene expression, PGE2 or Interleukin-2 secretion
Ara-C, Vincristine, Daunorubicin	HL-60, Molt-4 (T cell leukemia) P388 & U937 (leukemia)	Growth inhibition Cell cycle arrest, Apoptosis
Irradiation, Heat Shock, Serum deprivation	WEHI, Aortic endothelial cells, NIH 3T3, Molt-4	Apoptosis, Cell cycle arrest, αB-crystallin upregulation

[1]TNFα, tumor necrosis factor α; NGF, nerve growth factor; Ara-C, 1-β-D-Arabinofuranosylcytosine; PGE2, prostaglandin E2.
[2]See references in text.

CERAMIDE AS A SECOND MESSENGER

The role of ceramide as a putative second messenger was first described in promyelocytic leukemia HL-60 cells which have the ability to differentiate into monocytes, macrophages, or granulocytes depending on the specific stimulus. Inducers of monocytic differentiation of these cells, such as 1α,25-dihydroxyvitamin D_3 (vitamin D_3) TNFα, and γ-interferon, were found to induce hydrolysis of membrane sphingomyelin resulting in the generation of ceramide and phosphorylcholine whereas inducers of differentiation into macrophages or granulocytes (phorbol esters and retinoic acid, respectively) were without effect[15-17]. The hydrolysis of sphingomyelin resulted from the action of a neutral sphingomyelinase and peaked at 30 minutes following TNFα and at 2 hours following vitamin D_3 treatment. Thereafter, sphingomyelin was slowly regenerated and ceramide levels returned to baseline by 2-4 hours. Importantly, the addition of exogenous, cell-permeable ceramide analogs resulted in monocytic differentiation of HL-60 cells within 2 hours; thus, implicating ceramide as the effector molecule. These studies defined the "sphingomyelin cycle" which is analogous to the "phosphatidylinositol cycle" in glycerolipids.

Subsequently, a number of inducers of the sphingomyelin cycle and ceramide accumulation were described and are summarized in table 1. This group of inducers is dominated by cytokines and components of the immune system involved in inflammation and host defense (e.g. TNFα, IL-1, γ-interferon, CD28, Fas, and complement)[15-23], or cytotoxic agents or conditions (e.g. chemotherapeutic agents, irradiation, serum deprivation, heat shock, TNFα, and Fas)[11,24-30] which induce apoptosis or cell cycle arrest. The kinetics of ceramide generation vary with the inducer or cell type. For example, ceramide generation occurred within minutes following TNFα or IL-1β treatment of Jurkat cells[31] or murine thymocytes[21], respectively, or within several hours following serum deprivation of Molt-4 cells[11], or treatment of MCF-7 breast carcinoma cells with TNFα (Dbaibo et. al, submitted). The ability of cell-permeable ceramide analogs to reproduce the biologic effects of the specific inducer in these different systems solidified the role of ceramide as a second messenger. Importantly, analogs, of the naturally occurring D-*erythro*-ceramide, were the most potent of the ceramide stereoisomers in producing growth suppression. Additionally, D-*erythro*-dihydroceramide, the metabolic precursor of ceramide which differs structurally only by the absence of the *trans*-4,5 double bond, was completely inert in producing any effects attributable to ceramide[13,32,33]. These specific structural and stereochemical requirements suggested that a specific biochemical target is directly activated by ceramide to produce its biologic effects. Indeed, a ceramide-activated protein phosphatase (CAPP), a member of the 2A family of serine/threonine protein phosphatases, was identified[34,35]. This phosphatase was specifically activated by ceramide but not by sphingosine or dihydroceramide and was shown to be an effector molecule for ceramide-induced biology[36]. Moreover, this activity was conserved in the yeast *saccharomyces cerevisiae*[37] underscoring the central importance of this pathway. Additionally, other potential downstream effectors of ceramide including ceramide-activated protein kinase[38] and protein kinase ζ[38] were identified but their role in mediating the biologic effects of ceramide remains to be proven.

ROLE OF CERAMIDE IN CELL CYCLE ARREST

Molecular Control Of The Cell Cycle

Induction of cell cycle arrest is not only important in the stress response but is also a central mechanism of cell growth regulation, differentiation, development, and tumorigenesis. The retinoblastoma protein (Rb) is now recognized as a central regulator of the cell cycle[40-42]. The activity of Rb is determined by its phosphorylation status of the more than dozen distinct serine and threonine residues[42-45]. The hypophosphorylated form is the active form and predominates in G_0/G_1 while Rb becomes gradually phosphorylated towards the end of G_1 where it loses its growth-suppressive effects. The mechanism of action of Rb appears to be through the ability of the hypophosphorylated form to bind and inactivate a number of transcription factors which are crucial for the transcription of growth promoting genes. These include the E2F family of transcription factors as well as Elf-1, Myo-D, PU.1, ATF-2, and c-Abl[46,47]. The hypophosphorylated form of Rb is targeted by a number of viral proteins such as E1A of adenovirus, large T antigen of simian virus 40 (SV40), and E7 of papillomavirus[46-51]. These viral proteins bind and inactivate Rb allowing the release of transcription factors important for viral replication.

Rb is phosphorylated by cyclin-dependent kinases (cdk) particularly those regulated by the D and E cyclins. The activity of these cyclin-cdk complexes can be inhibited by a number of proteins including p21, p16, p15, and p27 which are induced by a number of growth-suppressive signals including transforming growth factor β and activation of the tumor suppressor p53[42]. Dephosphorylation of Rb is due to the action of the recently described Rb phosphatase which functions independently of p53, p21, and the cyclin-cdk complexes[52]. Therefore, the regulation of Rb phosphorylation, and hence cell cycle control, is achieved by a balance of signals which activate or inhibit the Rb kinases or phosphatase.

Activation of Rb by Ceramide

Initial studies using serum deprivation as a model revealed that Molt-4 cells underwent a specific G_0/G_1 cell cycle arrest when grown in the absence of serum[11]. These effects were first seen at 24 hours following serum deprivation but became maximal (> 96 % in G_0/G_1) by 48 hours. This was accompanied by a gradual rise in endogenous ceramide which reached 8- and 15-fold control levels at 48 and 96 hours, respectively. This raised the possibility that ceramide may mediate the effects of serum deprivation on cell cycle arrest. Indeed, treatment of Molt-4 cells with cell-permeable ceramides, but not dihydroceramide, resulted in accumulation of cells in G_0/G_1 by 16 hours. These were the first studies to implicate ceramide as a long-term modulator of cell function.

Additional evidence supporting a role for ceramide in the induction of cell cycle arrest came from studies utilizing the glucosylceramide synthase inhibitor, PDMP (threo-1-phenyl-2-decanoylamino-3-morpholino-1-propanol)[53]. Treatment of NIH 3T3 cells overexpressing the insulin-like growth factor-1 receptor with PDMP resulted in a time dependent accumulation of these cells in G_0/G_1 and G_2 which was attributed to the noted increase in endogenous ceramide. Cells were able to proceed through the cell cycle once PDMP was removed and ceramide levels normalized.

The importance of Rb in regulation of the cell cycle led us to examine the effects of serum deprivation on the phosphorylation state of Rb. We found that Rb became dephosphorylated when Molt-4 cells were grown in the absence of serum starting at 24 hours and being complete by 96 hours. Next, we examined the effects of ceramide on Rb phosphorylation. Treatment of Molt-4 cells with cell-permeable ceramides resulted in a dose- and time-dependent dephosphorylation of Rb with significant effects seen after 4-6 hours using 10 μM of ceramide[12]. In order to ascertain the importance of Rb in mediating ceramide-induced growth suppression, cell lines lacking functional Rb due to mutation or due to the expression of viral proteins which inactivate Rb were compared with cells expressing functional Rb. Rb-deficient cells and were found to be relatively resistant to ceramide. More specifically, we compared the effects of ceramide on G_0/G_1 cell cycle arrest in Molt-4 cells expressing the adenoviral E1A protein versus vector cells. We found that, in contrast to the vector cells, the Molt-4-E1A cells, which lack functional Rb, did not arrest in G_0/G_1. Therefore, the growth suppressive effects of ceramide appear to be partially mediated through the activation of Rb which, in turn, induces cell cycle arrest. Future studies aim at defining the molecular targets involved in this pathway; specifically, the effects of ceramide on the cyclin-cdk activities and the Rb phosphatase.

ROLE OF CERAMIDE IN APOPTOSIS

Apoptosis

All multicellular organisms have an inherent genetic program which encodes proteins that execute physiologic cell death. This process allows the organism to dispose of unwanted cells in order to achieve proper development and maintain homeostasis. Under stressful conditions such as infection, unrepairable genetic damage, or nutritional or growth factor deprivation, the sacrifice of a few cells allows the organism to survive and reproduce[54]. Morphologically, this process of cell suicide is often characterized by typical changes termed "apoptosis"[55]. Following an apoptotic signal, cells are packaged into small membrane bound "apoptotic bodies" containing various cellular organelles. The cellular chromatin becomes condensed and internucleosomal cleavage of DNA occurs resulting in multiples of 180 bp fragments which give the appearance of a ladder on DNA gel electrophoresis[56]. The resulting apoptotic bodies subsequently become engulfed by phagocytes thus sparing the organism potentially injurious reactions which could result from the release of intracellular molecules[57].

Molecular Regulation of Apoptosis

Despite significant progress in our understanding of the molecular regulators and executors of apoptosis over the past few years, many questions remain. It appears that two protein families are intimately involved in the regulation of apoptosis. The first, is the family of cysteine proteases related to the Interleukin-1β converting enzyme (ICE), generally believed to be involved in the execution phase of apoptosis where they are responsible for targeting a number of key proteins[58]. At least nine family members have been identified including the nematode *Caenorhabditis elegans* death protease, CED-3. These proteases may work sequentially, in a cascade analogous to that of the complement and clotting pathways, to promote cell death[59,60]. Bcl-2 is the founding member of a second family of apoptosis-regulating proteins. This expanding family of proteins includes both pro- and anti-apoptotic members which have the ability to dimerize with other members or with themselves. The ratio of death-promoting (Bax, Bad, Bak, and Bcl-x_S) versus death-inhibiting (e.g. Bcl-2 and Bcl-x_L) members in a cell may determine its fate[61]. However, the molecular mechanism of action of these proteins remains unknown.

Induction of Apoptosis by Ceramide

Early studies exploring the growth-suppressive functions of ceramide revealed that cytotoxicity was contributing to the antiproliferative effects[16,33,62]. Since a significant number of ceramide inducers were also known to activate the apoptotic program, it became important to evaluate the role of ceramide in regulating these effects. The leukemia cell line, U937, which undergoes apoptosis in response to TNFα was initially used. Treatment of these cells with the cell permeable ceramide, N-acetylsphingosine (C_2-ceramide), resulted in apoptosis characterized by typical DNA fragmentation on gel electrophoresis[13]. These effects occurred within 1-3 hours and were specific to C_2-ceramide since treatment with dihydroceramide or other amphiphilic lipids was ineffective. Later studies demonstrated the ability of ceramide to induce apoptosis in a wide variety of cell types mimicking the effects of a number of inducers including radiation, vincristine, antibody cross-linking, Fas, and serum deprivation (Table 1)[11,14,22-28].

The next important question was the nature of the molecular events triggered by ceramide to promote apoptosis. In this respect, the ICE-like proteases became possible targets for ceramide action. The most studied substrate for these proteases is the enzyme poly(ADP-ribose) polymerase (PARP) which is cleaved by some, but not all, members of this family of proteases[63,64]. PARP, which is involved in DNA repair, becomes cleaved during apoptosis to a signature 85 kDa fragment. The protease with greatest activity against PARP was recently identified as CPP32 (prICE/Yama/apopain)[65-67]. Cleavage of PARP after treatment of Molt-4 cells with C_2- or C_6-ceramide was found to occur within 3-4 hours yielding the specific apoptotic fragment[68]. These effects were specific to ceramide and were not seen after treatment with dihydroceramide or dioctanoylglycerol. Therefore these studies defined a novel downstream target of ceramide closely coupled to its apoptotic activities.

Targeting of The Ceramide Pathway By Anti-apoptotic Molecules

The rapidly accumulating information about the various proteins which inhibit apoptotic pathways and the emerging role of ceramide as a central mediator in apoptosis led to the examination of the possibility that some of these proteins may specifically target the ceramide pathway. Studies using cells overexpressing Bcl-2 showed that these cells were resistant to apoptosis induced by ceramide and many of its inducers[25,69,70]. For example, after treatment of Molt-4 cells with vincristine, a vinca alkaloid chemotherapeutic agent, ceramide accumulated and the cells subsequently underwent apoptosis[25]. Overexpression of Bcl-2 in these cells protected them from the apoptotic effects of vincristine *and* cell permeable ceramide. However, ceramide accumulation was not inhibited indicating that Bcl-2 was functioning downstream of ceramide. Specifically, Bcl-2 inhibited PARP cleavage induced by ceramide indicating that it was targeting an event either at, or proximal to, CPP32 activation but distal to ceramide accumulation.

Another anti-apoptotic protein that we studied was CrmA (cytokine response modifier), a cowpox virus protein, which is a potent inhibitor of ICE[71]. In addition to shutting down Interleukin-1β production, CrmA was found to be a potent inhibitor of apoptosis[72]. Its anti-apoptotic activity was presumed to be secondary to its ability to inhibit the ICE family of proteases although it was a 1000-fold more potent in inhibiting ICE compared to CPP32. Surprisingly, CrmA failed to inhibit apoptosis and PARP cleavage induced by ceramide although it potently inhibited these effects when the inducer was TNFα (Dbaibo et. al., submitted). Using MCF-7 breast carcinoma cells, treatment with TNFα resulted in a several fold accumulation of endogenous ceramide starting at 7-9 hours and reaching 4-5 fold by 24 hours. Expression of CrmA in these cells completely inhibited ceramide accumulation and subsequent apoptosis. In contrast, overexpression of Bcl-2 did not affect TNFα-induced ceramide accumulation significantly although it protected from apoptosis which confirmed the findings in Molt-4 cells with vincristine. Evaluation of PARP cleavage revealed it was completely inhibited by CrmA when the inducer was TNFα but not when it was ceramide. These results led to the suspicion that at least two ICE-like proteases were involved in this pathway: one upstream of ceramide generation inhibitable by CrmA, while the other is ceramide-activated and is not potently inhibited by CrmA (Figure 1). To test this hypothesis, the alkylating agent mitomycin C was utilized to induce apoptosis in MCF-7 cells. At concentrations that readily kill these cells, mitomycin C did not cause significant ceramide accumulation but resulted in complete PARP cleavage. Although Bcl-2 inhibited mitomycin C-induced PARP cleavage, CrmA was ineffective. These studies indicated that these two anti-apoptotic proteins achieve their function by targeting two distinct points along the TNFα-induced ceramide pathway: CrmA works proximally on a protease upstream of ceramide accumulation while Bcl-2 works distally to ceramide accumulation at or proximal to CPP32 (Figure 1). Moreover, they indicated that

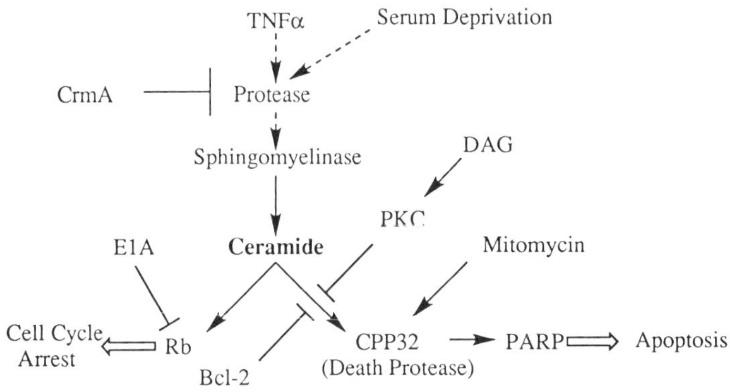

Figure 1. Schematic presentation of the pathways involved in the ceramide-regulated stress response.

proteases are not only important in the execution phase of apoptosis but also in the signaling phase represented by the generation of ceramide.

CERAMIDE AND THE STRESS-ACTIVATED KINASES

The stress-activated protein kinases (SAPKs) are presumably involved in the cellular response to intra- and extracellular stress such as heat shock, protein synthesis inhibition, ultraviolet light, and TNFα. A prototype of this family is the c-Jun N-terminal protein kinase (JNK) which preferentially phosphorylates the c-Jun and ATF2 transcription factors[73]. Since some of the effects of TNFα were known to be mediated by the sphingomyelin pathway, the role of sphingomyelinase and ceramide was investigated. Bacterial sphingomyelinase was found to reproduce the effects of TNFα on JNK activation in HepG2 cells[73]. C_2-ceramide and sphingomyelinase, but not dihydroceramide, activated JNK in HL-60 cells within 5 minutes[74]. Inhibition of the upstream kinases responsible for phosphorylating and activating JNK by expressing dominant negative mutants protected cells from apoptosis induced by NGF withdrawal[75], TNFα, or C_2-ceramide[76] although it did not interfere with ceramide generation. Therefore, these studies indicated that activation of the JNK pathway was potentially required for the ceramide-mediated stress response.

CONCLUSIONS

These studies support a role for ceramide as a central sensor/effector molecule in the stress response. Conditions which are stressful to the cell appear to converge on ceramide accumulation. However, this is not a terminal event as indicated by studies where apoptosis was prevented by Bcl-2 or inhibition of JNK activation. Additionally, elevation of diacylglycerol appeared to significantly attenuate ceramide-induced apoptosis indicating that elevation of ceramide may not be sufficient to initiate apoptosis[11,13]. Indeed, ceramide may function as a sensor which alerts the cell to impending danger. The fate of the cell is then determined by the degree of expression of proteins, such as Bcl-2, intracellular level of diglycerides, or the level of activation of the mitogen-activated protein kinase (MAPK) pathway which appears to counterbalance the JNK pathway[75]. Under these conditions, apoptosis may be inhibited and the cells will be driven by the elevated ceramide to undergo cell cycle arrest which is not affected by either Bcl-2 or increased diacylglycerol[11,25]. This allows the cell to repair damage and recover from the imposed stress. In this context, the ceramide-induced transcriptional activation of αB-crystallin, one of the heat shock proteins which are thought to function by repairing misfolded proteins, may represent an important function for ceramide in enhancing recovery from stress[29].

Future studies will focus on the events that occur immediately proximal and distal to ceramide accumulation. Better understanding of the enzymes that regulate ceramide generation as well as the immediate targets of ceramide will provide valuable tools in exploring the complex pathways by which cells respond to stress.

REFERENCES

1. W. Hilt, and D. H. Wolf, Stress-induced proteolysis in yeast, *Mol. Microbiol.* 6:2437-2442 (1992).
2. A. J. Fornace Jr, J. Jackman, M. C. Hollander, B. Hoffman-Liebermann, and D. A. Liebermann, Genotoxic-stress-response genes and growth-arrest genes. gadd, MyD, and other genes induced by treatments eliciting growth arrest, *Ann. N. Y. Acad. Sci.* 663:139-153 (1992).
3. C. E. Canman, C. Y. Chen, M. H. Lee, and M. B. Kastan, DNA damage responses: p53 induction, cell cycle perturbations, and apoptosis., *Cold Spring Harb. Symp. Quant. Biol.* 59:277-286 (1994).
4. S. Hakomori, Glycosphingolipids in cellular interaction, differentiation, and oncogenesis, *Annu. Rev. Biochem.* 50:733-764 (1981).
5. Y. A. Hannun, and R. M. Bell, Functions of sphingolipids and sphingolipid breakdown products in cellular regulation, *Science* 243:500-507 (1989).
6. Y. A. Hannun, The Sphingomyelin cycle and the second messenger function of ceramide, *J. Biol. Chem.* 269:3125-3128 (1994).
7. R. Kolesnick, and D. W. Golde, The sphingomyelin pathway in tumor necrosis factor and interleukin-1 signaling, *Cell* 77:325-328 (1994).
8. Y. A. Hannun, C. R. Loomis, A. H. Merrill, Jr., and R. M. Bell, Sphingosine inhibition of protein kinase C activity and of phorbol dibutyrate binding in vitro and human platelets, *J. Biol. Chem.* 261:12604-12609 (1986).
9. Y. A. Hannun, and C. M. Linardic, Sphingolipid breakdown products: anti-proliferative and tumor-suppressor lipids, *Biochim. Biophys. Acta Bio-Membr.* 1154:223-236 (1993).
10. M. Pushkareva, L. M. Obeid, and Y. A. Hannun, Ceramide: an endogenous regulator of apoptosis and growth suppression, *Immunol. Today* 16:294-297 (1995).
11. S. Jayadev, B. Liu, A. E. Bielawska, J. Y. Lee, F. Nazaire, M. Y. u. Pushkareva, L. M. Obeid, and Y. A. Hannun, Role for ceramide in cell cycle arrest., *J. Biol. Chem.* 270:2047-2052 (1995).
12. G. S. Dbaibo, M. Y. Pushkareva, S. Jayadev, J. K. Schwarz, J. M. Horowitz, L. M. Obeid, and Y. A. Hannun, Retinoblastoma gene product as a downstream target for a ceramide-dependent pathway of growth arrest., *Proc. Natl. Acad. Sci. U S A* 92:1347-1351 (1995).
13. L. M. Obeid, C. M. Linardic, L. A. Karolak, and Y. A. Hannun, Programmed cell death induced by ceramide, *Science* 259:1769-1771 (1993).
14. W. D. Jarvis, R. N. Kolesnick, F. A. Fornari, R. S. Traylor, D. A. Gewirtz, and S. Grant, Induction of apoptotic DNA damage and cell death by activation of the sphingomyelin pathway, *Proc. Natl. Acad. Sci. USA* 91:73-77 (1994).
15. T. Okazaki, R. M. Bell, and Y. A. Hannun, Sphingomyelin turnover induced by vitamin D_3 in HL-60 cells. Role in cell differentiation, *J. Biol. Chem.* 264:19076-19080 (1989).
16. T. Okazaki, A. Bielawska, R. M. Bell, and Y. A. Hannun, Role of ceramide as a lipid mediator of 1α,25-dihydroxyvitamin D_3-induced HL-60 cell differentiation, *J. Biol. Chem.* 265:15823-15831 (1990).
17. M. -Y Kim, C. Linardic, L. Obeid, and Y. Hannun, Identification of sphingomyelin turnover as an effector mechanism for the action of tumor necrosis factor alpha and gamma-interferon. Specific role in cell differentiation, *J. Biol. Chem.* 266:484-489 (1991).
18. F. Niculescu, H. Rus, S. Shin, T. Lang, and M. L. Shin, Generation of diacylglycerol and ceramide during homologous complement activation, *J. Immunol.* 150:214-224 (1993).
19. L. R. Ballou, C. P. Chao, M. A. Holness, S. C. Barker, and R. Raghow, Interleukin-1-mediated PGE2 production and sphingomyelin metabolism.

Evidence for the regulation of cyclooxygenase gene expression by sphingosine and ceramide, *J. Biol. Chem.* 267:20044-20050 (1992).
20. S. Mathias, A. Younes, C. -C Kan, I. Orlow, C. Joseph, and R. N. Kolesnick, Activation of the sphingomyelin signaling pathway in intact EL4 cells and in a cell-free system by IL-1β, *Science* 259:519-522 (1993).
21. L. M. Boucher, K. Wiegmann, A. Futterer, K. Pfeffer, T. Machleidt, S. Schütze, T. W. Mak, and M. Krönke, CD28 signals through acidic sphingomyelinase., *J. Exp. Med.* 181:2059-2068 (1995).
22. M. G. Cifone, R. De Maria, P. Roncaioli, M. R. Rippo, M. Azuma, L. L. Lanier, A. Santoni, and R. Testi, Apoptotic signaling through CD95 (Fas/Apo-1) activates an acidic sphingomyelinase., *J. Exp. Med.* 180:1547-1552 (1994).
23. C. G. Tepper, S. Jayadev, B. Liu, A. Bielawska, R. Wolff, S. Yonehara, Y. A. Hannun, and M. F. Seldin, Role of ceramide as an endogenous mediator of Fas-induced cytotoxicity, *Proc. Natl. Acad. Sci. USA* 92:8443-8447 (1995).
24. J. C. Strum, G. W. Small, S. B. Pauig, and L. W. Daniel, 1-β-D-arabinofuranosylcytosine stimulates ceramide and diglyceride formation in HL-60 cells, *J. Biol. Chem.* 269:15493-15497 (1994).
25. J. Zhang, N. Alter, J. C. Reed, C. Borner, L. M. Obeid, and Y. A. Hannun, Bcl-2 interrupts the ceramide-mediated pathway of cell death, *Proc. Natl. Acad. Sci., USA* (1996) *in press*.
26. R. Bose, M. Verheij, A. Haimovitz-Friedman, K. Scotto, Z. Fuks, and R. N. Kolesnick, Ceramide synthase mediates daunorubicin-induced apoptosis: an alternative mechanism for generating death signals, *Cell* 82:405-414 (1996).
27. A. Haimovitz-Friedman, C. C. Kan, D. Ehleiter, R. S. Persaud, M. McLoughlin, Z. Fuks, and R. N. Kolesnick, Ionizing radiation acts on cellular membranes to generate ceramide and initiate apoptosis., *J. Exp. Med.* 180:525-535 (1994).
28. J. Quintans, J. Kilkus, C. L. McShan, A. R. Gottschalk, and G. Dawson, Ceramide mediates the apoptotic response of WEHI 231 cells to anti-immunoglobulin, corticosteroids and irradiation, *Biochem. Biophys. Res. Commun.* 202:710-714 (1994).
29. Y. Chang, A. Abe, and J. A. Shayman, Ceramide formation during heat shock: A potential mediator of αB-crystallin transcription, *Proc. Natl. Acad. Sci., USA* 92:12275-12279 (1995).
30. G. J. Pronk, K. Ramer, P. Amiri, and L. T. Williams, Requirement of an ICE-like protease for induction of apoptosis and ceramide generation by REAPER, *Science* 271:808-810 (1996).
31. S. Schütze, K. Potthoff, T. Machleidt, D. Berkovic, K. Wiegmann, and M. Krönke, TNF activates NF-κB by phosphatidylcholine-specific phospholipase C-induced "acidic" sphingomyelin breakdown., *Cell* 71:765-776 (1992).
32. A. Bielawska, H. M. Crane, D. Liotta, L. M. Obeid, and Y. A. Hannun, Selectivity of ceramide-mediated biology: lack of activity of erythro-dihydroceramide, *J. Biol. Chem.* 268:26226-26232 (1993).
33. G. S. Dbaibo, L. M. Obeid, and Y. A. Hannun, TNFα signal transduction through ceramide: dissociation of growth inhibitory effects of TNFα from activation of NF-κB, *J. Biol. Chem.* 268:17762-17766 (1993).
34. R. T. Dobrowsky, and Y. A. Hannun, Ceramide stimulates a cytosolic protein phosphatase, *J. Biol. Chem.* 267:5048- 5051 (1992).
35. R. T. Dobrowsky, C. Kamibayashi, M. C. Mumby, and Y. A. Hannun, Caramide activates heterotrimeric protein phosphatase 2A, *J. Biol. Chem.* 268:15523-15530 (1993).
36. R. A. Wolff, R. T. Dobrowsky, A. Bielawska, L. M. Obeid, and Y. A. Hannun, Role of ceramide-activated protein phosphatase in ceramide-mediated signal transduction, *J. Biol. Chem.* 269:19605-19609 (1994).
37. J. D. Fishbein, R. T. Dobrowsky, A. Bielawska, S. Garrett, and Y. A. Hannun, Ceramide-mediated biology and CAPP are conserved in *Saccharomyces cerevisiae*, *J. Biol. Chem.* 268:9255-9261 (1993).
38. J. Liu, S. Mathias, Z. Yang, and R. N. Kolesnick, Renaturation and tumor necrosis factor-alpha stimulation of a 97-kDa ceramide-activated protein kinase, *J. Biol. Chem.* 269:3047-3052 (1994).

39. J. Lozano, E. Berra, M. M. Municio, M. T. Diaz-Meco, I. Dominguez, L. Sanz, and J. Moscat, Protein kinase C ζ isoform is critical for κB-dependent promoter activation by sphingomyelinase, *J. Biol. Chem.* 269:19200-19202 (1994).
40. R. A. Weinberg, The retinoblastoma gene and cell growth control, *Trends Biochem. Sci.* 15:199-202 (1990).
41. E. Harlow, For our eyes only, *Nature* 359:270-271 (1992).
42. R. A. Weinberg, The retinoblastoma protein and cell cycle control, *Cell* 81:323-330 (1995).
43. J. W. Ludlow, J. Shon, J. M. Pipas, D. M. Livingston, and J. A. DeCaprio, The retinoblastoma susceptibility gene product undergoes cell cycle-dependent dephosphorylation and binding to and release from SV40 large T, *Cell* 60:387-396 (1990).
44. D. W. Goodrich, N. P. Wang, Y. -W Qian, E. Y. -H.P. Lee, and W. -H Lee, The retinoblastoma gene product regulates progression through the G1 phase of the cell cycle, *Cell* 67:293-302 (1991).
45. J. A. DeCaprio, Y. Furukawa, F. Ajchenbaum, J. D. Griffin, and D. M. Livingston, The retinoblastoma-susceptibility gene product becomes phosphorylated in multiple stages during cell cycle entry and progression, *Proc. Natl. Acad. Sci. USA* 89:1795-1798 (1992).
46. J. R. Nevins, E2F: a link between the Rb tumor suppressor protein and viral oncoproteins, *Science* 258:424-429 (1992).
47. J. Y. Wang, E. S. Knudsen, and P. J. Welch, The retinoblastoma tumor suppressor protein, *Adv. Cancer Res.* 64:25-85 (1994).
48. S. Bagchi, P. Raychaudhuri, and J. R. Nevins, Adenovirus E1A proteins can dissociate heteromeric complexes involving the E2F transcription factor: a novel mechanism for E1A trans-activation, *Cell* 62:659-669 (1990).
49. S. Chellappan, V. B. Kraus, B. Kroger, K. Munger, P. M. Howley, W. C. Phelps, and J. R. Nevins, Adenovirus E1A, simian virus 40 tumor antigen, and human papillomavirus E7 protein share the capacity to disrupt the interaction between transcription factor E2F and the retinoblastoma gene product, *Proc. Natl. Acad. Sci. USA* 89:4549-4553 (1992).
50. J. W. Ludlow, Interactions between SV40 large-tumor antigen and the growth suppressor proteins pRB and p53, *FASEB J.* 7:866-871 (1993).
51. K. Vousden, Interactions of human papillomavirus transforming proteins with the products of tumor suppressor genes, *FASEB J.* 7:872-879 (1993).
52. Q. P. Dou, B. An, and P. L. Will, Induction of a retinoblastoma phosphatase activity by anticancer drugs accompanies p53-independent G1 arrest and apoptosis., *Proc. Natl. Acad. Sci. U S A* 92:9019-9023 (1995).
53. C. S. Rani, A. Abe, Y. Chang, N. Rosenzweig, A R. Saltiel, N. S. Radin, and J. A. Shayman, Cell cycle arrest induced by an inhibitor of glucosylceramide synthase. Correlation with cyclin-dependent kinases., *J. Biol. Chem.* 270:2859-2867 (1995).
54. D. L. Vaux, and A. Strasser, The molecular biology of apoptosis, *Proc. Natl. Acad. Sci., USA* 93:2239-2244 (1996).
55. J. F. R. Kerr, A. H. Wyllie, and A. R. Currie, Apoptosis: a basic biological phenomenon with wide-ranging implications in tissue kinetics, *Br. J. Cancer* 26:239-257 (1972).
56. A. H. Wyllie, Glucocorticoid-induced thymocyte apoptosis is associated with endogenous endonuclease activation, *Nature* 284:555-556 (1980).
57. L. E. Gerschenson, and R. J. Rotello, Apoptosis: a different type of cell death, *FASEB J.* 6:2450-2455 (1992).
58. S. J. Martin, and D. R. Green, Protease activation during apoptosis: death by a thousand cuts?, *Cell* 82:349-352 (1995).
59. A. M. Chinnaiyan, C. G. Tepper, M. F. Seldin, K. O'Rourke, F. C. Kischkel, S. Hellbardt, P. H. Krammer, M. E. Peter, and V. M. Dixit, FADD/MORT1 is a common mediator of CD 95 (Fas/APO-1) and tumor necrosis factor receptor-induced apoptosis, *J. Biol. Chem.* 271:4961-4965 (1996).

60. M. Enari, R. V. Talanian, W. W. Wong, and S. Nagata, Sequential activation of ICE-like and CPP32-like proteases during Fas-mediated apoptosis, *Nature* 380:723-726 (1996).
61. Z. N. Oltvai, and S. J. Korsmeyer, Checkpoints of dueling dimers foil death wishes, *Cell* 79:189-192 (1994).
62. A. Bielawska, C. M. Linardic, and Y. A. Hannun, Ceramide-mediated biology: determination of structural and stereospecific requirements through the use of N-acyl-phenylaminoalcohol analogs, *J. Biol. Chem.* 267:18493-18497 (1992).
63. S. H. Kaufmann, S. Desnoyers, Y. Ottaviano, N. E. Davidson, and G. G. Poirier, Specific proteolytic cleavage of poly(ADP-ribose) polymerase: an early marker of chemotherapy-induced apoptosis., *Cancer Res* 53:3976-3985 (1993).
64. Y. A. Lazebnik, S. H. Kaufmann, S. Desnoyers, G. G. Poirier, and W. C. Earnshaw, Cleavage of poly(ADP-ribose) polymerase by a proteinase with properties like ICE., *Nature* 371:346-347 (1994).
65. T. Fernandes-Alnemri, G. Litwack, and E. S. Alnemri, CPP32, a novel human apoptotic protein with homology to Caenorhabditis elegans cell death protein Ced-3 and mammalian interleukin-1 beta-converting enzyme., *J. Biol. Chem.* 269:30761-30764 (1994).
66. M. Tewari, L. T. Quan, K. O'Rourke, S. Desnoyers, Z. Zeng, D. R. Beidler, G. G. Poirier, G. S. Salvesen, and V. M. Dixit, Yama/CPP32 beta, a mammalian homolog of CED-3, is a CrmA-inhibitable protease that cleaves the death substrate poly(ADP-ribose) polymerase., *Cell* 81:801-809 (1995).
67. D. W. Nicholson, A. Ali, N. A. Thornberry, J. P. Vaillancourt, C. K. Ding, M. Gallant, Y. Gareau, P. R. Griffin, M. Labelle, Y. A. Lazebnik, N. A. Munday, S. M. Raju, M. E. Smulson, T. Yamin, V. L. Yu, and D. K. Miller, Identification and inhibition of the ICE/CED-3 protease necessary for mammalian apoptosis., *Nature* 376:37-43 (1995).
68. M. J. Smyth, D. K. Perry, J. Zhang, G. G. Poirier, Y. A. Hannun, and L. M. Obeid, prICE: a downstream target for ceramide-induced apoptosis and for the inhibitory action of bcl-2, *Biochem. J.* (1996) *in press*.
69. S. J. Martin, S. Takayama, A. J. McGahon, T. Miyashita, J. Corbeil, R. N. Kolesnick, J. C. Reed, and D. R. Green, Inhibition of ceramide-induced apoptosis by Bcl-2, *Cell Death Differ.* 2:253-257 (1995).
70. N. Karasavvas, R. K. Erukulla, R. Bittman, R. Lockshin, D. Hockenbery, and Z. Zakeri, BCL-2 suppresses ceramide-induced cell killing, *Cell Death Differ.* 3:149-151 (1996).
71. C. A. Ray, R. A. Black, S. R. Kronheim, T. A. Greenstreet, P. R. Sleath, G. S. Salvesen, and D. J. Pickup, Viral inhibition of inflammation: cowpox virus encodes an inhibitor of the interleukin-1β converting enzyme., *Cell* 69:597-604 (1992).
72. M. Tewari, and V. M. Dixit, Fas- and tumor necrosis factor-induced apoptosis is inhibited by the poxvirus crmA gene product., *J. Biol. Chem.* 270:3255-3260 (1995).
73. J. M. Kyriakis, P. Banerjee, E. Nikolakaki, T. Dai, E. A. Rubie, M. F. Ahmad, J. Avruch, and J. R. Woodgett, The stress-activated protein kinase subfamily of c-Jun kinases, *Nature* 369:156-160 (1994).
74. J. K. Westwick, A. E. Bielawska, G. Dbaibo, Y. A. Hannun, and D. A. Brenner, Ceramide activates the stress-activated protein kinases, *J. Biol. Chem.* 270:22689-22692 (1995).
75. Z. Xia, M. Dickens, J. Raingeaud, R. J. Davis, and M. E. Greenberg, Opposing effects of ERK and JNK-p38 MAP kinases on apoptosis., *Science* 270:1326-1331 (1995).
76. M. Verheij, R. Bose, X. H. Lin, B. Yao, W. D. Jarvis, S. Grant, M. J. Birrer, E. Szabo, L. I. Zon, J. M. Kyriakis, A. Haimovitz-Friedman, Z. Fuks, and R. N. Kolesnick, Requirement for ceramide-initiated SAPK/JNK signalling in stress-induced apoptosis, *Nature* 380:75-79 (1996).

SPHINGOSINE-1-PHOSPHATE - A NEW LIPID SIGNALING MOLECULE

Sarah Spiegel, Peter Burbelo, Olivier Cuvillier, Lisa Edsall, Elena Fuior, Burkhard Kleuser, Ana Olivera, Zhenxing Tu and Fang Wang

Department of Biochemistry and Molecular Biology
Georgetown University Medical Center
Washington, DC 20007

ABSTRACT

Sphingolipid metabolites have been suggested to play an important role in cellular signaling. Branching pathways of sphingolipid metabolism may regulate either apoptosis or mitogenic effects depending on the cell type and the nature of the stimulus. Ceramide produced by sphingomyelinase activation has been shown to induce apoptosis and cell growth arrest, whereas further metabolites of ceramide, sphingosine and sphingosine-1-phosphate (SPP), induce mitogenesis in many cell types and have been implicated as second messengers in cellular proliferation induced by PDGF and serum. Mitogens such as PDGF and serum, but not EGF, increase levels of sphingosine and SPP and also activate cytosolic sphingosine kinase activity. Dihydrosphingosine, an inhibitor of sphingosine kinase, reduces DNA synthesis induced by PDGF and serum but not by EGF. Thus elevation of sphingosine and SPP levels and activation of sphingosine kinase may have important biological roles in signal transduction pathways activated by specific growth factors. SPP mobilizes calcium from internal stores in an inositol trisphosphate-independent manner, increases phosphatidic acid levels by activation of PLD, and also activates extracellular signal regulated kinases (ERK), which are all prominent events in the control of cellular proliferation. SPP-induced activation of the transcription factor AP-1, one of the transcription factors activated by the ERK pathway, is the first link between the effects of sphingolipids metabolites on cellular proliferation and gene expression. Recently, we found that SPP prevents the appearance of the hallmarks of apoptosis, such as intranucleosomal DNA fragmentation and morphological changes, that result from elevations of ceramide. Our results suggest that the balance between intracellular levels of ceramide and SPP determines the fate of the cell.

INTRODUCTION

Much of the interest in sphingolipid second messengers stems from the seminal studies of Bell and Hannun which suggested that sphingosine can act as a negative regulator of protein kinase C (PKC) (Hannun and Bell, 1989). However, further studies revealed a more complex role for sphingosine. For example, in fibroblasts, sphingosine stimulates proliferation (Zhang et al., 1990a). This mitogenic effect is stereospecific in contrast to the lack of stereospecificity in inhibition of protein kinase C (Olivera et al., 1994). Moreover, fumonisin B1, a fungal inhibitor of ceramide synthase, increases the level of long-chain sphingoid bases and enhances cell growth of Swiss 3T3 fibroblasts (Schroeder et al., 1994). Sphingosine and SPP play a role in T cell proliferation, since cell growth inhibition induced by ISP-1/myriocin, a potent immuno-suppressant of structural similarity to sphingosine which inhibits serine palmitoyltransferase and sphingolipid metabolism, is completely abolished by the addition of sphingosine or SPP (Miyake et al., 1995). Sphingosine is rapidly taken up by cells and phosphorylated on its primary hydroxyl group by the action of a specific sphingosine kinase to form SPP (Zhang et al., 1991). Subsequent studies have suggested that the mitogenic effect of exogenous sphingosine is mediated, at least in part, via its conversion to SPP (Zhang et al., 1991).

ROLE OF SPHINGOSINE-1-PHOSPHATE IN GROWTH FACTOR SIGNALING

SPP was first described as a degradation product of long chain sphingoid bases. Our recent observation that intracellular levels of sphingosine, SPP, and sphingosine kinase activity are regulated by certain growth factors, such as PDGF, in quiescent Swiss 3T3 fibroblasts (Olivera and Spiegel, 1993), suggests that these breakdown products of cellular sphingolipids may be important mediators of the mitogenic action of PDGF (Olivera and Spiegel, 1993). These effects have also been observed in other cell types (Table 1). PDGF increases levels of sphingosine in vascular smooth muscle cells (Jacobs and Kester, 1993) and glomerular mesangial cells (Coroneos et al., 1995), and also increases SPP in arterial smooth muscle cells (Bornfeldt et al., 1995) and in airway smooth muscle cells (Pyne et al., 1996). Similarly, other mitogens such as the B subunit of cholera toxin and the phorbol ester 12-O-tetradecanoyl phorbol 13 acetate (TPA) significantly increases sphingosine kinase activity (Wang et al., 1996; Mazurek et al., 1994; Spiegel et al., 1994), whereas other mitogens like bombesin, bradykinin, insulin and EGF have little or no effect. Further support for the role of SPP in cellular proliferation arose from the use of a competitive inhibitor of sphingosine kinase, D,L-*threo*-dihydrosphingosine, which completely eliminates formation of SPP and selectively blocks cellular proliferation induced by PDGF (Olivera and Spiegel, 1993). It has been previously shown that several cytokines, including TNF-α, vitamin D_3, γ-interferon, and IL-1 stimulate the activity of sphingomyelinase resulting in the formation of ceramide which, in turn, mediates cell cycle arrest and apoptosis

(Kolesnick and Golde 1994; Hannun and Obeid, 1995). Collectively, these studies suggest that the dissociation of growth factor-induced mitogenesis from cytokine-mediated apoptosis may be due to distinct sphingolipid-derived second messengers. In agreement, PDGF, but not endothelin-1, mediates proliferation of glomerular mesangial cells, in part through ceramidase-regulated sphingosine formation, whereas inflammatory cytokines stimulate sphingomyelinase but not ceramidase activity leading to accumulation of ceramide (Coroneos et al., 1995).

Although the mitogenic effects of sphingosine and SPP have been shown to be widespread, in some cell types sphingosine inhibits cell growth either by inhibiting PKC or by inducing dephosphorylation of the retinoblastoma gene product leading to sequestration of E2F with the resultant loss of its DNA-binding and gene transcribing abilities (Dbaibo et al., 1995). In addition, it has been suggested that sphingosine may be an endogenous modulator of apoptosis during phorbol ester-induced differentiation of HL-60 cells (Ohta et al., 1995) and TNF-α-induced apoptosis in human neutrophils (Ohta et al., 1994). However, because ceramide and sphingosine are interconverted by simple acylation/deacylation reactions, further studies are needed to determine whether one or both are the active species or whether these are cell type specific responses.

Table 1. Changes in cellular levels of sphingolipid metabolites, sphingosine and sphingosine-1-phosphate by various stimuli.

Stimuli	Sphingolipid	Cell Type
Serum/plasma	Sphingosine	Human neutrophils
Serum/PDGF/TPA	Sphingosine/SPP	Swiss 3T3 fibroblasts
PDGF	Sphingosine	Vascular smooth muscle
PDGF	SPP	Arterial smooth muscle ells
PDGF	SPP	Airway smooth muscle cells
PDGF	Sphingosine	Mesangial cells
TPA	SPP	Balb/c 3T3 fibroblasts HL 60 , U937, Jurkat cells
Thrombin	SPP	Human platelets
Antigen clustering of FcεRI	SPP	Rat mast-cell line
Fumonisin B_1	Sphingosine Sphinganine	Many cell types, cattle, horses
Dexamethasone/ Phorbol dibutyrate	Sphingosine	Epstein Barr transformed human B lymphocytes

Calcium mobilization

Sphingosine and SPP play an important role in the mobilization of calcium from internal sources (Ghosh et al., 1990; Zhang et al., 1991; Mattie et al., 1994; Ghosh et al., 1994). This mobilization of calcium appears to involve an inositol trisphosphate-independent pathway and it should be noted that calcium release is the only early event in sphingosine-induced mitogenesis that has been demonstrated to be stereospecific. Calcium release is time- and temperature-dependent, implicating an intervening enzymatic step, consistent with the conversion of sphingosine to SPP (Ghosh et al., 1990). SPP mobilizes calcium from internal sources in many different cell types, including Swiss 3T3 fibroblasts (Mattie et al., 1994; Zhang et al., 1991), human foreskin fibroblasts (Chao et al., 1994) and REF52 fibroblasts and astrocytoma cells (Randriamampita and Tsien, 1993). The roles of sphingosine and SPP in calcium signaling following PDGF receptor stimulation were also investigated in oligodendrocytes (Fatatis and Miller, 1996). PDGF induced two distinct patterns of calcium signaling which were consequences of different modulatory roles played by sphingosine and SPP. The different delays associated with the two responses could result from the different times required for the production of sphingosine and SPP. PDGF is not only a mitogen for oligodendroglial cells but is also important for the development of oligodendrocyte progenitors. The different calcium mobilization patterns modulated by sphingosine and SPP may be related to the complex actions of PDGF on these cells. Recently, it was observed that calcium mobilization following FcεR1 engagement in mast cells is mediated by SPP (Choi et al., 1996). Inhibitors of sphingosine kinase such as DL-*threo*-dihydrosphingosine block the FcεR1-mediated calcium signal, but leave other kinase pathways intact. These finding suggest an exciting area in the development of SPP inhibitors as potential new antihistamine drugs. A sphingosylphosphorylcholine-gated calcium channel with unique pharmacological and electrophysiological properties has recently been characterized and cloned (Mao et al., 1996). However, it is not clear whether this channel is the target of SPP-induced calcium mobilization.

Phosphatidic acid

Sphingosine and SPP uniformly have a stimulatory effect on phosphatidic acid levels in many cell types (Spiegel and Milstien, 1995). In view of the potential roles of phosphatidic acid in cellular proliferation, signal transduction and activation of the MAP kinase cascade, a connection between long chain sphingoid bases and phosphatidic acid has important implications for the biological actions of sphingolipid metabolites. Therefore, it is not surprising that many studies have been directed toward examining these issues (reviewed in Spiegel and Milstien, 1995). Sphingosine and SPP have been shown to stimulate phospholipase D (PLD) activity in several cell types (Spiegel and Milstien, 1995). Phosphatidylcholine was the primary PLD substrate in sphingosine-treated NG108-15 neuroblastoma-glioma hybrid cells (Lavie and Liscovitch, 1990a) and in Swiss 3T3 fibroblasts (Zhang et al., 1990b) but a specific hydrolysis of phosphatidylethanolamine has also been observed in NIH 3T3 fibroblasts (Kiss and Anderson, 1990). A major pathway for sphingosine-induced increases in phosphatidic acid levels is the inhibition of phosphatidic acid phosphohydrolase (PAP). Sphingosine inhibited PAP in cell lysates from NG108-15 cells (Lavie et al., 1990b), neutrophils (Mullmann et al., 1991), in plasma membranes and endoplasmic reticulum fractions from rat liver (Jamal et al., 1991), and also in preparations from the yeast *Saccharomyces cerevisiae*. (Wu et al., 1993). However, in contrast to sphingosine, SPP has no effect on PAP activity in Swiss 3T3 cells, indicating that the most likely mechanism for SPP-induced phosphatidic acid accumulation in Swiss 3T3 fibroblasts is the activation of PLD (Desai et al., 1992). Similar results were obtained with bovine pulmonary artery endothelial cells in which SPP stimulated PLD but had no effect on membrane-associated PAP activity (Natarajan et al., 1994). It was suggested that sphingosine and SPP share a similar PLD stimulating activity but differ in their PAP inhibitory activity (Natarajan et al., 1994). Sphingosine can also activate an 80 kD form of diacylglycerol kinase *in vitro* , while inhibiting another isoform of 150 kD (Sakane et al., 1989). At least 4 different isoforms of diacylglycerol kinase have been detected in Jurkat T cells, of which two in the cytosol are activated by sphingosine (Yamada and Sakane, 1993). Thus, the effect of sphingosine on diacylglycerol kinase, and hence on phosphatidic acid production, could be different depending on the diacylglycerol kinase isozyme present in a particular cell. In contrast to sphingosine and SPP, ceramide inhibits PLD activation (Gomez-Munoz et al., 1994, Venable et al., 1994). Increased levels of ceramide may exhibit growth arresting effects by blocking the PLD signaling pathway, while the mitogenic effects of sphingosine and SPP may be due to the stimulation of PLD (Gomez-Munoz et al., 1994). The regulation of phosphatidic acid levels by sphingolipid metabolites indicates that the sphingolipid cycle could regulate the glycerophospholipid cycle. Cross-talk between these pathways adds another level of complexity to signaling pathways utilizing lipid metabolites and additional sites of regulation.

c-AMP

Both sphingosine (Zhang et al., 1990b) and SPP (Goodemote et al., 1995), rapidly and drastically decrease cellular cAMP levels in Swiss 3T3 fibroblasts. Pertussis toxin (PT) treatment prevents the inhibitory effect of sphingosine and SPP on cAMP accumulation, suggesting that a PT-sensitive G_i protein may be involved. In agreement, SPP suppresses cAMP accumulation in airway smooth muscle cells (Pyne et al., 1996). However, sphingosine has the opposite effect of cAMP in these cells. In contrast, in human smooth muscle cells, SPP increases cAMP levels, induces actin filament disassembly and PIP_2 turnover, and calcium mobilization (Bornfeldt et al., 1995). The relevance of changes in cAMP levels to mitogenic effects of sphingosine and SPP is still unclear.

Extracellular signal-regulated kinases

Activation of the mitogen-activated protein kinases (MAP kinases), also known as extracellular-signal-regulated kinases, is a key step of the signaling cascade to the nucleus initiated by binding of a growth factor to its receptor and subsequent tyrosine phosphorylation. The two best characterized ERK kinases are the 42 kDa p42mapk/ERK2 and the 44 kDa p44mapk/ERK2. Both enzymes are activated by dual specificity MAP kinase kinases (MKKs or MEKs) which are in turn regulated by serine phosphorylation by several kinases, including MEK kinase and Raf-1. Recent studies have demonstrated that SPP, similar to PDGF, rapidly and transiently activates Raf, MEK, and ERK kinases (Wu et al., 1995). In vitro experiments have not revealed any direct effects of SPP on Raf, MEK, or ERK kinase, implying action(s) upstream of Raf. The ability of SPP to stimulate PLD leading to an increase in cellular phosphatidic acid could be important since phosphatidic acid has been implicated in this pathway. Similarly, SPP induces the activation of ERK2 in airway smooth muscle cells (Pyne et al., 1996). AP-1, a transcription factor complex, consisting of homo and/or heterodimers of the fos and jun gene products, becomes activated following stimulation of the ERK signaling pathway and stimulates the transcription of several growth promoting genes. One link between sphingolipids metabolites, cellular proliferation, and gene expression involves the activation of AP-1 (Su et al., 1994). SPP not only stimulates the Raf/MKK/ERK pathway in 3T3 fibroblasts (Wu et al, 1995) it also markedly activates the transcriptional activity of AP-1 (Su et al., 1994). AP-1 may be a convergence point where signals carried by sphingolipid-derived messengers are integrated with the well-known intracellular second messages derived from glycerophospholipids, e.g. diacylglycerol.

SPHINGOSINE-1-PHOSPHATE INHIBITS CERAMIDE-INDUCED APOPTOSIS

Paradoxically, key regulatory components that play crucial roles in cell growth, such as p53, Rb-1, cyclin D1, c-fos, c-myc and $p34^{cdc2}$ kinase, have also recently been implicated in programmed cell death (Shi et al., 1994) and it has been suggested that the pathways of apoptosis and mitosis may be mechanistically related or tightly coupled (Harrington et al., 1994). Several studies suggest that activation of PKC by TPA or diacylglycerol analogs prevents the effects of TNF-α, Fas ligation and cell permeable ceramides on DNA fragmentation, suggesting that protein kinase C can antagonize the ceramide-mediated apoptotic pathway (Jarvis et al., 1994; Obeid et al., 1993). Moreover, like TNF-α, ionizing radiation induces rapid sphingomyelin hydrolysis to ceramide leading to apoptosis of bovine aortic endothelial cells and activation of protein kinase C blocks both radiation-induced apoptosis and the increase in the level of ceramide (Haimovitz-Friedman et al., 1994). Collectively, these results support the notion that protein kinase C can activate a signaling pathway that opposes the ceramide pathway. Recently, we found that the growth stimulating sphingolipid SPP acts similar to the PKC signaling pathway by opposing programmed cell death induced by the related sphingolipid, ceramide (Cuvillier et al., 1996). SPP prevents the appearance of the hallmarks of apoptosis, intranucleosomal DNA fragmentation and morphological changes, resulting from elevations of ceramide induced by TNF-α and Fas ligation, sphingomyelinase treatment, or by cell permeable ceramide analogs (Cuvillier et al., 1996). Furthermore, we established that stimulation of sphingosine kinase and concomitant increase in cellular SPP levels are responsible for the inhibition of ceramide-induced apoptosis resulting from activation of PKC (Cuvillier et al., 1996). The concentration of SPP required for these cytoprotective effects is similar to that previously shown to stimulate proliferation of various cell types and to activate AP-1 (Zhang et al., 1991; Su et al., 1994). However, in different cell types undergoing apoptosis it might be anticipated that the effective concentration of SPP could be different depending on basal levels of ceramide, the increases in ceramide induced by external stimuli, and the rate of degradation of SPP. Furthermore, high concentrations of SPP might be able to induce apoptosis in some cell types due to conversion to sphingosine, which might induce apoptosis (Ohta et al., 1995), or by re-acylation to ceramide.

It appears that the relative intracellular levels of ceramide and SPP are determining factors in the fate of the cell. This relationship may be analogous to the Bcl-2/Bax rheostat switch in which the signal must reach a critical threshold for a response and the extent of the response is dictated by a set point. In a similar manner, the balance between cellular levels of ceramide that favor death to levels of SPP that inhibit death is critical. Although the ceramide/SPP rheostat may be an inherent characteristic of each cell type, external stimuli can reset this ratio. Thus, the regulation of sphingomyelinase, ceramidase, and sphingosine kinase activities may determine whether signals flow through ceramide or SPP, altering the balance between these mitogenic and apoptotic sphingolipid signals. Ceramide levels are markedly elevated in human immunodeficiency virus (HIV)-infected T lymphocytes (Van Veldhoven et al., 1992) which are known to undergo apoptosis, and ceramide also induces HIV replication in chronically infected HL-60 cells (Rivas et al., 1993). These studies suggest a role for ceramide in the progression of AIDS (Hannun, 1994; Rivas et al., 1993). The effects of SPP in preventing apoptosis might be important for treatment of AIDS and other diseases in which apoptosis may contribute to the pathology, such as neurodegenerative diseases and ischemic stroke.

ACKNOWLEDGMENTS

This work was supported by research grants 1RO1 GM43880 and CA61774 from the National Institutes of Health. Peter Burbelo was supported by a Hitchings-Elion fellowship from the Wellcome Trust. Olivier Cuvillier was supported by a fellowship from the Association pour la Recherche contre le Cancer (France). Burkhard Kleuser was supported by a Deutsche Forschungsgemeinschaft Research fellowship.

The abbreviations used are: SPP, sphingosine-1-phosphate; TPA, 12-O-tetradecanoylphorbol 13-acetate; PDGF, platelet derived growth factor; BSA, bovine serum albumin; FCS, fetal calf serum; EGF, epidermal growth factor; AP 1, activator protein 1; MAPK, mitogen-activated protein kinase; ERK, extracellular signal-regulated kinase; TNF-α, tumor necrosis factor α.

REFERENCES

Bornfeldt, K. E., Graves, L.M., Raines, E.W., Igarashi, Y., Wayman, G., Yamamura, S., Yatomi, Y., Sidhu, J. S., Krebs, E. G., Hakomori, S. and Ross, R., 1995, Sphingosine-1-phosphate inhibits PDGF-induced chemotaxis of human arterial smooth muscle cells: spatial and temporal modulation of PDGF chemotactic signal transduction, *J. Cell. Biol.* 130:193.

Chao, C. P., Laulederkind, S. J. and Ballou, L. R., 1994, Sphingosine-mediated phosphatidylinositol metabolism and calcium mobilization, *J. Biol. Chem.* 269:5849.

Choi, O. Kim, J.H., and Kinet, J.P., 1996, Calcium mobilization by sphingosine kinase in signalling by the FcεRI antigen receptor, *Nature.* 380: 634.

Coroneos, E., Martinez, M., McKenna, S., and Kester, M., 1995, Differential regulation of sphingomyelinase and ceramidase activities by growth factors and cytokines: Implication for cellular proliferation and differentiation, *J. Biol. Chem.* 270: 23305.

Cuvillier, O., Pirianov, G., Kleuser, B., Vanek, P.G., Coso, O.A., Gutkind, J.S., and Spiegel, S., 1996, Suppression of ceramide-mediated programmed cell death by sphingosine-1-phosphate, *Nature.* in press.

Dbaibo, G. S., Wolff, R. A., Obeid, L M. and Hannun, Y. A., 1995, Activation of a retinoblastoma protein-dependent pathway by sphingosine, *Biochem. J.* 310:453.

Desai, N. N., Zhang, H., Olivera, A., Mattie, M. E. and Spiegel, S., 1992, Sphingosine-1-phosphate, a metabolite of sphingosine, increases phosphatidic acid levels by phospholipase D activation, *J. Biol. Chem.* 267:23122.

Fatatis, A. and Miller, R. J., 1996, Sphingosine and sphingosine-1-phosphate differentially modulate platelet-derived growth factor-BB-induced Ca^{2+} signaling in transformed oligodendrocytes, *J. Biol. Chem.* 271:295.

Ghosh, T. K., Bian, J. and Gill, D. L., 1990, Intracellular calcium release mediated by sphingosine derivatives generated in cells, *Science.* 248:1653.

Ghosh, T. K., Bian, J. and Gill, D. L., 1994, Sphingosine 1-phosphate generated in the endoplasmic reticulum membrane activates release of stored calcium, *J. Biol. Chem.* 269:22628.

Goodemote, K. A., Mattie, M. E., Berger, A. and Spiegel, S., 1995, Involvement of a pertussis toxin sensitive G protein in the mitogenic signaling pathways of sphingosine-1-phosphate, *J. Biol. Chem.* 270:10272.

Gomez-Munoz, A., Martin, A., O'Brien, L. and Brindley, D. N., 1994, Cell-permeable ceramides inhibit the stimulation of DNA synthesis and phospholipase D activity by phosphatidate and lysophosphatidate in rat fibroblasts, *J. Biol. Chem.* 269:8937.

Haimovitz-Friedman, A., Kan, C. C., Ehleiter, D., Persaud, R. S., McLoughlin, M., Fuks, Z. and Kolesnick, R. N., 1994, Ionizing radiation acts on cellular membranes to generate ceramide and initiate apoptosis, *J. Exp. Med.* 180:525.

Hannun, Y. A. and Bell, R. M., 1989, Functions of sphingolipids and sphingolipid breakdown products in cellular regulation, *Science.* 243:500.

Hannun, Y. A., 1994, The sphingomyelin cycle and the second messenger function of ceramide, *J. Biol. Chem.* 269:3125.

Hannun, Y. A. and Obeid, L. M., 1995, Ceramide: an intracellular signal for apoptosis. *TIBS.* 20:73.

Harrington, E. A., Fanadi, A. and Evan, G. I., 1994, Oncogenes and cell death. *Curr. Biol.* 4:120.

Jacobs, L. S. and Kester, M. ,1993, Sphingolipids as mediators of effects of platelet-derived growth factor in vascular smooth muscle cells, *Am. J. Physiol.* 265:c740.

Jamal, Z., Martin, A., Gomez-Munoz, A. , and Brindley, D. N., 1991, Plasma membrane fraction from rat liver contain a phosphatidate phosphohydrolase distinct from that in the endoplasmic reticulum and cytosol, *J. Biol. Chem.* 266:2988.

Jarvis, W. D., Fornari, F. A., Browning, J. L., Gewirtz, D. A., Kolesnick, R. N. and Grant, S., 1994, Attenuation of ceramide-induced apoptosis by diglyceride in human myeloid leukemia cells, *J. Biol. Chem.* 269:31685.

Kiss, Z. and Anderson, W. B., 1990, ATP stimulates the hydrolysis of phosphatidylethanolamine in NIH 3T3 cells. Potentiating effects of guanosine triphosphates and sphingosine, *J. Biol. Chem.* 265:7345.

Kolesnick, R. and Golde, D. W., 1994, The sphingomyelin pathway in tumor necrosis factor and interleukin-1 signaling, *Cell.* 77:325.

Lavie, Y. and Liscovitch, M., 1990a, Activation of phospholipase D by sphingoid bases in

NG108-15 neural-derived cells, *J. Biol. Chem.* 265:3868.
Lavie, Y., Piterman, O. and Liscovitch, M., 1990b, Inhibition of phosphatidic acid phosphohydrolase activity by sphingosine. Dual action of sphingosine in diacylglycerol signal termination, *FEBS Lett.* 277:7.
Mao, C., Kim, S. H., Almenoff, J. S., Rudner, X. L., Kearney, D. M. and Kindman, L. A., 1996, Molecular cloning and characterization of SCaMPER, a sphingolipid Ca^{2+} release-mediating protein from endoplasmic reticulum, *Proc. Natl. Acad. Sci. USA.* 93:1993.
Mattie, M. E., Brooker, G. and Spiegel, S., 1994, Sphingosine-1-phosphate, a putative second messenger, mobilizes calcium from internal stores via an inositol trisphosphate-independent pathway, *J. Biol. Chem.* 269:3181.
Mazurek, N., Megidish, T., Hakomori, S. and Igarashi, Y., 1994, Regulatory effect of phorbol esters on sphingosine kinase in BALB/C 3T3 fibroblasts (variant A31): demonstration of cell type-specific response -a preliminary note, *Biochem. Biophys. Res. Commun.* 198:1.
Miyake, Y. Kozutsumi, Y., Nakamura, S., Fujita, T. and Kawasaki, T., 1995, Serine palmitoyltransferase is the primary target of a sphingosine-like immunosuppressant, ISP-1/myriocin, *Biochem Biophys Res Commun.* 211:396.
Natarajan, V., Jayaram, H. N., Scribner, W. M. and Garcia, J. G. N., 1994, Activation of endothelial cell phospholipase D by sphingosine and sphingosine-1-phosphate, *Am. J. Respir. Cell Mol. Biol.* 11:221.
Obeid, L. M., Lenardic, C. M., Karolak, L. A. and Hannun, Y. A., 1993, Programmed cell death induced by ceramide, *Science.* 259:1769.
Ohta, H., Yatomi, Y., Sweeney, E.A., Hakomori, S and Igarashi, Y., 1994, A possible role of sphingosine in induction of apoptosis by tumor necrosis factor-alpha in human neutrophils, *FEBS Lett.* 355:267.
Ohta, H., Sweeney, E.A., Masamune, A., Yatomi, Y., Hakomori, S. and Igarashi, Y., 1995, Induction of apoptosis by sphingosine in human leukemic HL-60 cells: a possible endogenous modulator of apoptotic DNA fragmentation occurring during phorbol ester-induced differentiation, *Cancer Res.* 55:691.
Olivera, A., and Spiegel, S., 1993, Sphingosine-1-phosphate as second messenger in cell proliferation induced by PDGF and FCS mitogens, *Nature.* 365:557.
Olivera, A., Zhang, H., Carlson, R. O., Mattie, M. E., Schmidt, R. R. and Spiegel, S., 1994, Stereospecificity of sphingosine-induced intracellular calcium mobilization and cellular proliferation, *J. Biol. Chem.* 69:17924.
Pyne, S., Chapman, J., Steele, L., and Pyne, N.J., 1996, Sphingomyelin-derived lipids differentially regulate the extracellular signal-regulated kinases 2 (ERK-2) and c-Jun N-terminal kinase (JNK) signal cascades in airway smooth muscle, *Eur. J. Biochem.* 237:819.
Rivas, C. I., Golde, D. W., Vera, J. C. and Kolesnick, R. N., 1993, Involvement of the sphingomyelin pathway in TNF signaling for HIV poroduction in chronically infected HL-60 cells, *Blood.* 83:2191.
Ramdriamampita, C., and Tsien, R.Y., 1993, Emptying of intracellular Ca2+ stores releases a novel small messenger that stimulates Ca2+ influx, *Nature.* 364:809.
Sakane, F., Yamada, K. and Kanoh, H., 1989, Different effects of sphingosine, R59022 and anionic amphiphiles on two diacylglycerol kinase isozymes purified from porcine thymus cytosol, *FEBS Lett.* 255:409.
Schroeder, J. J., Crane, H. M., Xia, J., Liotta, D. C. and Merrill, A. H., 1994, Disruption of sphingolipid metabolism and stimulation of DNA synthesis by fumonisin B1. A molecular mechanism for carcinogenesis associated with Fusarium moniliforme, *J. Biol. Chem.* 269:3475.
Shi, L., Nishioka, W. K., Th'ng, J., M., B. E., Litchfield, D. W. & Greenberg, A. H., 1994, Premature p34cdc2 activation required for apoptosis, *Science.* 263:1143.
Spiegel, S., Olivera, A., Zhang, H., Thompson, E. W., Su,Y., and Berger, A., 1994, Sphingosine-1-phosphate, a novel second messenger involved in cell growth regulation and signal transduction, affects growth and invasiveness of human breast cancer cells, *Breast Canc. Res. Treat.* 31:95.
Spiegel, S. and Milstien S., 1995, Sphingolipid metabolites: members of a new class of lipid second messengers, *J. Membr. Biol.* 146:225.
Su, Y., Rosenthal, D., Smulson, M. and Spiegel, S., 1994, Sphingosine 1-phosphate, a novel signaling molecule, stimulates DNA binding activity of AP-1 in quiescent Swiss 3T3 fibroblasts, *J. Biol. Chem.* 269:16512.
Van Veldhoven, P. P., Matthews, T. J., Bolognesi, D. P. and Bell, R. M., 1992, Changes in

bioactive lipids, alkylglycerol and ceramide, occur in HIV-infected cells, *Biochem. Biophys. Res. Commun.* 187:209.

Venable, M. E., Blobe, G. C. and Obeid, L. M., 1994, Identification of a defect in the phospholipase D/diacylglycerol pathway in cellular senescence, *J. Biol. Chem.* 269:26040.

Wang, F., Buckley, N.E., Olivera, A., Kimberly A. Goodemote, K.A., Su, Y. and Spiegel, S., 1996, Involvement of sphingolipids metabolites in cellular proliferation modulated by ganglioside GM1, *Glycoconjugate J.* in press.

Wu, W. I., Lin, Y. P., Wang, E., Merrill, A. H. and Carman, G. M., 1993, Regulation of phosphatidate phosphatase activity from the yeast *Saccharomyces cerevisiae* by sphingoid bases, *J. Biol. Chem*, 268:13830.

Wu, J., Spiegel, S. and Sturgill, T.W., 1995, Sphingosine-1-phosphate rapidly activates the MAP kinase pathway by a G-protein dependent mechanism, *J. Biol. Chem.* 270:11484.

Yamada, K. and Sakane, F., 1993, The different effects of sphingosine on diacylglycerol kinase isozymes in Jurkat cells, a human T-cell line, *Biochim. Biophys. Acta.* 1169:211.

Zhang, H., Buckley, N. E., Gibson, K. and Spiegel, S., 1990a, Sphingosine stimulates cellular proliferation via a protein kinase C-independent pathway, *J. Biol. Chem.* 265:76.

Zhang, H., Desai, N. N., Murphey, J. M. and Spiegel, S., 1990b, Increases in phosphatidic acid levels accompany sphingosine-stimulated proliferation of quiescent Swiss 3T3 cells, *J. Biol. Chem.* 265:21309.

Zhang, H., Desai, N. N., Olivera, A., Seki, T., Brooker, G. and Spiegel, S., 1991, Sphingosine-1-phosphate, a novel lipid, involved in cellular proliferation, *J. Cell. Biol.* 114:155.

USE OF BODIPY™-LABELED SPHINGOLIPIDS TO STUDY MEMBRANE TRAFFIC IN ANIMAL CELLS

Richard E. Pagano and Chii-shiarng Chen

Mayo Clinic and Foundation
200 First Street, S.W.
Rochester, MN 55905-0001

INTRODUCTION

Fluorescent lipid analogs have been used to study membrane lipid traffic in animal cells for a number of years (reviewed in Pagano and Sleight, 1985; Koval and Pagano, 1991; Hoekstra and Kok, 1992; Koval, 1993; Rosenwald and Pagano, 1993). In this approach, one of the naturally occurring fatty acids of a lipid is replaced with a short-chain fluorescent fatty acid. Most of the resulting fluorescent lipid analogs can be readily integrated into cellular membranes by spontaneous lipid transfer from exogenous sources. The intracellular distribution of the labeled molecules can then be observed in living cells by high resolution fluorescence microscopy, and temporal changes in the distribution of a given lipid and its metabolites can be correlated with changes in its metabolism. This approach has been used to study the uptake, metabolism, and transport of phosphatidylcholine (Sleight and Pagano, 1984), phosphatidylserine and phosphatidylethanolamine (Tanaka and Schroit, 1983; Martin and Pagano, 1987), phosphatidylinositol (Ting and Pagano, 1990, 1991), phosphatidic acid (Pagano and Longmuir, 1985), ceramide (reviewed in Pagano, 1990; Rosenwald and Pagano, 1993), and various other sphingolipids (Koval and Pagano, 1989, 1990; Mayor et al., 1993; Kok et al., 1989; 1991; 1995; Schwarzmann and Sandhoff, 1990) analogs labeled with *N*-[6-[(7-nitrobenz-2-oxa-1,3-diazol-4-yl)amino]caproic acid (C_6-NBD-).

In the present Chapter we summarize recent studies on sphingolipid trafficking using lipids labeled with *N*-[5-(5,7-dimethyl BODIPY™)-1-pentanoic acid (C_5-DMB-). This probe is a particularly attractive alternative to NBD because the BODIPY™ fluorophore (i) has a higher fluorescence yield and is more photostable than NBD (Johnson et al., 1991), and (ii) is less polar than NBD and therefore is probably better anchored in the membrane bilayer than NBD which "loops back" to the membrane/water interface (Chattopadhyay and London, 1987; Wolf et al, 1992). The structures of the sphingolipids which will be discussed are given in Fig. 1 and correspond to C_5-DMB-labeled ceramide (Cer), glucosylceramide (GlcCer), and sphingomyelin (SM).

Frontiers in Bioactive Lipids
Edited by Vanderhoek, Plenum Press, New York, 1996

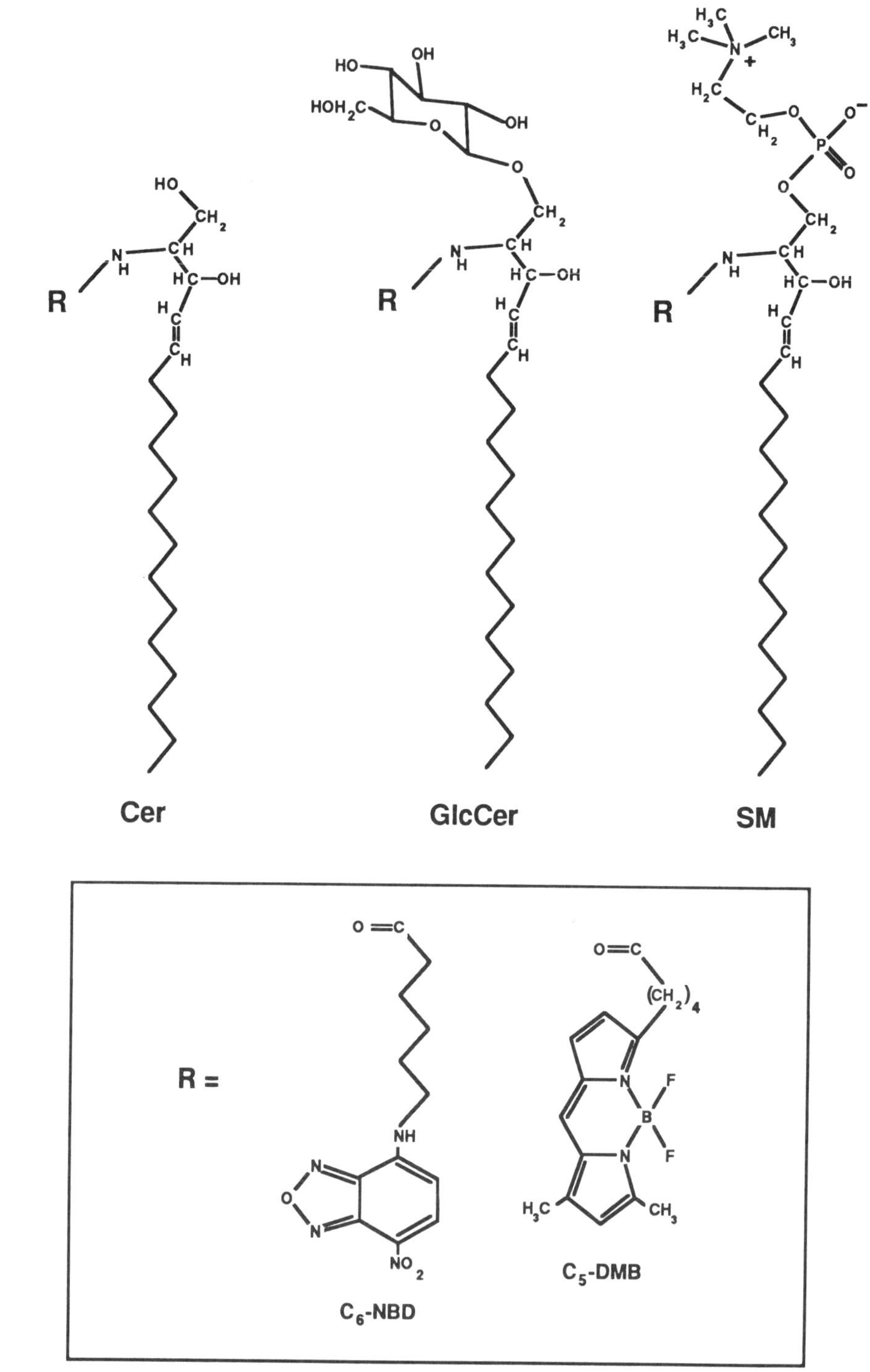

Figure 1. Structures of fluorescent analogs of ceramide (Cer), glucosylceramide (GlcCer), and sphingomyelin (SM). The R group (in box) corresponds to either *N*-[6-[(7-nitrobenz-2-oxa-1,3- diazol-4-yl)amino]caproic acid (C_6-NBD-) or to *N*-[5-(5,7-dimethyl BODIPY™)-1-pentanoic acid (C_5-DMB-).

SPECTRAL PROPERTIES OF BODIPY™-LIPIDS

The fluorescence emission spectra of lipid vesicles comprised of 1-palmitoyl, 2-oleoyl phosphatidylcholine (POPC) and various amounts of C_5-DMB-Cer are shown in Fig. 2. At low mole fractions, a single fluorescence emission peak at ~515 nm is observed. However, with increasing mole fractions this fluorescence decreases in intensity while a second peak of fluorescence appears at ~620 nm. The ratio of fluorescence intensities at 620 nm and 515 nm increases approximately linearly from 2 to 50 mol % C_5-DMB-Cer (Fig. 2, inset). Virtually identical results have been obtained using POPC vesicles containing C_5-DMB-SM in place of C_5-DMB-Cer, demonstrating that the shift in fluorescence emission is not limited to the ceramide analog. In addition, identical spectral shifts were obtained when C_5-DMB-Cer was incorporated into lipid vesicles prepared from dioleoylphosphatidylcholine (DOPC) or from lipids extracted from rat liver Golgi membranes, suggesting that the shift in fluorescence emission was not dependent on the presence of POPC (Pagano et al., 1991).

A number of control experiments have been carried out to examine the effect of various parameters on the fluorescence emission properties of C_5-DMB-lipids. No effect of pH on C_5-DMB-SM fluorescence in lipid vesicles was detected between pH 5 and 7.4, consistent with previous studies demonstrating that the BODIPY™ fluorophore is not pH sensitive (Karolin et al., 1994; Kang and Haugland, 1989). The fluorescence emission spectra of C_5-DMB-SM in artifical lipid vesicles was also not affected when the membrane potential across the vesicle bilayer was varied from 0 to 100 mV (data not shown). Finally, no differences in fluorescence emission spectra were detected among lipid vesicles ranging in size from ~30 to ~90 nm in diameter which were formed from DOPC and 20 mol % C_5-DMB-lipid, demonstrating that vesicle curvature did not affect the fluorescence emission of

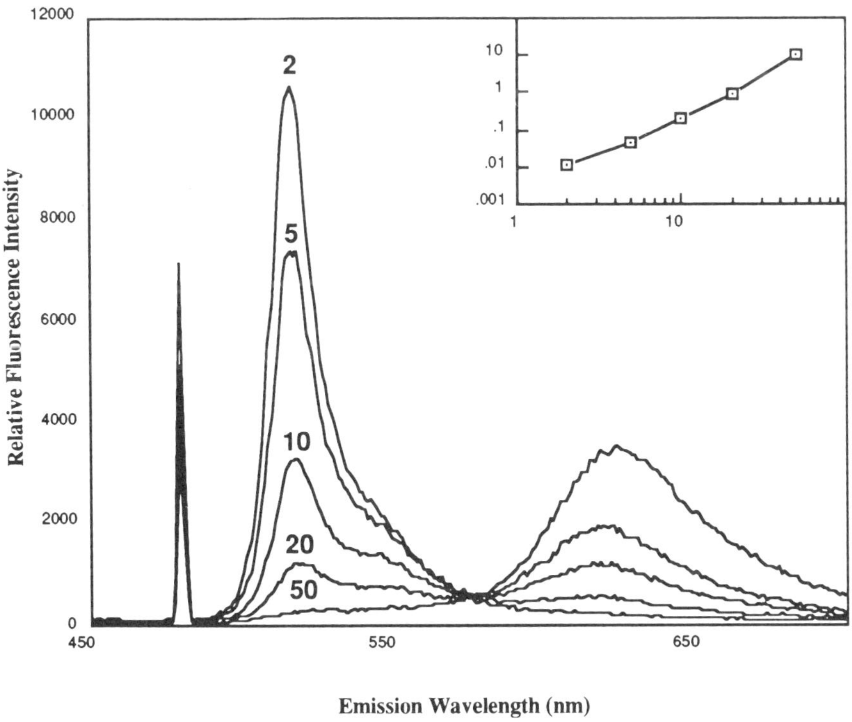

Figure 2. Fluorescence emission spectra of lipid vesicles containing various amounts of C_5-DMB-Cer. Emission spectra of small unilamellar vesicles formed from POPC and 2, 5, 10, 20, or 50 mol% C_5-DMB-Cer (*Total* lipid concentration was 31 μ*M*). Excitation was at 480 nm. *Inset:* ratio of relative fluorescence intensities at 620 nm and 515 nm *vs* mol% C_5-DMB-Cer. Reproduced from Pagano et al., 1991 by permission from The Rockefeller University Press.

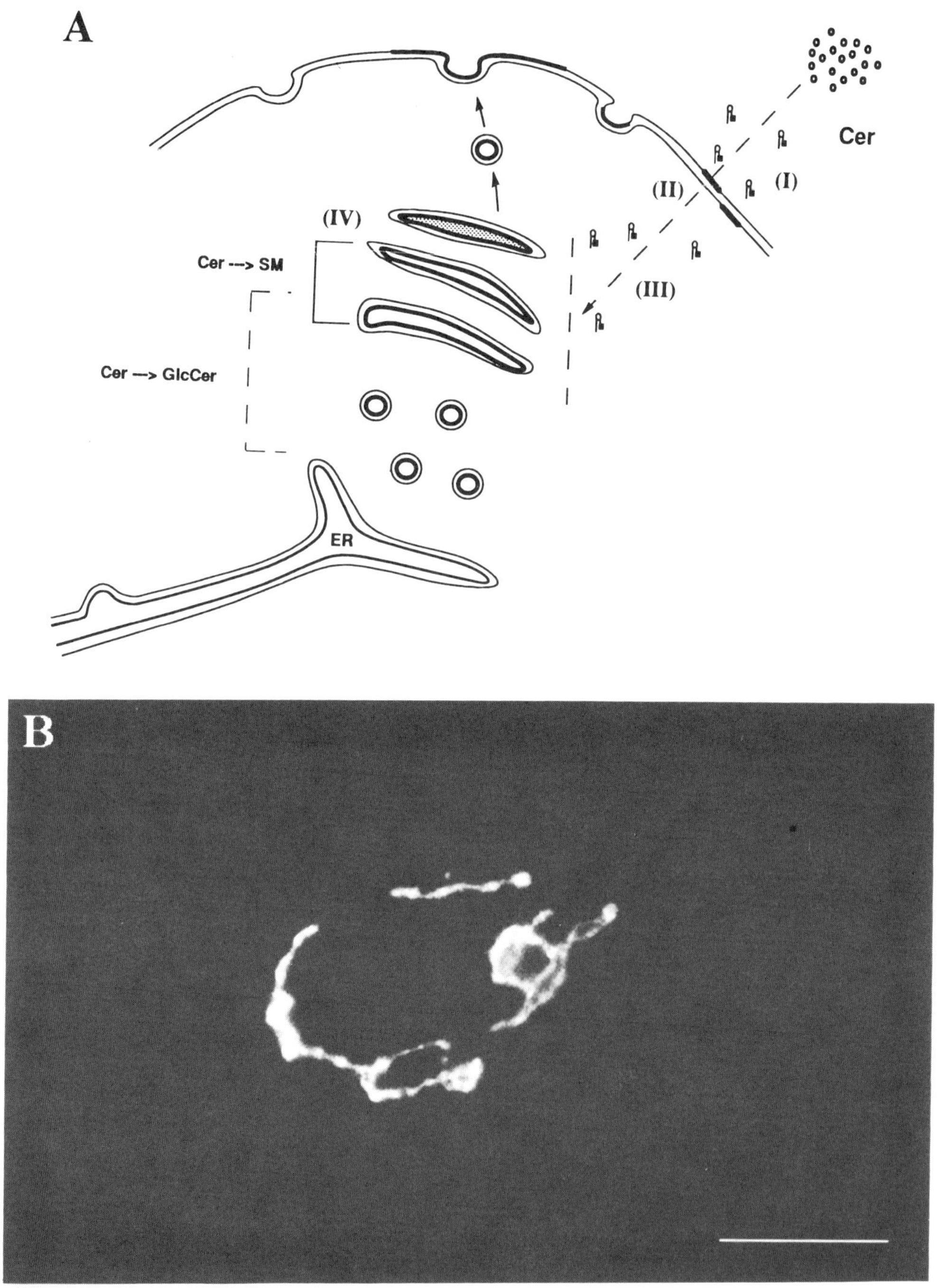

Figure 3. Uptake, distribution, and transport of fluorescent ceramide in cultured mammalian cells. (A) Working model (see text for details). Thick lines represent portions of the bilayer containing fluorescent lipid. (B) Fluorescence micrograph of human skin fibroblasts which were incubated with C_5-DMB-Cer for 30 min at 4°C, washed, and warmed to 37°C for 30 min. Samples were excited with blue light (450-490 nm) and observed in the red region of the spectrum (≥ 590 nm). Bar, 20 μm.

C_5-DMB-SM (data not shown). These results indicate that pH, membrane potential, and vesicle size have no effect on the spectral properties of C_5-DMB-lipids.

The spectral properties of C_5-DMB-lipids documented above can be used to differentiate membranes within living cells which contain high concentrations of the fluorescent lipid and consequently fluoresce red, from those which contain lower concentrations and fluoresce green. Typically, specimens are excited with blue light (450-470 nm) and visualized in the green (520-560 nm), green + red (≥ 520 nm), or red (≥ 590 nm) regions of the spectrum. Quantitative fluorescence microscopy and ratio imaging of specimens can also be used to estimate the concentration of the BODIPY™-lipid in cellular membranes. For details of these procedures the reader is referred to Pagano et al., 1991.

CELLULAR UPTAKE, TRANSPORT, AND METABOLISM OF C_5-DMB-CER (Pagano et al., 1991)

When cells are incubated with C_6-NBD-Cer or C_5-DMB-Cer, a striking concentration of fluorescence is seen in a perinuclear structure that corresponds to the Golgi apparatus (Lipsky and Pagano, 1983, 1985; Pagano et al., 1991). Subsequent studies have demonstrated that this distribution of intracellular fluorescence is the result of a series of steps as shown in Fig. 3A: (I) monomer transfer of the fluorescent lipid from an appropriate donor (e.g., liposomes containing fluorescent Cer or complexes of the fluorescent Cer with defatted BSA) to the plasma membranes of the recipient cells, (II) transbilayer movement from the outer to the inner leaflet of the plasma membrane, (III) spontaneous transfer of the fluorescent lipid to intracellular membranes, and finally (IV) concentration of the lipid at the Golgi complex.

The accumulation of C_5-DMB-Cer at the Golgi apparatus of living cells results in a striking change in fluorescence emission from green to red wavelengths (Pagano et al., 1991). Based on fluorescence ratio imaging microscopy of human skin fibroblasts treated with C_5-DMB-Cer, it was estimated that the concentration of this lipid and its metabolites at the Golgi apparatus is approximately 5-10 mol %. With time at 37°C, the fluorescent ceramide is metabolized to the corresponding fluorescent analogs of SM and GlcCer. Previous studies have shown that SM and GlcCer synthases are localized principally at the Golgi complex (Coste et al., 1986; Futerman et al., 1990; Jeckel et al., 1990; Futerman and Pagano, 1991). Unlike C_6-NBD-Cer, C_5-DMB-Cer is a relatively poor substrate for GlcCer synthase so that the major product formed in most cell types is C_5-DMB-SM (Pagano et al., 1991). The intracellular distribution of C_5-DMB-Cer and its metabolites is also observed to change with time at 37°C. The fluorescent lipid first accumulates in the Golgi apparatus (Fig. 3B), but at later times the plasma membrane becomes fluorescent, indicative of transport of fluorescent SM to the plasma membrane (data not shown). This latter process is temperature and energy dependent.

CELLULAR UPTAKE, TRANSPORT, AND METABOLISM OF C_5-DMB-SM (Martin and Pagano, 1994; Chen et al., 1995)

Figure 4A shows a model for the movement of C_5-DMB-SM from the plasma membrane to intracellular membranes. This lipid is integrated into the plasma membrane bilayer during incubation at low temperature. During subsequent warming to 37°C, the fluorescent lipid is internalized into numerous fluorescent vesicles which are scattered throughout the cytoplasm (Fig. 4B), and, by analogy to previous studies with C_6-NBD-SM, correspond to labeled endosomes (Koval and Pagano, 1990). This internalization process is ATP-dependent and is inhibited at low temperature. C_5-DMB-SM can be almost completely removed from the cell surface after low temperature labeling by incubation with defatted BSA ("back-exchange"), suggesting that the fluorescent lipid is present in the *outer leaflet* of the plasma membrane bilayer. The presence of discrete, "punctate dots" of fluorescence, rather than a more generalized labeling of many different intracellular membranes, suggests that C_5-DMB-SM is restricted to the *lumenal leaflet* of endocytic vesicles (topologically equivalent to the *outer leaflet* of the plasma membrane bilayer) during internalization and therefore unable to equilibrate and label other cellular membranes. This idea is consistent with our finding that C_5-DMB-SM (as well as other C_5-DMB-lipids) labels many different intracellular membranes in perforated cells where the plasma membrane was disrupted and a barrier to transmembrane movement of lipids no longer exists.

Recently we found that the spectral properties of C_5-DMB-SM can be used to detect unique populations of very early endosomes in cultured human skin fibroblasts (Chen et al., 1995). When cells are incubated with C_5-DMB-SM at 4°C, washed, briefly (≤1 min) warmed to 37°C to allow internalization to occur, and then "back-exchanged" with defatted BSA, C_5-DMB-SM is found to be distributed in a punctate pattern throughout the cytoplasm. Interestingly, within the same cell, some endosomes exhibit green fluorescence, while others emit red/orange fluorescence. Furthermore, the red/orange endosomes are usually seen at the periphery of the cell, while the green endosomes are more uniformly distributed throughout the cytoplasm. This mixed population of endosomes is seen after internalization times as short as 7 sec, and is also seen over a wide range of C_5-DMB-SM concentrations (1-25 μM). Quantitative fluorescence microscopy demonstrated that the apparent concentration of the lipid analog in the red/orange endosomes is several fold higher than its initial concentration at the plasma membrane, suggesting selective internalization ("sorting") of the lipid into a subset of early endosomes. Co-localization studies with anti-Transferrin receptor antibodies further demonstrated that this subpopulation of endosomes results from receptor-mediated endocytosis. We conclude that the spectral properties of C_5-DMB-SM can be used to distinguish unique populations of early endosomes from one another and to record dynamic changes in their number and distribution within living cells.

In addition to endosomes, some labeling of the Golgi apparatus is also when incubations are carried out over longer periods of time at 37°C. This most likely results from hydrolysis of C_5-DMB-SM at the plasma membrane and/or lysosomes (see Fig. 4) and subsequent transport of the resulting fluorescent ceramide to the Golgi apparatus which has a high affinity for this lipid (reviewed in Pagano, 1990; Rosenwald and Pagano, 1993). Finally it should be noted that internalized SM "recycles" back to the plasma membrane over time (see Koval and Pagano, 1989, 1990).

INTERNALIZATION AND SORTING OF C_5-DMB-GlcCer FROM THE PLASMA MEMBRANE TO THE GOLGI APPARATUS (Martin and Pagano, 1994)

Incubation of human skin fibroblasts with C_5-DMB-GlcCer/BSA results in a markedly different pattern of intracellular fluorescence than that seen with C_5-DMB-SM. Namely, when cells are incubated with C_5-DMB-GlcCer/BSA at low temperature, washed, and warmed to 37°C, the fluorescent lipid redistributes into punctate dots of fluorescence throughout the cytoplasm and rapidly labels other intracellular membranes (e.g., the nuclear envelope and Golgi apparatus) (Fig. 5).

A working model for the intracellular transport of C_5-DMB-GlcCer is given in Fig. 5A. During low temperature incubation, C_5-DMB-GlcCer labels the plasma membrane. However, 25-30% of the C_5-DMB-GlcCer incorporated into the plasma membrane at low temperature is resistant to back-exchange suggesting that a portion of the C_5-DMB-GlcCer is transported to the *inner leaflet* of the plasma membrane bilayer where it is inaccessible to back-exchange (Fig. 5, Step I). Once present on the *cytoplasmic leaflet* of the plasma membrane, C_5-DMB-GlcCer can move by spontaneous or protein-mediated diffusion to label other intracellular membranes (Fig. 5, Step II). During warming to 37°C, some C_5-DMB-GlcCer also enters the cell by an endocytic route (Fig. 5, Step III) and evidence has been presented that transbilayer movement of C_5-DMB-GlcCer also occurs in endosomes (Fig. 5, Step IV). The process of C_5-DMB-GlcCer transport from the plasma membrane and endosomes to label other intracellular membranes is inhibited at 10°C, but not by a wide variety of drugs, inhibitors, and enzymatic treatments (Martin and Pagano, 1994). Thus it seems likely that the transbilayer movement of C_5-DMB-GlcCer at the plasma membrane and endosomes and subsequent labeling of intracellular membranes by this lipid is not the result of a protein-mediated process.

The mechanism for accumulation of C_5-DMB-GlcCer at the Golgi apparatus (Fig. 5, Step V) is unknown. One intriguing possibility for this finding is related to the synthesis of

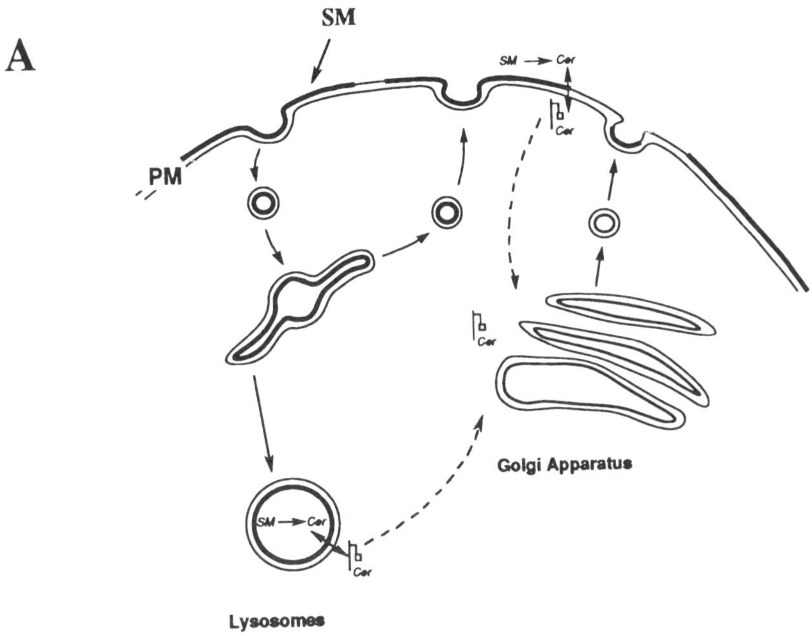

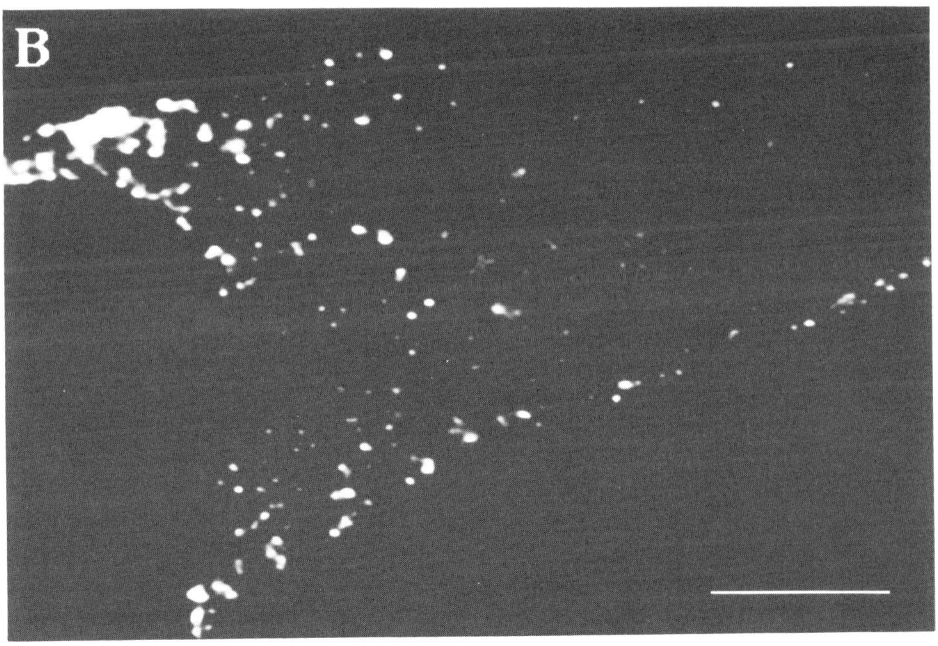

Figure 4. Internalization of fluorescent sphingomyelin from the plasma. (A) Working model (see text for details). Thick lines represent portions of the bilayer containing fluorescent lipid. (B) Fluorescence micrograph of human skin fibroblasts which were incubated with C_5-DMB-SM for 30 min at 4°C, washed, warmed to 37°C for 30 sec, and back-exchanged. Samples were excited with blue light (450-490 nm) and observed in the red region of the spectrum (≥ 590 nm). Bar, 20 μm.

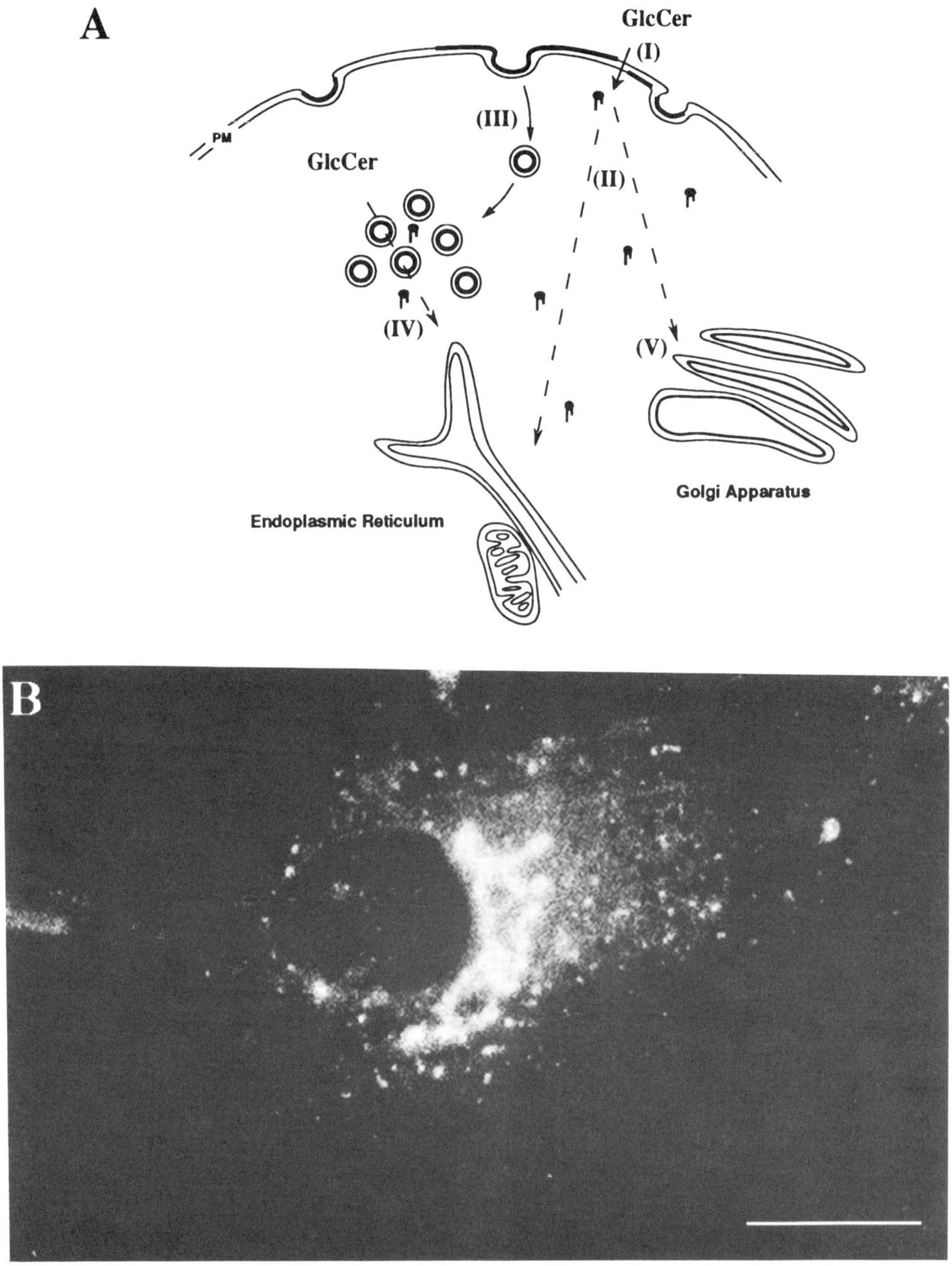

Figure 5. internalization and sorting of C_5-DMB-GlcCer from the plasma membrane to the Golgi apparatus. Working model (see text for details). Thick lines represent portions of the bilayer containing fluorescent lipid. (B) Cells were incubated with C_5-DMB-GlcCer/BSA for 30 min at 2°C, washed, warmed for 30 min at 37°C, and subsequently "back-exchanged." Samples were excited with blue light (450-490 nm) and observed in the green + red region of the spectrum (≥ 520 nm). Note the accumulation of fluorescence at the Golgi apparatus. Bar, 20 μm.

endogenous GlcCer. Previous studies have demonstrated that GlcCer synthase is enriched at the Golgi complex, and that synthesis of GlcCer occurs on the cytosolic surface of membrane vesicles derived from this organelle (Coste et al., 1985, 1986; Futerman and Pagano, 1991; Jeckel et al., 1992; Trinchera et al., 1991). These findings imply that after its synthesis, GlcCer must undergo transbilayer movement to the lumenal surface to account for the known topology of higher-order glycosphingolipids within the Golgi apparatus and plasma membrane. Thus, it is possible that C_5-DMB-GlcCer accumulates at the Golgi apparatus because it utilizes the same molecular machinery present at the Golgi apparatus for transbilayer movement of newly-synthesized endogenous GlcCer.

CONCLUSIONS

In this Chapter we have summarized some of the properties of the BODIPY™-fluorophore and highlighted known differences in the uptake, metabolism, and distribution of C_5-DMB-Cer, -GlcCer, and -SM in human skin fibroblasts. The unique spectral properties of the BODIPY™ fluorophore allow the investigator to discriminate among different labeled organelles within single living cells and, in conjunction with quantitative fluorescence microscopy, to determine the concentration of a given lipid and its metabolites at a given organelle. This methodology can easily be applied to other cell systems and it is anticipated that important new information about sphingolipid trafficking will be forthcoming using this approach.

ACKNOWLEDGEMENTS

Supported by U.S.P.H. Grant R37 GM-22942.

REFERENCES

Chattopadhyay, A. and E. London. 1987. Parallax method for direct measurement of membrane penetration depth utilizing fluorescence quenching by spin-labeled phospholipids. *Biochemistry* **26**:39-.

Chen, C.S., Martin, O.C., and R.E. Pagano. 1995. Internalization of a plasma membrane lipid in human skin fibroblasts: Insights from a sphingomyelin analog which reports concentration changes during endocytosis. *Molecular Biology of the Cell.* **6**: 231a.

Coste, H., M.B. Martel, and R. Got. 1986. Topology of glucosylceramide synthesis in Golgi membranes from porcine submaxillary glands. *Biochim. Biophys. Acta.* **858**: 6.

Futerman, A.H., Stieger, B., Hubbard, A.L., and Pagano, R.E. 1990. Sphingomyelin synthesis in rat liver occurs predominantly at the cis and medial cisternae of the Golgi apparatus. *J. Biol. Chem.* **265**: 8650.

Futerman, A.H. and Pagano, R.E. 1991. Determination of the intracellular sites and topology of glucosylceramide synthesis in rat liver. *Biochem. J.* **280**: 295.

Hoekstra, D. and J. W. Kok. 1992. Trafficking of glycosphingolipids in eukaryotic cells; sorting and recycling of lipids. *Biochim. Biophys. Acta.* **1113**: 277.

Jeckel, D. A. Karrenbauer, R. Birk, R.R. Schmidt, and F. Wieland. 1990. Sphingomyelin is synthesized in the cis Golgi. FEBS (Fed. Eur. Biochem. Soc.) Lett. **261**: 155.

Jeckel, D., A. Karrenbauer, K.N.J. Burger, G. Van Meer, and F. Wieland. 1992. Glucosylceramide is synthesized at the cytosolic surface of various Golgi subfractions. *J. Cell Biol.* **117**: 259.

Johnson, I. D., H. C. Kang, and R. P. Haugland. 1991. Fluorescent membrane probes incorporating dipyrrometheneboron difluoride fluorephores. *Anal. Biochem.* **198**: 228.

Kang, H. C and R. P. Haugland. 1989. Spectral properties of 4-sulfonato-3,3',5,5'-tetramethyl-2,2'-pyrromethen-1,1'-borondifluoride complex (Bodipy), its sodium slat, and protein derivatives. *Proc. SPIE Int. Soc. Opt. Eng.* **1063**: 68.

Karolin, J., L. B.-A. Johansson, L. Strandberg, and T. Ny. 1994. Fluorescence and adsorption spectroscopic properties of dipyrrometheneboron difluordie (BODIPY) derivatives in liquids, lipid membranes, and proteins. *J. Am. Chem. Soc.* **116**: 7801.

Kok, J. W., T. Babia, K. Klappe, and D. Hoekstra. 1995. Fluorescent, short-chain C_6-NBD-sphingomyelin, but not C_6-NBD-glucosylceramide, is subject to extensive degradation in the plasma membrane: implications for signal transduction related to cell differentiation. *Biochem. J.* **309**: 905.

Kok, J. W., T. Babia, D. Hoekstra. 1991. Sorting of sphingolipids in the endocytic pathway of HT29 cells. *J. Cell Biol.* **114**: 231.

Kok, J. W., S. Eskelinen, K. Hoekstra, and D. Hoekstra. 1989. Salvage of glucosylceramide by recycling after internalization along the pathway of receptor-mediated endocytosis. *Proc. Natl. Acad. Sci. USA.* **86**: 9896.

Koval, M. and R. E. Pagano. 1991. Intracellular transport and metabolism of sphingomyelin. *Biochim. Biophys. Acta* **1082**:113.

Koval, M. and R. E. Pagano. 1990. Sorting of an internalized plasma membrane lipid between recycling and degradative pathways in normal and Niemann-Pick, type A fibroblasts. *J. Cell Biol.* **111**: 429.

Koval, M. and R. E. Pagano. 1989. Lipd recycling between the plasma membrane and intracellular compartments: Transport and metabolism of fluorescent sphingomyelin analogues in cultured fibroblasts. *J. Cell Biol.* **108**: 2169.

Koval, M. 1993. Plasma membrane lipid transport in cultured cells: Studies using lipid analogs and model systems. *Advances in Cell and Molecular Biology of Membranes* **1**: 199.

Martin, O.C. and Pagano, R.E. 1987. Transbilayer movement of fluorescent analogs of phosphatidylserine and phosphatidylethanolamine at the plasma membrane of cultured cells: Evidence for a protein-mediated and ATP-dependent process(es). *J. Biol. Chem.* **262**: 5890.

Martin, O. C. and R. E. Pagano. 1994. Internalization and sorting of a fluorescent analog of Glucosylceramide to the Golgi apparatus of human skin fibroblasts: Utilization of endocytic and nonendocytic transport mechanisms. *J. Cell Biol.* **125**: 769.

Mayor, S., J. F. Presley, and F. R. Maxfield. 1993. Sorting of membrane components from endosomes and subsequent recycling to the cell surface occurs by a bulk flow process. *J. Cell Biol.* **121**: 1257.

Nichols, J. W. and R. E. Pagano. 1982. Use of resonance energy transfer to study the kinetics of amphiphile transfer between vesicles. *Biochmeistry* **21**: 1721.

Pagano, R.E. 1990. The Golgi apparatus: Insights from lipid biochemistry. *Biochem. Soc. Trans.* **18**: 361.

Pagano, R.E. and Longmuir, K.J. 1985. Phosphorylation, Transbilayer Movement, and Facilitated Intracellular Transport of Diacyl-glycerol are Involved in the Uptake of a Fluorescent Analog of Phosphatidic Acid by Cultured Fibroblasts. *J. Biol. Chem.* **260**: 1909.

Pagano, R. E. and R. G. Sleight. 1985. Defining lipid transport pathways in animal cells. *Science* **229**: 1051.

Pagano, R. E. and O. C. Martin. 1988. A series of fluorescent *N*-(Acyl)-sphingosines: Synthesis, physical properties, and studies in cultured cells. *Biochemistry* **27**: 4439.

Pagano, R. E., O.C. Martin, H.C. Kang, and R.P. Haugland. 1991. A novel fluorescent ceramide analog for studying membrane traffic in animal cells: Accumulation at the Golgi apparatus results in altered spectral properties of the sphingolipid precursor. *J. Cell Biol.* **113**: 1267.

Rosenwald, A. G. and R. E. Pagano. 1993. Intracellular transport of ceramide and its metabolites at the Golgi complex: insights from short-chain analogs. *Adv. Lipid Res.* **26**: 101.

Schwarzmann, G. and K. Sandhoff. 1990. Metabolism and intracellular transport of glycosphingolipids. *Biochemistry* **29**: 10865.

Sleight, R. G. and R. E. Pagano. 1984. Transport of a fluorescent phosphatidylcholine analog from the plasma membrane to the Golgi apparatus. *J. Cell Biol.* **99**: 742.

Struck, D. K. and R. E. Pagano. 1980. Insertion of fluorescent phospholipids into the plasma membrane of a mammalian cell. *J. Biol. Chem.* **255**: 5404.

Tanaka, Y. And A.J. Schroit. 1983. Insertion of fluorescent phosphatidylserine into the plasma membrane of red blood cells. Recognition by autologous macrophages. *J. Biol. Chem.* **258**: 11335.

Ting, A.E. and Pagano, R.E. 1990. Detection of a phosphatidylinositol-specific phospholipase C at the surface of Swiss 3T3 cells and its potential role in the regulation of cell growth. *J. Biol. Chem.* **265**: 5337.

Ting, A.E. and Pagano, R.E. 1991. Density-dependent inhibition of cell growth is correlated with the activity of a cell surface phosphatidylinositol-specific phospholipase C. *Eur. J. Cell Biol.* **56**: 401.

Trinchera, M., A. Fabbri, and R. Ghidoni. 1991. Topography of glycosyltransferases involved in the initial glycosylations of gangliosides. *J. Biol. Chem.* **266**: 20907.

Wolf, D. E., A. P. Winiski, A. E. Ting, K. M. Bocian, and R. E. Pagano. 1992. Determination of the transbilayer distribution of fluorescent lipid analogues by nonradiative fluorescence resonance energy transfer. *Biochemistry* **31**: 2865-2873.

BIOLOGICAL SIGNIFICANCE OF TUMOR GANGLIOSIDES: SHEDDING, TRANSFER, AND IMMUNOSUPPRESSION

Stephan Ladisch

Center for Cancer and Transplantation Biology
Children's Research Institute
Children's National Medical Center
111 Michigan Avenue, N.W.
Washington, D.C. 20010-2970

INTRODUCTION

Tumor formation is a complex, multiple-step process influenced by many factors. Local interactions between the tumor cell and the host are critical to tumor formation, and it is becoming recognized that the release of soluble factors by tumor cells, into the local tumor microenvironment, may dramatically alter the functions of host cells (1).

One of such factors is cell surface gangliosides, an important class of tumor-derived molecules, which may act as intercellular signaling molecules, according to the following hypothesis: Gangliosides synthesized by tumor cells and expressed in their cell membranes are shed in substantial quantities into the tumor cell microenvironment (2-5). Subsequently, these bioactive molecules bind to host immunocytes in this same microenvironment, and suppress the cellular anti-tumor immune responses (6). In this paper, new findings of our laboratory which support this hypothesis will be reviewed.

GANGLIOSIDE SHEDDING

It has been proposed since a number of years that the release of molecules by tumor cells may be important in evasion of host immune responses by tumors (3). It is now clear that ganglioside shedding is very common in tumor systems, and that the alterations in circulating gangliosides in tumor-bearing hosts are in fact caused by shedding by tumors (7). Studies have shown a correlation between increased circulating ganglioside concentrations and tumor burden (2) and the presence of specific tumor-derived gangliosides in the circulation (4, 8, 9). Our recent work, discussed, has shown a direct relationship between quantitative aspects of ganglioside shedding in human neuroblastoma and tumor progression in vivo (10). Since gangliosides have immunosuppressive and other biological properties, the study of these molecules has a high potential for elucidating a mechanism by which tumors escape host immune responses.

Shedding of human neuroblastoma tumor gangliosides in vitro. Because gangliosides shed by tumor cells are implicated in modulating tumor formation (see below), we have characterized both the quantitative and the qualitative aspects of ganglioside shedding by the human neuroblastoma cell line LAN-5 (8) in vitro. Having previously shown a high degree of ceramide structural heterogeneity among tumor gangliosides, and markedly different immunosuppressive activity according to ceramide structure of gangliosides, it became important to definitively document that these biologically active molecules are shed, as we hypothesized. We discovered that neuroblastoma ganglioside shedding is selective according to ceramide structure. In studying the shedding process in LAN-5 human neuroblastoma cells in vitro (5), we found that not only G_{D2}, but also the two other major gangliosides G_{M2} and G_{T1b} (mean total concentration=38 nmol LBSA/10^8 cells) were shed. These molecules are shed very rapidly (1-3 pmol/10^6 cells/h, or 0.5% of the total cell gangliosides/h). Every major carbohydrate species was shed in proportion to its concentration in the cell. However, selectively greater shedding of ceramide subspecies containing shorter chain fatty acids was found, and these are the same species with the most potent immunosuppressive activity (11). Confirmed in 3 other neuroblastoma cell lines (LAN-1, IMR, KCNR), high expression, rapid shedding, and selective shedding of certain ceramide species are therefore characteristic of human neuroblastoma tumors, and may result in significant in vivo accumulation of these biologically active molecules during tumor growth. The independence of shedding from cell density and proliferative state implies that in vivo even a tumor with a low proliferative index may have a high shedding rate.

Tumor progression is highly linked to neuroblastoma ganglioside shedding. Several earlier studies have shown a high level of G_{D2} ganglioside in the plasma of patients with neuroblastoma (12). In one study (9), the concentration of G_{D2} in patients of all clinical stages, studied prior to the initiation of treatment, was elevated to 545 $\pm$ 136 pmol/ml, in comparison to 198 $\pm$ 54 in patients with localized disease (Stages I and II), suggesting a relationship with widespread tumor (8). These findings raise the question of whether a direct relationship exists between the presence of this ganglioside and the outcome in neuroblastoma.

We therefore performed a clinical study to test this hypothesis and to answer the important question of the biological significance of ganglioside shedding we performed the clinical study relating tumor evolution to pretreatment circulating ganglioside concentration (10). More specifically, this study was designed to determine whether significant ganglioside shedding influences tumor progression. Ganglioside shedding was measured as circulating tumor-derived G_{D2}, and the outcome of 74 patients with advanced stage (III and IV) disease was determined. By Kaplan Meier analysis, progression-free survival (PFS) was found to be inversely related to circulating G_{D2} levels at the time of diagnosis ($p=0.018$). The outcome of the quartile of patients having the highest circulating G_{D2} levels (>568 pmol/ml) is seen to be strikingly different from that of the quartile of patients with the lowest (<103 pmol/ml) G_{D2} levels ($p=0.013$): median PFS was shorter (9 vs 28 months), and the long-term survival rate lower (2-year PFS of 24% vs 70%).

We conclude that more rapid disease progression and lower survival of patients with neuroblastoma are associated with high circulating G_{D2} levels at diagnosis. This study provided a first line of evidence in humans for a relationship between ganglioside shedding and tumor progression, and suggested a biologic relationship between shed tumor gangliosides (which are associated with circulating lipoproteins in vivo (13)) and tumor progression.

GANGLIOSIDE TRANSFER

Transport of immunosuppressive human neuroblastoma gangliosides. The mechanism of transport of gangliosides is a significant question that needs to be addressed regarding the shedding process by human neuroblastoma cells in vivo. To investigate this question in vivo we traced the serum localization of shed G_{D2}. Sera from patients with neuroblastoma tumors were separated into the lipoprotein fractions (VLDL, LDL, HDL) and lipoprotein-depleted serum. 73% of the total G_{D2} was present in the LDL fraction, while VLDL and HDL contained 21 and 6% respectively. Significantly, lipoprotein-depleted serum was devoid of G_{D2}. Thus, shed neuroblastoma tumor gangliosides are exclusively associated with the serum lipoprotein (and predominantly LDL) fractions in vivo (13). These findings have implications for the immunological detection of these molecules and ultimately for the development of approaches to their removal from the circulation of patients with cancer.

Transfer of tumor gangliosides. Having demonstrated that tumor gangliosides reach the circulation in vivo (and are transported by lipoproteins), we sought to complement this work by determining whether cell to cell transfer of tumor gangliosides could be documented (14). To answer this question we developed a dual chamber culture system in which metabolically labeled lymphoma cells (which shed gangliosides) were placed in one chamber and fibroblast target cells were placed in the other chamber. Thus in this system, the lymphoma cells shed radiolabeled gangliosides into the conditioned medium of the contralateral chamber containing the fibroblasts. The striking finding of these experiments was that the shed lymphoma cell gangliosides bound avidly to the target fibroblasts in a trypsin-resistant manner ($1x10^6$ and $7x10^6$ molecules/fibroblast in 24 and 48 hours). Since these are similar to binding rates of purified lymphoma gangliosides added exogenously, these results suggest that cell to cell ganglioside transfer is an efficient process in a system which models the in vivo microenvironment. Thus, these results document a process which has long been inferred to exist, the transfer of intact ganglioside molecules from one independent, viable cell population to another. The fact that the transfer of gangliosides from the lymphoma cells to the fibroblasts occurred at a relatively low concentration of shed gangliosides (0.1μM) underscores the potential biologic importance of the interaction of these bioactive molecules with target cells in the tumor microenvironment (14).

IMMUNOSUPPRESSION BY GANGLIOSIDES

Immunosuppression by gangliosides. The study of gangliosides as molecules which may regulate the cellular immune response has recently been reviewed (15). First studies documented inhibition of lymphoproliferation by normal brain gangliosides (16-18). The next phase in the study of gangliosides began with our initial work in the YAC-1 lymphoma, in which we demonstrated that *tumor* gangliosides markedly suppress the cellular immune response in vitro (6). These findings have been confirmed by other laboratories (19), leading to increasing acceptance of the hypothesis (6) that tumor-derived gangliosides may adversely affect host immune responses in vivo (20,21).

Gangliosides inhibit several steps in the cellular immune response, *in* vitro, including antigen processing/presentation (7), lymphocyte proliferation (6, 17, 18, 22), and cytotoxic effector function of NK cells (12, 23-25). Gangliosides also suppress major histocompatibility complex class I and class II genes in cultured astrocytes (26), and induce a selective and complete modulation of CD4 from the surface of T cells by endocytosis (27). However, in the past it has been difficult to delineate immunoregulatory effects of tumor gangliosides in vivo, partly due to the lack of large quantities of highly purified tumor gangliosides. To overcome this problem, we have studied both natural and chemically synthesized human neuroblastoma tumor gangliosides, all HPLC-purified, and tested their inhibitory effects on the antitumor cellular immune response in vivo.

Mechanisms of immunosuppression. Characteristic effects on two cell populations have been reported. First, gangliosides inhibit the accessory function (28), TNF production (29), and other functions (30) of adherent macrophage-enriched peripheral blood mononuclear cells. Gangliosides induce a reversible state of unresponsiveness of antigen processing cells to antigenic stimulation and it is quite likely that such a mechanism could be operative in the local environment of the tumor in vivo (28). Second, gangliosides impair the responsiveness of human lymphocytes to activation by a direct interference with the proliferation of mitogen-activated lymphocytes (31,32). This has been demonstrated to be due to an interference with IL-2 dependent cell proliferation, probably by direct binding of gangliosides to IL-2 (33). Gangliosides also inhibit cytotoxic effector function (34) and B cell immunoglobulin production (35). Most recently, modulatory effects of gangliosides on the expression of T-cell surface membrane proteins have been described (e.g., the modulation of CD4 by exogenous gangliosides; 27, 36).

We formulate these multiple effects of gangliosides into the concept that gangliosides mediate intercellular signaling. As intercellular signaling molecules, gangliosides released by tumor cells act by several mechanisms: they can associate with antigen-processing/ presenting cells to inhibit antigen presentation, interact with lymphokines (such as IL-2) to inhibit cytokine-mediated lymphocyte proliferation, alter the expression of cell surface receptor molecules such that the cell becomes less responsive to proliferative stimuli, or, as most recently suggested, affect intracellular signal transduction after insertion into the target cell membrane (37, 38). A major issue of controversy is the unequivocal demonstration of tumor ganglioside-induced immunosuppression in vivo.

Tumor gangliosides inhibit the murine allogeneic immune response in vivo. With the demonstration of potent in vitro inhibition of human cellular immune responses by tumor gangliosides, we therefore sought to determine the inhibitory effects of tumor gangliosides on the allogeneic immune response in vivo. We used a murine model of the allogeneic immune response developed by Kroczek et al. (39). This experimental system measures the cellular immune response in the lymph node draining the injection site of a foreign antigen in the form of allogeneic cells. (i) Inhibition of murine MHC class II-restricted allogeneic immune response by human neuroblastoma gangliosides (40). Four days after injection of allogeneic (C3H) spleen cells into the hind footpad of BALB/c mice, the cellular immune response in the draining popliteal lymph nodes is evidenced as an increase in lymph node mass (2-fold), lymphocyte number (6-fold) and lymphocyte DNA synthesis (6-fold). Purified human neuroblastoma gangliosides (10 nmols) co-injected with the stimulating allogeneic cells significantly suppressed this in vivo immune response. The increase in the lymph node mass was reduced by 65% (0.66 vs 1.89 mg), the increase in lymphocyte number (4.0×10^6 cells/node) was almost completely inhibited (1.1×10^6 cells/node), and in vitro [^{3}H]thymidine uptake by the lymphocytes recovered in vivo was reduced by 80% (40).

We also examined potential systemic toxicity of gangliosides administered in vivo. When gangliosides were coinjected in the footpad of mice with the allogeneic cells (inhibiting the immune response in vivo, Table 1), the injection site appeared normal, and no erythema was observed 24 hours after the ganglioside injection. The mice behaved normally and the ganglioside injection did not affect the body weight of mice. At the age of 10 weeks old, the body weight was 24.0±0.4 g (n=5) before the ganglioside injection, and 24.3±0.3 g 4 days later. The control group (n=5) weighed 24.2±1.2 g before the injection vs 24.8±1.1 g four days later. Thus, there was no short-term systemic toxicity associated with tumor ganglioside treatment.

Finally, to determine whether lipid-cell interactions of a nonspecific nature could cause inhibition of the in vivo allogeneic immune response attributed to gangliosides specifically, liposomes of cholesterol:lecithin (molar ratio 1:1) were prepared. Mice received either human gangliosides (5 nmol) or the liposomes (5 or 50 nmol) admixed with the allogeneic cells. As in the previous experiments, human neuroblastoma tumor gangliosides inhibited the allogeneic stimulation. In contrast, the liposomes had no inhibitory effect on either lymph node mass or lymph node lymphocyte number, supporting a specific inhibitory effect on the murine immune response by gangliosides. Thus, the results taken together provide strong evidence for an immunoregulatory effect of tumor gangliosides in vivo.

CONCLUSIONS AND FUTURE DIRECTIONS

As delineated above, the shedding, transport and immunologic activity of the shed tumor gangliosides makes it important to understand and possibly alter (downregulate) their metabolism, a future direction of our work. As a first approach, we have studied ganglioside metabolism in LAN-5 neuroblastoma cells and used retinoic acid as a probe (because of its known differentiating effects on neuroblastoma cells), characterizing its effect upon ganglioside synthesis and shedding (41). Three surprising observations were made: (i) 20 μM retinoic acid caused a marked (2-fold) increase in cellular ganglioside content, with a slight relative enhancement in G_{D1a} and G_{T1b} synthesis, (ii) ganglioside shedding increased in parallel with increased cellular ganglioside content, and (iii) retinoic acid caused a quantitatively similar increase in cell membrane phospholipids, which are also shed. We conclude that enhanced ganglioside synthesis and shedding by retinoic acid are part of a previously undescribed generalized stimulatory effect of retinoic acid on membrane lipid metabolism. The results suggest that retinoic acid may not be effective in downregulating ganglioside metabolism. Current studies are demonstrating that another molecule, PDMP, which inhibits glucosylceramide synthase, may be highly effective in inhibiting ganglioside synthesis and shedding (42). Studies such as these may yield new understanding of ganglioside metabolism, and result in abrogation of the shedding of immunosuppressive gangliosides by tumor cells.

ACKNOWLEDGMENTS

The author's work cited here was supported by grants CA61010 and CA52361 from the National Cancer Institute, and by the Stewart Trust, Phi Beta Psi Sorority, and The Children's Cancer Foundation.

REFERENCES

1. J. Travis, Do tumor-altered T cells depress immune responses? Science, *258*: 1732-1733, 1992.
2. T.M. Kloppel, T.W. Keenan, M.J. Freeman, and D.J. Morre, Glycolipid-bound sialic acid in serum: Increased levels in mice and humans bearing mammary carcinomas, Proc. Natl. Acad.Sci. USA, *74*: 3011-3013, 1977.
3. P.H. Black,. Shedding from the cell surface of normal and cancer cells, Adv. Cancer Res., *32*:75-199, 1980.
4. S. Ladisch, and Z.-L. Wu, Detection of a tumor-associated ganglioside in plasma of patients with neuroblastoma, Lancet, *i*: 136-138, 1985.
5. R. Li, and S. Ladisch, Shedding of human neuroblastoma gangliosides, Biochim. Biophys. Acta, *1083*: 57-64, 1991.
6. S. Ladisch, B. Gillard, C. Wong, and L. Ulsh, Shedding and immunoregulatory activity of YAC-1 lymphoma cell gangliosides, Cancer Res., *43*: 3808-3813, 1983.

7. S. Ladisch, and Z.-L. Wu, Circulating gangliosides as tumor markers, Prog. Clin. Biol. Res. *175*:277,1985.
8. J. Portoukalian, G. Zwingelstein, N. Abdul-Malak, and J.-F. Dore, Alteration of gangliosides in plasma and red cells of humans bearing melanoma tumors, Biochem. Biophys. Res. Commun., 85: 916-920, 1978.
9. S. Ladisch, Z.-L. Wu, S. Feig, L. Ulsh, E. Schwartz, G. Floutsis, F. Wiley, C. Lenarsky, and R. Sceger, Shedding of G_{D2} ganglioside by human neuroblastoma. Int. J. Cancer, 39: 73-76, 1987.
10. L., Valentino, T. Moss, E. Olson, H.-J. Wang, R. Elashoff, and S. Ladisch, Shed tumor gangliosides and progression of human neuroblastoma. Blood, *75*: 1564-1567, 1990.
11. S. Ladisch, R. Li, and E. Olson, Ceramide structure predicts tumor ganglioside immunosuppressive activity, Proc. Nat'l Acad. Sci., *91*:1974-1978, 1994.
12. G. Grayson, and S. Ladisch, Immunosuppression by human gangliosides: II. Carbohydrate structure and the inhibition of natural cytotoxicity, Cellular Immunol., *139*: 18-29,1992.
13. L.Valentino, and S. Ladisch, Localization of shed human tumor gangliosides: association with serum lipoproteins, Cancer Research, *52*: 810-814, 1992.
14. R.Olshefski, and S.Ladisch, Intercellular transfer of shed tumor cell gangliosides, FEBS Letters, in press.
15. E. V. Dyatlovitskaya, and L.D. Bergelson, Glycosphingolipids and antitumor immunity, Biochim. Biophys. Acta., 907: 125-143, 1987.
16. H.C.Miller, and W.J. Esselman, Modulation of the immune response by antigen reactive lymphocytes after cultivation with gangliosides, J. Immunol., *115*: 839-843, 1975.
17. J. Ryan, and M. Shinitzky, Possible role for glycosphingolipids in the control of immune responses, Eur. J. Immun., 9: 171-175, 1979.
18. R. Whisler, and A. Yates, Regulation of lymphocyte responses by human gangliosides. I. characteristics of inhibitory effects and the induction of impaired activation, J. Immunol., 125: 2106-2111, 1980.
19. T.A. Gonwa, M.A. Westrick, and B.A. Macher, Inhibition of mitogen- and antigen-induced lymphocyte activation by human leukemia cell gangliosides, Cancer Res., 44: 3467-3470, 1984.
20. S. Ladisch, S. Kitada, and E.F. Hays, Gangliosides shed by tumor cells enhance tumor formation in mice, J. Clin. Invest., 79: 1879-1882, 1987.
21. M. Hachida, R. Irie, and D.L. Morton, Significant immunosuppressive effect of ganglioside G_{M3} in organ transplantation, Transplantation Proceedings, 22: 1663-1665, 1990.
22. G. Floutsis, L.Ulsh, and S. Ladisch, Immunosuppressive activity of human neuroblastoma tumor gangliosides, Int. J. Cancer, 43: 6-9, 1988.
23. L.D. Bergelson, E.V. Dyatlovitskaya, T.E. Klyuchareva, E.V. Kryukova, A.F. Lemenovskaya, V.A. Matveeva, and E.V. Sinitsyna, The role of glycosphingolipids in natural immunity. Gangliosides modulate the cytotoxicity of natural killer cells, Eur. J. Immunol., 19: 1979-1983,1989.
24. I. Ando, D. Hoon, Y. Suzuki, R. Saxton, S.Golub, and R. Irie, Ganglioside G_{M2} on the K562 cell line is recognized as a target structure by human natural killer cells, Int. J. Cancer,40: 12-17, 1987.
25. K. Bezouska, Oligosaccharide ligands for NKR-P1 protein activate NK cells and cytotoxicity, Nature, 372:150-157, 1994.
26. P.T. Massa, Specific suppression of major histocompatibility complex class I and class II genes in astrocytes by brain-enriched gangliosides, J. Exp. Med., 178: 1357-1363, 1993.

27. H. Repke, E. Barber, S. Ulbricht, K. Buchner, F. Hucho, R. Kopp, H. Scholz, et al, Ganglioside-induced CD4 endocytosis occurs independent of serine phosphorylation and is accompanied by dissociation of P56(lck), J. Immunol., *149*: 2585-2591, 1992.
28. S. Ladisch, L. Ulsh, B.Gillard, and C.Wong, Modulation of the immune response by gangliosides: Inhibition of adherent monocyte accessory function in vitro, J. Clin. Invest., 74: 2074-2081, 1984.
29. H.W.L. Ziegler-Heitbrock, E. Kafferlein, J.G. Haas, M.S. NicoletteMeyer, C. Weber, and D. Flieger, Gangliosides suppress tumor necrosis factor production in human monocytes, J. Immunol., 148: 1753-1758, 1992.
30. D.S.B. Hoon, T. Jung, J. Naungayan, A.J. Cochran, D.L. Morton, and W.H. McBride, Modulation of human macrophage functions by gangliosides, Immunology Letters, *20*: 269-276, 1989.
31. W.D. Merritt, M. Bailey, and D.H. Pluznik, Inhibition of interleukin-2 dependent cytotoxic T-lymphocyte growth by gangliosides, Cell. Immunol., 89: 1-10, 1984.
32. F. Sharom, A. Chiu, and J. Chu, Membrane gangliosides modulate interleukin-2-stimulated T-lymphocyte proliferation, Biochim. Biophys. Acta, 1094: 35-42, 1991.
33. R.J. Robb, The suppressive effect of gangliosides upon IL 2-dependent proliferation as a function of inhibition of IL 2-receptor association, J. Immunol., 136: 971-976, 1986.
34. N.V. Prokazova, E.V. Dyatlovitskaya, and L.D. Bergelson, Sialylated lactosylceramides. Possible inducers of non-specific immunosuppression and atherosclerotic lesions, Eur. J. Biochem., 172: 1-6, 1988.
35. H. Kimata,, and A. Yoshida, Differential effects of gangliosides on Ig production and proliferation by human B cells, Blood, 84: 1193-1200, 1994.
36. H. Offner, T. Thieme, and A. Vandenbark, Gangliosides induce selective modulation of CD4 from helper T lymphocytes, J. Immunol., 139: 3295-3305, 1987.
37. S.-I. Hakomori, New directions in cancer therapy based on aberrant expression of glycosphingolipids: anti-adhesion and ortho-signaling therapy, Cancer Cells, 3: 461-470, 1991.
38. C.B. Zeller, and R.B. Marchase, Gangliosides as modulators of cell function, Invited Review, C1341-C1355, 1992.
39. Induction of IL-2 receptor expression in vivo. Response to allogeneic cells. Transplantation, 44: 547-553, 1987.
40. R. Li, S. Ladisch, and N. Villacreses,Tumor gangliosides inhibit the murine allogeneic immune response in vivo, Cancer Research 55:211-214, 1995.
41. R. Li, and S. Ladisch, Alteration of neuroblastoma ganglioside metabolism by retinoic acid, J. Neurochem., *59*: 2297-2303, 1992.
42. R. Li, and S. Ladisch, Abrogation of shedding of immunosuppressive neuroblastoma gangliosides, International Journal of Cancer, submitted for publication.

LYSOPHOSPHATIDIC ACID-INDUCED SIGNALS IN ASTROCYTES

Jeffrey N. Keller,[1] Marion R. Steiner,[2] Mark P. Mattson[3] and Sheldon M. Steiner[1]

[1] Molecular and Cell Biology Group, School of Biological Sciences
[2] Department of Microbiology and Immunology
[3] Sanders-Brown Center on Aging and Department of Anatomy and Neurobiology
University of Kentucky
Lexington, KY 40506

INTRODUCTION

Lysophosphatidic acid (LPA) is a potent lipid biomediator which evokes a broad array of responses in a diverse group of cell types (reviewed in Moolenaar, 1995; Tokumura, 1995). A sampling of LPA induced effects include stimulation of cell motility in cells ranging from *Dictyostelium* amoebae to monocytes, vasoactive responses, and platelet aggregation (Jalink et al., 1993; Zhou et al., 1995; Tokumura, 1995; Tokumura et al., 1981). The cell-type specificity of LPA-induced effects is illustrated by the LPA-induced stimulation of proliferation of cells, such as fibroblasts and myoblasts; inhibition of proliferation of some tumor cells; and stimulation of differentiation of keratinocytes (van Corven et al., 1989; Yoshida et al., 1996; Tigyi et al., 1994; Piazza et al., 1995). LPA stimulates many intracellular signal transduction pathways leading to an increase in the concentration of intracellular calcium ($[Ca^{2+}]_i$), and stimulation of the activities of serine/threonine protein kinases, tyrosine kinases and phospholipases (e.g., Jalink et al., 1995; Amano et al., 1996; Howe and Marshall, 1993; van Corven et al., 1989; Ha et al., 1994). One LPA signal transduction pathway involves the small G protein rho since ADP-ribosylation of rho by *Clostridium botulinum* C3 exoenzyme inhibits some LPA-induced signals (Kumagai et al., 1993). Information about the cells and stimuli which induce the formation of LPA is limited, although it is well established that platelets produce LPA (Mauco et al., 1978). Platelets are a likely source of the LPA (bioassay) observed in the cerebrospinal fluid following hemorrhagic injury; in contrast, no LPA is detectible in the cerebrospinal fluid prior to injury (Tigyi et al., 1995). This finding is but one indication that LPA is likely to be a significant biomediator in the brain. The brain is a rich source of both LPA and LPA receptors, and enzymes for the biosynthesis and degradation of LPA have been isolated from the brain (Das and Hajra, 1989; van der Bend et al., 1992; Thomson et al., 1994; Thompson and Clark, 1994; Thomson and Clark, 1995). Moreover, brain tumor-derived cells, glioma and neuroblastoma cells, are responsive to LPA (Smalheiser and Ali, 1994; Koschel and Tas, 1993). Thus, studies were initiated in this laboratory to define the functions of LPA in normal cells of the brain. This study focuses on the LPA-induced responses of astrocytes derived from neonatal rat brains.

RESULTS

LPA Induces Increases in $[Ca^{2+}]_i$, Lipid Peroxidation and DNA Synthesis in Astrocytes

Given the LPA stimulated increase in $[Ca^{2+}]_i$ in multiple cell types, the effect of LPA on astrocyte $[Ca^{2+}]_i$ was determined. There was a rapid increase in $[Ca^{2+}]_i$, when astrocytes were treated with 1 μM LPA (1-oleoyl-*sn*-glycero-3-phosphate) (Fig. 1A). The $[Ca^{2+}]_i$ returned to the basal level in approximately 5-6 min. Since LPA has been reported to affect the release of calcium from intracellular stores and the uptake of extracellular calcium (Jalink et al., 1990; Tigyi et al., 1996), the importance of these two sources of calcium in the astrocyte response to LPA was examined. In cells incubated in buffer without any added calcium, LPA still induced an increase in $[Ca^{2+}]_i$; however, there was a more rapid return to the basal level (Fig. 1B).

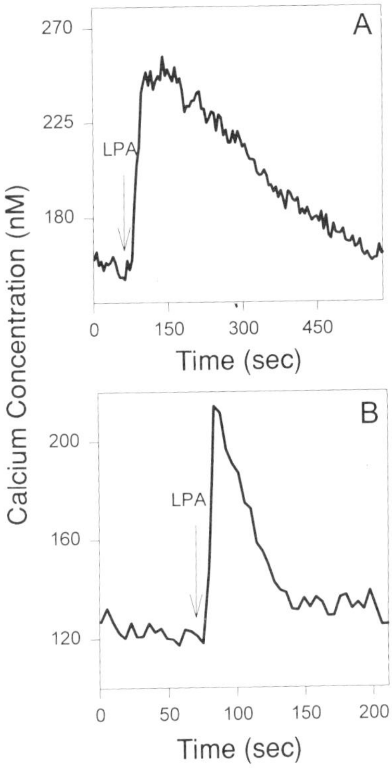

Figure 1. LPA stimulates an increase in $[Ca^{2+}]_i$ in astrocytes incubated in Locke's solution (A) or in Locke's solution without calcium chloride (B). Cells were loaded with the acetoxymethyl ester form of fura-2, then washed and incubated in Locke's solution with/without calcium chloride. LPA plus fatty acid free bovine serum albumin (BSA) were added to cells, final concentration of BSA, 0.5 mg/ml. All controls without LPA contained 0.5 mg/ml BSA.

Recently, a new and distinctive LPA-mediated response was observed in HeLa cells, namely an LPA-induced production of reactive oxygen species (Chen et al., 1995). The possibility of LPA stimulation of the production of reactive oxygen species in astrocytes was examined with an *in situ* thiobarbituric acid reactivity assay. LPA treatment of astrocytes resulted in an increase in thiobarbituric acid reactive substances; this LPA-mediated increase in thiobarbituric acid reactivity was sensitive to pretreatment of cells with the anti-oxidant, propyl gallate (Table 1).

Table 1. LPA stimulates an increase in thiobarbituric acid reactive substances *in situ*.

Treatment	Thiobarbituric acid reactivity (*in situ* relative intensity)
None	9.4 ± 0.4
LPA	28.4 ± 3.8
Propyl gallate	9.1 ± 0.6
LPA + propyl gallate	9.3 ± 1.0

Where indicated, cells were pretreated with 5 μM propyl gallate for 30 min and then they were incubated with 25 μM LPA for 1 h. The cells were next treated/fixed with thiobarbituric acid in acetic acid-trichloroacetic acid. Fluorescence was analyzed by confocal laser scanning microscopy and intensity was quantified using Molecular Dynamics software. $\rho < 0.001$ sample with LPA versus sample without LPA, Student's *t* test.

Next the possibility that LPA might affect astrocyte proliferation was examined. LPA stimulated DNA synthesis in astrocytes preincubated for 3 days in Locke's solution (Fig. 2).

LPA Inhibits Glutamate and Glucose Uptake in Astrocytes

Glutamate uptake by astrocytes has multiple functions, e.g., regulation of synaptic transmission, protection from excitotoxicity (Storm-Mathisen et al., 1995). Thus it was of interest to determine if LPA might affect glutamate uptake by astrocytes. LPA treatment resulted in a 24% inhibition of glutamate uptake by astrocytes (Table 2). This may be indicative of a marked effect of LPA on a single class of glutamate transporters or a partial effect on one or more classes of transporters. Since some LPA responses can be inhibited by pretreatment with *Clostridium botulinum* C3 exoenzyme, the effect of this exoenzyme on the LPA-mediated inhibition of glutamate uptake was evaluated. *Clostridium botulinum* C3 exoenzyme had no effect on either the basal or LPA-inhibited glutamate uptake (Table 2).

Table 2. LPA inhibits glutamate uptake.

Treatment	[^{3}H]glutamate uptake (% of control)
None	100 ± 3
LPA	76 ± 3
C3 exoenzyme	98 ± 2
LPA + C3 exoenzyme	76 ± 2

Cells were preincubated for 3 h with 1 μg/ml *Clostridium botulinum* C3 exoenzyme and then incubated with 10 μM LPA for 30 min. [^{3}H]glutamate was added for the last 7 min of the incubation and then the radioactivity and protein content of each sample were determined. The value of the glutamate uptake for the control sample, was defined as 100%; mean ± SEM of two experiments performed in triplicate. $\rho < 0.001$ sample with LPA versus sample without LPA, Student's *t* test.

Modulation of glutamate uptake in astrocytes can indirectly affect glucose uptake, i.e., glutamate uptake leads to increased intracellular Na^+ and this results in increased Na^+/K^+ ATPase activity which results in increased glucose uptake (Pellerin and Magistretti, 1994). Therefore, the possibility that LPA might inhibit glucose uptake was explored. Incubation of cells with 10 μM LPA resulted in an inhibition of glucose uptake (Fig. 3). Pretreatment of cells with ouabain, an inhibitor of Na^+/K^+ ATPase, resulted in a marked reduction in glucose uptake which was not significantly affected by LPA (Fig. 3). Conversely, in the presence of the ionophore, gramicidin, glucose uptake was markedly stimulated and LPA had no effect on the gramicidin-stimulated uptake (Fig. 3). These data are consistent with the LPA-mediated inhibition of glucose uptake occurring via an indirect mechanism involving a change in glutamate uptake leading to a change in Na^+/K^+ ATPase activity.

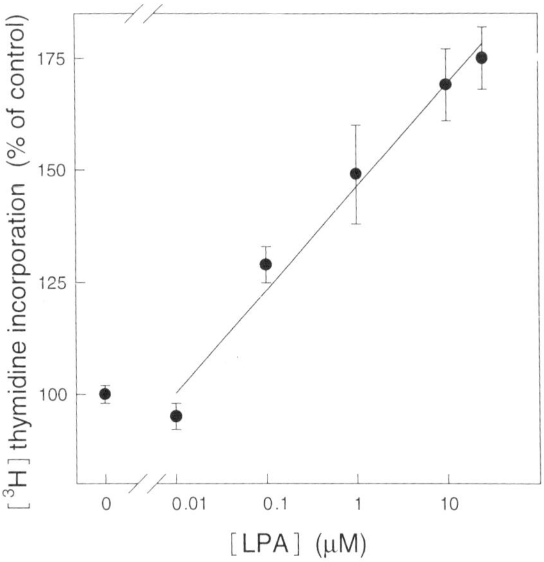

Figure 2. LPA-induces an increase in DNA synthesis. Astrocytes were incubated in Locke's solution for three days and then LPA, as indicated, plus [^{3}H]thymidine were added. After an 18 h incubation, trichloroacetic acid precipitable radioactivity and protein were measured. Data are the mean ± SEM ($n \geq 12$ per LPA concentration). 100% = response of the cells without LPA. First order linear regression indicated, correlation coefficient 0.87.

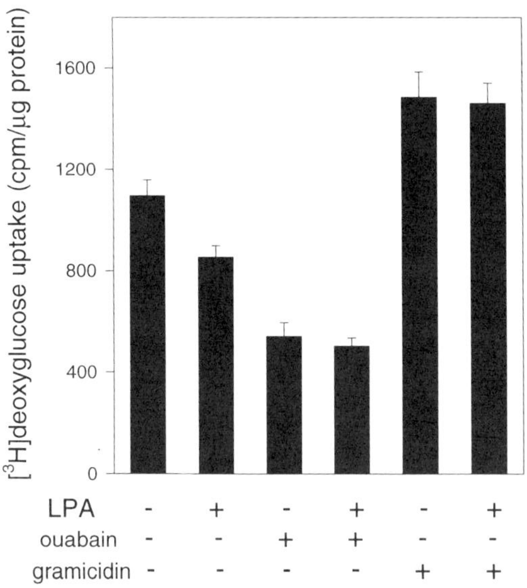

Figure 3. LPA partially inhibits glucose uptake. Astrocytes were pretreated, for 30 min, with 10 μM ouabain or 5 μM gramicidin, as indicated, and then the cells were incubated with 50 μM glutamate plus 10 μM LPA for 30 min. [^{3}H]deoxyglucose was added to the cultures for the last 7 min of the incubation and then cellular radioactivity and protein were determined. The results are the mean ± SEM of three analyses each performed in quadruplicate.

SUMMARY

The combined findings indicate that LPA evokes multiple responses in astrocytes, including increases in $[Ca^{2+}]_i$, lipid peroxidation and DNA synthesis, and inhibition of glutamate uptake and glucose uptake. LPA is likely to have an even broader impact on astrocytes given that the possibility of synergy between LPA and other biomediators is yet to be explored in these cells. Significant synergy is observed in smooth muscle cells in response to a variety of biomediators (Tokumura et al., 1994; Ohata et al., 1995; Schachter et al., 1995). The effects of LPA on astrocytes can also affect neurons in a variety of ways. Astrocytes have protective and support functions for neurons; hence, there can be changes relating simply to increased numbers of astrocytes. As described above, regulation of extracellular glutamate concentrations affects both neurotransmission and the potential for excitotoxicity. In addition, astrocyte energy metabolism can modulate neuronal energy metabolism; thus, the reduced glucose uptake by astrocytes can lead to a reduction in lactate available for neurons (Dringen et al., 1993). Thus, LPA is a potent biomediator which is likely to have multiple roles in the functions of astrocytes in normal and pathological conditions.

REFERENCES

Amano, M., Mukai, H., Ono, Y., Chihara, K., Matsui, T., Hamajima, Y., Okawa, K., Iwamatsu, A and Kaibuchi, K. (1996) Identification of a putative target for rho as the serine-threonine kinase protein kinase N. *Science* 271:648-650.

Chen, Q., Olashaw, N. and Wu, J. (1995) Participation of reactive oxygen species in the lysophosphatidic acid-stimulated mitogen-activated protein kinase kinase activation pathway. *J. Biol. Chem.* 270:28499-28502.

Das, A.K. and Hajra, A.K. (1989) Quantification, characterization and fatty acid composition of lysophosphatidic acid in different rat tissues. *Lipids* 24:329-333.

Dringen, R., Wiesinger, H. and Hamprecht, B. (1993) Uptake of L-lactate by cultured rat brain neurons *Neurosci. Lett.* 163: 5-7.

Ha, K.S., Yeo, E.J. and Exton, J.H. (1994) Lysophosphatidic acid activation of phosphatidylcholine-hydrolysing phospholipase D and actin polymerization by a pertussis toxin-sensitive mechanism. *Biochem. J.* 303:55-59.

Howe, L. and Marshall, C.J. (1993) Lysophosphatidic acid stimulates mitogen-activated protein kinase activation via a G-protein-coupled pathway requiring p21ras and p74raf-1. *J. Biol. Chem.* 268:20717-20720.

Jalink, K., van Corven, E.J. and Moolenaar, W.H. (1990) Lysophosphatidic acid, but not phosphatidic acid, is a potent Ca^{2+}-mobilizing stimulus for fibroblasts. Evidence for an extracellular site of action. *J. Biol. Chem.* 265:12232-12239.

Jalink, K., Moolenaar, W.H., and Van Duijn, B. (1993) Lysophosphatidic acid is a chemoattractant for *Dictyostelium discoideum* amoebae. *Proc. Natl. Acad. Sci. USA* 90:1857-1861.

Jalink, K., Hengeveld, T., Mulder, S., Postma, F.P., Simon, M.F., Chap, H., van der Marel, G.A., van Boom, J.H., van Blitterswijk, W.J. and Moolenaar, W.H. (1995) Lysophosphatidic acid-induced Ca^{2+} mobilization in human A431 cells: structure-activity analysis. *Biochem. J.* 307:609-616.

Koschel, K. and Tas, P.W.L. (1993) Lysophosphatidic acid reverts the β-adrenergic agonist-induced morphological response in C6 glioma cells. *Exp. Cell Res.* 206:162-166.

Kreps, D.M., Whittle, S., Hoffman, J.M. and Toews, M.L. (1993) Lysophosphatidic acid mimics serum-induced sensitization of cyclic AMP accumulation. *FASEB J.* 7:1376-1380.

Kumagai, N., Morii, N., Fujisawa, K., Nemoto, Y. and Narumiya, S. (1993) ADP-ribosylation of rho p21 inhibits lysophosphatidic acid-induced protein tyrosine phosphorylation and phosphatidylinositol 3-kinase activation in cultured Swiss 3T3 cells. *J. Biol. Chem.* 268:24535-24538.

Mauco, G., Chap, H., Simon, M.-F. and Douste-Blazy, L. (1978) Phosphatidic and lysophosphatidic acid production in phospholipase C-and thrombin-treated platelets. Possible involvement of a platelet lipase. *Biochimie* 60:653-661.

Moolenaar, W.H. (1995) Lysophosphatidic acid, a multifunctional phospholipid messenger. *J. Biol. Chem.* 270:12949-12952.

Ohata, H., Seito, N., Aizawa, H., Nobe, K. and Momose, K. (1995) Sensitizing effect of lysophosphatidic acid on mechanoreceptor-linked response in cytosolic free Ca^{2+} concentration in cultured smooth muscle cells. *Biochem. Biophys. Res. Commun.* 208:19-25.

Pellerin, L. and Magistretti, P.J. (1994) Glutamate uptake into astrocytes stimulates aerobic glycolysis: a mechanism coupling neuronal activity to glucose utilization. *Proc. Natl. Acad. Sci. USA* 91:10625-10629.

Piazza, G.A., Ritter, J.L. and Baracka, C.A. (1995) Lysophosphatidic acid induction of transforming growth factors α and β: modulation of proliferation and differentiation in cultured human keratinocytes and mouse skin. *Exp. Cell Res.* 216:51-64.

Schachter, J.B., Yasuda, R.P. and Wolfe, B.B. (1995) Adenosine receptor activation potentiates phosphoinositide hydrolysis and arachidonic acid release in DDT_1-MF_2 cells: putative interrelations. *Cell. Signal.* 7:659-668.

Smalheiser, N.R. and Ali, J.Y. (1994) Acute neurite retraction triggered by lysophosphatidic acid: timing of the inhibitory effects of genistein. *Brain Res.* 660:309-318.

Storm-Mathisen, J., Danbolt, N.C. and Ottersen, O.P. (1995) Localization of glutamate and its membrane transport proteins, in: *CNS Neurotransmitters and Neuromodulators Glutamate*, T.W. Stone, ed., CRC Press, Boca Raton, pp. 1-18.

Thompson, F.J. and Clark, M.A. (1994) Purification of a lysophosphatidic acid-hydrolysing lysophospholipase from rat brain. *Biochem. J.* 30: 457-461.

Thomson, F.J., Perkins, L., Ahern, D. and Clark, M. (1994) Identification and characterization of a lysophosphatidic acid receptor. *Mol. Pharmacol.* 45:718-723.

Thomson, F.J. and Clark, M.A. (1995) Purification of a phosphatidic-acid-hydrolysing phospholipase A_2 from rat brain. *Biochem. J.* 306:305-309.

Tigyi, G., Dyer, D.L. and Miledi, R. (1994) Lysophosphatidic acid posses dual action in cell proliferation. *Proc. Natl .Acad. Sci. USA* 91:1908-1912.

Tigyi, G., Hong, L., Yakubu, M., Parfenova, H., Shibata, M. and Leffler, C.W. (1995) Lysophosphatidic acid alters cerebrovascular reactivity in piglets. *Am. J. Physiol.* 268:H2048-H2055.

Tigyi, G., Fischer, D.J., Sebök, Á., Yang, C., Dyer, D.L. and Miledi, R. (1996) Lysophosphatidic acid-induced neurite retraction in PC12 cells: control by phosphoinositide-Ca^{2+}signaling and rho. *J. Neurochem.* 66:537-548.

Tokumura, A., Fukuzawa, K., Isobe, J. and Tsukatani, H. (1981) Lysophosphatidic acid-induced aggregation of human and feline platelets: structure-activity relationship. *Biochem. Biophys. Res. Commun.* 99:391-396.

Tokumura, A., Iimori, M., Nishioka, Y., Kitahara, M., Sakashita, M. and Tanaka, S. (1994) Lysophosphatidic acids induce proliferation of cultured vascular smooth muscle cells from rat aorta. *Am. J. Physiol.* 267:C204-C210.

Tokumura, A. (1995) A family of phospholipid autacoids: occurrence, metabolism and bioactions. *Prog. Lipid Res.* 34:151-184.

van Corven, E.J., Groenink, A., Jalink, K., Eichholtz, T. and Moolenaar, W.H. (1989) Lysophosphatidate-induced cell proliferation: identification and dissection of signaling pathways mediated by G proteins. *Cell* 59:45-54.

van der Bend, R.L., Brunner, J., Jalink, K., van Corven, E.J., Moolenaar, W.H. and van Blitterswijk, W.J. (1992) Identification of a putative membrane receptor for the bioactive phospholipid, lysophosphatidic acid. *EMBO J.* 11:2495-2501.

Yoshida, S., Fujisawa-Sehara, A., Taki, T., Arai, K., Nabeshima, Y. (1996) Lysophosphatidic acid and bFGF control different modes in proliferating myoblasts. *J. Cell Biol.* 132:181-193.

Zhou, D., Luini, W., Bernasconi, S., Diomede, L., Salmona, M., Mantovani, A., Sozzani, S. (1995) Phosphatidic acid and lysophosphatidic acid induce haptotactic migration of human monocytes. *J. Biol. Chem.* 270:25549-25556.

GLYCEROPHOSPHOINOSITOL-4-PHOSPHATE IN INTRACELLULAR SIGNALLING

Christopher P. Berrie, Marco Falasca, Angela Carvelli, Cristiano Iurisci, and Daniela Corda

Istituto di Ricerche Farmacologiche "Mario Negri Sud"
Consorzio Mario Negri Sud
Department of Cell Biology and Oncology
66030 Santa Maria Imbaro (Chieti)
Italy

INTRODUCTION

Cell Membrane Lipids and Signal Transduction

Over the past two decades, the build-up of a complex understanding of the metabolism of cell membrane lipids has seen the development of the roles of these lipids from solely "passive" structural components of the membrane bilayer to "active" regulators of cell function. These roles have included the fields of lipid-protein interaction, membrane biogenesis and membrane transport (for reviews see Casey, 1995; Liscovitch and Cantley, 1995; Milligan et al., 1995), but it is in that of signal transduction (for reviews see Bell and Burns, 1991; Nishizuka, 1992, 1995; Billah, 1993; Divecha and Irvine, 1995; Liscovitch and Cantley, 1995; Milligan et al., 1995; Corda and Falasca, 1996) where novel lipid second messengers continue to be described (Panayotou and Waterfield, 1992; Cuadrado et al., 1993; Hannun, 1994; Su et al., 1994; Moolenaar, 1995; Pushkareva et al., 1995; Falasca et al., 1996a). Thus membrane lipids are acted upon by intracellular phospholipases (phospholipase A_2 (PLA_2), phospholipase C (PLC) and phospholipase D), leading to the production of lipid second messengers that are involved in the modulation of different functional cellular responses (Exton, 1990; Dennis et al., 1991; Cockcroft and Thomas, 1992; Nishizuka, 1992; Berridge, 1993; Billah, 1993; Glaser et al., 1993; Iacovelli et al., 1993; Mayer and Marshall, 1993; Dennis, 1994; Divecha and Irvine, 1995; Lee and Rhee, 1995). Activation of one of these phospholipases, PLA_2, leads to the mobilisation of arachidonic acid from an *sn-2* linkage of phosphatidylcholine (PtdCho), phosphatidylethanolamine (PtdEth) and phosphatidylinositol (PtdIns) in numerous mammalian cell types (Van den Bosch, 1980; Irvine, 1982), which, in turn, serves as a substrate for the production of other cellular second messengers, such as the prostaglandins and leukotrienes (Piomelli and Greengard, 1990; Piomelli, 1993; Khan et al., 1995). In the case of PLA_2 activation, this is also accompanied by the formation of biologically active lysolipids that have been shown to be mitogens in several cell lines, including lysophosphatidic acid (LysoPtdCOOH; Moolenaar, 1995), lysophosphatidylinositol (LysoPtdIns; Falasca and Corda, 1994) and lysophosphatidylcholine (LysoPtdCho; Asaoka et al., 1991). These lysolipids themselves are the substrates for lysophospholipases that release the fatty acid from the *sn-1* linkage, thus producing glycerophospholipids.

The Glycerophosphoinositols

As indicated above, the glycerophosphoinositols are lipid metabolites that can be derived from membrane phosphoinositides by the successive action of a phospholipase and a lysophospholipase, and the observation that their levels in cells are greatly augmented by ras transformation led to the proposal that they might have a role in the mechanism of action of ras (Alonso et al., 1988; Alonso and Santos, 1990; Valitutti et al., 1991; Iacovelli et al., 1993; Corda et al., 1994; Corda and Falasca, 1996). Under these conditions, the constitutive production of the glycerophosphoinositols is accompanied by a concomitant elevation of the levels of arachidonic acid and LysoPtdIns, hence confirming that the transformation process involves a PLA_2 and a lysophospholipase action on the membrane phosphoinositides (Valitutti et al., 1991; Falasca et al., 1995). Furthermore, the specificity for the phosphoinositides of this constitutive PLA_2 activation in ras transformed cells was confirmed by the evaluation of lysophospholipid levels, where those of LysoPtdIns showed a three-fold increase over the parent cell lines, while those of LysoPtdCho and LysoPtdEth either remained unchanged, or were slightly decreased (Falasca et al., 1995).

Glycerophosphoinositol-4-phosphate

While glycerophosphoinositol (GroPIns) is the most abundant product in ras transformed cells (Alonso et al., 1988; Alonso and Santos, 1990; Valitutti et al., 1991; Corda et al., 1994), increases in the levels of its phosphorylated derivatives, glycerophosphoinositol-4-phosphate (GroPIns-4-P) and glycerophosphoinositol-4,5-bisphosphate (GroPIns-4,5-P_2) were also seen (Iacovelli et al., 1993; Corda et al., 1994). Indeed, investigations into the possible physiological roles of the glycerophosphoinositols demonstrated that GroPIns-4-P selectively exerts an inhibitory effect on the GTP-binding protein that stimulates adenylyl cyclase in a number of cell lines (Iacovelli et al., 1993). Furthermore, estimates of the intra-cellular concentrations of GroPIns-4-P by the metabolic labelling of the cellular inositol pool under these conditions are in the range of those seen for GroPIns-4-P inhibition of adenylyl cyclase (Iacovelli et al., 1993). Hence, we have investigated the production and possible roles of GroPIns-4-P in normal cells under physiological conditions.

HORMONE STIMULATION OF GroPtdIns-4-P ACCUMULATION

One of the major criteria for the establishing of a physiological role for GroPIns-4-P in normal cell function is the demonstration of its production under hormone stimulation. We have therefore monitored the levels of GroPIns-4-P in Swiss 3T3 cells exposed to a number

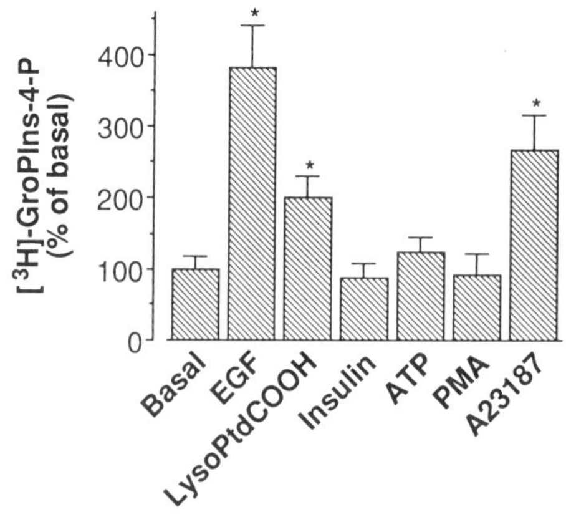

Figure 1. GroPIns-4-P accumulation in Swiss 3T3 cells stimulated for 45s with 10ng/ml EGF, 100μM ATP, 100nM PMA, 1μM insulin, 10μM A23187 or 100μM LysoPtdCOOH.

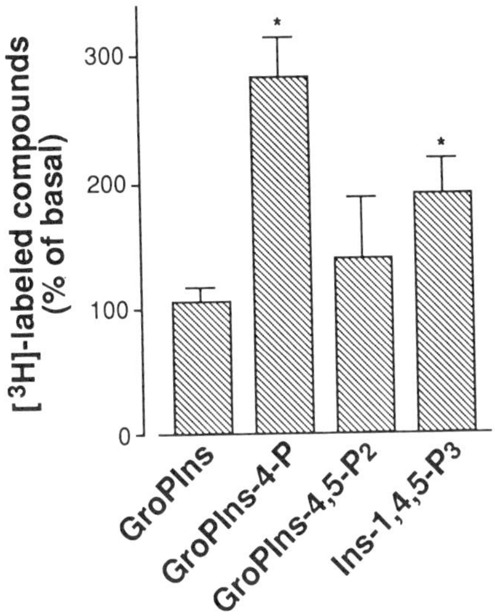

Figure 2. EGF stimulation (10ng/ml; 30s) of glycerophosphoinositides and Ins-1,4,5-P_3 accumulation in Swiss 3T3 cells.

of stimulatory agents, using ^{3}H-inositol equilibrium-loading and HPLC separation techniques (Falasca and Corda, 1994; Falasca et al., 1995; Falasca et al., 1996a). As shown in Figure 1, the growth factor EGF, the mitogen LysoPtdCOOH, and the ionophore A23187 (which is known to stimulate the PLA_2 involved in GroPIns-4-P formation in ras-transformed cells) all induced an increase in the GroPIns-4-P levels, which were unaffected by insulin, ATP and the phorbol ester phorbol myristate acetate (PMA). Therefore, the formation of GroPIns-4-P requires the activation of specific receptors, such as the EGF receptor in fibroblasts (Figure 1), and the IgE receptor in RBL cells and the TSH receptor in thyrocytes (Falasca et al., 1996a), corroborating that this compound might function as a novel intracellular messenger.

Further investigations of this EGF stimulation of GroPIns-4-P accumulation in these cells have demonstrated a rapid increase in its levels within the first 30s of stimulation (Figure 2), which decreases to twice the basal levels within 5m of stimulation (a level that is maintained for some 30m; Falasca et al., 1996a). This data would thus suggest pathway(s) for the rapid catabolism of GroPIns-4-P. One such mechanism could involve its hydrolysis to inositol phoshates and glycerol; alternatively, GroPIns-4-P might be reacylated to form PtdIns-4-P, dephosphorylated to form GroPIns, or phosphorylated to form GroPIns-4,5-P_2. However, as also illustrated in Figure 2, EGF stimulation was unable to demonstrate any significant increases in either GroPIns or GroPIns-4,5-P_2 within 30s (although GroPIns-4,5-P_2 does show a small, but significant, transient increase around 60s of stimulation; Falasca et al., 1996a), indicating that the catabolism of GroPIns-4-P is unlikely to involve either dephosphorylation or further phosphorylation. Furthermore, as EGF stimulation in these cells results in both a rapid decrease in the levels of PtdIns-4-P, which return to basal within 30s (Falasca et al., 1996a), and an increase in the levels of inositol-4-phosphate formation that closely follows GroPIns-4-P formation (Falasca et al., 1996a), these could represent the principal catabolic pathways of GroPIns-4-P. Figure 2 also demonstrates that EGF stimulation was also able to induce the accumulation of inositol (1,4,5)-trisphosphate (Ins-1,4,5-P_3) within 30s; this accumulation rapidly returns to basal levels within 1m (Falasca et al., 1996a). Furthermore, significant EGF-induced formation of GroPIns-4-P was apparent at an EGF concentration of 0.5ng/ml, at which concentration the levels of Ins-1,4,5-P_3 remained at basal (Falasca et al., 1996a). Thus both the kinetics and the dose-dependency of GroPIns-4-P and Ins-1,4,5-P_3 formation indicate that EGF stimulates two independent pathways, and that the increase in GroPIns-4-P is not secondary to a phospholipase C-induced increase in cytosolic calcium.

INHIBITION OF EGF-STIMULATED GroPIns-4-P ACCUMULATION

In order to further investigate the mechanism of action of EGF-stimulated GroPIns-4-P accumulation in these cells, we have tested the effects of a number of specific inhibitors. As shown in Figure 3, the tyrosine kinase inhibitor genistein completely inhibited the EGF-stimulated GroPIns-4-P accumulation. Furthermore, wortmannin, an inhibitor of PtdIns 3-kinase and PD98059, an inhibitor of MAP kinase kinase (MEK; Pang et al., 1995) were also able to inhibit this GroPIns-4-P accumulation by 80% and 50%, respectively (Figure 3). Conversely, while it has been shown that PKC is able to phosphorylate and activate a purified form of cytosolic PLA_2 (Nemenhoff et al., 1993), a PKC activator, PMA, and a PKC inhibitor, calphostin, were both unable to induce a significant formation of GroPIns-4-P (Figure 3). Taken together, these results indicate that EGF-stimulated accumulation of GroPIns-4-P requires phosphorylation of the specific tyrosine kinase receptor and activation of PtdIns 3-kinase. One of the substrates of PtdIns 3-kinase is the small GTP binding protein Rac, which has recently been demonstrated to be directly involved in the activation of PLA_2 (Hawkins et al., 1995; Parker, 1995; Peppelenbosch et al., 1995). We have shown that EGF stimulation of fibroblasts expressing the dominant-negative form of the Rac protein (Qiu et al., 1995) was unable to induce GroPIns-4-P accumulation, and, conversely, that those cells expressing a constitutively active Rac protein demonstrated higher basal levels of GroPIns-4-P (Falasca et al., 1996a). Hence it is the ras/Rac cascade that appears to be involved in the EGF stimulation of PLA_2 activation in these cells, although the partial inhibition by the MEK inhibitor, PD98059, also indicates the possible involvement of MAP kinase activation of PLA_2. Thus, the PtdIns 3-kinase and MAP kinase pathways both appear to be involved in the modulation of PLA_2 activity either via an action on Rac, or via an as yet unclear cross-talk mechanism. In either case, however, we can conclude that the enzyme responsible for the formation of GroPIns-4-P is downstream from the activation of ras, in agreement with the high levels of this compound found in ras transformed cells (Valitutti et al., 1991).

GroPIns-4-P INHIBITION OF ADENYLYL CYCLASE FUNCTION

As we have previously demonstrated that exogenously added GroPIns-4-P can inhibit adenylyl cyclase both *in vitro* and *in vivo* (Iacovelli et al., 1993; Corda et al., 1994), we have investigated the effects of EGF on the cAMP levels in these Swiss 3T3 cells (Falasca et al., 1996a). As shown in Table 1, EGF was indeed able to cause a decrease not only in the basal cAMP levels, but also in those stimulated by forskolin and cholera toxin, with the maximal

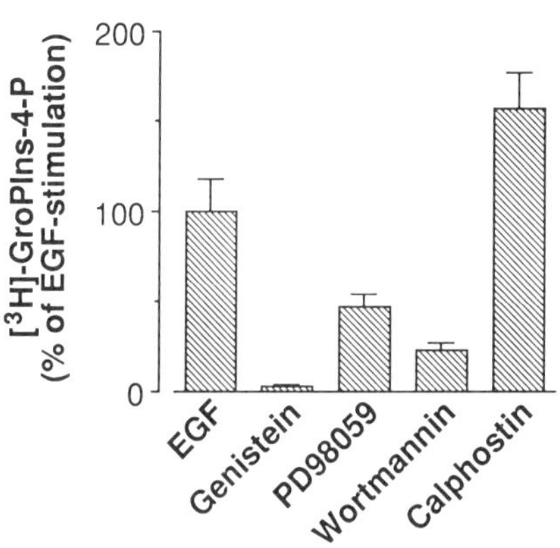

Figure 3. Modulation of basal and EGF-stimulated GroPIns-4-P accumulation by pharmacological agents. Cells were pretreated for 10m with 1μM calphostin C, 100nM wortmannin, 10μM PD98059 or 25μM genistein, followed by 45s of stimulation with 10ng/ml EGF, 100nM PMA or 100μM A23187.

Table 1. Effects of EGF on basal, and cholera toxin (CT) and forskolin-induced increases in cAMP levels in Swiss-3T3 cells.

	cAMP levels (pmol/ml)					
	Time (min)					
Stimulants	1	% Inhibition	3	% Inhibition	30	% Inibition
Control	2.50±0.40		1.64±0.47		1.10±0.2	
EGF	1.42±0.20*	(43%)	1.41±0.31		1.95±0.13*	
Forskolin	5.94±1.02		3.79±0.75*		77.2±7.2*	
Forskolin+EGF	1.51±0.31**	(75%)	1.71±0.48**	(55%)	40.5±6.24**	(52%)
CT	2.83±0.33		2.12±0.15		88.83±16.91*	
CT+EGF	1.05±0.16***	(63%)	0.83±0.2***	(39%)	53.15±18.24***	(40%)

Swiss-3T3 cells were stimulated with EGF (10 ng/ml) and/or cholera toxin (10 nM) or forskolin (10 μM) for the indicated times. The cAMP assay was then performed as described in Iacovelli et al., 1993. Data are the mean of 5 separate experiments performed in triplicate.
*, $p<0.05$ vs control; **, $p<0.05$ vs forskolin; ***, $p<0.05$ vs cholera toxin.

inhibition occurring in parallel with the formation of GroPIns-4-P, and to an extent that was similar to that seen with exogenously added GroPIns-4-P (Iacovelli et al., 1993; Falasca et al., 1996a). This effect was not inhibited by pertussis toxin pretreatment of the cells (Falasca et al., 1996a), indicating that the classical G_i-mediated inhibition of adenylyl cyclase is not involved in this phenomenon. Thus, an alternative mechanism for the modulation of adenylyl cyclase takes place in these cells. Interestingly, it has been reported previously that in a different fibroblast cell line (NRK-CL2-T), EGF could decrease the prostaglandin E_1-induced stimulation of adenylyl cyclase, and that this effect could not be observed in crude membrane preparations, indicating the requirement for cytosolic components (Anderson et al., 1979). Similarly, it has been proposed that in hepatocytes, EGF could decrease cAMP levels by activating a phosphodiesterase, although this has yet to be proven (Bosch et al., 1986). Our proposal is thus that the receptor-induced formation of GroPIns-4-P might lead to an inhibitory regulation of adenylyl cyclase, and we have previously proposed that this effect could be mediated by a direct action on the G_s protein (Iacovelli et al., 1993). This was initially based on the observations that in cell lines such as FRTL5 thyroid cells, smooth muscle cells, pituitary cells (AtT20) and pneumocytes (A549), GroPIns-4-P inhibition of adenylyl cyclase was only evident when the G_s protein was activated directly by cholera toxin, aluminium fluoride or receptor coupling, and not when forskolin was used as a direct stimulator of the enzyme (Iacovelli et al., 1993). Furthermore, GroPIns-4-P was also able to inhibit the cholera toxin-induced ADP-ribosylation of G_s in the same cell systems (Iacovelli et al., 1993). It has also been reported that in some cases the interaction of forskolin with adenylyl cyclase requires an activated G_s protein (Bender and Neer, 1983), with G_s and forskolin thus inducing synergistic activation of adenylyl cyclase types II, V and VI (Sutkowski et al., 1994). In these cases, an inhibition of G_s would prevent the formation of the ternary complex (G_s, enzyme, forskolin) that has been proposed to be important for the high affinity binding site for forskolin (Nelson and Seamon, 1986). Thus, by acting on G_s, GroPIns-4-P would indirectly modulate this association, and thus the effect of forskolin, although it cannot be formally ruled out that GroPIns-4-P might have an additional mechanism of action, involving a direct interaction with particular adenylyl cyclase isoforms that could be present in the cells examined in these studies.

The glycerophosphoinositides have also recently been shown to act on another G protein-coupled enzyme, PLA_2. Thus, GroPInsP and GroPIns-4-P were both able to inhibit (by up to 80%) the ATP- and aluminium fluoride-induced arachidonic acid release in FRTL5 cells, but not the stimulation induced by either melittin (Sharma, 1992) or high calcium levels (Corda and Falasca, 1996). This suggests that also in this case the action of these glycerophosphoinositols is exerted at the level of the G proteins, and thus further indicates that they may act as specific modulators of G protein-mediated function.

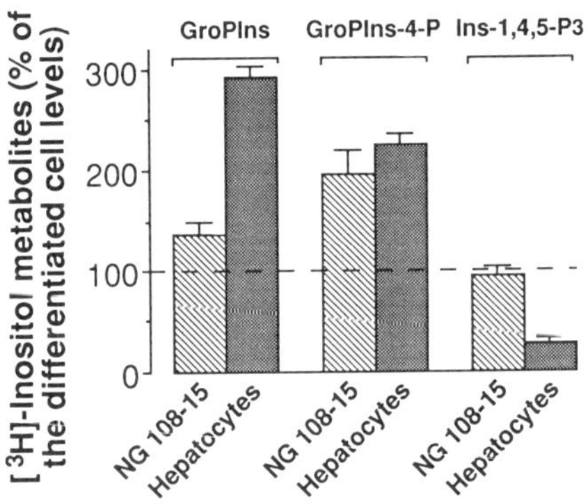

Figure 4. Levels of ^{3}H-inositol-labeled metabolites in undifferentiated and differentiated NG-108-15 cells, and in rat hepatocytes during development (modified from Falasca et al., 1996b).

GroPIns-4-P AND CELL DIFFERENTIATION

In order to further determine the role of GroPIns-4-P production in the activation of the ras/MAP kinase cascade, we have also investigated two cell systems that are models of neuronal and hepatic cell differentiation (Leoni et al., 1987; Falasca et al., 1996b). As is illustrated in Figure 4, in NG-108-15 cells that can be induced to differentiate by the addition of dibutyryl cAMP, it is in the undifferentiated state where high levels of GroPIns, GroPIns-4-P (and LysoPtdIns; see Falasca et al., 1996b) are seen. This involvement of the PLA_2/lyso-phospholipase pathway as a characteristic of the undifferentiated state of these cells is further supported by the observation that Ins-1,4,5-P_3 levels do not change during this process (Figure 4). Furthermore, hepatocytes that were isolated during liver development demonstrated higher levels of GroPIns and GroPIns-4-P in the undifferentiated state, with the converse being seen in the case of Ins-1,4,5-P_3 levels (Figure 4 and Falasca et al., 1996b). This thus indicates that these changes in the glycerophosphoinositide, LysoPtdIns (Falasca et al., 1996b) and Ins-1,4,5-P_3 levels are in fact a result of the activation of a specific PLA_2/lysophospholipase pathway, and not just a result of non-specific alterations in membrane lipid synthesis. The concept that these changes are a characteristic of, rather than a controlling factor in, the activation of the ras pathway has also been supported by the demonstration of higher levels of GroPIns and GroPIns-4-P in the differentiated forms of both PC12 and HL60 cells (Falasca et al., 1996b), with further evidence having been seen in the induction of various haematopoietic cells, including HL60 cells and myeloblasts (French et al., 1991; Bunce et al., 1993; Mountford et al., 1994). These results would thus indicate that increased PLA_2 activity is not only associated with oncogenic ras, but also a characteristic of physiological events that are modulated by ras dependent pathways.

CONCLUDING COMMENTS

Recent reports that have analysed other biological activities of GroPIns-4-P (and GroPIns-4,5-P_2), such as the *in vitro* inhibition of phosphoinositide specific PLC (Cruz-Rivera et al., 1990) and the inhibition of the cardiac sarcolemmal sodium/calcium exchanger (Luciani et al., 1995), have indicated that these effects are independent of the modulation of adenylyl cyclase; instead, they appear to be related to the role of the glycerophosphoinositols as competitors of PtdIns-4-P (and PtdIns-4,5-P_2) for those effects that involve a direct interaction of the polar head-groups of these molecules with the protein substrates.

Furthermore, in the yeast *Saccharomyces cerevisiae*, where glucose stimulation has been shown to cause a rapid elevation in the levels of GroPIns-4-P and GroPIns-4,5-P_2 derived from the deacylation of PtdIns-4-P and PtdIns-4,5-P_2, respectively, these effects have been related to the ability of these yeast to utilise these glycerophosphoinositols as

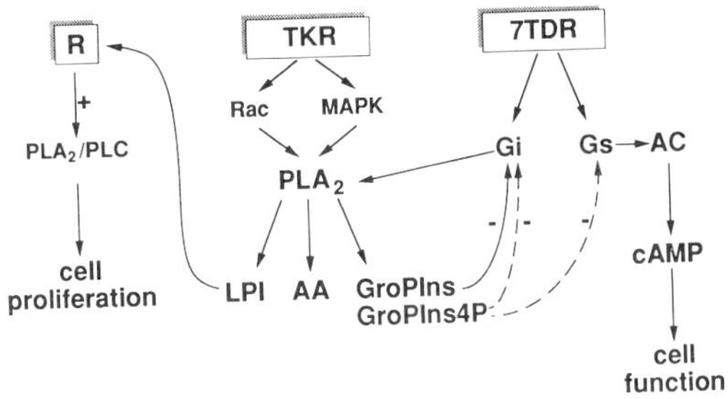

Figure 5. Scheme of the signalling pathways involving GroPIns and GroPIns-4-P.

alternative sources for the metabolism of the phosphatidylinositols (Angus and Lester, 1975; Hawkins et al., 1993; Patten et al., 1995). It would therefore appear that these glycerophosphoinositols are able to interact with several different cellular targets, with the elucidation of the different mechanisms that could be involved remaining to be completed.

However, if the molecular mechanisms of the actions of GroPIns and GroPIns-4-P discussed here are to be summarised at this stage (see Figure 5), we can consider that the formation of GroPIns-4-P in normal cells that are activated by tyrosine kinase receptors provides a novel cross-talk mechanism between activation of the ras pathway and the inhibition of adenylyl cyclase. Thus one of the potential targets for the glycerophosphoinositols, with specific reference to the formation of GroPIns-4-P, is adenylyl cyclase, and we can further conclude that GroPIns-4-P is not only a ubiquitous novel second messenger that is able to mediate cross-talk between cytosolic PLA_2 and adenylyl cyclase, but also that these high levels of GroPIns-4-P characterise phenomena that are downstream from the ras pathway, and can be activated by hormone stimulation, by oncogenic ras, and by the active form of the ras protein.

Acknowledgements. This study was supported in part by the Italian Association for Cancer Research (AIRC) and the Italian National Research Council (N°95.00558.PF39, ACRO and Convenzione CNR-Consorzio Mario Negri Sud). M.F. was the recipient of a fellowship from the Centro di Formazione e studi per il Mezzogiorno (FORMEZ). C.P.B. is the recipient of a European Commission TMR Programme Research Training Grant (N° ERB4001GT953337).

REFERENCES

Alonso, T., and Santos, E., 1990, Increased intracellular glycerophosphoinositol is a biochemical marker for transformation by membrane-associated and cytoplasmic oncogenes, *Biochem. Biophys. Res. Commun.*, 171:14.

Alonso, T., Morgan, R.O., Marvizon, J.C., Zarbl, H., and Santos, E., 1988, Malignant transformation by ras and other oncogenes produces common alterations in inositol phospholipid signaling pathways, *Proc. Natl. Acad. Sci. USA*, 85:4271.

Anderson, W.B., Gallo, M., Wilson, J., Lovelace, E., and Pastan, I., 1979, Effect of epidermal growth factor on prostaglandin E_1-stimulated accumulation of cyclic AMP in fibroblastic cells, *FEBS Lett.*,102:329.

Angus, W.W., and Lester, R.L., 1975, The regulated catabolism of endogenous and exogenous phosphatidylinositol by *Saccharomyces cerevisiae* leading to extracellular glycerophosphorylinositol and inositol, *J. Biol. Chem.*, 250:22.

Asaoka, Y., Oka, M., Yoshida, K., and Nishizuka, Y., 1991, Lysophosphatidylcholine as a possible second messenger synergistic to diacylglycerol and calcium ions for T-lymphocyte activation, *Biochem. Biophys. Res. Commun.*, 178:1378.

Bell, R.M., and Burns, D.J., 1991, Lipid activation of protein kinase C, *J. Biol. Chem.*, 266:4661.

Bender, J.L., and Neer, E.J., 1983, Properties of the adenylate cyclase catalytic unit from caudate nucleus, *J. Biol. Chem.*, 258:2432.

Berridge, M.J., 1993, Inositol trisphosphate and calcium signalling, *Nature*, 361:315.

Billah, M.M., 1993, Phospholipase D and cell signaling, *Curr. Opin. Immunol.*, 5:114.

Bosch, F., Bouscarel, B., Slaton, J., Blackmore, P.F., and Exton, J.H., 1986, Epidermal growth factor mimics insulin effects in rat hepatocytes, *Biochem. J.*, 239:523.

Bunce, C.M., French, P.J., Allen, P., Mountford, J.D., Moor, B., Greaves, M.F., Michell, R.H., and Brown, G., 1993, Comparison of the levels of inositol metabolites in transformed haemopoietic cells and their normal counterparts, *Biochem. J.*, 289:667.

Casey, P.J., 1995, Protein lipidation in cell signaling, *Science*, 268:221.

Cockcroft, S., and Thomas, G.M.H., 1992, Inositol-lipid-specific phospholipase C isoenzymes and their differential regulation by receptors, *Biochem. J.*, 288:1.

Corda, D., and Falasca, M., 1996, Glycerophosphoinositols as potential markers of ras-induced transformation and novel second messengers, *Anticancer Res.*, 16:in press.

Corda, D., Falasca, M., Iacovelli, L., Carvelli, A., D'Arcangelo, D., and Ramakrishna, V., 1994, Novel cellular activities of phosphoinositide metabolites originated by the action of phospholipase A_2, in: *GTPase-Controlled Molecular Machines*, D. Corda, H. Hamm, and A. Luini, eds., Ares Serono Symposia Publications, Rome.

Cruz-Rivera, M., Bennet, C.F., and Crooke, S.T., 1990, Glycerol-3-phospho-D-myo-inositol 4-phosphate (Gro-PIP) is an inhibitor of phosphoinositide-specific phospholipase C, *Biochim. Biophys. Acta*, 1042:113.

Cuadrado, A., Carnero, A., Dolfi, F., Jiménez, B., and Lacal, J.C., 1993, Phosphorylcholine: a novel second messenger essential for mitogenic activity of growth factors, *Oncogene*, 8:2959.

Dennis, E.A., 1994, Diversity of group types, regulation, and function of phospholipase A_2, *J. Biol. Chem.*, 269:13057.

Dennis, E.A., Rhee, S.G., Billah, M.M., and Hannun, Y.A., 1991, Role of phospholipases in generating lipid second messengers in signal transduction, *FASEB J.*, 5:2068.

Divecha, N., and Irvine, R.F., 1995, Phospholipid signaling, *Cell*, 80:269.

Exton, J.H., 1990, Signaling through phosphatidylcholine breakdown, *J. Biol. Chem.*, 265:1.

Falasca, M., and Corda, D., 1994, Elevated levels and mitogenic activity of lysophosphatidylinositol in k-ras transformed epithelial cells, *Eur. J. Biochem.*, 221:383.

Falasca, M., Carvelli, A., Iurisci, C., Qiu, R.-G., Symons, M.H., and Corda, D., 1996a, Growth factor-induced formation of glycerophosphoinositol-4-phosphate, a putative novel intracellular messenger that inhibits adenylyl cyclase activity, submitted.

Falasca, M., Marino, M., Carvelli, A., Iurisci, C., Leoni, S., and Corda, D., 1996b, Changes in the levels of glycerophosphoinositols during differentiation of hepatic and neuronal cells, *Eur. J. Biochem.*, in press.

Falasca, M., Silletta, M.G., Carvelli, A., Di Francesco, A.L., Fusco, A., Ramakrishna, V., and Corda, D., 1995, Signalling pathways involved in the mitogenic action of lysophosphatidylinositol, *Oncogene*, 10:2113.

French, P.J., Bunce, C.M., Stephens, L.R., Lord, J.M., McConnell, F.M., Brown, G., Creba, J.A., and Michell, R.H., 1991, Changes in the levels of inositol lipids and phosphates during the differentiation of HL-60 promyelocytic cells towards neutrophils or monocytes, *Proc. R. Soc. Lond. B Biol. Sci.*, 245:193.

Glaser, K.B., Mobilio, D., Chang, J.Y., and Senko, N., 1993, Phospholipase A_2 enzymes: regulation and inhibition, *Trends Pharmacol. Sci.*, 14:92.

Hannun, Y.A., 1994, The sphingomyelin cycle and the second messenger function of ceramide, *J. Biol. Chem.*, 269:3125.

Hawkins, P.T., Eguinoa, A., Qiu, R.-G., Stokoe, D., Cooke, F.T, Walters, R., Wennström, S., Claesson-Welsh, L., Evans, T., Symons, M., and Stephens, L., 1995, PDGF stimulates an increase in GTP-Rac via activation of phosphoinositide 3-kinase, *Curr. Biol.*, 5:393.

Hawkins, P.T., Stephens, L.R., and Piggot, J.R., 1993, Analysis of inositol metabolites produced by *Saccharomyces cerevisiae* in response to glucose stimulation, *J. Biol. Chem.*, 268:3374.

Iacovelli, L., Falasca, M., Valitutti, S., D'Arcangelo, D., and Corda, D., 1993, Glycerophosphoinositol 4-phosphate, a putative endogenous inhibitor of adenylylcyclase, *J. Biol. Chem.*, 268:20402.

Irvine, R.F., 1982, How is the level of free arachidonic acid controlled in mammalian cells? *Biochem. J.*, 204:3.

Khan, W.A., Blobe, G.C., and Hannun, Y.A., 1995, Arachidonic acid and free fatty acids as second messengers and the role of protein kinase C, *Cell Signal.*, 7:171.

Lee, S.B., and Rhee, S.G., 1995, Significance of phosphatidylinositol hydrolysis and regulation of phospholipase C isozymes. *Curr. Opin. Cell Biol.*, 7:183.

Leoni, S., Spagnuolo, S., Dini, L., and Conti Devirgiliis, L., 1987, Regulation of amino acid transport in isolated hepatocytes during development, *J. Cell. Physiol.*, 130:103.

Liscovitch, M., and Cantley, L.C., 1995, Signal transduction and membrane traffic: the PITP/phosphoinositide connection, *Cell*, 81:659.

Luciani, S., Antolini, M., Bova, S., Cargnelli, G., Cusinato, G., Bebetto, P., Trevisi, L., and Varotto, R., 1995, Inhibition of cardiac sarcolemmal sodium-calcium exchanger by glycerophosphoinositol 4-phosphate and glycerophosphoinositol 4,5-bisphosphate, *Biochem. Biophys. Res. Commun.*, 206:674.

Mayer, R.J., and Marshall, L.A., 1993, New insights on mammalian phospholipase A_2 (s); comparison of arachidonoyl-selective and -nonselective enzymes, *FASEB J.*, 7:339.

Milligan, G., Parenti, M., and Magee, A.I., 1995, The dynamic role of palmitoylation in signal transduction, *Trends Biochem Sci.*, 20:181.

Moolenaar, W.H., 1995, Lysophosphatidic acid, a multifunctional phospholipid second messenger, *J. Biol. Chem.*, 270:12949.

Mountford, J.C., Bunce, P.J., French, P.J., Michell, R.H., and Brown, G., 1994, Intracellular concentrations of inositol, glycerophosphoinositol and inositol pentakisphosphate increase during haemopoietic cell differentiation, *Biochim. Biophys. Acta*, 1222:101.

Nelson, C.A., and Seamon, K.B., 1986, Binding of [^{3}H]-forskolin to human platelet membranes. Regulation by guanyl-5'-yl imidodiphosphate, NaF, and prostaglandins E_1 and D_2, *J. Biol. Chem.*, 261:13469.

Nemenoff, R.A., Winitz, S., Qian, N.-X., Van Putten, V., Johnson, G.L. and Heasley, L.E., 1993, Phosphorylation and activation of a high molecular weight form of phospholipase A_2 by p42 microtubule-associated protein 2 kinase and protein kinase C, *J. Biol. Chem.*, 268:1960.

Nishizuka, Y., 1992, Intracellular signaling by hydrolysis of phospholipids and activation of protein kinase C, *Science*, 258:607.

Nishizuka, Y., 1995, Protein kinase C and lipid signaling for sustained cellular responses, FASEB. J., 9:484.

Panayotou, G., and Waterfield, M.D., 1992, Phosphatidylinositol 3-kinase: a key enzyme in diverse signalling processes, *Trends Cell. Biol.*, 2:358.

Pang, L., Sawada, T., Decker, S.J., and Saltiel, A.R., 1995, Inhibition of MAP kinase kinase blocks the differentiation of PC-12 cells induced by nerve growth factor, *J. Biol. Chem.*, 270:13585.

Parker, P.J., 1995, *Curr. Biol.*, 5:577.

Patten, J.L., Pessoa-Brandao, L., and Henry, S.A., 1995, Production and reutilization of an extracellular phosphatidylinositol catabolite, glycerophosphoinositol, by *Saccharomyces cerevisiae*, *J. Bacteriol.*,177: 3379.

Peppelenbosch, M.P., Qiu, R.-G., de Vries-Smits, A.M.M., Tertoolen, L.G.J., de Laat, S.W., McCormick, F., Hall, A., Symons, M.H., and Bos, J.L., 1995, Rac mediates growth factor-induced arachidonic acid release, *Cell*, 81:849.

Piomelli, D., 1993, Arachidonic acid in cell signaling, *Curr. Opin. Cell Biol.*, 5:274.

Piomelli, D., and Greengard, P., 1990, Lipoxygenase metabolites of arachidonic acid in neuronal transmembrane signalling, *Trends Pharmacol. Sci.*, 11:367.

Pushkareva, M., Obeid, L.M., and Hannun, Y.A., 1995, Ceramide: an endogenous regulator of apoptosis and growth suppression. *Immunol. Today*, 16:294.

Qiu, R.-G., Chen, J., Kirn, D., McCormick, F., and Symons, M., 1995, An essential role for Rac in Ras transformation, *Nature*, 374:457.

Sharma, S.V., 1992, Melittin resistance: a counterselection for ras transformation, *Oncogene*, 7:193.

Su, Y., Rosenthal, D., Smulson, M., and Spiegel, S., 1994, Sphingosine 1-phosphate, a novel signaling molecule, stimulates DNA binding activity of AP-1 in quiescent Swiss 3T3 fibroblasts, *J. Biol. Chem.*, 269:16512.

Sutkowski, E.M., Tang, W.-J., Broome, C.W., Robbins, J.D., and Seamon, K.B., 1994, Regulation of forskolin interactions with type I, II, V, and VI adenylyl cyclases by G_s alpha, *Biochemistry*, 33:12852.

Valitutti, S., Cucchi, P., Colletta, G., Di Filippo, C., and Corda, D., 1991, Transformation by the k-ras oncogene correlates with increases in phospholipase A_2 activity, glycerophosphoinositol production and phosphoinositol lipid synthesis in thyroid cells, *Cell. Signal.*, 3:321.

Van den Bosch, H., 1980, Intracellular phospholipases, *Biochem. Biophys. Acta*, 604:191.

BIOSYNTHESIS OF PLATELET ACTIVATING FACTOR AND RELATED ACETYLATED LIPIDS

Fred Snyder, Ten-ching Lee, Merle L. Blank, Koji Shinozaki, Ming-che Ou, and Veronica Fitzgerald

Medical Sciences Division
Oak Ridge Associated Universities
P.O. Box 117
Oak Ridge, TN 37831

INTRODUCTION

Platelet-activating factor (PAF) is synthesized by either de novo or remodeling enzymes (Snyder, 1995a,b). The remodeling of a preexisting membrane phospholipid, 1-alkyl-2-acyl-*sn*-glycero-3-phosphocholine (alkylacyl-GPC), to form PAF can occur via the direct action of a phospholipase A_2 (PLA_2) or via a stepwise reaction sequence catalyzed by CoA-independent (CoA-ind) or CoA-dependent (CoA-dep) transacylases. Once PAF is synthesized, its acetate moiety can be transferred to other lysophospholipids and fatty alcohols as well as to sphingosine by a CoA-independent transacetylase. The C_2-ceramides that are produced are of special interest in view of their known biological activities.

A brief summary of the properties of these enzymes and some newly identified selective inhibitors of the enzymes in the remodeling pathway of PAF biosynthesis are the subject of this report. These selective inhibitors have little, if any, effect on the de novo enzymes. Details of our original publications relating to most of these studies can be found in the references cited in the "Methods" section.

METHODS

Earlier work from our labortory have described the enzymatic systems discussed in the subsequent section. These enzymes include acetyl-CoA:lyso-PAF acetyltransferase (Wykle et al., 1980), the CoA-ind transacylase (Uemura et al., 1991; Blank et al., 1995), the CoA-dep tranacylase (Blank et al., 1995), the de novo enzymes (Lee etal., 1986,1988; Woodard et al., 1987; Renooij and Snyder, 1981), PAF:lysoplasmalogen transacetylase (Lee et al., 1992), and PAF:sphingosine transacetylase (Lee et al., 1996).

Potential inhibitors (up to 20µM) dissolved in dimethylsulfoxide or ethanol were incubated with rat spleen microsomes (100 micrograms of protein) for a period of 10 minutes; enymatic reactions were initiated by substrate additions. Control samples contained the vehicle without inhibitor. All calculations were based on data obtained over the linear range of the reaction kinetics. Although a variety of inhibitors were tested, only sanguinarine and the N-methylcarbamyl-PAF analog are discussed in this report.

RESULTS AND DISCUSSION

De Novo Synthesis of PAF

The three enzymes of the de novo pathway for PAF synthesis are (I) acetyl-CoA:alkylglycero-P acetyltransferase, (II) alkylacetylglycero-P phosphohydrolase, and (III) CDP-choline:alkylacetylglycerol cholinephosphotransferase (Lee et al., 1986,1988; Renooij and Snyder, 1981). These enzymes catalyze the following reaction steps:

$$\text{alkylglycero-P} \xrightarrow[\text{(I)}]{} \text{alkylacetylglycero-P} \xrightarrow[\text{(II)}]{} \text{alkylacetylglycerol} \xrightarrow[\text{(III)}]{} \text{PAF}$$

All of the de novo enzymes appear to be distributed ubiquitously in mamalian tissues and this route is thought to provide essential physiological levels of PAF. The enzyme activities are not affected by stimuli that induce hypersensitivity reactions or by arachidonic acid but do respond to certain physiological factors e.g., neurotransmitters. The selective inhibitors of the remodeling pathway, discussed later in this section, do not significantly affect the de novo enzymatic activities.

Remodeling Pathways of PAF Synthesis

General Aspects. The remodeling of a membrane lipid (alkylacyl-GPC) via the hydrolysis of the acyl group to form lyso-PAF as an intermediate in the synthesis of PAF is the pathway responsible for the cellular hypersensitivity responses induced by PAF. Involvement of a specific PLA_2 and two different types of transacylases have been implicated as the source of the lyso-PAF intermediate. A number of studies have indicated that the preferred alkylacyl-GPC substrate for any of these enzymes contains an arachidonoyl or other polyunsaturated acyl moiety at the *sn*-2 position. The lyso-PAF (alkylglycerophosphocholine) produced in this hydrolytic reaction serves as the substrate for the acetyltransferase that synthesizes PAF (Wykle et al., 1980).

Phospholipase A_2 (PLA_2). For a long time, the direct hydrolysis of the acyl group of alkylacyl-GPC had been thought to be catalyzed by a specific PLA_2. However, a clear linkage between the agonist-stimulation of PAF synthesis and PLA_2 activity in this pathway has never been demonstrated and therefore the role of the direct action of a specific PLA_2 in PAF biosynthesis remains to be established.

$$\text{alkylacyl-GPC} \xrightarrow[PLA_2]{} \text{lyso-PAF} \xrightarrow[\textit{acetyltransferase}]{} \text{PAF}$$

CoA-ind and CoA-dep Transacylases. These two different types of enzymatic activities catalyze a transacylation reaction between alkylacyl-GPC and an acyl acceptor. The CoA-ind transacylase (Uemura et al., 1991; Nieto et al., 1991,;Venable et al., 1991; Suguira et al., 1990) uses a lysophospholipid, e.g. lysoplasmalogen (alk-1-enylglycero-phosphoethanolamine or alk-1-enyl-GPE) as the acyl acceptor, whereas the CoA-dep transacylase (Blank et al., 1995; Suguira et al., 1995) utilizes CoA as the acceptor molecule. The reactions for these two routes are shown below:

CoA-ind transacylase

alkylacyl-GPC	+	lysoplasmalogen (alk-1-enyl-GPE)	→	lyso-PAF	+	plasmalogen (alk-1-enylacyl-GPE)

CoA-dep transacylase

alkylacyl-GPC + CoA → lyso-PAF + acyl-CoA

The lyso-PAF formed via transacylation in both of the above reactions can then be acetylated by the acetyl-CoA:lyso-PAF acetyltransferase (Wykle et al., 1980). Although both transacylase activities are widely distributed among mammalian cells and tissues, the CoA-ind transacylase is thought to play the most important role in PAF biosynthesis.

In addition to their significance in PAF biosynthesis, the transacylases appear to be the major way in which the *sn*-2 position of phospholipids are remodeled. The high degree of selectivity the transacylases exhibit for polyunsaturated moieties such as arachidonic acid accounts for how these highly unsaturated acyl groups can be transferred from one phospholipid molecule to another in the absence of acyl-CoAs. The reader is referred to two reviews that cover the extensive literature on the subject of transacylases (MacDonald and Sprecher, 1991; Snyder et al., 1992)

PAF Transacetylase

A novel reaction involving the transfer of the acetate group to a variety of other lipids has recently been characterized by our laboratory (Lee et al., 1992, 1996). The PAF transacetylase does not require CoA and has properties differing from those of the transacylases that transfer long-chain acyl moieties among lipids. The different reactions and various types of acetyl acceptor molecules utilized by the PAF transacetylase are illustrated below.

PAF + lysoglycerophospholipids → *sn*-2-acetylated phospholipid + lyso-PAF

PAF + fatty alcohols → fatty alcohol acetates + lyso-PAF

PAF + sphingosine → C_2-ceramides + lyso-PAF

The significance of the acetylated lipids produced by the PAF transacetylase remains obscure, but the formation of the C_2-ceramides (Lee et al., 1996) suggests the possibility of an important physiological role because of the diverse biological properties of the ceramides it can produce. It is also obvious that the PAF transacetylase could occupy a strategic pivitol position as an enzymatic regulator of the cellular functions of sphingosine as well as other acetylated lipid mediators.

Selective Inhibitors of Enzymes in the PAF Remodeling Pathway

Our recent investigation of a number of potential inhibitors of the different enzymes responsible for PAF biosynthesis have identified some that selectively inhibit the enzymes (acetyltransferase and transacylases) of the remodeling pathway of PAF biosynthesis with little or no influence on any of the de novo enymes. Sanguinarine (an isoquinoline alkaloid) and an *sn*-2 carbamyl analog of PAF selectively inhibited the lyso-PAF acetyltransferase and the transacylases associated with the remodeling pathway without seriously affecting any of the enzyme activities of the de novo pathway of PAF biosynthesis. Detailed results of these enzyme inhibitor studies will be presented in a separate paper.

ACKNOWLEDGEMENTS

Work supported by the Office of Energy Research, US Department of Energy (DE-AC05-760R00033) and the National Heart, Lung, and Blood Institute (HL27109-14).

REFERENCES

Blank, M.L., Fitzgerald, V., Smith, Z.L., and Snyder, F., 1995, Generation of the precursor (lyso-PAF) of platelet-activating factor via a CoA-dependent transacylase, Biochem. Biophys. Res. Commun. 210:1995.

Lee, T-c., Malone, B., and Snyder, F., 1986, A new de novo pathway for the formation of 1-alkyl-2-acetyl-*sn*-glycerols, a precursor of platelet-activating factor, J. Biol. Chem. 261:537.

Lee, T-c., Malone, B., and Snyder, F., 1988, Formation of 1-alkyl-2-acetyl-*sn*-glycerols via the de novo biosynthetic pathway for platelet-activating factor- characterization of 1-alkyl-2-acetyl-*sn*-glycero-3-phosphate phosphohydrolase in rat spleens, J. Biol. Chem. 263:1755.

Lee, T-c., Ou, M-c., Shinozaki, K., Malone, B., and Snyder, F., 1996, Biosynthesis of N-acetylsphingosine by platelet-activating factor: sphingosine CoA-independent transacetylase in HL-60 cells, J. Biol. Chem. 271:209.

Lee, T-c., Uemura Y., and Snyder, F., 1992, A novel CoA-independent transacetylase produces ethanolamine plasmalogen and acyl analogs of platelet-activating factor (PAF) with PAF as the acetate donor in HL-60 cells, J. Biol. Chem. 267:19992.

Nieto, M.L., Venable, M.E., Bauldry, S.A., Greene, G.G., Kennedy, M., Bass, D.A., and Wykle, R.L., 1991, Evidence that hydrolysis of ethanolamine plasmalogens triggers synthesis of platelet-activating factor via a transacylation reaction, J. Biol. Chem. 266:18699.

Rennoij, W. and Snyder, F., 1981, Biosynthesis of 1-alkyl-2-acetyl-*sn*-glycero-3-phosphocholine (platelet-activating factor and a hypotensive lipid) by cholinephosphotransferase in various rat tissues, Biochim. Biophys. Acta 663:545.

Snyder, F., 1995a, Platelet-activating factor and its analogs: metabolic pathways and related intracellular processes, Biochim. Biophys. Acta, 1254:231.

Snyder, F., 1995b, Platelet-activating factor: the biosynthetic and catabolic enzymes, Biochem. J. 305:689.

Sugiura, T., Fukuda, T., Masuzawa, Y., and Waku, K., 1990, Ether lysophospholipid-induced production of platelet-activating factor in human polymorphonuclear leukocytes, Biochim. Biophys. Acta, 1047:223.

Sugiura, T., Kudo, N., Ojima, T., Mabuchi-Itoh, K., Yamashita, A., and Waku, K., 1995, Coenzyme A-dependent cleavage of membrane phospholipids in several rat tissues: ATP-independent acyl CoA synthesis and the generation of lysophospholipids, Biochim. Biophys. Acta, 1255:167.

Uemura, Y., Lee, T-c., 1991, A coenzyme A-independent transacylase is linked to the formation of platelet-activating factor (PAF) by generating the lyso-PAF intermediate in the remodeling pathway, J. Biol. Chem., 266:8268.

Venable, M.E., Nieto, M.L., Schmidt, J.D., and Wykle, R.L., Conversion of 1-O-[^{3}H]-alkyl-2- arachidonoyl-*sn*-glycero-3-phosphorylcholine to lyso platelet-activating factor by the CoA-independent transacylace in membrane fractions of human neutrophils, J.Biol.Chem. 266:18691.

Woodard, D.S., Lee, T-c, and Snyder, F., 1987, The final step in the de novo biosynthesis of platelet-activating factor -properties of a unique CDP-choline:1-alkyl-2-acetylglycerol cholinephosphotransferase in microsomes from the renal inner medulla of rats, J. Biol. Chem., 262:2520.

Wykle, R.L., Malone, B., and Snyder, F., 1980, Enzymatic synthesis of 1-alkyl-2-acetyl-*sn*-glycero-3-phosphocholine, a hypotensive and platelet-aggregating lipid, J. Biol. Chem. 255:10256.

ANTIPROLIFERATIVE PHOSPHOLIPID ANALOGUES ACT VIA INHIBITION OF PHOSPHATIDYLCHOLINE BIOSYNTHESIS

Christoph C. Geilen,[1] Thomas Wieder, [1,2] and Constantin E. Orfanos[1]

[1]Department of Dermatology and
[2]Department of Molecular Biology and Biochemistry
University Medical Center Benjamin Franklin, The Free University of Berlin
Hindenburgdamm 30, D-12200 Berlin, Germany

INTRODUCTION

In current anticancer therapies most cytostatic agents impair cell division by crosslinking DNA (e.g. cis-platin or alkylating agents), disrupting the cytoskeleton (e.g. vinblastine) or rectifying the cytoskeleton (e.g. taxol). In a new approach to cancer chemotherapy, the cell membrane was described as a target for cytostatic agents[1,2] and it has been proposed that inhibition of cell signaling pathways represents their mode of action[3,4]. Alkyllysophospholipids, as promising candidates for this kind of cytostatics, possess antineoplastic properties in vitro and in vivo[5], leading to the development of another class of antiproliferative phospholipid analogues, the alkylphosphocholines. The prototype of these phospholipid analogues, hexadecylphosphocholine (HePC), has been shown to inhibit cell proliferation and tumor growth [6-8]. Using radiolabelled HePC[9] it was possible to show that more than 95% of this compound is membrane-bound[10] thereby confirming cellular membranes as the primary target of HePC. Since the exact mechanism(s) of alkylphospholipid actions still remained obscure we determined their influence on phosphatidylcholine biosynthesis. Phosphatidylcholine is the main phospholipid of cellular membranes of eucaryotic cells and its biosynthesis is mainly governed by the rate-limiting enzyme of the CDP-choline pathway, CTP:phosphocholine cytidylyltransferase (EC 2.7.7.15)[11,12]. The activity of cytidylyltransferase is regulated by reversible translocation to membranes and the translocation process has been shown to be influenced by various lipids[13]. Furthermore, reversible phosphorylation seems to be involved in enzyme regulation[14,15] and cytidylyltransferase is a substrate for mitogen-activated protein kinases [16,17]. However, the following paragraph will concentrate on cytidylyltransferase/lipid interactions.

Frontiers in Bioactive Lipids
Edited by Vanderhock, Plenum Press, New York, 1996

PHOSPHATIDYLCHOLINE BIOSYNTHESIS AS A TARGET OF PHOSPHOLIPID ANALOGUES

Alkylphosphocholines

It was demonstrated that HePC (for chemical structure see Fig. 1) inhibits the biosynthesis of phosphatidylcholine by disturbing the translocation process of the rate limiting enzyme, CTP: phosphocholine cytidylyltransferase (EC 2.7.7.15), and that this inhibition correlates with the antiproliferative effects of this analogue in different cell lines[18-21]. Therefore, the effects of different phospholipid analogues on cell proliferation and phosphatidylcholine biosynthesis were systematically investigated. Using alkylphosphocholines, alkylphosphoethanolamines and

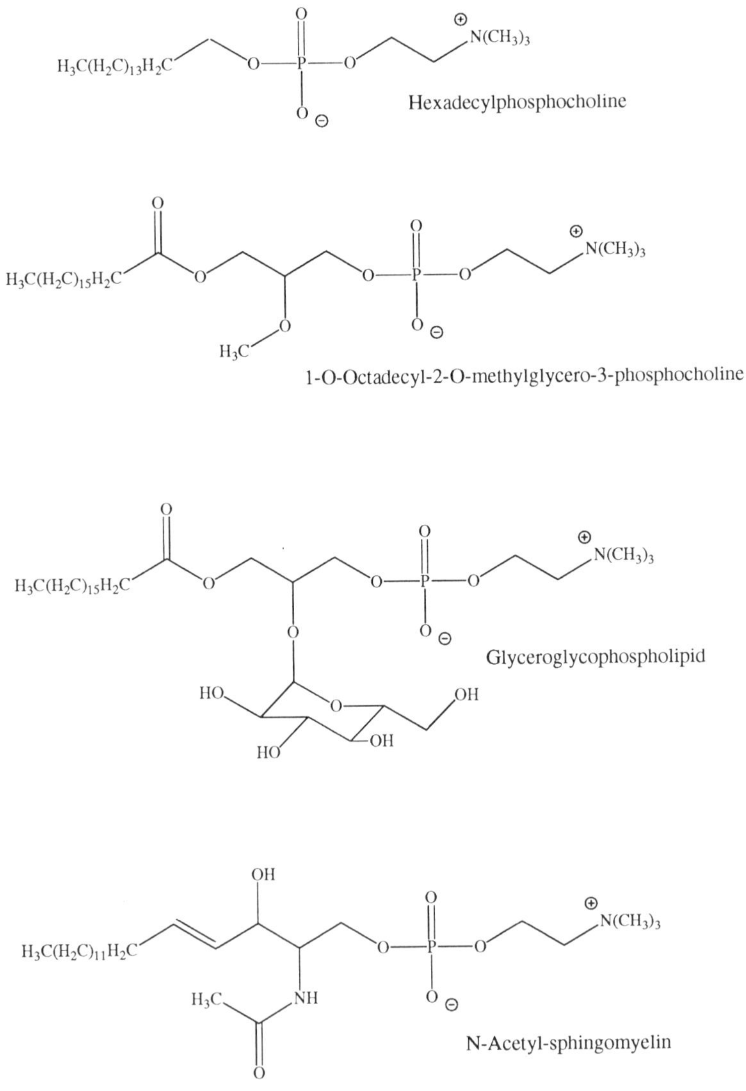

Figure 1. Chemical structures of different bioactive phospholipid analogues.

alkylphosphoserines, we could show that all biologically active phospholipid analogues contained a phosphocholine head group [22].

Etherlipids

In further experiments the etherlipid, 1-O-octadecyl-2-O-methylglycero-3-phosphocholine (ET-18-OCH_3), was used and its effects on cell proliferation and phosphatidylcholine biosynthesis were compared with the effects of HePC. This study clearly revealed that inhibition of phosphatidylcholine biosynthesis by both analogues preceded cell death. Although ET-18-OCH_3 was a more effective inhibitor of phosphatidylcholine biosynthesis as compared with HePC this compound showed strong toxic side effects[23]. In this context, it was shown by another group that circumventing the inhibited *de novo* biosynthesis of phosphatidylcholine by addition of lyso-phosphatidylcholine was sufficient to abolish the cytotoxic effect of ET-18-OCH_3[24].

Glyceroglycophospholipid (Glc-PC)

Recently, the synthesis of a new glucose-containing phospholipid analogue with antiproliferative properties has been published[25]. This compound did not inhibit protein kinase C *in vitro*, even at cytotoxic concentrations (>100μM). On the other hand, a dose- and time-dependent inhibition of phosphatidylcholine biosynthesis with a half-inhibitory concentration of 10 μM was observed (unpublished results). Hence, GlcPC is the first phospholipid analogue which specifically inhibits phosphatidylcholine biosynthesis without accompanying inhibitory effects on protein kinase C (Table 1).

Sphingomyelin analogues

To investigate the influence of choline head group-containing sphingolipids on phosphatidylcholine biosynthesis, the cell-permeable sphingomyelin analogue, N-acetyl-sphingosine-1-phosphocholine (AcSM), was used to modify the sphingomyelin content of

Table 1. Effect of different phospholipid analogues on cell proliferation, phosphatidylcholine biosynthesis, CTP:phosphocholine cytidylyltransferase and protein kinase C

Lipid analogue[1]	Cell Proliferation	PC Biosynthesis	Membrane-bound CT	PKC Activity
HePC	⇓	⇓	⇓	⇓
Et-18-O-CH_3	⇓	⇓	⇓	⇓
GlcPC	⇓	⇓	n.d.	=
AcSM	⇓	⇓	⇓	n.d.

[1] Abbreviations: Hexadecylphosphocholine (HePC); 1-O-octadecyl-2-O-methylglycero-3-phosphocholine (ET-18-OCH_3); glyceroglucophospholipid (GlcPC); *N*-acetyl-sphingomyelin (AcSM); phosphatidylcholine (PC); CTP:phosphocholine cytidylyltransferase (CT); protein kinase C (PKC); inhibition (⇓); not determined (n.d.)

cellular membranes. Treatment of HaCaT cells with AcSM led to a fast incorporation of this compound into the cells and subsequently to inhibition of phosphatidylcholine biosynthesis and translocation of CTP:phosphocholine cytidylyltransferase[26].

In conclusion, interfering with the cell membrane metabolism of phospholipids by the use of membrane-active agents may be a useful, novel approach to treatment of cancer and hyperproliferative skin disorders, e.g. psoriasis or keloids.

OTHER MECHANISMS OF ACTION

The mechanisms of action of phospholipid analogues has been a subject of research for many years and different targets of these antiproliferative agents have been proposed. Among these targets were kinases, such as protein kinase C[27,28], membrane-bound proteins, such as sodium pump[27] and different enzymes of the phospholipid metabolism, such as phospholipase C[28] and acyltransferases[29]. Although there are many reports in the literature about the potential targets of phospholipid analogues, none of these targets has been systematically investigated with different analogues sharing common structural features. On the contrary, protein kinase C which has been thought to be the main target of phospholipid analogues has been shown to be of minor importance for the antiproliferative effect of ET-18-OCH_3 and HePC[30]. Recently, a totally new mechanism of action of ET-18-OCH_3 was proposed. It was shown that the ether lipid induced the expression of fos and jun proto-oncogenes and activated AP-1 transcription factor[31]. However, the signaling pathways finally leading to enhanced gene expression still need to be elucidated.

Acknowledgment

This study was supported by a grant of the Deutsche Forschungsgemeinschaft (Ge 641/3-3), research grants from the University Medical Center Benjamin Franklin and the Berliner Krebsgesellschaft. The authors want to thank Dr. D. Arndt and Dr. R. Zeisig (Max-Delbrück Center of Molecular Medicine, Berlin-Buch, Germany), Dr. W. Reutter and Dr. R.T.C. Huang (Institute of Molecular Biology and Biochemistry, Berlin-Dahlem, Germany), Dr. J. Mulzer and Dr. M. Mickeleit (Institute of Organic Chemistry, Berlin-Dahlem, Germany) and ASTA Medica AG (Frankfurt a. M., Germany) for providing phospholipid analogues. HaCaT cells were a gift from Dr. N.E. Fusenig (German Cancer Research Center, Heidelberg, Germany).

REFERENCES

1. W. E. Berdel, Membrane-interactive lipids as experimental anticancer drugs, *Br. J. Cancer* 64: 208(1991).
2. H. H.Grunicke, The cell membrane as a target for cancer chemotherapy, *Eur. J. Cancer* 27: 281 (1991).
3. V. G. Brunton and P. Workman, Cell-signaling targets for antitumor drug development, *Cancer Chemother. Pharmacol.* 32: 1 (1993).
4. S. P. Langdon and J. F. Smyth, Inhibition of cell signalling pathways, *Cancer Treat. Rev.* 21: 65 (1995).

5. W. E. Berdel, R. Andreesen, and P. G. Munder, Synthetic alkyllysophospholipid analogues: a new class of antitumor agents, in: *Phospholipids and Cellular Regulation*, J. F. Kuo, ed., CRC Press, Boca Raton (1985).
6. C. Unger, E. A. M. Fleer., J. Kötting, W. Neumüller, and H. Eibl, Antitumoral activity of alkylphosphocholines and analogues in human leukemia cell lines, *Prog. Exp. Tumor Res.* 34: 25 (1992).
7. C. C. Geilen, R. Haase, K. Buchner, Th. Wieder, F. Hucho, and W. Reutter, The phospholipid analogue, hexadecylphosphocholine, inhibits protein kinase C in vitro and antagonises phorbol ester-stimulated cell proliferation, *Eur. J. Cancer* 27, 1650 (1991).
8. C. Unger, H. Eibl, H. W. Breiser, H. W. van Heyden, J. Engel, P. Hilgard, H. Sindermann, M. Peukert, and G. S. Nagel, Hexadecylphosphocholine (D 18506) in the topical treatment of skin metastases: a phase-I trial, *Onkologie* 11: 295 (1988).
9. C. C. Geilen, A. Samson, Th. Wieder, H. Wild, and W. Reutter, Synthesis of hexadecylphospho[methyl-^{14}C]-choline, *J. Labelled Comp. Radiopharm.* 31: 1071 (1992).
10. C. C. Geilen, Th. Wieder, A. Haase, W. Reutter, D. M. Morré, and D. J. Morré, Uptake, subcellular distribution and metabolism of the phospholipid analogue hexadecylphosphocholine in MDCK cells, *Biochim. Biophys. Acta* 1211: 14 (1994).
11. D. E. Vance, Phosphatidylcholine metabolism: masochistic enzymology, metabolic regulation, and lipoprotein assembly, *Biochem. Cell Biol.* 68: 1151 (1990).
12. C. Kent, Regulation of phosphatidylcholine biosynthesis, *Prog. Lipid Res.* 29: 87 (1990)
13. R. B. Cornell, Regulation of CTP:phosphocholine cytidylyltransferase by lipids. 1. Negative surface charge dependence for activation, *Biochemistry* 30: 5873 (1991).
14. S. L. Pelech and D. E. Vance, Regulation of rat liver cytosolic CTP:phosphocholine cytidylyltransferase by phosphorylation and dephosphorylation, *J. Biol. Chem.* 257: 14198 (1982).
15. J. D. Watkins and C. Kent, Phosphorylation of CTP:phosphocholine cytidylyltransferase in vivo, *J. Biol. Chem.* 265: 2190 (1990).
16. M. Wieprecht, Th. Wieder, C. C. Geilen, and C. E. Orfanos, Growth factors stimulate phosphorylation of CTP:phosphocholine cytidylyltransferase in HeLa cells, *FEBS Lett.* 353: 221 (1994).
17. M. Wieprecht, Th. Wieder, C. Paul, C. C. Geilen, and C. E. Orfanos, Evidence for phosphorylation of CTP:phosphocholine cytidylyltransferase by multiple proline-directed protein kinases, *J. Biol. Chem.*, in press.
18. R. Haase, Th. Wieder, C.C. Geilen, and W. Reutter, The phospholipid analogue hexadecylphosphocholine inhibits phosphatidylcholine biosynthesis in Madin-Darby canine kidney cells, *FEBS Lett.* 288: 129 (1991).
19. C.C. Geilen, Th. Wieder, and W. Reutter, Hexadecylphosphocholine inhibits translocation of CTP:phosphocholine cytidylyltransferase in Madin-Darby canine kidney cells, *J. Biol. Chem.* 267: 6719 (1992).
20. Th. Wieder, C. C. Geilen, and W. Reutter, Antagonism of phorbol ester-stimulated phosphatidylcholine biosynthesis by the phospholipid analogue, hexadecylphosphocholine, *Biochem. J.* 291: 561 (1993).
21. M. Detmar, C.C. Geilen, Th. Wieder, C.E. Orfanos, and W. Reutter, The phospholipid analogue hexadecylphosphocholine inhibits proliferation and phosphatidylcholine biosynthesis of human epidermal keratinocytes in vitro, *J. Invest. Dermatol.* 102: 490 (1994).
22. C.C. Geilen, A. Haase, Th. Wieder, D. Arndt, R. Zeisig, and W. Reutter, Phospholipid analogues - side chain and polar head group-dependent effects on phosphatidylcholine biosynthesis, *J. Lipid Res.* 35: 625 (1994).
23. Th. Wieder, A. Haase, C.C. Geilen, and C.E. Orfanos, The effect of two synthetic phospholipids on cell proliferation and phosphatidylcholine biosynthesis in Madin-Darby canine kidney cells, *Lipids* 30: 389 (1995).
24. K.P. Boggs, C.O. Rock, and S. Jackowski, Lysophosphatidylcholine attenuates the cytotoxic effects of the antineoplastic phospholipid 1-O-octadecyl-2-O-methyl-rac-glycero-3-phosphocholine, *J. Biol. Chem.* 270: 11612 (1995).
25. M. Mickeleit, Th. Wieder, K. Buchner, C. C. Geilen, J. Mulzer, and W. Reutter, Glyceroglycophospholipid (Glc-PC), ein neuartiges glucosidisches Phospholipid, *Angew. Chem.* 107: 2879 (1995).
26. Th. Wieder, C. Perlitz, M. Wieprecht, R.T.C. Huang, C.C. Geilen, and C.E. Orfanos, Two new sphingomyelin analogues inhibit phosphatidylcholine biosynthesis by decreasing membrane-bound CTP: phosphocholine cytidylyltransferase levels in HaCaT cells, *Biochem. J.* 311: 873 (1995).
27. B. Zheng, K. Oishi, M. Shoji, H. Eibl, W. E. Berdel, J. Hajdu, W. R. Vogler, and J. F. Kuo, Inhibition of protein kinase C, (sodium plus potassium)-activated adenosine triphosphatase, and sodium pump by synthetic phospholipid analogues, *Cancer Res.* 50: 3025 (1990).
28. F. Überall, H. Oberhuber, K. Maly, J. Zaknun, L. Demuth, and H. H. Grunicke, Hexadecylphosphocholine inhibits inositol phosphate formation and protein kinase C activity, *Cancer Res.* 51: 807 (1991).

29. W. R. Vogler, E. Whigham, W. D. Bennett, and A. C. Olson, Effect of alkyl-lysophospholipids on phosphatidylcholine biosynthesis in leukemic cell lines, *Exp. Hematol.* 13: 629 (1985).
30. D. Berkovic, K. Berkovic, E. A. M. Fleer, H. Eibl, and C. Unger, Inhibition of calcium-dependent protein kinase C by hexadecylphosphocholine and 1-O-octadecyl-2-O-methyl-rac-glycero-3-phosphocholine do not correlate with inhibition of proliferation of HL60 and K562 cell lines, *Eur. J. Cancer* 30A: 509 (1994).
31. F. Mollinedo, C. Gajate, and M. Modolell, The ether lipid 1-O-octadecyl-2-O-methyl-rac-glycero-3-phosphocholine induces expression of fos and jun proto-oncogenes and activates AP-1 transcription factor in human leukaemic cells, *Biochem. J.* 302: 325 (1994).

LIPOPOLYSACCHARIDE-, LIPOSOME-ENCAPSULATED MTP-PE- AND DEXAMETHASONE- REGULATED PROSTAGLANDIN RELEASE IN RAT LIVER MACROPHAGES: ROLE OF CYTOSOLIC PHOSPHOLIPASE A_2 AND PROSTAGLANDIN H SYNTHASE-2

P. Ambs[1], E. Fitzke[1], C. Créminon[2], J. MacLouf[2], and P. Dieter[1]

[1]Institut für Biochemie und Molekularbiologie, Klinik für Tumorbiologie, Breisacherstr. 117, D-79106 Freiburg, Germany,
[2]INSERM U. 348, Hôpital Lariboisière 8, rue Guy Patin, 75475 Paris cedex 10, France,

INTRODUCTION

Rat liver macrophages (RLM), activated by a variety of agents, are known to release various eicosanoids including prostaglandin (PG)E_2 and PGD_2[1]. The formation of prostanoids is strictly dependent on Ca^{2+} and is thought to be determined by the activation of a phospholipase (PL)A_2[2] and the expression of prostaglandin H synthase (PGHS)-2. It has been shown that RLM express an intracellular $cPLA_2$ which is strictly Ca^{2+}-dependent in the range from 10^{-7}-10^{-6} M and which shows a Ca^{2+}-dependent translocation to cellular membranes[3,4]. $cPLA_2$ phosphorylation has been shown to be most likely catalysed by mitogen activated protein (MAP) kinase[4]. Further evidence has been provided that in addition to regulation by $cPLA_2$, release of prostanoids in RLM is also dependent on the expression of PGHS-2[5].

The synthesis and release of prostanoids in RLM is elicited by a number of agents including two components of the outer membrane of bacteria, lipopolysaccharide (LPS) and N-acetyl-muramyl-L-alanyl-D-isoglutaminyl-L-alanine-2-[1',2'-dipalmitoyl-sn-glycero-3']-ethylamide (MTP-PE) incorporated into liposomes[6]. Although these two compounds share many properties, they seem to use dirferent signal transduction pathways to activate RLM[6].

The aim of this study was to investigate the possible mechanisms by which LPS and liposome-encapsulated MTP-PE induce an increased release of PGE_2 and PGD_2.

MATERIALS AND METHODS

RLM from male Wistar rats were isolated and cultured as described previously[1]. PLA_2 activity was determined according to the method described previously[3,4]. Release of PGE_2 and PGD_2 was measured by ELISA[6] and RIA[6], respectively. SDS-PAGE and Western-blot analysis were performed as described elsewhere[4,7]. The blots were either probed with polyclonal antibodies raised against U937 $cPLA_2$[8] (generous gift from Dr. J. D. Clark; Genetics Institute, Massachusetts, USA) or with polyclonal antibodies raised against PGHS-2[9].

RESULTS

It is known that LPS induces a release of prostanoids in RLM with a more pronounced stimulation of PGE_2 compared to PGD_2[10]. In contrast to LPS, liposome-encapsulated MTP-PE induces a similar release of PGE_2 and PGD_2, thereby not affecting the ratio of released PGE_2/PGD_2 measured in unstimulated cells (Table 1).

It has been shown that incubation of RLM with dexamethasone prior to stimulation with LPS decreases the release of PGE_2 and PGD_2 to levels measured in unstimulated cells[1]. Surprisingly, dexamethasone inhibits only the release of PGD_2, but not of PGE_2, in response to liposome-encapsulated MTP-PE (Table 1).

Table 1. Formation of PGE_2 and PGD_2 in untreated, LPS- and liposome-encapsulated MTP-PE-treated cells and the effects of pre-incubation with dexamethasone

treatment	Release of PGE_2 [pmol/10^6 cells]		Release of PGD_2 [pmol/10^6 cells]		Ratio of PGE_2/D_2	
	- dex	+ dex	- dex	+ dex	- dex	+ dex
None	6 ± 2	5 ± 2	6 ± 3	6 ± 3	0.9	0.9
LPS	41 ± 6	6 ± 2	17 ± 5	5 ± 3	2.4	0.8
MTP-PE	42 ± 12	41 ± 9	53 ± 14	8 ± 3	0.9	6

RLM (48 h in primary culture) were treated without (-dex) or with (+dex) dexamethasone (1 μM) for 24 h. Thereafter LPS (0.5 μg/ml; from *Salmonella minnesota* R595, provided by Dr. Galanos, Freiburg, FRG) or liposome-encapsulated MTP-PE-liposomes (25 μg/ml; provided by Dr. Tarcsay, Ciba-Geigy, Basel, Switzerland) were added for 24 h. The media were removed and formation of PGE_2 and PGD_2 was determined. Results are means of ± SD of 3 independent experiments.

The LPS-induced release of arachidonic acid (AA)[4] and prostaglandins (Table 1) is paralleled by an about 2fold increased PLA_2 activity (Table 2) and by an increased expression of $cPLA_2$ (Figure 1). We have recently described[4] a LPS-induced $cPLA_2$ phosphorylation which is observed already 10 min after LPS-treatment (Figure 1). Phosphorylation of $cPLA_2$ can be detected by a shift of the enzyme to a higher apparent molecular weight on SDS-PAGE[4]. The LPS-induced phosphorylation of $cPLA_2$ is almost complete and remains unchanged up to 8 h. Thereafter $cPLA_2$ becomes slowly dephosphorylated. 24 h to 48 h after LPS-exposure the minor part of $cPLA_2$ is in the phosphorylated state (Figure 1).

In contrast to LPS, liposome-encapsulated MTP-PE does not induce an increase of the expression of $cPLA_2$. Liposome-encapsulated MTP-PE induces a phosphorylation of $cPLA_2$, however, with a complete different time-course than LPS. Phosphorylation of $cPLA_2$ can be detected at first after 16-24 h, 48 h after addition of liposome-encapsulated MTP-PE phosphorylation of $cPLA_2$ is almost complete (Figure 1). Treatment of RLM with liposome-encapsulated MTP-PE leads also to a small increase of $cPLA_2$ activity (Table 2).

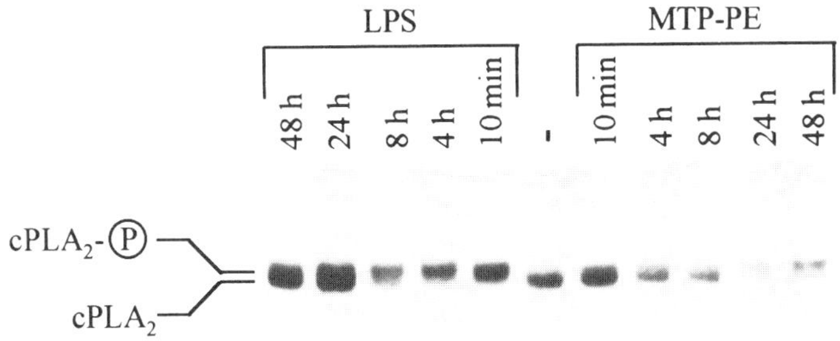

Figure 1. Expression and phosphorylation of $cPLA_2$ in response to LPS and liposome-encapsulated MTP-PE. RLM (48 h in primary culture) were treated without (-) or with LPS (0.5 µg/ml) or liposome-encapsulated MTP-PE (25 µg/ml) for the indicated times. The lysates (20 µg) were immunoblotted with anti-$cPLA_2$ antibodies.

Table 2. Effects of LPS, liposome-encapsulated MTP-PE and dexamethasone on PLA_2-activity.

Treatment	PLA_2 activity [% of control]
None	100
LPS	179 ± 27
MTP-PE	129 ± 6
dexamethasone	51 ± 3

RLM (48 h in primary culture) were incubated without or with LPS (0.5 µg/ml), liposome-encapsulated MTP-PE (25 µg/ml) or dexamethasone (1 µM) for 24 h. Thereafter PLA_2-activity of lysates (2.5 µg) was determined. A value of 100% corresponds to a PLA_2 activity of 0.12 ± 0.02 nmol AA/min x mg protein. Results are means ± SD of 3 independent experiments.

The induction of PGHS-2 is another mechanism by which inflammatory agents are able to stimulate a formation of prostanoids in macrophages[11]. In RLM the expression of PGHS-2 is observed after stimulation with LPS. PGHS-2 expression becomes detectable 4-8 h after addition of LPS and reaches a maximum after 24 h. Thereafter the amount of PGHS-2 decreases but can still be detected after 48 h (Figure 2). Liposome-encapsulated MTP-PE induces PGHS-2 expression with a time-course similar to LPS: PGHS-2 is detectable after 8 h and reaches maximum levels after 24 h. But in contrast to LPS, 48 h after the addition of liposome-encapsulated MTP-PE no decrease of the levels of PGHS-2 was observed (Figure 2).

Dexamethasone seems to act on both levels, $cPLA_2$ and PGHS-2. Addition of dexamethasone for 24 h decreases the amount and activity of $cPLA_2$ (Table 2, Figure 3). However, it does not affect the LPS-induced phosphorylation of $cPLA_2$, in contrast to the phosphorylation induced by liposome-encapsulated MTP-PE which is completely inhibited by dexamethasone (Figure 3). The expression of PGHS-2 induced by LPS and liposome-

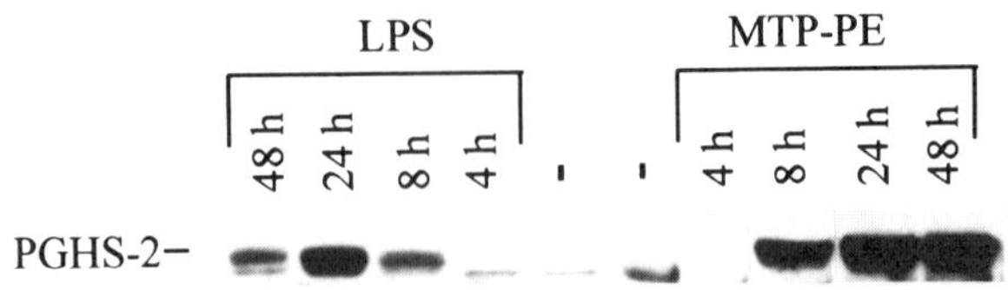

Figure 2. PGHS-2 expression in response to LPS and liposome-encapsulated MTP-PE. RLM (48 h in primary culture) were treated without (-) or with LPS (0.5 μg/ml) or liposome-encapsulated MTP-PE (25 μg/ml) for the indicated times. The lysates (20 μg) were immunoblotted with anti-PGHS-2 antibodies.

encapsulated MTP-PE is completely abolished if the cells are pretreated with dexamethasone (Figure 4).

It is known that PGE_2 modulates many acivities of RLM[10]. Here we show that addition of PGE_2 24 h prior to stimulation with LPS or liposome-encapsulated MTP-PE leads to a markedly enhanced expression of PGHS-2 (Figure 4).

DISCUSSION

The LPS-induced release of prostanoids in RLM (Table 1) seems to be regulated by two mechanisms. Firstly, LPS induces an increase of the expression of $cPLA_2$ (Figure 1) which is accompanied by an increased PLA_2 activity (Table 2). LPS also induces a phosphorylation of $cPLA_2$ (Figure 1). The role of $cPLA_2$ phosphorylation in RLM, however, is unclear because phosphorylation can be detected already after 10 min while at this time-point almost no release of AA and prostanoids can be measured[4]. Furthermore, only a small increase of $cPLA_2$ activity (about 1.4 fold) is observed after phosphorylation of the enzyme[4]. We have recently provided evidence that Ca^{2+} seems to be more important for the activation of $cPLA_2$ in RLM: Ca^{2+} is absolutely required[2] and seems also to be sufficient, even without any measurable phosphorylation of $cPLA_2$[4], for maximal activation

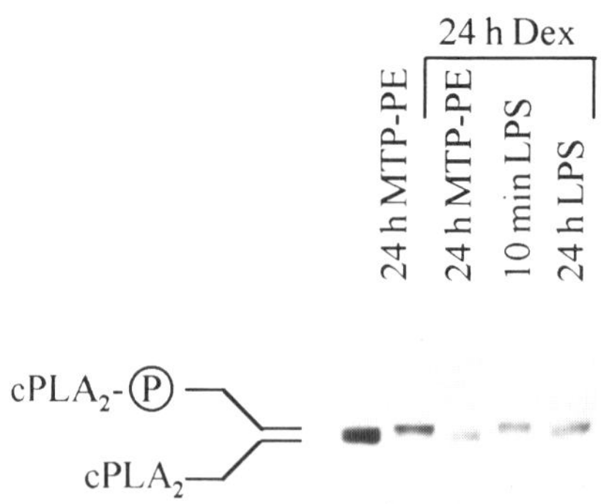

Figure 3. Effects of incubation of RLM with dexamethasone on the expression and phosphorylation of $cPLA_2$ induced by LPS- and liposome-encapsulated MTP-PE. RLM (48 h in primary culture) were treated without (-) or with dexamethasone (Dex, 1 μM). Thereafter LPS (0.5 μg/ml) or liposome-encapsulated MTP-PE (25 μg/ml) were added for 10 min or 24 h as indicated. The lysates (20 μg) were immunoblotted with anti-$cPLA_2$ antibodies.

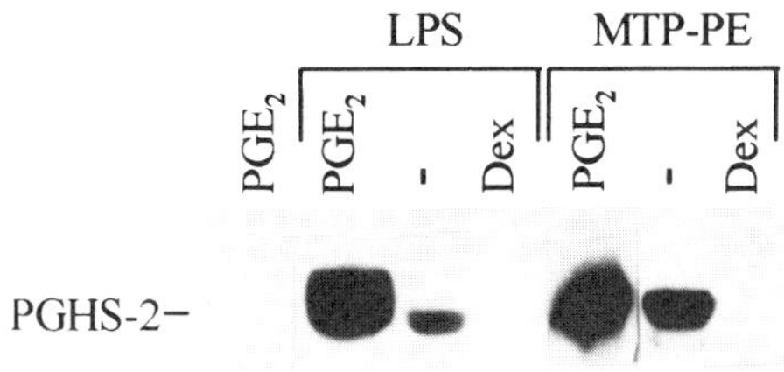

Figure 4. Effects of incubation of RLM with dexamethasone or PGE_2 on the expression of PGHS-2 induced by LPS- and liposome-encapsulated MTP-PE-liposomes. RLM (48 h in primary culture) were treated without (-), with dexamethasone (Dex, 24h, 1 µM) or with PGE_2 (24h, 1 µM). Thereafter LPS (0.5 µg/ml) or liposome-encapsulated MTP-PE-liposomes (25 µg/ml) were added for another 24 h. The lysates (20 µg) were immunoblotted with anti-PGHS-2 antibodies.

of the AA cascade. A second mechanism of LPS to induce release of AA in RLM seems to be an induction of PGHS-2 expression (Figure 2). The time-course of the expression of $cPLA_2$ and PGHS-2 by LPS correlates well to the time-course of LPS-induced release of AA and PGE_2[4,10].

Like LPS, liposome-encapsulated MTP-PE induces an expression of PGHS-2, but in contrast to LPS, a more prolonged expression of $cPLA_2$ is observed (Figure 2). This correlates with a prolonged increase of the release of PGE_2[6]. In contrast to LPS, liposome-encapsulated MTP-PE does not induce an increased expression of $cPLA_2$ (Figure 1). However, a small increase of PLA_2 activity is measured after liposome-encapsulated MTP-PE (Table 2). This increase might be based on phosphorylation of $cPLA_2$ which is observed 24 h after addition of liposome-encapsulated MTP-PE (Figure 1). Phosphorylation of $cPLA_2$ in RLM has been shown to be mediated by PKC-ß and MAP-kinase or by tyrosine kinases and MAP-kinase for stimulation with zymosan and phorbol ester or with LPS, respectively[4]. In all these cases phosphorylation is rapid and can be detected within the first 10 min after addition of the agents. In the case of liposome-encapsulated MTP-PE, phosphorylation of $cPLA_2$ is slow and can be detected at first after 24 h (Figure 1). A mechanism which involves gene expression might therefore be involved in liposome-encapsulated MTP-PE-induced phosphorylation of $cPLA_2$. PKC-δ might be a candicate for mediating this effect because PKC-δ expression is enhanced by liposome-encapsulated MTP-PE with a similar time-course as phosphorylation of $cPLA_2$[6].

Dexamethasone has been shown to inhibit almost completely the release of AA and prostanoids upon stimulation with zymosan, phorbol ester, A23187 or LPS in RLM[1] (Table 1). This effect might be based on the inhibition of the expression of $cPLA_2$ (Figure 3; Table 2) and/or by the inhibition of the expression of PGHS-2 (Figure 4). Surprisingly, the liposome-encapsulated MTP-PE- induced release of PGE_2 is not inhibited by dexamethasone while the release of PGD_2 is inhibited (Table 1). The reason for this "specific" effect of dexamethasone on PGE_2 release is not known. An effect of dexamethasone on PGE_2 synthase in RLM, however, is most likely.

PGE_2 has been shown to inhibit the LPS-induced release of prostanoids in RLM[10]. Here we show that PGE_2 leads to an increase of the LPS- and liposome-encapsulated MTP-PE-. induced expression of PGHS-2 (Figure 4). Further investigations including the effect of PGE_2 on the expression of $cPLA_2$ and other enzymes of the AA cascade are necessary to clarify these two contradictionary findings.

In conclusion we have shown that RLM express different transduction pathways to trigger and regulate the AA cascade: Stimulation of pre-existing enzymes in case of zymosan, phorbol ester or A23187, most probably by an increase in intracellular Ca^{2+} and involving $cPLA_2$ and PLC/DAG lipase[4,12], changes in the expression of $cPLA_2$ and/or

PGHS-2 in case of stimulation with LPS and liposome-encapsulated MTP-PE[4]. Furthermore, we provided evidence that LPS and liposome encapsulated MTP-PE might use different mechanisms to stimulate release of prostanoids in RLM.

REFERENCES

1. P.Dieter, A.Schulze-Specking, and K.Decker, Differential inhibition of prostaglandin and superoxide production by dexamethasone in primary cultures of rat Kupffer cells, *Eur. J. Biochem.* 159: 451 (1986)
2. P.Dieter, A.Schulze-Specking, and K.Decker, Ca^{2+} requirement of prostanoid but not superoxide production by rat Kupffer cells, *Eur. J. Biochem.* 177: 61 (1993)
3. H.Krause, P.Dieter, A.Schulze-Specking, A.Ballhorn, and K.Decker. Ca^{2+}- induced reversible translocation of PLA_2 between the cytosol and the membrane fraction of rat liver macrophages, *Eur. J. Biochem.* 199: 355 (1991).
4. P.Ambs, M.Baccarini, E.Fitzke, and P.Dieter, Role of cytosolic phospholipase A_2 in arachidonic acid release of rat-liver macrophages: regulation by Ca^{2+} and phosphorylation, *Biochem. J.* 311: 189 (1995)
5. F.Zhang, U.Warskulat, M.Wettstein, R.Schreiber, H.Henninger, K.Decker, and D.Häussinger, Hyperosmolarity stimulates prostaglandin synthesis and cyclooxygenase-2 expression in activated rat liver macrophages, *Biochem. J.* 312: 135 (1995)
6. P.Dieter, P.Ambs, E.Fitzke, H.Hidaka, and H.Schwende, Comparative studies of cytotoxicity and the release of TNF-α, nitric oxide, and eicosanoids in liver macrophages treated with lipopolysaccharide and liposome encapsulated MTP-PE, *J. Immunol.* 155: 2595 (1995)
7. J.Duyster, H.Schwende, E.Fitzke, H.Hidaka, and P.Dieter, Different roles of PKC-ß and PKC-δ in AA-cascade, superoxide formation and phosphoinositide hydrolysis, *Biochem. J.* 292: 203 (1993).
8. J.D.Clark, L.L.Lin, R.W.Kriz, S.R.Chakkodabylu, L.A.Sultzman, A.Y.Lin, N.Milona, and J.L.Knopf, A novel arachidonic acid-selective $cPLA_2$ contains a Ca^{2+}-dependent translocation domain with homology to PKC and GAP, *Cell* 65: 1043 (1991).
9. A.Habib, C.Créminon, Y.Frobert, J.H.Grassi, P.Pradelles, and J.MacLouf, Demonstration of an inducible cyclooxygenase in human endothelial cells using antibodies raised against the carboxyl-terminal region of the cyclooxygenase-2, *J. Biol. Chem.*, 268: 23448 (1993).
10. Th.Peters, U.Karck, and K.Decker, Interdependence of tumor necrosis factor, prostaglandin E_2, and protein synthesis in lipopolysaccharide-exposed kupffer-cells, *Eur. J. Biochem.* 191: 583 (1990)
11. H.R.Herschman, Prostaglandin Synthase 2, *Biochim. Biophys. Acta*, 1299: 125 (1996)
12. P.Dieter and E.Fitzke, Formation of diacylglycerol, inositol phosphates, arachidonic acid and its metabolites in macrophages, *Eur. J. Biochem.* 218: 753 (1993).

REGULATION OF PHOSPHOLIPASE C-β ISOENZYMES

T. Kendall Harden, Theresa M. Filtz, Andrew Paterson,
Marie-Christine Galas, José L. Boyer, and Gary L. Waldo

Department of Pharmacology
School of Medicine
University of North Carolina
Chapel Hill, NC 27599

INTRODUCTION

A broad range of hormones, neurotransmitters, growth factors, and chemoattractants produce their physiological effects through phospholipase C (PLC)-catalyzed initiation of the inositol lipid signaling cascade (Berridge and Irvine, 1987). Although tyrosine phosphorylation of the SH2- and SH3-domain-containing PLC-γ isoenzymes accounts for interface of a number of growth factors with this signaling pathway, the majority of extracellular stimuli act through seven-transmembrane-spanning receptors that activate G-proteins that in turn activate isoenzymes of the PLC-β family (Rhee and Choi, 1992). The focus of this chapter is on this latter group of signaling proteins.

G-protein-mediated regulation of effector proteins was elucidated initially for hormone-regulated adenylyl cyclase and for the rhodopsin-mediated phototransduction cascade. However, G-protein-regulated plasma membrane effector proteins now are known to include phospholipases and ion channels, and additional targets for G-proteins recently have been proposed (Hepler and Gilman, 1992; Gilman, 1995). G-proteins are heterotrimers that consist of a guanine nucleotide-binding α-subunit and β- and γ-subunits. Activation involves receptor-promoted exchange of GTP for GDP and supposed dissociation of the G-protein heterotrimer into a GTP-liganded α-subunit and free βγ-dimer. Although effector protein stimulation historically has been associated with the activity of GTP-liganded α-subunits, G-protein βγ-subunits also are important direct regulators of several effector proteins (Clapham and Neer, 1993; Boyer et al., 1994b).

The G-protein regulators of PLC-β isoenzymes include the four α-subunits, i.e. α_q, α_{11}, α_{14}, and α_{16}, that comprise the Gq family of pertussis- and cholera toxin-insensitive G-proteins (Simon et al., 1991; Hepler and Gilman, 1992). However, several receptors that couple to pertussis toxin-sensitive G-proteins, i.e. Gi and Go promote inositol lipid hydrolysis in certain target cells. The mechanism of this form of G-protein-mediated regulation of PLC is not unambiguously known. However, βγ-subunits activate certain of the PLC-β isoenzymes (Camps et al., 1992; Boyer et al., 1992; Blank et al., 1992), and receptor-promoted release of βγ-subunits from pertussis toxin substrate G-proteins, which typically are present in relatively high amounts in most cells, is thought to account for pertussis toxin-sensitive inositol lipid signaling. Offermanns and Simon (1995) recently have reported that $G\alpha_{16}$ does not exhibit the fidelity of receptor coupling that has been shown to be the case with other G-proteins including those of the Gq class. That is, receptors that were previously shown to have high selectivity for $G\alpha_s$ or $G\alpha_i$ nonetheless markedly promoted inositol lipid hydrolysis in cells expressing $G\alpha_{16}$. The physiological significance of this phenomenon is uncertain since $G\alpha_{16}$ is primarily expressed in cells of hematopoietic origin instead of in the cell types that express many of the Gs- or Gi-linked receptors.

Our realization of the remarkable inositol lipid signaling response to guanine nucleotides led to adoption of the turkey erythrocyte membrane as a model to study G-protein-regulated PLC (Harden et al., 1987; Harden et al., 1988; Boyer et al., 1989a; Boyer et al., 1989b). A soluble PLC (PLC-βt) was purified from turkey erythrocytes (Morris et al., 1990a) and subsequently was the first purified PLC isoenzyme to be shown to be regulated by a G-protein (Morris et al., 1990b). Amino acid sequence obtained from PLC-βt illustrated that it exhibited sequence similarity to a PLC isoenzyme (PLC-β1) that Rhee and coworkers had previously purified from rat brain (Ryu et al., 1987). Nucleotide sequence encoding this mammalian isoenzyme was eventually cloned (Suh et al., 1988) and cDNA for three additional mammalian PLC-β isoenzymes (PLC-β2, PLC-β3, and PLC-β4) subsequently were isolated (Park et al., 1992; Jhon et al., 1993; Lee et al., 1993). We have isolated a cDNA that encodes PLC-βt (Waldo et al., 1996), and a Xenopus PLC-β (PLC-βX; Ma et al., 1993) and two drosophila PLC-β (Bloomquist et al., 1988; Shortridge et al., 1991) isoenzymes have been cloned. These PLC-β isoenzymes have in common a relatively short amino terminal domain, so-called X- and Y-domains that likely comprise the catalytic domain, a domain that separates the X- and Y-domains, and a relatively long carboxy terminal domain. The overall percent amino acid identities of these isoenzymes are compared to PLC-βt in Fig. 1.

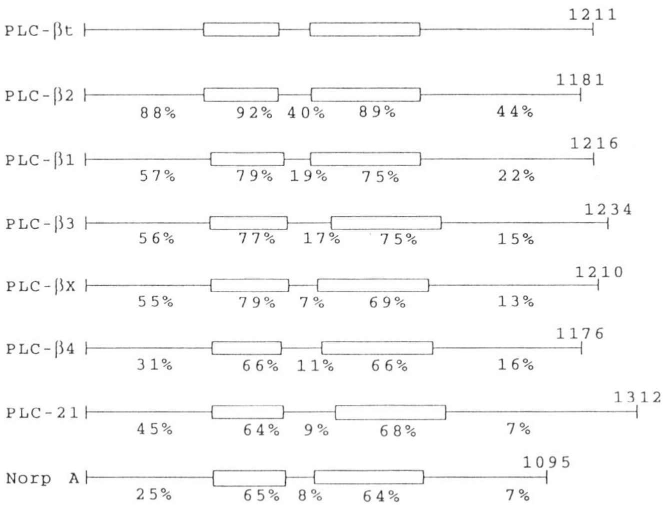

Figure 1. PLC-β isoenzymes. The amino acid sequences of the PLC-β isoenzymes were aligned and a linear schematic representation comparing PLC-βt with the seven other PLC-β isoenzymes is presented. Open boxes represent the highly conserved X- and Y-domains. The percent of amino acid sequence identity of each isoenzyme with the corresponding regions of PLC-βt is indicated below each diagram. The total amino acid length is indicated on the right.

EXPRESSION OF PLC-β ISOENZYMES USING THE BACULOVIRUS/INSECT CELL SYSTEM

Although the existence of PLC-β isoenzymes and the identity of their regulating G-proteins has been recognized for several years, knowledge of the molecular details of the regulation of these signaling proteins has been slow to develop. This lack of progress has followed at least in part from unavailability of the PLC-β isoenzymes in purified form. Alternative approaches for studying isoenzyme-selective or -specific signaling properties of these proteins have been used, but problems in interpretation are inherent in a number of these approaches. For example, cellular expression of PLC-β isoenzymes and/or G-protein subunits from mammalian expression vectors usually has involved measurement of signaling activities at nonphysiologically high concentrations of proteins and under conditions in which the contribution of endogenous signaling proteins is ill-defined.

To circumvent some of the problems associated with study of PLC-β isoenzymes, our laboratory has developed procedures for expression of large amounts of recombinant proteins using the baculovirus/Sf9 insect cell expression system. This methodology has the advantage of production of large amounts of recombinant enzyme that is indistinguishable in regulated activities from the enzymes purified from native sources. Availability of these recombinant proteins has permitted our laboratory to begin to address the following questions: a) what is the selectivity of G-protein α- and βγ-subunits for regulation of the various PLC-β isoenzymes? b) what is the mechanism of α- and βγ-subunit-promoted activation of PLC-β? c) what is the localization of the binding domains in PLC-β isoenzymes for α- and βγ-subunits? d) what is the selectivity and what are the sites of regulation of PLC-β isoenzymes by protein phosphorylation and what is the identity of the modified amino acid residues? Our progress on several of these topics is summarized in this chapter.

To date we have expressed and purified large amounts of five isoenzymes including rat PLC-β1, human PLC-β2, PLC-βt, a PLC-β (PLC-βX) that was cloned by Iyengar and coworkers from Xenopus oocytes (Ma et al., 1993), and a deletion mutant of PLC-β2 that lacks approximately 400 amino acids of the carboxyl terminus. We have described elsewhere (Paterson and Harden, 1995) the salient features of PLC-β isoenzyme expression and purification from infected Sf9 insect cells. Briefly, up to 10 liters of cells (3×10^6 cells/liter) have been infected as suspension cultures and isoenzyme expression optimized versus time in culture, e.g. 48 hours. The PLC-β isoenzymes are expressed as soluble proteins to high levels, e.g. up to 5% of total Sf9 cell protein, and these recombinant proteins can be easily purified to homogeneity by conventional chromatography. Purification of up to 30 mg of homogeneous PLC-β isoenzyme has been accomplished.

There is little evidence for post-translational modification of PLC-β isoenzymes and, likewise, no suggestion of post-translational modification of baculovirus-expressed PLC-β isoenzymes has been obtained. The physical properties of the purified recombinant enzymes are indistinguishable from those of isoenzymes purified from native sources. They co-migrate in various chromatographic or electrophoresis systems and share immunoreactivity with isoenzyme specific antisera. The PLC-β isoenzymes are not phosphoproteins after expression in Sf9 cells. We also have compared the enzyme activities of recombinant to native PLC-β isoenzymes. This has been most extensively accomplished with PLC-β1 (Paterson et al., 1995), PLC-β2 (Paterson et al., 1995) and PLC-βt (Waldo et al., 1996). The dependence on PtdIns(4,5)P_2 substrate and Ca^{++} concentration, and the maximal activity observed against substrate presented in a cholate/phospholipid vesicle assay were indistinguishable between native and recombinant enzymes. Moreover, the maximal activation and concentration dependence of activation by α_{11} and by βγ-subunits were the same for three different native/recombinant isoenzyme pairs, i.e. PLC-β1, PLC-β2, and PLC-βt.

SELECTIVITY OF ACTIVATION OF PLC-β ISOENZYMES BY G-PROTEIN SUBUNITS

Preparations of PLC-β1 obtained from native sources have been predictably contaminated with other proteins as well as by degradation products of the enzyme that retain catalytic but not α-subunit-regulated activity; moreover, homogeneous preparations of PLC-β2 have not been readily available. Production of large amounts of recombinant PLC-β isoenzymes has allowed us to address the absolute selectivity of activation of PLC-β1, PLC-β2, and PLC-βt by G-protein subunits to an extent that has been unattainable with enzymes purified from native sources. Assay conditions were established in which maximal reaction rates were obtained with respect to PLC-β isoenzyme concentration (Paterson et al., 1995). Under these conditions PLC-β2 required greater than 10-fold higher concentrations of both $G\alpha_{11}$ and of isoenzyme for activation than did PLC-β1. In contrast, PLC-β2 was much more effectively activated by G-protein βγ-subunits than was PLC-β1 (Paterson et al., 1995). This differential activation was not a consequence of different concentration dependencies on βγ-subunit for activation. Instead, the efficacy of βγ-subunits for activation of PLC-β2 was 5-10 fold greater than was the efficacy of βγ-subunits for activation of PLC-β1. Thus, it is the extent to which the reaction velocities of PLC-β2 and PLC-β1 are elevated that discriminates their response to βγ-subunits.

The regulation of PLC-βt by α_{11} and βγ-subunits has been compared to the regulation of PLC-β1 and PLC-β2 by these G-protein subunits. PLC-βt is unique because, in contrast to PLC-β1 and PLC-β2, it is effectively activated by both α- and βγ-subunits. Indeed, PLC-βt is essentially as sensitive to activation by α_{11} as is PLC-β1 and is more sensitive to activation by βγ-subunits than is PLC-β2.

Genes encoding 5 different G-protein β-subunits and approximately ten γ-subunits have been identified. Since most potential combinations of these β- and γ-subunits form functional βγ-dimers, the possibility of signaling diversity among these βγ-subunits is considerable. In collaboration with Garrison and coworkers we have defined the specificity of activation of PLC-β isoenzymes by βγ-dimers of defined composition purified after baculovirus-promoted expression in Sf9 cells (Boyer et al., 1994a). Most βγ-dimers, i.e. $\beta_1\gamma_2$, $\beta_2\gamma_2$, $\beta_1\gamma_3$ activated PLC-βt with a concentration dependence and maximal effect similar to that previously observed with βγ-subunits purified from bovine brain or turkey erythrocytes. However, the maximal effect observed with $\beta_2\gamma_3$ was only 50% of that observed with the other βγ-subunit preparations. Moreover, $\beta_1\gamma_1$, which is the predominant form of βγ-subunit in retina was much less effective than were other forms of βγ-subunit. Similarly low activities of βγ-subunit have been observed by others (Camps et al., 1992; Dietrich et al., 1994). One possible explanation for this difference in activity involves the prenyl group substituent at the carboxy terminus of the various γ-subunits. The γ_1-subunit is substituted with a farnesyl (C15) group whereas the other γ-subunits are geranylgeranyl (C20)-substituted. The question of whether the presence of different prenyl moieties confers different activities to βγ-subunits is being directly examined by J.C. Garrison and colleagues.

PHOSPHORYLATION OF PLC-β ISOENZYMES

In contrast to the substantial body of information on agonist-induced desensitization of adenylyl cyclase-activating receptor-signaling pathways, e.g. for the β-adrenergic receptor (Lefkowitz, 1993; Lohse, 1993), very little is known about the molecular details of agonist-induced desensitization of receptor-regulated PLC. The extensive cross-talk that occurs among the various second messenger cascades also has not been clearly defined for the inositol lipid signaling cascade. As such, we have begun to apply the avian erythrocyte as a model to address the molecular details of agonist-induced desensitization of a receptor-regulated inositol lipid signaling pathway. This should prove a very useful system for study of desensitization since the three proteins that comprise this PLC signaling cohort have been identified and their cDNA cloned. They include the P_{2Y1}-purinergic receptor (Filtz et al., 1994), avian $G\alpha_{11}$ (Maurice et al., 1993), and PLC-βt (Waldo et al., 1996).

Preincubation of turkey erythrocytes with P_{2Y}-purinergic receptor agonists (Martin and Harden, 1989), with phorbol ester activators of protein kinase C (Galas and Harden, 1996b), or with molecules that promote elevation of intracellular cyclic AMP (Galas and Harden, 1996a), all resulted in a reduction in agonist-stimulated inositol lipid hydrolysis in membranes isolated from these cells. The properties of P_{2Y}-purinergic receptor agonist-induced desensitization of this pathway suggest a role for a protein kinase, e.g. a G-protein-coupled receptor kinase, that is not regulated by the second messengers produced as a consequence of inositol lipid breakdown (Galas and Harden, 1996b). However, our data with direct activators of protein kinase C (PKC) and of cyclic AMP-dependent protein kinase (PKA) also strongly suggest that either $G\alpha_{11}$ or PLC-βt are directly phosphorylated and their activities regulated by protein phosphorylation.

We have taken advantage of the availability of large amounts of homogeneous recombinant PLC-βt to initiate a study of phosphorylation of this isoenzyme in vitro. PLC-βt is an excellent substrate for PKA. We have established conditions for study of enzyme phosphorylation under initial rate conditions, and have shown that PLC-βt is phosphorylated at a reaction rate similar to that of the best protein substrates, e.g. phosphorylase kinase, or of an optimal histone substrate, histone H2b. PKA-catalyzed phosphorylation of PLC-βt was to a stoichiometry of 2.5 - 3.0 mole/mole. In comparative experiments, PLC-β1 was a very poor substrate for PKA while PLC-β2 was phosphorylated at a rate approximately one-half that of PLC-βt. The effect of phosphorylation of PLC-βt on regulated activity of this isoenzyme is under investigation, and the sites of phosphorylation are being identified.

Activation of PKC has been shown in many cell types, including turkey erythrocytes (Galas and Harden, 1996b), to result in loss of capacity of hormones to promote inositol lipid hydrolysis (Rittenhouse and Sasson, 1985; Orellana et al., 1985; Leeb-Lundberg et al., 1985). Although PLC-β1 has been reported to be phosphorylated as a result of activation of PKC in intact cells (Ryu et al., 1990), the stoichiometry of this phosphorylation was not established. In our hands PLC-β1 is a very poor substrate for PKC in vitro, whereas purified PLC-βt is a very good substrate. The physiological relevance of this PKC-promoted phosphorylation of PLC-βt is being established.

CONCLUSIONS AND FUTURE DIRECTIONS

The baculovirus expression system provides a powerful approach for obtaining large amounts of PLC-β isoenzymes in homogeneous form, and availability of these purified proteins is leading to considerable insight into the regulatory properties of PLC-β isoenzymes. Although inroads have been made in definition of the selectivity of interaction of G-protein subunits with individual PLC-β isoenzymes, the full range of selectivities among the various proteins remains to be established. An important aspect of cellular signaling involves the cross-talk that occurs among the various second messenger response cascades. Very little is known about homologous or heterologous regulation of PLC-associated signaling, and our studies of phosphorylation of PLC-β isoenzymes should lead to identification of amino acid residues that play key roles in the regulated function of these isoenzymes. Both of these research directions should lead naturally to delineation of important regulatory domains within the PLC-β isoenzymes, including not only those involved in α- and βγ-subunit interaction and in protein phosphorylation, but also those involved in the binding of Ca^{++}, the binding of phosphoinositide substrate, and the binding of PLC-β isoenzymes to the phospholipid bilayer. To date essentially all studies of G-protein/PLC-β interaction have been carried out after reconstitution of proteins into substrate-containing phospholipid vesicles. Although this general assay system has provided results that are consistent with those obtained from studies in which these proteins have been overexpressed in mammalian cells, they may very poorly represent the interaction of PLC-β with phosphoinositide substrates in natural lipid bilayers. As such, assay systems, e.g. lipid monolayers (James et al., 1995; Carman et al., 1995), that more closely approximate physiologically relevant lipid compositions and interfacial surface pressures ultimately need to be applied to questions about the regulatory properties of PLC-β isoenzymes. Finally, some insight into regulatory domains of PLC-β isoenzymes have been made from mutational analyses. However, these domains have been defined in only the broadest of terms, and more precise definition is needed. The current availability of large amounts of homogeneous PLC-β isoenzymes should progress to solution of their crystal structure.

REFERENCES

Berridge, M.J. and Irvine, R.F. 1987, Inositol trisphosphate and diacylglycerol: two interacting second messengers, *Ann. Rev. Biochem.* 56:159.

Blank, J.L., Brittain, K.A., and Exton, J.H. 1992, Activation of cytosolic phosphoinositide phospholipase C by G-protein βγ subunits, *J.Biol.Chem.* 267:23069.

Bloomquist, B.T., Shortridge, R.D., Schneuwly, S., Perdew, M., Moutell, C., Steller, H., Rubiu, G., and Pak, W.L. 1988, Isolation of a putative phospholipase C gene of Drosophila, norpA, and its role in phototransduction, *Cell,* 54:723.

Boyer, J.L., Downes, C.P., and Hardem T.K. 1989a, Kinetics of activation of phospholipase C by P_2 purinergic receptor agonists and guanine nucleotides, *J.Biol.Chem.* 264:884.

Boyer, J.L., Waldo, G.L., Evans, T., Northup, J.K., Downes, C.P., and Harden, T.K. 1989b, Modification of AlF_4^--and receptor-stimulated phospholipase C by G-protein βγ-subunits, *J.Biol.Chem.* 264:13917.

Boyer, J.L., Waldo, G.L., and Harden, T.K. 1992, βγ-Subunit activation of G-protein-regulated phospholipase C, *J.Biol.Chem.* 267:25451.

Boyer, J.L., Graber, S.G., Waldo, G.L., Harden, T.K., and Garrison, J.C. 1994a, Selective activation of phospholipase C by recombinant G-protein α- and βγ-subunits, *J.Biol.Chem.* 269:2814.

Boyer, J.L., Paterson, A., and Harden, T.K. 1994b, G-protein-mediated regulation of phospholipase C: involvement of βγ subunits, *Trends Cardiovascular Med.* 4:88.

Camps, M., Hou, C., Sidiropoulos, D., Stock, J.B., Jakobs, K.H., and Gierschik, P. 1992, Stimulation of phospholipase C by guanine-nucleotide-binding protein βγ subunits, *Eur.J.Biochem.* 206:821.

Carman, G.M., Deems, R.A., and Dennis, E.A. 1995, Lipid signaling enzymes and surface dilution kinetics, *J.Biol.Chem.* 270:18711.

Clapham, D.E. and Neer, E.J. 1993, New roles for G-protein βγ-dimers in transmembrane signalling, *Nature,* 365:403.

Dietrich, A., Meister, M., Brazil, D., Camps, M., and Gierschik, P. 1994, Stimulation of phospholipase C-β_2 by recombinant guanine-nucleotide-binding protein βγ dimers produced in a baculovirus/insect cell expression system, *Eur.J.Biochem.* 219:171.

Filtz, T.M., Li, Q., Boyer, J.L., Nicholas, R.A., and Harden, T.K. 1994, Expression of a cloned P_{2Y}-purinergic receptor that couples to phospholipase C, *Mol.Pharmacol.* 46:8.

Galas, M.C. and Harden, T.K. 1996a, Cyclic AMP-induced desensitization of G-protein-regulated phospholipase C, *Eur. J. Pharmacol.* in press.

Galas, M.C. and Harden, T.K. 1996b, Receptor-induced heterologous desensitization of receptor-regulated phospholipase C: lack of involvement of protein kinase C, *Eur.J.Pharmacol.Mol.Pharmacol.* in press.

Gilman, A.G. 1995, G proteins and regulation of adenylate cyclase, *Ang. Chem. Int. Ed. Eng.* 34:1406.

Harden, T.K., Stephens, L., Hawkins, P.T., and Downes, C.P. 1987, Turkey erythrocyte membranes as a model for regulation of phospholipase C by guanine nucleotides, *J.Biol.Chem.* 262:9057.

Harden, T.K., Hawkins, P.T., Stephens, L., Boyer, J.L., and Downes, C.P. 1988, Phosphoinositide hydrolysis by guanosine 5'-0-(3-thiotriphosphate)-activated phospholipase C of turkey erythrocyte membranes, *Biochem.J.* 252:583.

Hepler, J.R. and Gilman, A.G. 1992, G proteins, *Trends Biochem.Sci.* 17:383.

James, S.R., Paterson, A., Harden, T.K., and Downes, C.P. 1995, Kinetic analysis of phospholipase Cβ isoforms using phospholipid-detergent mixed micelles, *J.Biol.Chem.* 270:11872.

Jhon, D.-Y., Lee, H.-H., Park, D., Lee, C.-W., Lee, K.-H., Yoo, O.J., and Rhee, S.G. 1993, Cloning, sequencing, purification, and G_q-dependent activation of phospholipase C-β3, *J.Biol.Chem.* 268:6654.

Lee, C.-W., Park, D.J., Lee, K.-H., Kim, C.G., and Rhee, S.G. 1993, Purification, molecular cloning, and sequencing of phospholipase C-β4, *J.Biol.Chem.* 268:21318.

Leeb-Lundberg, L.M.F., Cotecchia, S., Lomasney, J.W., Bernardis, J.F., Lefkowitz, R.J., and Caron, M.G. 1985, Phorbol esters promote α_1-adrenergic receptor phosphorylation and receptor uncoupling from inositol phosphate metabolism, *Proc. Natl. Acad. Sci. USA,* 82:5651.

Lefkowitz, R.J. 1993, G protein-coupled receptor kinases, *Cell,* 74:409.

Lohse, M.J. 1993, Molecular mechanisms of membrane receptor desensitization, *Biochimica et Biophysica Acta,* 1179:171.

Ma, H.-W., Blitzer, R.D., Healy, E.C., Premont, R.T., Landau, E.M., and Iyengar, R. 1993, Receptor-evoked Cl^- current in *Xenopus* oocytes is mediated through β-type phospholipase C, *J.Biol.Chem.* 268:19915.

Martin, M.W. and Harden, T.K. 1989, Agonist-induced desensitization of P_{2Y}-purinergic receptor-and guanine nucleotide-regulated phospholipase C, *J.Biol.Chem.* 264:19535.

Maurice, D.H., Waldo, G.L., Morris, A.J., Nicholas, R.A., and Harden, T.K. 1993, Identification of $G\alpha_{11}$ as the phospholipase C-activating G-protein of turkey erythrocytes, *Biochem.J.* 290:765.

Morris, A.J., Waldo, G.L., Downes, C.P., and Harden, T.K. 1990a, A receptor and G-protein-regulated polyphosphoinositide specific phospholipase C. 1. Purification and properties, *J.Biol.Chem.* 265:13501.

Morris, A.J., Waldo, G.L., Downes, C.P., and Harden, T.K. 1990b, A receptor and G-protein-regulated polyphosphoinositide specific phospholipase C. 2. P_{2Y}-purinergic receptor and G-protein-mediated regulation of the purified enzyme, *J.Biol.Chem.* 265:13508.

Orellana, S.A., Solski, P.A., and Brown, J.H. 1985, Phorbol ester inhibits phosphoinositide hydrolysis and calcium mobilization in cultured astrocytoma cells, *J.Biol.Chem.* 260:5236.

Park, D., Jhon, D.-Y., Kriz, R., Knopf, J., and Rhee, S.G. 1992, Cloning, sequencing, expression, and G_q-independent activation of phospholipase C-β2, *J.Biol.Chem.* 267:16048.

Paterson, A., Boyer, J.L., Watts, V.J., Morris, A.J., Price, E.M., and Harden, T.K. 1995, Expression and purification of recombinant phospholipase C-β1 and phospholipase C-β2: differential activation by $G\alpha_{11}$ and βγ-subunits, *Cell.Signaling,* 7:709.

Paterson, A., and Harden, T.K. 1996, Expression, purification, and functional reconstitution of recombinant phospholipase C-β isoenzymes, *Methods in Neurosciences,* 29:246.

Rhee, S.G. and Choi, K.D. 1992, Regulation of inositol phospholipid-specific phospholipase C isozymes, *J.Biol.Chem.* 267:12393.

Rittenhouse, S.E. and Sasson, J.P. 1985, Mass changes in myoinositol trisphosphate in human platelets stimulated by thrombin: inhibitory effects of phorbol ester, *J.Biol.Chem.* 260:8657.

Ryu, S.H., Cho, K.S., Lee, K.Y., Suh, P.G., and Rhee, S.G. 1987, Purification and characterization of two immunologically distinct phosphoinositide-specific phospholipase C from bovine brain, *J.Biol.Chem.* 262:12511.

Ryu, S.H., Kim, U.-H., Wahl, M.I., Brown, A.B., Carpenter, G., Huang, K.-P., and Rhee, S.G. 1990, Feedback regulation of phospholipase C-β by protein kinase C, *J.Biol.Chem.* 265:17941.

Shortridge, R.D., Yoon, J., Lending, C.R., Bloomquist, B.T., Perdew, M.H., and Pak, W.L. 1991, A drosophila phospholipase C gene that is expressed in the central nervous system, *J.Biol.Chem.* 266:12474.

Simon, M.I., Strathmann, M.P., and Gautam, N. 1991, Diversity of G proteins in signal transduction, *Science,* 252:802.

Suh, P.-G., Ryu, S.H., Moon, K.H., Suh, W., and Rhee, S.G. 1988, Cloning and sequence of multiple forms of phospholipase C, *Cell,* 54:161.

Waldo, G.L., Paterson, A., Boyer, J.L., Nicholas, R.A., and Harden, T.K. 1996, Molecular cloning, expression, and regulatory activity of $G\alpha_{11}$- and βγ-subunit-stimulated PLC-β from avian erythrocytes, *Biochem.J.* 314:917.

PHOSPHOLIPASE D

John H. Exton

Howard Hughes Medical Institute and Department of Molecular Physiology and Biophysics and Pharmacology, Vanderbilt University School of Medicine Nashville, TN

INTRODUCTION

Phospholipase D (PLD) is a ubiquitous enzyme that hydrolyses mainly phosphatidylcholine (PC) to yield phosphatidic acid (PA) and choline. It is regulated by a variety of hormones, neurotransmitters, growth factors, cytokines and related molecules and is believed to play a role in signal transduction in many cell types. The signalling molecule produced by PLD is considered to be PA since resting levels of choline are high in most cells.

The products of the metabolism of PA are 1,2-diacylglycerol (DAG) produced by the action of phosphatidate phosphohydrolase, and lysophosphatidic acid (LPA) produced by the action of a specific phospholipase A_2. Many functions for PA have been proposed, namely, promotion of mitogenesis in fibroblasts, stimulation of the respiratory burst in neutrophils, the regulation of small G proteins and the activation of several protein kinases and enzymes of phosphoinositide metabolism. In addition, much of the DAG that accumulates in stimulated cells arises from PA, and an important role for DAG arising from PC hydrolysis is activation of Ca^{2+}-independent isoforms of protein kinase C (PKC). LPA is a major product of PA metabolism in platelets and is an important intercellular signal. Many cells respond to LPA through a G protein-coupled membrane receptor.

ENZYMOLOGY OF PHOSPHOLIPASE D

PLD activity has been detected in all mammalian tissues examined and there are probably several isozymes (Billah & Anthes, 1990; Exton 1994; Bocckino & Exton, 1996; Provost et al. 1996), but efforts to purify the enzyme have met with limited success. PLD activity has been found in the plasma membranes and cytosol of most tissues, and also in the endoplasmic reticulum, Golgi and microsomes of certain tissues (Bocckino & Exton, 1996; Provost et al. 1996). The enzyme has been partially purified from rat brain microsomes (Taki & Kanfer, 1979; Kobayashi & Kanfer, 1991) and completely purified from porcine lung microsomes (Okamura & Yamashita, 1994). The enzyme from lung has an M_r of 190,000 and is specific for PC. It is stimulated by Ca^{2+} and Mg^{2+} ions and unsaturated fatty acids. Brown et al. (1995) have solubilized PLD from HL60 cell membranes using NaCl. The enzyme is potently activated by phosphatidylinositol 4,5 bisphosphate (PIP_2) and by the small G protein ADP-ribosylation factor (ARF). In further work, Brown et al. (1995) have partially purified PLD from porcine brain membranes. Like the HL60 enzyme, it is markedly stimulated by PIP_2 and ARF. However, the stimulation by ARF is increased further by the addition of undefined cytosolic factors (Singer et al. 1995).

The presence of two forms of PLD has been demonstrated in rat brain membranes using Triton solubilization and heparin chromatography (Massenburg et al. 1994). One form is dependent on oleate, whereas the other is markedly activated by ARF and PIP_2. Soluble forms of PLD have been identified in eosinophils (Kater et al. 1976), HL60 cells (Siddiqi et al. 1995) and in several mammalian tissues (Wang et al. 1991; Provost et al. 1996).

An ARF-responsive human PLD has recently been cloned from a HeLa cDNA library (Hammond et al. 1995). It has a Mr of 120,000 and has many blocks of amino acids that are conserved in plant (rice, corn, castor bean) and yeast (S. cerevisiae) PLDs. Expression of the PLD in Sf9 insect cells results in a large enhancement of PLD activity that is selective for PC and catalyzes transphosphatidylation (Hammond et al. 1995). The expressed enzyme is activated by PIP_2 and ARF, but inhibited by oleate.

REGULATION OF PHOSPHOLIPASE D

A very large number of agonists increase the activity of PLD in many cells and tissues (Bocckino & Exton, 1996). Many of these agonists act through receptors coupled to the heterotrimeric G proteins G_q and G_i, whereas others are growth factors or cytokines that act through receptors with tyrosine kinase activity or which activate soluble tyrosine kinases. However, it is not clear how the signals arising from activation of G_q or G_i or from enhanced tyrosine kinase activity are communicated to PLD isozymes.

ROLE OF PROTEIN KINASE C IN THE ACTIVATION OF PHOSPHOLIPASE D

Tumor-promoting phorbol esters stimulate PLD in a large number of tissues and cells (Exton 1990, 1994; Billah & Anthes 1990; Bocckino & Exton, 1996). This indication that the enzyme is regulated by PKC is supported by many studies employing PKC inhibitors and down-regulation of the enzyme (Exton 1990, 1994; Bocckino & Exton 1996). Some of the agents that have been employed to inhibit PKC lack specificity insofar as they inhibit other protein kinases. Nevertheless, studies with a variety of these inhibitors generally support the involvement of PKC in the actions of many G protein-linked agonists and growth factors on PLD. On the other hand, negative or partial effects have been observed in some studies and there are also studies in which PLD and PKC activities are dissociated. Except for a few reports, down-regulation of PKC by long-term phorbol ester treatment has been found to completely or partly inhibit the activation of PLD by a variety of growth factors and G-protein-linked agonists (Bocckino & Exton 1996). In summary, the overall results of inhibitor and down-regulation studies indicate that agonist activation of PLD involves both PKC-dependent and -independent mechanisms.

There is much evidence that the PKC-dependent mechanism involves hydrolysis of PIP_2 to yield inositol trisphosphate and DAG since most agonists that activate PLD also activate phosphoinositide phospholipase C (PI-PLC) (Bocckino & Exton 1996). A clear example of the involvement of PI-PLC in the activation of PLD by growth factors comes from the work of Yeo et al. (1994). Using TRMP cells it was shown that PDGF receptors with mutation of Tyr-1021 to Phe, which results in a loss of activation of PI-PLCγ, did not mediate PDGF activation of PLD. On the other hand, retention of Tyr-1021, but mutation of other Tyr residues involved in signalling, permitted PLD activation (Yeo et al. 1994). Furthermore, treatment of the cells with a PKC inhibitor or down-regulation of PKC greatly impaired the PLD response. These results indicate that PLD activation by PDGF in these cells is dependent on PI-PLC and PKC. This conclusion is supported by Lee et al. (1994) who found that NIH 3T3 fibroblasts stably overexpressing PI-PLCγ1 have a greatly enhanced PLD response to PDGF, and that treatment with a tyrosine kinase inhibitor greatly decreases PLD activation and also tyrosine phosphorylation of PI-PLCγ1.

Enhancement of PLD activation by endothelin, thrombin and PDGF has been observed in fibroblasts overexpressing the α- and β_I-isozymes of PKC (Pai et al. 1991; Pachter et al. 1992; Eldar et al. 1993). On the other hand, depletion of PKCα by antisense methods decreases the activation of the phospholipase by purinergic agonists (Balboa et al. 1994). Thus these studies support the idea that Ca^{2+}-dependent PKC isozymes play an important role in the regulation of PLD.

Regulation of PLD by PKC could be direct or involve another protein(s). There have been several reports that PKC enhances membrane-associated PLD activity (Conricode et al. 1992; Ohguchi et al. 1995, 1996; Lopez et al. 1995). Although one of these studies supported a phosphorylation-dependent mechanism (Lopez et al. 1995), the others indicated that phosphorylation was not involved. For example, activation of PLD occurred in the demonstrated absence of ATP (Conricode et al. 1992) and was dissociated from kinase activity (Conricode et al. 1992; Ohguchi et al. 1995). The activation showed a concentration dependence on both phorbol ester and Ca^{2+} (Conricode et al. 1992, 1994). This indication of the involvement of Ca^{2+}-dependent PKC isozymes was supported by experiments with individual PKC isozymes, which showed that only the α, ßI and ßII forms were effective (Conricode et al. 1994; Ohguchi et al. 1996).

In studies of the regulation of brain PLD, Singer et al. (1995) identified a factor that activated the enzyme in a nucleotide-independent manner. This factor was later shown to be PKCα (Singer et al. 1996). In accord with the findings of Conricode et al. (1992), the activation of PLD by PKCα was enhanced by phorbol ester and Ca^{2+}, but occurred in the absence of ATP. Furthermore, staurosporine and protein phosphatase treatment inhibited the kinase activity of the enzyme without altering its effect on PLD. The results of Singer et al. (1996) therefore strongly support a mechanism of activation of PLD that is independent of phosphorylation. Ohguchi et al. (1996) also showed that a cytosolic factor required for the activation of PLD in HL60 membranes by phorbol ester was PKC. Although MgATP was required, this was attributed to its stimulation of the resynthesis of PIP_2, a cofactor required for PLD activity. The α- and ß-isozymes of PKC were able to stimulate PLD, but not the γ-isozyme.

ROLE OF CALCIUM IONS IN THE REGULATION OF PHOSPHOLIPASE D

Studies with Ca^{2+} ionophores indicate that a rise in cytosolic Ca^{2+} can stimulate PLD (Bocckino & Exton, 1996). There is also evidence that depletion of cellular Ca^{2+} by chelators inhibits the stimulation of the enzyme by agonists linked to G_q and G_i (Halenda & Rehm 1990; Kessels et al. 1991; Huang et al. 1991; Lin & Gilfillan 1992; Wu et al. 1992; Balboa et al. 1994). A major question is whether these findings stem from the involvement of Ca^{2+}-dependent PKC isozymes or indicate another role for Ca^{2+} ions. Studies of the effects of Ca^{2+} on partially purified PLD have usually shown stimulation, but the concentrations required for stimulation have varied widely and high concentrations have generally been inhibitory (Taki & Kanfer 1979; Cockcroft et al. 1994; Okamura & Yamashita 1994; Brown et al. 1995; Siddiqi et al. 1995; Bocckino & Exton 1996). Thus it is unclear whether or not there is direct control of PLD by changes in cytosolic Ca^{2+}.

ROLE OF THE SMALL G PROTEIN ARF IN THE REGULATION OF PHOSPHOLIPASE D

A new development in the study of the regulation of PLD is the involvement of small G proteins of the ARF and Rho families. Early studies showed that cytosol was required for the stimulatory effect of GTPγS on PLD in membranes of neutrophils or HL60 cells (Anthes et al. 1991; Olson et al. 1991). A similar conclusion was drawn from experiments with permeabilized HL60 cells (Geny & Cockcroft, 1992; Geny et al. 1993). In simultaneous studies, Brown et al. (1993) and Cockcroft et al. (1994) purified the cytosolic factor to homogeneity and showed that it had amino acid sequences corresponding to ARF1 and ARF3. Both groups also showed that recombinant ARF was effective and that myristoylation greatly enhanced its potency. Activation of PLD has been observed with myristoylated ARF1, 3, 5 and 6, with little difference in their relative potency (Massenburg et al. 1994; Brown et al. 1995).

Although it is now well accepted that ARF is required for the activation of PLD in neutrophils and HL60 cells by GTP analogues, recent evidence indicates the requirement for additional cytosolic factors (Lambeth et al. 1995; Bourgoin et al. 1995; Singer et al. 1995, 1996). One of these factors has been identified as a 50 kDa protein (Lambeth et al. 1995; Bourgoin et al. 1995) but its structure and function remain unknown. As described above, another factor has been shown to be the PKCα (Singer et al. 1996). This kinase can activate the enzyme directly or synergize with ARF or Rho to produce a marked activation (Singer et al. 1996).

Houle et al. (1995) have observed that membrane translocation of ARF is associated with activation of PLD in HL60 cells. Thus in membranes from cells treated with f-Met-Leu-Phe or 4ß-phorbol 12-myristate 13-acetate (PMA), the response of PLD to guanosine 5′-0-(3-thiotriphosphate) (GTPγS) is enhanced and there is an increase in membrane-associated ARF. The findings support the view that association of ARF with the membrane as a necessary step for activation of PLD.

Interestingly, the domain of ARF involved in activating PLD is different from that involved in cholera toxin activation (Zhang et al. 1995). Although most studies of the regulation of PLD have utilized plasma membrane preparations, the enzyme is also present in Golgi membranes where it is activated by ARF in a brefeldin A-sensitive manner (Ktiskakis et al. 1995; Provost et al. 1996). Brefeldin A is an inhibitor of the activation of ARF by GTP/GDP exchange (Donaldson et al. 1992; Helms & Rothman, 1992; Randazzo et al. 1993).

The majority of investigations of the regulation of PLD by ARF have been carried out in vitro. However, there is evidence that this regulation occurs in intact cells. Using human embryonic kidney (HEK) cells stably expressing M3 muscarinic receptors, Rümenapp et al. (1995) found that the activation of PLD by carbachol was inhibited by Brefeldin A in a dose-dependent manner. Translocation of ARF from cytosol to membranes was also observed in cells treated with carbachol (Rümenapp et al. 1995). In addition to indicating a role for ARF in the activation of PLD by a G protein-linked agonist, these results suggest that activation of ARF is associated with its membrane translocation.

In unpublished studies, I. Guillemain-Raddassi and J. Exton have found that treatment of HL60 cells with brefeldin A inhibits the activation of PLD by f-Met-Leu-Phe, ATP and PMA. These results thus resemble those of Rümenapp et al. (1995) and support a role for ARF in the regulation of PLD in vivo.

ROLE OF SMALL G PROTEINS OF THE RHO FAMILY IN THE REGULATION OF PHOSPHOLIPASE D

In a study of the effects of GTPγS on the PLD activity of neutrophil lysates, Bowman et al. (1993) found that protein factors in the cytosol and membranes were needed. The GTPγS-binding protein was found to be associated with the membranes and its properties indicated that it was a small G protein. This was supported by the fact that a GDP dissociation stimulatory protein (GDS) for small G proteins increased the stimulation of membrane-associated PLD by GTP. When a GDP dissociation inhibitor (GDI) that was selective for members of the Rho family was tested, it inhibited the activation of the enzyme by GTPγS (Bowman et al. 1993).

The findings of Bowman et al. (1993) have been confirmed using membranes or homogenates from HL60 cells, neutrophils or rat liver. Thus the stimulation of PLD by GTPγS was inhibited by Rho-GDI (Malcolm et al. 1994; Siddiqi et al. 1995; Kwak et al. 1995; Ohguchi et al. 1995) and in the homogenates (Bourgoin et al. 1995). Furthermore, the addition of recombinant RhoA, Rac1 or CDC42 to membranes previously treated with Rho-GDI restored GTPγS stimulation of the phospholipase (Malcolm et al. 1994; Siddiqi et al. 1995; Kwak et al. 1995). RhoA was the most effective Rho family member in both HL60 and liver membranes. As expected, ARF promoted the stimulation of PC-PLD in HL60 membranes, and the combination of ARF and RhoA produced a synergistic effect (Siddiqi et al. 1995). A similar synergism has also been reported by Singer et al. (1995, 1996) and Kuribara et

Although in some studies, addition of Rho proteins alone is capable of inducing GTPγS-stimulated membrane PLD activity (Malcolm et al. 1994; Siddiqi et al. 1995; Ohguchi et al. 1995), Kwak et al. (1995) found that the effect of RhoA on neutrophil membrane PLD was greatly enhanced by a 50 kDa protein partially purified from neutrophil cytosol.

Many reports have shown a synergism between PMA and GTPγS in the activation of membrane PLD (Quian & Drewes, 1989; Olson et al. 1991; Geny & Cockcroft, 1992; Coorssen & Haslam, 1993; Ohguchi et al. 1995), and recent evidence indicates that it involves PKC and both Rho and ARF. Thus Ohguchi et al. (1995, 1996) observed that RhoGDI inhibited the stimulatory effect of PKC on HL60 membrane PLD activity and that the addition of RhoA plus PKCα produced a synergistic activation in the presence of PMA and GTPγS. Singer et al. (1996) also showed that PKCα produced a synergistic activation when added with either ARF or RhoA to a PLD preparation from brain. The mechanisms involved in the interactions of PKC with the two small G proteins are undefined.

A key issue concerning the role of ARF and Rho family members in the regulation of PLD is how these small G proteins fit into the general scheme by which agonists control this enzyme in vivo. Recent studies have indicated that Rho family members are involved in the signalling pathways initiated by receptors linked to G proteins and those with tyrosine kinase activity. For example, there is much evidence that the Rho family member Rac mediates the effects of growth factors on membrane ruffling or formation of lamellipodia in fibroblasts (Ridley et al. 1992; Ridley, 1994; Nobes & Hall, 1995) and that another member Cdc42Hs is involved in filopodia and actin microspike formation in Swiss 3T3 cells (Kozma et al. 1995; Nobes & Hall, 1995). Furthermore, Rho is involved in the effects of the G protein-linked agonists on the formation of actin stress fibers and assembly of focal adhesions in fibroblasts (Paterson et al. 1990; Ridley & Hall, 1992; Takai et al. 1995).

In addition to the evidence that Rho and Rac mediate some of the cytoskeletal changes induced by G-protein-linked agonists and growth factors, there are data implicating their involvement in the regulation of PLD. Schmidt et al. (1994) have reported that treatment of HEK cells with C. difficile toxin B blocks carbachol stimulation of PLD. The toxin also inhibits its activation by GTPγS and AlF_4^- in intact or permeabilized cells. Malcolm et al. (1996) have also shown that scrape-loading of C3 exoenzyme into Rat1 fibroblasts greatly attenuates the activation of PLD by LPA, endothelin-1 and PMA without altering the activation of PI-PLC. Activation of PLD was associated with translocation of RhoA to membranes. In unpublished studies of the effects of PDGF and EGF on PLD in fibroblasts, J. Hess, A. Ross and J. Exton have also found inhibition of the PDGF effect by C3 exoenzyme. They have also shown that the effect of EGF on PLD is greatly enhanced in fibroblasts overexpressing Rac1 and abolished in cells expressing dominant negative N17Rac1. These studies indicate a role for Rho in the activation of PLD by agonists linked to G proteins and PKC and a role for both Rho and Rac in the activation induced by growth factors. Based on other data (Ridley et al. 1992) Rho appears to be downstream from Rac.

The above results indicate that growth factors and G protein-linked agonists must be capable of controlling one or more of the proteins that regulate the activation state (GTP/GDP ratio) of the Rho family of small G proteins. Nobes et al. (1995) have shown that wortmannin, an inhibitor of phosphatidylinositol 3-kinase, blocks the membrane ruffling induced in Swiss 3T3 cells by PDGF, but not that due to microinjected activated V12Rac1, and does not alter the effects of LPA and activated V14RhoA. On the other hand, the tyrosine kinase inhibitor tyrphostin inhibits the effects of LPA, but not those of activated RhoA. These data indicate that phosphatidylinositol 3-kinase is involved in the activation of Rac by growth factors, whereas tyrosine phosphorylation mediates the activation of Rho by LPA. Tyrosine kinase activity may also be needed for formation of stress fibers and focal adhesions induced by LPA, bombesin or microinjected V14RhoA, since genistein blocked these responses (Ridley & Hall, 1992). Rozengurt and associates (Rankin et al. 1994; Seckl et al. 1995) have reported that C3 exoenzyme blocks the tyrosine phosphorylation of focal adhesion kinase and paxillin induced by bombesin, endothelin or GTPγS in Swiss 3T3 cells, supporting a role for tyrosine kinase in downstream signalling from Rho.

In unpublished studies, I. Fleming and J. Exton have observed that treatment of Swiss 3T3 cells with LPA or bombesin causes a rapid translocation of RhoA from the cytosol to a membrane fraction. This is not induced by PDGF or accompanied by translocation of Rac1. Its relationship to the presumed activation of RhoA by these G-protein linked agonists remains to be defined. The idea that the translocation is controlled by membrane-associated GDS proteins is supported by findings of Bokoch et al. (1994). They showed that in membranes plus cytosol from neutrophils, GTP analogs induced rapid membrane association of Rho, Rac and Cdc42 and that this was associated with dissociation of the Rho proteins from their complexes with GDI.

There is evidence of coupling between the Ras and Rho signalling pathways. Some groups have shown that transformation of cells by oncogenic Ras is critically dependent upon Rac or Rho (Qiu et al. 1995a, 1995b; Prendergast et al. 1995; Khosravi-Far et al. 1995). Examination of actin polymerization in these studies indicated that the effects of Ras on transformation were mediated by two pathways, one involving Raf and the MAP kinase cascade, and the other involving Rac and Rho, with the latter also controlling the actin cytoskeleton (Qiu et al. 1995a; Khosravi-Far et al. 1995; Prendergast et al. 1995). The effects of injection of C3 exoenzyme or expression of dominative negative or constitutively active forms of Rac or Cdc42 also indicate a need for Rho family members for cell cycle progression (Olson et al. 1995). In support of the conclusions of Qiu et al. (1995a) and Prendergast et al. (1995), Olson et al. (1995) found that the active forms of Rho, Rac and Cdc42 did not activate MAP kinase, but that Rac and Cdc42 activated Jun kinase.

The mechanism of coupling between the Ras and Rho signalling pathways is not understood. However, Rac activation by growth factors is blocked by antibodies to Grb2, suggesting that this adaptor protein, which binds SOS a GDS for Ras, is involved (Matuoka et al. 1993). Association of Rho with the PDGF ß receptor is also induced in fibroblasts by PDGF treatment (Zubiaur et al. 1995), suggesting that Rho may be a component of the signalling complex. In addition, expression of the N-terminal domain of Ras-GAP disrupts actin stress fibers and focal adhesions in Rat 2 cells, consistent with impairment of Rho function (McGlade et al. 1993).

ROLE OF SOLUBLE TYROSINE KINASES AND NOVEL SERINE/THREONINE KINASES IN THE REGULATION OF PHOSPHOLIPASE D

Activation of the T-cell antigen receptor and of the Fc_{ϵ} receptor for IgE results in stimulation of PLD (Stewart et al. 1991; Dinh & Kennerly, 1991; Lin & Gilfillan, 1992). These receptors do not possess intrinsic tyrosine kinase activity, but their activation increases the activity of soluble tyrosine kinases of the Src family (Weiss, 1993). The activation of PLD in the RBL-2H3 mast cell line is abolished by tyrphostin and genistein, implying the involvement of these tyrosine kinases (Kumada et al. 1993; Lin et al. 1994). There is also a report of PLD activation in v-Src transformed cells (Song et al. 1991), although later work has shown a dependence on Ras (Jiang et al. 1995a) and a GDS for the related small G protein Ral (Jiang et al. 1995b).

A role for a tyrosine kinase(s) in the activation of PLD by G protein-mediated agonists is indicated by other studies. Dubyak et al. (1993) and Kusner et al. (1993) observed that ATP markedly augmented the stimulation of PLD by GTPγS in permeabilized U937 cell. The effect was mimicked by vanadate, an inhibitor of protein tyrosine phosphatases, and blocked by tyrosine kinase inhibitors. These inhibitors have also been found to impair PLD activation in platelets (Martinson et al. 1994), neutrophils (Uings et al. 1992), vascular smooth muscle cells (Wilkes et al. 1993), fibroblasts (Briscoe et al. 1995), and HEK cells transfected with M3 muscarinic receptors (Schmidt et al. 1994). However, in most of these studies, high doses of inhibitors were employed and their specificity of action was not established. In one study that used many inhibitors (Martinson et al. 1994), the inhibition of PLD did not correlate well with changes in tyrosine phosphorylation.

There is additional evidence for the involvement of tyrosine kinase(s) in the regulation of PLD. For example, in permeabilized HL60 cells, Bourgoin & Grinstein (1992) found a strong correlation between the effects of vanadyl hydroperoxide on tyrosine phosphorylation and PLD activity. Vanadate plus H_2O_2 activated PLD in neutrophils (Uings et al. 1992) and platelets (Martinson et al. 1994), and stimulatory effects of vanadate have been reported in other studies (Ward et al. 1995; Schmidt et al. 1994). It is not known whether PLD is subject to direct phosphorylation and activation by a receptor-associated or soluble tyrosine kinase or whether a signalling cascade is involved.

As described in a previous section, there is evidence that tyrosine kinase activity is involved in upstream and downstream signalling involving Rho (Nobes et al. 1995; Ridley & Hall 1994; Rankin et al. 1994; Seckl et al. 1995). However, the specific involvement of tyrosine kinase(s) in Rho-mediated activation of PLD has not been defined. There is now much evidence that protein serine/threonine kinases are targets of Rho family proteins (Manser et al. 1994; Martin et al. 1995; Kumagai et al. 1995; Knaus et al. 1995; S. Zhang et al. 1995; Bagrodia et al. 1995). These have been designated p21-activated protein kinases (PAKs) and are related to the STE20 kinase of S. cerevisiae (Martin et al. 1995). Recently, a serine/threonine kinase that specifically binds to GTP-RhoA has been identified (Leung et al. 1995). Whether or not any of these serine/threonine kinases is involved in PLD regulation is presently unknown.

CONCLUDING REMARKS

The cloning of human PLD (Hammond et al. 1995) and of other isozymes should provide much needed molecular information concerning the structure-function relationships and regulation of these enzymes. Thus the catalytic domain(s) should be defined and the interaction sites and mechanisms of regulation by ARF, Rho, PIP_2 and oleate and by PKC and other protein kinases should be identified. Molecular biological and immunological approaches should also help define the functions and the tissue and subcellular distribution of the various isozymes.

A related critical issue is to define the mechanisms by which Rho and ARF family members are activated by growth factors and other agonists. Another point of interest will be to discover what role PLD plays in the presently known cellular changes induced by activation of Rho and ARF. It is anticipated that the next few years will yield exciting developments concerning the cellular role of PLD.

Anthes, J.C., Wang, P., Siegel, M.I., Egan, R.W., and Billah, M.M., 1991, Granulocyte phospholipase D is activated by a guanine nucleotide dependent protein factor, Biochem. Biophys. Res. Commun. 175:236.

Bagrodia, S., Taylor, S.J., Creasy, C.L., Chernoff, J., and Cerione, R.A., 1995, Identification of a mouse $p32^{Cdc42/Rac}$ activated kinase, J. Biol. Chem. 270:22731.

Balboa, M.A., Firestein, B.L., Godson, C., Bekk, K.S., and Insel, P.A., 1994, Protein kinase C_α mediates phospholipase D activation by nucleotides and phorbol ester in Madin-Darby canine kidney cells. Stimulation of phospholipase D is independent of activation of polyphosphoinositide-specific phospholipase C and phospholipase A_2, J. Biol. Chem. 269:10511.

Billah, M.M., and Anthes, J.C., 1990, The regulation and cellular functions of phosphatidylcholine hydrolysis, Biochem. J. 269:281.

Bocckino, S.B. and Exton, J.H., 1996, Handbook of Lipid Research, Plenum, New York.

Bokoch, G.M., Bohl, B.P., and Chuang, T.-H., 1994, Guanine nucleotide exchange regulates membrane translocation of Rac/Rho GTP-binding proteins, J. Biol. Chem. 269:31674.

Bourgoin, S. and Grinstein, S., 1992, Peroxides of vanadate induce activation of phospholipase D in HL-60 cells. Role of tyrosine phosphorylation, J. Biol. Chem. 267:11908.

Bourgoin, S., Harbour, D., Desmarais, Y., Takai, Y., and Beaulieu, A., 1995, Low molecular weight GTP-binding proteins in HL-60 granulocytes. Assessment of the role of ARF and of a 50-kDa cytosolic protein in phospholipase D activation, J. Biol. Chem. 270:3172.

Bowman, E.P., Uhlinger, D.J., and Lambeth, J.D., 1993, Neutrophil phospholipase D is activated by a membrane-associated Rho family small molecular weight GTP-binding protein, J. Biol. Chem. 268:21509.

Briscoe, C.P., Martin, A., Cross, M., and Wakelam, M.J.O., 1995, The roles of multiple pathways in regulating bombesin-stimulated phospholipase D activity in Swiss 3T3 fibroblasts, Biochem. J. 306:115.

Brown, H.A., Gutowski, S., Moomaw, C.R., Slaughter, C., and Sternweis, P.C., 1993, ADP-ribosylation factor, a small GTP-dependent regulatory protein, stimulates phospholipase D activity, Cell 75:1137.

Brown, H.A., Gutowski, S., Kahn, R.A., and Sternweis, P.C., 1995, Partial purification and characterization of Arf-sensitive phospholipase D from porcine brain, J. Biol. Chem. 270.14935.

Cockcroft, S., Thomas, G.M.H., Fensome, A., Geny, B., Cunningham, E., Gout, I., Hiles, I., Totty, N.F., Truong, O., and Hsuan, J.J., 1994, Phospholipase D: A downstream effector of ARF in granulocytes, Science 263:523.

Conricode, K.M., Brewer, K.A., and Exton, J.H., 1992, Activation of phospholipase D by protein kinase C. Evidence for a phosphorylation-independent mechanism, J. Biol. Chem. 267:7199.

Conricode, K.M., Smith, J.L., Burns, D.J., and Exton, J.H., 1994, Phospholipase D activation in fibroblast membranes by the α and ß isoforms of protein kinase C, FEBS Lett. 342:149.

Coorssen, J.R., and Haslam, R.J., 1993, GTPγS and phorbol ester act synergistically to stimulate both Ca^{2+}-independent secretion and phospholipase D activity in permeabilized human platelets. Inhibition by BAPTA and analogues, FEBS Lett. 316:170.

Dinh, T.T., and Kennerly, D.A., 1991, Assessment of receptor-dependent activation of phosphatidylcholine hydrolysis by both phospholipase D and phospholipase C, Cell Regul. 2:299.

Donaldson, J.G., Finazzi, D., and Klausner, R.D., 1992, Brefeldin A inhibits Golgi membrane-catalysed exchange of guanine nucleotide onto ARF protein, Nature 360:350.
Dubyak, G.R., Schomisch, S.J., Kusner, D.J., and Xie, M., 1993, Phospholipase D activity in phagocytic leucocytes is synergistically regulated by G-protein- and tyrosine kinase-based mechanisms, Biochem. J. 292:121.
Eldar, H., Ben-Av, P., Schmidt, U.-S., Livneh, E., and Liscovitch, M., 1993, Up-regulation of phospholipase D activity induced by overexpression of protein kinase C-α. Studies in intact swiss/3T3 cells and in detergent-solubilized membranes in vitro, J. Biol.Chem. 268:12560.
Exton, J.H., 1990, Signaling through phosphatidylcholine breakdown, J. Biol. Chem. 265:1.
Exton, J.H., 1994, Phosphatidylcholine breakdown and signal transduction, Biochim. Biophys. Acta 1212:26.
Geny, B. and Cockcroft, S., 1992, Synergistic activation of phospholipase D by protein kinase C- and G-protein-mediated pathways in streptolysin O-permeabilized HL60 cells, Biochem. J. 284:531.
Geny, B., Fensome, A., and Cockcroft, S., 1993, Rat brain cytosol contains a factor which reconstitutes guanine-nucleotide-binding- protein-regulated phospholipase-D activation in HL60 cells previously permeabilized with streptolysin O, Eur. J. Biochem. 215:389.
Halenda, S.P., and Rehm, A.G., 1990, Evidence for the calcium-dependent activation of phospholipase D in thrombin-stimulated human erythroleukemia cells, Biochem. J. 267:479.
Hammond, S.M., Altshuller, Y.M., Sung, T.-C., Rudge, S.A., Rose, S., Engebrecht, J.A., Morris, A.J., and Frohman, M.A., 1995, Human ADP-ribosylation factor-activated phosphatidylcholine-specific phospholipase D defines a new and highly conserved gene family, J. Biol. Chem. 270:29640.
Helms, J.B., and Rothman, J.E., 1992, Inhibition by brefeldin A of a Golgi membrane enzyme that catalyses exchange of guanine nucleotide bound to ARF, Nature 360:352.
Houle, M.G., Kahn, R.A., Naccache, P.H., and Bourgoin, S., 1995, ADP-ribosylation factor translocation correlates with potentiation of GTPγS-stimulated phospholipase D activity in membrane fractions of HL-60 cells, J. Biol. Chem. 270:22795.
Huang, R., Kucera, G.L., and Rittenhouse, S.E., 1991, Elevated cytosolic Ca^{2+} activates phospholipase D in human platelets, J. Biol. Chem. 266:1652.
Jiang, H., Lu, Z., Luo, J.-Q., Wolfman, A., and Foster, D.A., 1995a, Ras mediates the activation of phospholipase D by v-Src, J. Biol. Chem.270:6006.
Jiang, H., Luo, J.-Q., Urano, T., Frankel, P., Lu, Z., Foster, D.A., and Feig, L.A., 1995b, Involvement of Ral GTPase in v-Src-induced phospholipase D activation, Nature 378:409.
Kater, L.A., Goetzl, E.J., and Austen, K.F., 1976, Isolation of human eosinophil phospholipase D, J. Clin. Invest. 57:1173.
Kessels, G.C.R., Roos, D., and Verhoeven, A.J., 1991, fMet-leu-phe-induced activation of phospholipase D in human neutrophils. Dependence on changes in cytosolic free Ca^{2+} concentration and relation with respiratory burst activation, J. Biol. Chem. 266:23152.
Khosravi-Far, R., Solski, P.A., Clark, G.J., Kinch, M.S., and Der, C.J., 1995, Activation of Rac1, RhoA, and mitogen-activated protein kinases is required for Ras transformation, Mol. Cell. Biol. 15:6443.

Knaus, U.G., Morris, S., Dong, H.-J., Chernoff, J., and Bokoch, G.M., 1995, Regulation of human leukocyte p21-activated kinases through G protein-coupled receptors, Science 269:221.
Kobayashi, M., and Kanfer, J.N., 1991, Solubilization and purification of rat tissue phospholipase D, Meth. Enzymol. 197:575.
Kozma, R., Ahmed, S., Best, A., and Lim, L., 1995, The Ras-related protein Cdc42Hs and bradykinin promote formation of peripheral actin microspikes and filopodia in Swiss 3T3 fibroblasts, Molec. Cell. Biol. 15:1942.
Ktistakis, N.T., Brown, H.A., Sternweis, P.C., and Roth, M.G., 1995, Phospholipase D is present on Golgi-enriched membranes and its activation by ADP-ribosylation factor is sensitive to brefeldin A, Proc. Natl. Acad. Sci. USA 92:4952.
Kumada, T., Miyata, H., and Nozawa, Y., 1993, Involvement of tyrosine phosphorylation in IgE receptor-mediated phospholipase D activation in rat basophilic leukemia (RBL-2H3) cells, Biochem. Biophys. Res. Commun. 191:1363.
Kumagai, N., Morii, N., Ishizaki, T., Watanabe, N., Fukisawa, K., Saito, Y., and Narumiya, S., 1995, Lysophosphatidic acid-induced activation of protein Ser/Thr kinases in cultured rat 3Y1 fibroblasts. Possible involvement in rho p21-mediated signalling, FEBS Lett. 366:11.
Kuribara, H., Tago, K., Yokozeki, T., Sasaki, T., Takai, Y., Morii, N., Narumiya, S., Katada, T., and Kanaho, Y., 1995, Synergistic activation of rat brain phospholipase D by ADP-ribosylation factor and rhoA p21, and its inhibition by Clostridium botulinum C3 exoenzyme, J. Biol. Chem. 270:25667.
Kusner, D.J., Schomisch, S.J., and Dubyak, G.R., 1993, ATP-induced potentiation of G-protein-dependent phospholipase D activity in a cell-free system from U937 promonocytic leukocytes, J. Biol. Chem. 268:19973.
Kwak, J.-Y., Lopez, I., Uhlinger, D.J., Ryu, S.Y., and Lambeth, J.D., 1995, RhoA and a cytosolic 50-kDa factor reconstitute GTPγS-dependent phospholipase D activity in human neutrophil subcellular fractions, J. Biol. Chem. 270:27093.
Lambeth, J.D., Kwak, J.-Y., Bowman, E.P., Perry, D., Uhlinger, D.J., and Lopez, I., 1995, ADP-ribosylation factor functions synergistically with a 50-kDa cytosolic factor in cell-free activation of human neutrophil phospholipase D, J. Biol. Chem. 270:2431.
Lee, Y.H., Kim, H.S., Pai, J.-K., Ryu, S.H., and Suh, P.-G., 1994, Activation of phospholipase D induced by platelet-derived growth factor is dependent upon the level of phospholipase C-γ1, J. Biol. Chem. 269:26842.
Leung, T., Manser, E., Tan, L., and Lim, L., 1995, A novel serine-threonine kinase binding the ras-related RhoA GTPase which translocates the kinase to peripheral membranes, J. Biol. Chem. 270:29051.
Lin, P., and Gilfillan, A.M., 1992, The role of calcium and protein kinase C in the IgE-dependent activation of phosphatidylcholine-specific phospholipase D in a rat mast (RBL 2H3) cell line, Eur. J. Biochem. 207:163.
Lin, P., Fung, W.-J.C., Li, S., Chen, T., Repetto, B., Huang, K.-S., and Gilfillan, A.M., 1994, Temporal regulation of the IgE-dependent 1,2-diacylglycerol production by tyrosine kinase activation in a rat (RBL2H3) mast-cell line, Biochem. J. 299:109.

Lopez, I., Burns, D.J., and Lambeth, J.D., 1995, Regulation of phospholipase D by protein kinase C in human neutrophils. Conventional isoforms of protein kinase C phosphorylate a phospholipase D-related component in the plasma membrane, J. Biol. Chem. 270:19465.
Malcolm, K.C., Elliott, C.M., and Exton, J.H., 1996, Evidence for Rho-mediated agonist stimulation of phospholipase D in Rat1 fibroblasts. Effects of Clostridium botulinum C3 exoenzyme, J. Biol. Chem. (In press).
Malcolm, K.C., Ross, A.H., Qiu, R.-G., Symons, M., and Exton, J.H., 1994, Activation of rat liver phospholipase D by the small GTP binding protein rhoA, J. Biol. Chem. 269:25951.
Manser, E., Leung, T., Salihuddlin, H., Zhao, Z.-s., and Lim, L., 1994, A brain serine/threonine protein kinase activated by Cdc42 and Rac1, Nature 367:40.
Martin, G.A., Bollag, G., McCormick, F., and Abo, A., 1995, A novel serine kinase activated by rac1/CDC42Hs-dependent autophosphorylation is related to PAK65 and STE20, EMBO J. 14:1970.
Martinson, E., Scheible, S., and Presek, P., 1994, Inhibition of phospholipase D of human platelets by protein tyrosine kinase inhibitors, Cell. Mol. Biol. 40:627.
Massenburg, D., Han, J.-S., Liyanage, M., Patton, W.A., Rhee, S.G., Moss, J., and Vaughan, M., 1994, Activation of rat brain phospholipase D by ADP-ribosylation factors 1, 5, and 6: Separation of ADP-ribosylation factor-dependent and oleate-dependent enzymes, Proc. Natl. Acad. Sci. USA 91:11718.
Matuoka, L., Shibasaki, F., Shibata, M., and Takenawa, T., 1993, Ash/Grb-2, a SH2/SH3-containing protein, couples to signaling for mitogenesis and cytoskeletal reorganization by EGF and PDGF, EMBO J. 12:3467.
McGlade, J., Brunkhorst, B., Anderson, D., Mbamalu, G., Settleman, J., Dedhar, S., Rozakis-Adcock, M., Chen, L.B., and Pawson, T., 1993, The N-terminal region of GAP regulates cytoskeletal structure and cell adhesion, EMBO J. 12:3073.
Nobes, C.D., and Hall, A., 1995, Rho, Rac, and Cdc42 GTPases regulate the assembly of multimolecular focal complexes associated with actin stress fibers, lamellipodia, and filopodia, Cell 81:53.
Nobes, C.D., Hawkins, P., Stephens, L., and Hall, A., 1995, Activation of the small GTP-binding proteins rho and rac by growth factor receptors, J.Cell Sci. 108:225.
Ohguchi, K., Banno, Y., Nakashima, S., and Nozawa, Y., 1995, Activation of membrane-bound phospholipase D by protein kinase C in HL60 cells: synergistic action of a small GTP-binding protein RhoA, Biochem. Biophys. Res. Commun. 211:306.
Ohguchi, K., Banno, Y., Nakashima, S., and Nozawa, Y., 1996, Regulation of membrane-bound phospholipase D by protein kinase C in HL60 cells. Synergistic action of small GTP-binding protein RhoA, J. Biol. Chem. 271:4366.
Okamura, S.-L., and Yamashita, S., 1994, Purification and characterization of phosphatidylcholine phospholipase D from pig lung, J. Biol.Chem. 269:31207.
Olson, M.F., Ashworth, A., and Hall, A., 1995, An essential role for Rho, Rac, and Cdc42 GTPases in cell cycle progression through G_1, Science 269:1270.

Olson, S.C., Bowman, E.P., and Lambeth, J.D., 1991, Phospholipase D activation in a cell-free system from human neutrophils by phorbol 12-myristate 13-acetate and guanosine 5'-O-(3-thiotriphosphate). Activation is calcium dependent and requires protein factors in both the plasma membrane and cytosol, J. Biol. Chem. 266:17236.
Pachter, J.A., Pai, J.-K., Mayer-Ezell, R., Petrin, J.M., Dobek, E., and Bishop, W.R., 1992, Differential regulation of phosphoinositide and phosphatidylcholine hydrolysis by protein kinase C-β1 overexpression. Effects on stimulation by α-thrombin, guanosine 5'-O-(thiotriphosphate), and calcium, J. Biol. Chem. 267:9826.
Pai, J.-K., Dobek, E.A., and Bishop, W.R., 1991, Endothelin-1 activates phospholipase D and thymidine incorporation in fibroblasts overexpressing protein kinase $C_{\beta 1}$, Cell Regul. 2:897.
Paterson, H.F., Self, A.J., Garrett, M.D., Just, I., Aktories, K., and Hall, A., 1990, Microinjection of recombinant $p21^{rho}$ induces rapid changes in cell morphology, J. Cell Biol. 111:1001.
Prendergast, G.C., Khosravi-Far, R., Solski, P.A., Kurzawa, H., Lebowitz, P.F., and Der, C.J., 1995, Critical role of Rho in cell transformation by oncogenic Ras, Oncogene 10:2289.
Provost, J.J., Fudge, J., Israelit, S., Siddiqi, A.R., and Exton, J.H., 1996, Tissue specific distribution and subcellular distribution of phospholipase D in rat. Evidence for distinct RhoA- and ARF-regulated isozymes, Biochem. J. (submitted).
Qian, Z., and Drewes, L.R., 1989, Muscarinic acetylcholine receptor regulates phosphatidylcholine phospholipase D in canine brain, J. Biol. Chem. 264:21720.
Qiu, R.-G., Chen, J., Korn, D., McCormick, F., and Symons, M., 1995, An essential role for Rac in Ras transformation, Nature 374:457.
Qiu, R.-G., Chen, J., McCormick, F., and Symons, M., 1995, A role for Rho in Ras transformation, Proc. Natl. Acad. Sci. USA 92:11781.
Randazzo, P.A., Yang, Y.C., Rulka, C., and Kahn, R.A., 1993, Activation of ADP-ribosylation factor by Golgi membranes. Evidence for a brefeldin A-and protease-sensitive activating factor on Golgi membranes, J. Biol.Chem. 268:9555.
Rankin, S., Morii, N., Narumiya, S., and Rozengurt, E., 1994, Botulinum C3 exoenzyme blocks the tyrosine phosphorylation of $p125^{FAK}$ and paxillin induced by bombesin and endothelin, FEBS Lett. 354:315.
Ridley, A.J., 1994, Membrane ruffling and signal transduction, BioEssays 5:321.
Ridley, A.J., and Hall, A., 1992, The small GTP-binding protein rho regulates the assembly of focal adhesions and actin stress fibers in response to growth factors, Cell 70:389.
Ridley, A.J., Paterson, H.F., Johnston, C.L., Diekmann, D., and Hall, A., 1992, The small GTP-binding protein rac regulates growth factor-induced membrane ruffling, Cell 70:401.
Rümenapp, U., Geiszt, M., Wahn, F., Schmidt, M., and Jakobs K.H., 1995, Evidence for ADP-ribosylation-factor-mediated activation of phospholipase D by m3 muscarinic acetylcholine receptor, Eur. J. Biochem. 234:240.

Schmidt, M., Huwe, S.M., Fasselt, B., Homann, D., Rümenapp, U., Sandmann, J., and Jakobs, K.H., 1994, Mechanisms of phospholipase D stimulation by m3 muscarinic acetylcholine receptors. Evidence for involvement of tyrosine phosphorylation, Eur. J. Biochem. 225:667.
Seckl, M.J., Morii, N., Narumiya, S., and Rozengurt, E., 1995, Guanosine 5'-3-O-(Thio)triphosphate stimulates tyrosine phosphorylation of $p125^{FAK}$ and paxillin in permeabilized Swiss 3T3 cells. Role of $p21^{rho}$, J. Biol.Chem. 270:6984.
Siddiqi, A.R., Smith, J.L., Ross, A.H., Qiu, R.-G., Symons, M., and Exton, J.H., 1995, Regulation of phospholipase D in HL60 cells. Evidence for a cytosolic/phospholipase D, J. Biol. Chem. 270:8466.
Singer, W.D., Brown, H.A., Bokoch, G.M., and Sternweis, P.C., 1995, Resolved phospholipase D activity is modulated by cytosolic factors other than Arf, J. Biol. Chem. 270:14944.
Singer, W.D., Brown, H.A., Jiang, X., and Sternweis, P.C., 1996, Regulation of phospholipase D by protein kinase C is synergistic with ADP-ribosylation factor and independent of protein kinase activity, J. Biol. Chem. 271:4504.
Song, J., Pfeffer, L.M., and Foster, D.A., 1991, v-Src increases diacylglycerol levels via a type D phospholipase-mediated hydrolysis of phosphatidylcholine, Molec. Cell. Biol. 11:4903.
Stewart, S.J., Cunningham, G.R., Strupp, J.A., House, F.S., Kelley, L.L., Henderson, G.S., Exton, J.H., and Bocckino, S.B., 1991, Activation of phospholipase D: a signaling system set in motion by perturbation of the T lymphocyte antigen receptor/CD3 complex, Cell Regul. 2:841.
Takai, Y., Sasaki, T., Tanaka, K., and Nakanishi, H., 1995, Rho as a regulator of the cytoskeleton, Trends Biochem. Sci. 20:227.
Taki, T., and Kanfer, J.N., 1979, Partial purification and properties of a rat brain phospholipase D, J. Biol. Chem. 254:9761.
Uings, I.J., Thompson, N.T., Randall, R.W., Spacey, G.D., Bonser, R.W., Hudson, A.T., and Garland, L.G., 1992, Tyrosine phosphorylation is involved in receptor coupling to phospholipase D but not phospholipase C in the human neutrophil, Biochem. J. 281:597.
Wang, P., Anthes, J.C., Siegel, M.I., Egan, R.W., and Billah, M.M., 1991, Existence of a cytosolic phospholipase D. Identification and comparison with membrane-bound enzyme, J. Biol. Chem. 266:14877.
Ward, D.T., Ohanian, J., Heagerty, A.M., and Ohanian, V., 1995, Phospholipase D-induced phosphatidate production in intact small arteries during noradrenaline stimulation: involvement of both G-protein and tyrosine-phosphorylation-linked pathways, Biochem. J. 307:451.
Weiss, A., 1993, T cell antigen receptor signal transduction: A tale of tails and cytoplasmic protein-tyrosine kinases, Cell 73:209.
Wilkes, L.C., Patel, V., Purkiss, J.R., and Boarder, M.R., 1993, Endothelin-1 stimulated phospholipase D in A10 vascular smooth muscle derived cells is dependent on tyrosine kinase, FEBS Lett. 322:147.
Wu, H., James-Kracke, M.R., and Halenda, S.P., 1992, Direct relationship between intracellular calcium mobilization and phospholipase D activation in prostaglandin E-stimulated human erythroleukemia cells, Biochem. 31:3370.

Yeo, E.-J., Kazlauskas, A., and Exton, J.H., 1994, Activation of phospholipase C-γ is necessary for stimulation of phospholipase D by platelet-derived growth factor, J. Biol. Chem. 269:27823.

Zhang, G.-F., Patton, W.A., Lee, F.-J.S., Liyanage, M., Han, J.-S., Rhee, S.G., Moss, J., and Vaughan, M., 1995, Different ARF domains are required for the activation of cholera toxin and phospholipase D, J.Biol. Chem. 270:21.

Zhang, S., Han, J., Sells, M.A., Chernoff, J., Knaus, U.G., Olevitch, R.J., and Bokoch, G.M., 1995, Rho family GTPases regulate p38 mitogen-activated protein kinase through the downstream mediator Pak1, J. Biol.Chem. 270:23934.

Zubiaur, M., Sancho, J., Terhorst, C., and Faller, D.V., 1995, A Small GTP-binding protein, Rho, associates with the platelet-derived growth factor type-ß receptor upon ligand binding, J. Biol. Chem. 270:17221.

DISCRIMINATION BETWEEN VARIOUS PHOSPHOLIPASE D ACTIVITIES IN THE HUMAN NEUTROPHIL AND THEIR RELATIVE INVOLVEMENT IN OXIDATIVE BURST

Valérie PLANAT *, Michel RECORD, Hélène TRONCHERE, Gérard RIBBES, and Hugues CHAP.

Unité INSERM 326 « Phospholipides Membranaires, Signalisation Cellulaire et Lipoproteines », C.H.R de PURPAN, 31059, TOULOUSE Cedex, FRANCE

(*) to whom correspondance should be adressed

Abreviations : CB, cytochalasin B; DG, diglyceride; fMLP, formylMethionyl-Leucyl-Phenylalanine; GF, GF 109203X, protein kinase C inhibitor; PMA, phorbol myristate acetate; DMSO, dimethylsulfoxyde; PA, phosphatidic acid; PAO, phenylarsine oxyde; PEt, phosphatidylethanol; TLC, thin-layer chromatoraphy; WT, wortmannin.

INTRODUCTION

The fMLP-activated phospholipase D (fMLP-PLD) was distinguished from the PMA-PLD on their relative sensitivity to calcium chelation and tyrosine kinase inhibitor. The receptor-mediated activity was abolished by calcium deprivation and genistein treatment, at variance with the activity trigggered by the non-receptor agonist PMA.

With each agonist a maximum activity could be obtained: with fMLP, in the presence of cytochalasin B (5µg/ml), the highest amount of phosphatidic acid was observed. This activity was called fMLP-PLD_{max}. With PMA as agonist, the PMA-PLD_{max} was observed in presence of phenylarsine oxide (PAO) or wortmannin. Each of these maximum activities could be subdivided into fMLP-PLD and PMA-PLD ones, obtained by just adding the respective agonist to cell preparations. A further subdivision of the activities was obtained by their insensitivity (fMLP-PLD1) or sensitivity (fMLP-PLD2) to wortmannin and PAO in the case of

Our datas indicated that the role of phospholipase D was not an all-or-none process, but that only one of the activities (fMLP-PLD2) was involved in the oxidative burst. Our results suggested that the various phospholipase D activities we have characterized might have distinct targets in the cell.

The present work report multiple regulation pathway of the overall phospholipase D activity in the human neutrophil.

MATERIAL AND METHODS:

Materials

Agonists were bought at Sigma, France.Genistein was from Gibco, France. The protein kinase C inhibitor, GF 109203 X, was obtained from Calbiochem, USA. Wortmannin and hydroxy-wortmannin were supplied by Dr G.Payne, Sandoz Laboratory, Bale, Switzerland. Phenylarsine oxide was from Aldrich Chemical. Luminol and horseradish peroxidase was obtained Sigma, France.

Methods

Neutrophil preparation and labelling were performed as previously reported (1,2,3). Monitoring of phospholipase D activity in total cell and calcium chelation have been already described (4). Mathematical calculations for the best fits between tracings of H2O2-dependent chemiluminescence and phosphatidic acid generation were performed by Dr V. von Tscharner (1) according to Deranleau (5). Oxydative burst was determined as indicated in ref 4 or 1.

RESULTS

The tyrosine kinase dependent (fMLP-PLD) and independent (PMA-PLD) phospholipases D:

Control		GF 109203X		Genistein	
		-	+	-	+
0.4	**fMLP**	**1.5**	**1.5**	*1.5*	*0.4*
0.25	**PMA**	*1.75*	*0.75*	**0.85**	**0.9**

Comparative effect of protein kinase C (10μM) and tyrosine kinase (100μM) inhibitors on phospholipase D activity after 1 minute cell stimulation by 1μM fMLP or 100 nM PMA. Results represented the amount of phosphatidic acid formed in % of total radioactivity in the sample. "Control " indicated the value at the PA spot on TLC of untreated cells.

From results plotted in the table, it was clear that fMLP-PLD absolutely required tyrosine kinase for its activation at variance with PMA-PLD. The effect of the protein kinase C inhibitor was just the opposite: it decreased the PMA-PLD while leaving unaffected the fMLP-PLD one. Whereas the tyrosine kinase inhibitor fully abolished the fMLP-PLD, maximum effect of GF109203X was with 10 mM of the compound and levelled off around 50% inhibition at the plateau, at 5 min. Similarly, whereas calcium chelation fully inhibited the fMLP-PLD it decreased only by half the PMA-PLD at 1 min (4). Whether the GF109203X-sensitive activity is the one which is calcium dependent has to be checked.

Therefore, the PKC-dependent phospholipase D was termed PMA-PLD2, and the PKC-independent PMA-PLD1.

<u>The PI-3 kinase-activated (fMLP-PLD2) and PI-3 kinase-inhibited (PMA-PLD3) phospholipase D activities:</u>

Phospholipase D and PI-3 kinase share regulatory pathways via the the small g protein RhoA, which has been reported to activate both enzymes (6,7). This prompted us to examine the effect of wortmannin, well-characterized as a PI-3 kinase inhibitor in the human neutrophil (8). With fMLP as agonist, inhibition was found to be delayed always by a time-lag of 20 to 30 seconds, as compared with the beginning of the phospholipase D activation, and levelled off around 50 % inhibition (Fig.1A). The dose-dependency showed that maximum effect of wortmannin occurred at 100 nM, with an IC_{50} of 10 nM (Fig.1C). The IC_{50} two times higher than that reported for PI-3 kinase in the human neutrophil (9). Although this inhibitor has been reported to act also on phospholipase A2 activation pathway (10), this effect has not been observed in the human neutrophil (11). We have also checked an anolog of the compound, the 17-hydroxywortmannin, reported to be the most potent in the initial observations (12) . Results obtained with this analog confirmed the time-delayed and maximum inhibition of fMLP-induced phospholipase D (Fig.2). Therefore, the fMLP-PLD was actually the sum of two activities, a fMLP-"earlyPLD" or fMLP-PLD1 insensitive to wortmannin, and a fMLP-"delayedPLD" or fMLP-PLD2 sensitive to the compound.

Since genistein treatment abolished fMLP-PLD activation immediatly (4) and wortmannin after a time-lag, it was concluded that fMLP-PLD activation required tyrosine-kinase as an early event and PI-3 kinase as a delayed event.

Noteworthy, the effect of wortmannin was exactly the opposite when PMA was used as agonist, since it enhanced the PMA-PLD (Fig. 1B) by two times at one minute. PMA has been reported to activate also PI-3 kinase (13). The effect of wortmannin reported in Fig.1B indicated that in normal conditions, part of the overall PMA-PLD (about 50 %) could not be observed because it was repressed by PI-3 kinase activation. In that case, activation of the lipid kinase exerted a negative effect on phospholipase D activation: this part of the PMA-PLD was termed PMA-PLD3.

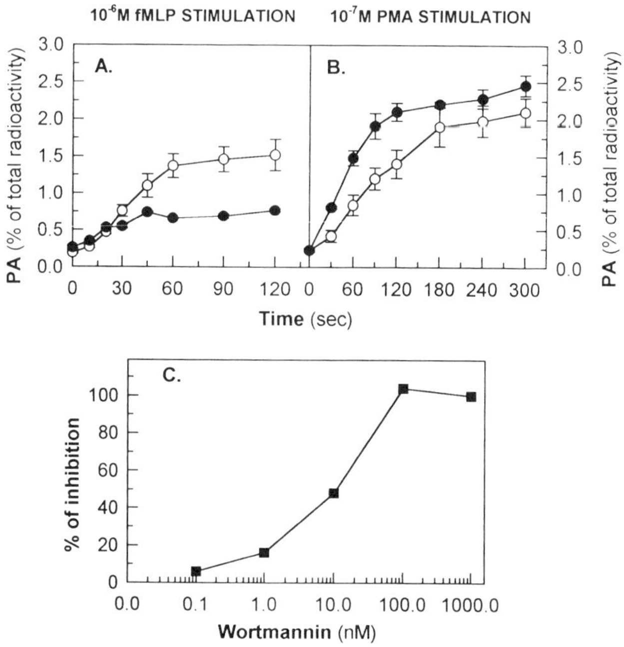

Figure 1: Effect of wortmannin on fMLP and PMA-activated phospholipase D.

Neutrophils were labelled on their alkyl-phoshatidylcholine subclass as indicated in Methods, then incubated for 15 min with the indicated concentrations of wortmannin (●) or not (O) prior to stimulation by either 1μM of fMLP (A) or 100 nM PMA (B). Concentration of the inhibitor was of 1μM in A and B. In C, dose-dependency was checked at 1 min and maximum inhibition of the fMLP-activated phospholipase D to wortmannin was normalized to 100 %; i.e results from pannel A with wortmannin represented the maximum inhibition.

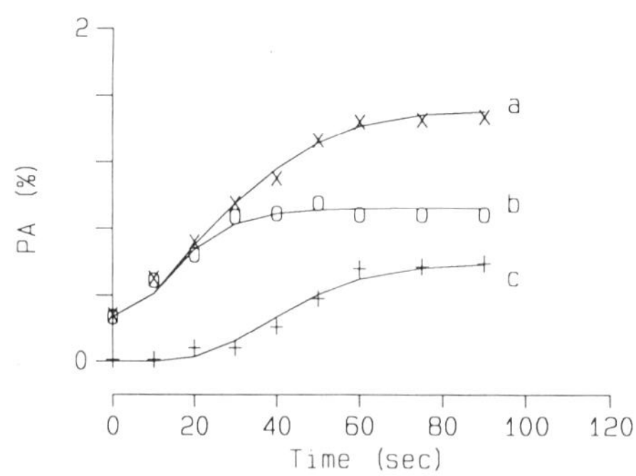

Figure 2: The 17-hydroxywortmannin (HWT) analog also discriminated between an early-insensitive phospholipase D activity and a delayed-sensitive one.

The compound at a concentration of 1 μM was used in similar conditions as in Fig 1, on cells stimulated by fMLP. Curve *a*: positive control (X); curve *b*: residual activity in HWT treated cells (0); curve *c*: difference between curves *a* and *b* at each time point (+).

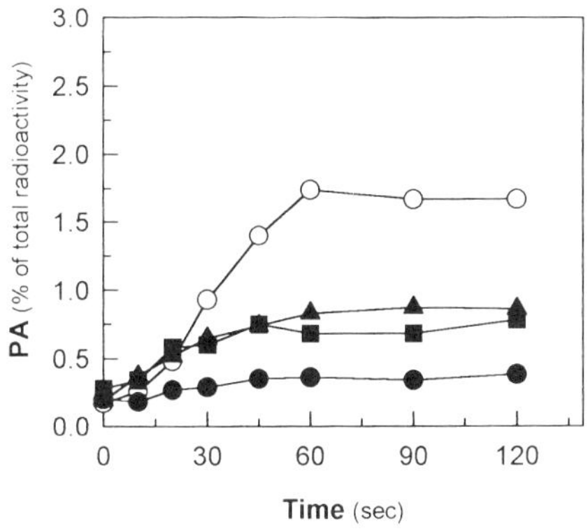

Figure 3: Phenylarsine oxide and wortmannin displayed the same intermediate inhibition on fMLP-activated phospholipase D as compared with genistein. Concentration of 1µM of either agonist (wortmannin ▲, PAO ■, genistein ●) was used on the same cell preparation stimulated with 1µM fMLP.

Alltogether, our results demonstrated that PI-3 kinase displayed an opposite action on phospholipase D activation depending whether the agonist interacted with a membranous receptor or not.

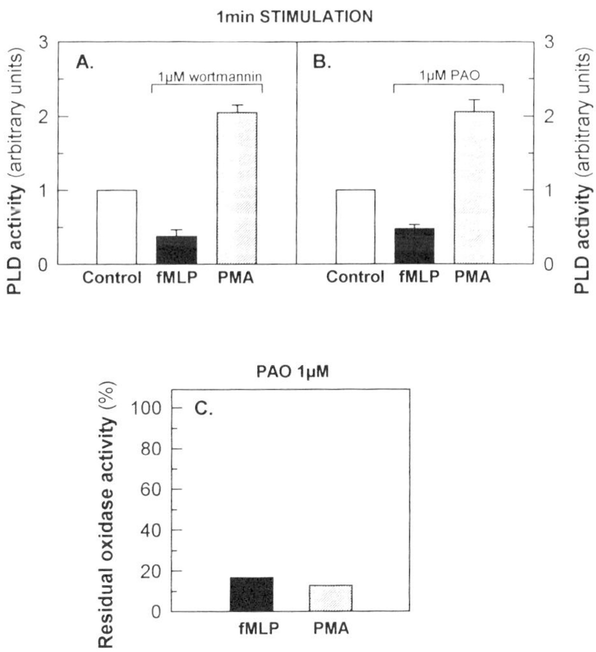

Figure 4: Comparative effect of PAO and wortmannin on phospholipase D and NADPH oxidase activities. The figure reported the phospholipase activities and the amount of $O_2^{\cdot -}$ after 1 min cell stimulation by either agonist, following cell treatment with wortmannin (A) or PAO (B,C).

The phosphatase-activated (fMLP-PLD2) and phosphatase-inhibited (PMA-PLD3) phospholipase D activities:

The similitude between the kinetics in presence of wortmannin and those previously observed with phenylarsine oxyde (PAO) (see ref.4) prompted us to verify the effect of both compounds on the same cell preparation, in the same conditions. Results plotted in Fig.3 demonstrated that the two inhibitors had exactly the same time-dependent effect on fMLP-PLD.

The same inhibitory effect of PAO was maintained at concentrations as low as 1nM (not shown), indicating the compound had a strong affinity for a precise target in the cell. We have controlled that PAO did not inhibit PI-3 kinase in-vitro, using immunoprecipitated PI-3 kinase from platelet. Therefore PAO interacted with a target closely related to PI-3 kinase activation. If we assume that the PI-3 kinase involved is the one which is activated by tyrosine phosphorylation, we can postulate that a *src* kinase could be involved in this process, and that PAO would inhibit activation of this *src* kinase by preventing its dephosphorylation at the level of tyrosine 527.

Consistent with the effect of wortmannin (Fig.4A), PAO which inhibited fMLP-PLD2 led to an opposite response when PMA was the agonist (Fig4B), unravelling a new part of the PMA-PLD, which was called PMA-PLD3.

Thus PAO and wortmannin displayed exactly the same effect, which was opposite on recepor-mediated and non-receptor mediated agonists.

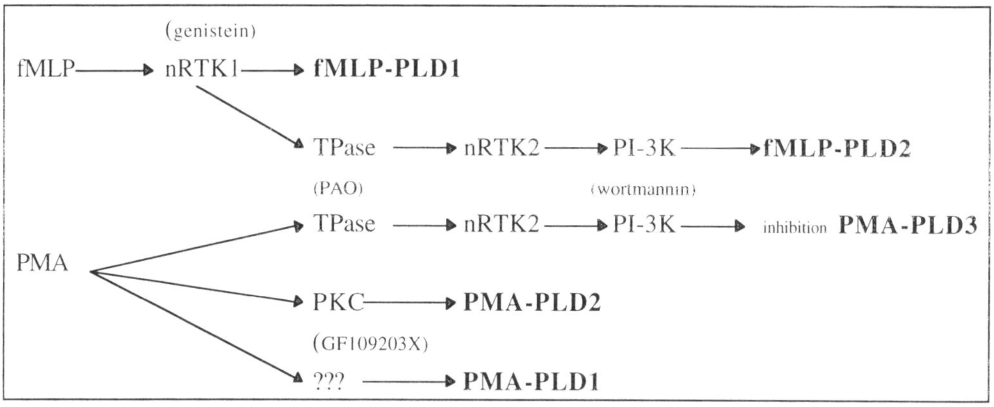

Scheeme 1: Possible pathways regulating the various phospholipase D activities. nRTK1= non-receptor tyrosine kinase 1; TPase=tyrosine phosphatase; nRTK2= non-receptor tyrosine kinase 2; PI-3K= PI-3 kinase; PKC=protein kinase C. The targets of the inhibitors used are indicated.

The relative involvement of the various phospholipases D in the oxydative burst.

Treatment of neutrophils with PAO led to an almost complete inhibition of superoxyde anion production at 1 min (and complete at 5 min;not shown) when cell were stimulated either by PMA or fMLP (Fig.4C). If one assume that phospholipases D activities and oxidative burst are related, results from Fig.4 indicated that inhibition of fMLP-PLD2 (sensitive to wortmannin and PAO) was involved in the inhibition oxidative burst, whereas with PMA, burst inhibition was a consequence of activation of a phospholipase D activity, namely the PMA-PLD3, activated by wortmannin and PAO. On the reverse, since PMA-PLD3 is normally repressed in cells, it would favor burst activation in physiological conditions . In that respect, the PKC-dependent PLD (PMA-PLD2) should be involved in superoxyde anion production, since PKC inhibitors affected both generation of phosphatidic acids and $O_2^{\cdot-}$ (14).

However the physiological agonist which is mimicked by PMA has to be established, since when neutrophils were activated by fMLP, PKCs were not involved in phospholipase D

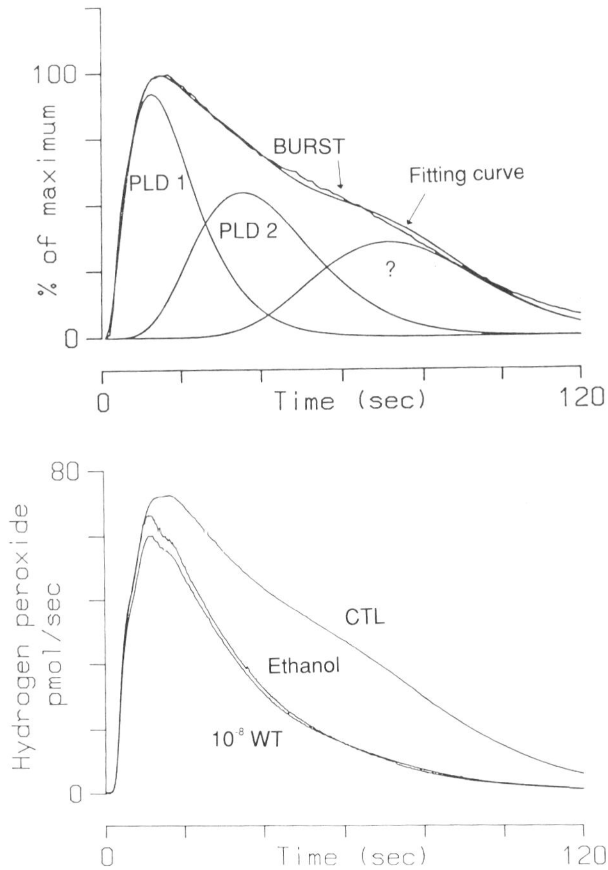

Figure 5: The fMLP-PLD2 only is involved in NADPH oxidase activation. In A, each activity was normalized in percent of total activity across the time. The fitting curve represented the mathematical addition of curves PLD1, PLD2 and "?".The oxidative burst was represented by the experimental tracing of H_2O_2-dependent chemiluminescence. In B, the H_2O_2 formation was reported in stimulated control cells, or in cells stimulated either in presence of ethanol (0.5%v/v) or 10 nM wortmannin.

activation (see ref.4 fig.1). Although fMLP triggered phospholipase Cβ generating diglycerides which activated proteine kinases C (15), those diglycerides were not involved in fMLP-dependent phospholipase D activation (see ref.4). Thus we focused on the relationship between phospholipase D and oxidative burst when fMLP was used as agonist.

The oxidative burst was monitored in our experimental conditions by chemiluminescence, which reported the instantaneous NADPH oxidase activity. Instead phospholipase D activity had been monitored by accumulation of phosphatidic acid (Fig.1-3). Mathematical calculations were performed from the results of Fig.2, to obtained the corresponding instantaneous activities of the fMLP-PLD1 and 2 from the curves *b* and *c*.

Such activities were then compared with the NADPH activity, in Fig. 5A. This figure showed that maximum PLD1 activation was reached after 15 sec, and after 40 sec cell stimulation for PLD2. The sum of areas under the curves of PLD1 and PLD2, plus the area under the curve of a so far unidentified process indicated as "?" on the figure, resulted in a curve which matched perfectly the experimental tracing of the NADPH oxidase activity (Fig 5A). This observation indicated that activation of the oxidase could be the result of 3 events: fMLP-PLD1, fMLP-PLD2 and a third one to be characterized.

To better analyse this hypothesis, we monitored the oxidase activity upon fMLP-induced cell activation in presence of 0.5 % ethanol, which allowed phosphatidylethanol formation competing with phosphatidic acid generation. We had controlled that the amount of phosphatidic acid made both by fMLP-PLD1 and 2 were decreased in presence of ethanol. Under these experimental conditions however, the early activation of the oxidase, during the first 20 seconds, was practically unaffected by the decrease in phosphatidic acid formation (Fig 5B). Instead, the oxidase activity dropped faster, since the duration of the oxidative burst shifted from 120 seconds and more to 60 seconds. Exactly the same tracing of oxidase activity was obtained with wortmannin treatment, which selectively affected fMLP-PLD2, instead of both PLD1 and 2 with ethanol (Fig.5B). These results indicated that only fMLP-PLD2 was involved in the oxidative burst, and more precisely in the duration of the burst, by controlling the decrease in the oxidase activity.

Conclusion: Phospholipase D activity exhibited various levels in stimulated human neutrophils.

Results reported herein indicated multiple regulatory pathways for phospholipase D activity in the human neutophil, depending upon the agonist; each of these pathways being involved in a specific level of second messenger formation.

With each agonist a maximum activity could be obtained: with fMLP, in the presence of cytochalasin B (5μg/ml), the highest amount of phosphatidic acid was observed (3.5%; Fig.6). This activity was called fMLP-PLD$_{max}$. With PMA as agonist, the PMA-PLD$_{max}$ was observed in presence of PAO or wortmannin. Each of these maximum activities could be subdivided into fMLP-PLD and PMA-PLD ones, obtained by just adding the respective agonist to cell preparations (Fig.6). A further subdivision of the activities was obtained by their sensitivity or

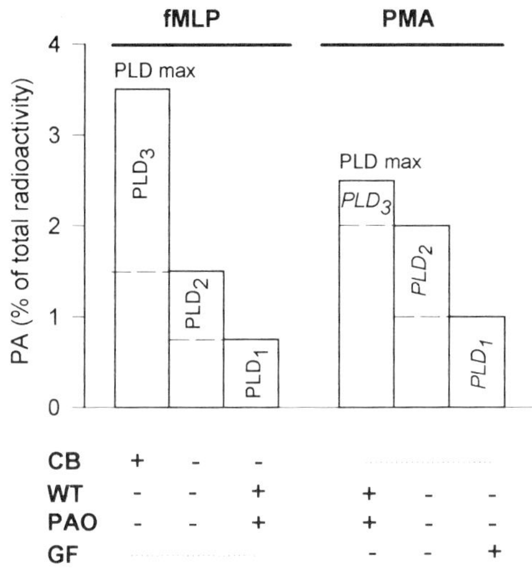

Figure 6: Various levels of phospholipase D activity in the human neutrophil;

Results represented typical amounts of phosphatidic acids formed (in % of total radioactivity in the sample) at the plateau for each agonist, i.e at 1 min for fMLP and 5 min for PMA.

With each agonist, the overall phospholipase D activity could be subdivided in three parts.

With fMLP, maximum activity was obtained in presence of cytochalasin B [CB] (5μg/ml). Omission of the compound led to suppression of fMLP-PLD3 activity. In presence of either wortmannin (WT) or phenylarsine oxide (PAO), the fMLP-PLD2 was inhibited, resulting in the minimum activity, fMLP-PLD1. The PKC inhibitor, GF109203X (GF) was ineffective, whereas calcium chelation or tyrosine kinase inhibitor completely suppressed the three activities (not represented).

With PMA, maximum activity was obtained in presence of WT and PAO. Omission of these compound suppressed the PMA-PLD3 activity. The PKC inhibitor abolished the PMA-PLD2, resulting in the minimum activity, PMA-PLD1. The conditions for complete inhibition were not found: calcium chelation decreased by 20 % at 5 min (50% at 1 min) the activity obtained in presence of PMA alone, whereas genistein had no effect (not represented; see ref 4)

insensitivity to wortmannin in the case of fMLP, and by the sensitivity / insensitivity to the protein kinase C inhibitor in the case of PMA.

Concerning the involvement of phospholipase D in neutrophil physiological responses, the question has been so far to establish whether this phospholipase was part of signalling pathways leading to either oxidative burst or/and degranulation. Results from Fig.5 indicated that the role of phospholipase D was not an all-or-none process, but that only one of the activities was involved in the oxidative burst. Our results suggested that the various phospholipase D activities we have characterized might have distinct targets in the cell.

DISCUSSION

Taken alltogether, the present datas reported two major pathways for phospholipase D activition depending upon the nature of the agonist. Effectively, the protein kinase C is not involved when the agonist is the fMLP; on the opposite, tyrosine kinases are not involved when the agonist is PMA. Two phospholipase D activities, topographically distinct, have been reported in HL-60 cells (16): a membranous one, calcium independent, activated essentially by RhoA, and a cytosolic one, calcium dependent and activated by Arf. Also RhoA-activated phospholipase D synergizes with the PMA-dependent activity (17). Whereas fMLP-activated phospholipase D requires a cytosolic 50 KD factor for full activation (18), the PMA-dependent activity requires a membranous one (19).

These respective factors have not been identified so far. From our results about in-vivo regulation, the full activation of fMLP-PLD would require the sequential involvement of tyrosine kinase 1 / tyrosine phosphatase / tyrosine kinase 2 / PI-3 kinase (scheeme 1). Recently, antibodies against the tyrosine kinase *lyn* have been shown to immunoprecipitate also PI-3 kinase (20). Therefore the non-receptor tyrosine kinase 2 we propose in the scheeme 1 could be *lyn*. It has been shown in platelets that the tyrosine phophatase SHPTP1 is activated by a non-receptor tyrosine kinase (21). We propose that another *src* kinase accounting for the non-receptor tyrosine kinase 1 would activate the SHPTP1. The full pathway between fMLP and PLD2 would involve sequentially: a *src* kinase/SHPTP1/*lyn*/PI-3 kinase. Most of the in-vitro PI-3 kinase activity is dependent upon $\beta\gamma$ for its activation (22). However this does not rule out the involvement of tyrosine phosphorylated PI-3 kinase in the activation process of the fMLP-PLD2. In that respect, the IC_{50} of fMLP-PLD2 towards wortmannin (10 µM), was closer to that reported for the tyrosine-dependent PI-3 kinase (5µM) than that of the $\beta\gamma$-dependent PI-3 kinase (40 µM)(23).

The proposed pathway for activating fMLP-PLD2, *src* kinase/SHPTP1/*lyn*/PI-3 kinase, is constituted by effectors which are immediatly activated upon cell stimulation, and would not explained the time-lag of 20-30 seconds we have always observed when comparing fMLP-PLD 1 and 2 activation (Fig.1,2). A possible explanation is that fMLP-PLD1 would be membranous phospholipase activated by RhoA/nRTK1 (or nRTK1/RhoA) whereas the fMLP-PLD2 would be a cytosolic enzyme requiring some delay to translocate on the membrane, namely the plasma membrane (24). We have shown that upon neutrophil activation, the plasma membrane was the site of action for the phospholipase (24). In HL-60 cells, the phospholipase D activity correlates with the amount of translocated Arf on the plasma membrane. Similarly in fibroblasts, the cytosolic phospholipase first interacts with the small g protein *ral* in the cytosol, before translocating to the membrane for intercating with *ras* (25). Therefore the "*src* kinase/SHPTP1/*lyn*/PI-3 kinase " pathway would also involved a translocation process of Arf/PLD2 requiring about 20 seconds.

This "*src* kinase/SHPTP1/*lyn*/PI-3 kinase" pathway appears as a route also used by the PMA-PLD3 activity (scheeme 1), but in a negative way since PMA led to inhibition of this phospholipase activity. The interest for the cell of such a regulation requires further investigations. From in-vitro studies, it has been shown that only the classical calcium-dependent PKCs were involved in PMA-activated phospholipase D (19). Our results from in-vivo studies indicated that a noticeable part of the overall PMA-PLD was calcium-independent. Therefore, the calcium independent PKCs of the neutrophil (δ and ζ) could be involved in part of PMA-PLD activity, in a way which remains to be defined. However the following pathway:

$$\text{agonist} \longrightarrow \text{PI-3 kinase} \xrightarrow{\text{PIP3}} \delta,\zeta\ \text{PKCs} \rightarrow \left|\begin{array}{l}\text{PMA-PLD3}\\ \text{fMLP-PLD2}\end{array}\right.$$

is ruled out for two reasons: 1/ because PMA-PLD3 requires inhibition and not activation of the PI-3 kinase, and 2/ because PKCs are not at all involved in the fMLP-PLD2.

From all our observations, the PI-3 kinase appears to play a central role. Whether its mode of action is really through PIP3 formation has to be checked, since PIP2 were initially described as activators of phospholipase D (26), and since the IC_{50} of wortmannin differs between PI-3 kinase and phospholipase D (9, and Fig1C). However PIP3 appears as potent as PIP2 in another myeloïd cell line, U937 cells (27). Further experiments using in-vitro determinations of phospholipase D activity in presence of PIP3 will have to be designed.

The SHPTP1 tyrosine phosphatase appears also as a strong candidate in the activation pathway of part of the fMLP-PLD and PMA-PLD (scheeme 1). However the dual regulation on tyrosine and serine residues induced by thrombin and PMA respectively, as reported by Li et al (21), deals with the isoform SHPTP1C, which is poorly present in the human neutrophil (F.Gaits, unpublished results). More probably, the isoform SHPTP2C could be the target of PAO. This is a major point to be clarified in the next future, since phospholipase D regulation by tyrosine phosphatases has received a poor attention so far.

REFERENCES

1.Gélas, P., von Tscharner, V., Record, M., Baggiolini, M., Chap, H.(1992) Biochem.J., 287, 67-72

2.Tronchère, H., Planat, V., Record, M., Tercé, F., Ribbes, G., Chap, H.(1995) J.Biol.Chem. 270, 13138-13146

3.Planat, V., Lanau, J.M., Record,M., Schauer, M., Ribbes, G., Müller-Peddinghaus, R., Chap, H. (1996) Biochem., J. *Submitted*

4. Planat, V., Tronchère, H., Record, M., Ribbes, G., Chap, H. (1996) Biochem.Biophys.Res.Commun., 218, 847-853

5.Deranleau, D.A.(1982) Experentia , 38, 661-662

6. Zhang, J., King, W.G., Dillon, S., Hall, A., Feig, L. and Rittenhouse, S.E. (1993) J. Biol. Chem., 268, 22251-22254

7.Malcom, K., Ross, A., Qiu, R., Symons, M., Exton, J. (1994) J.Biol.Chem., 269, 25951-25954

8. Thelen, M., Wymann, M.P. and Langen, H. (1994) Proc. Natl. Acad. Sci., 91, 4960-4964.

9. Arcado, A., Wymann, M. (1993) Biochem.J., 296, 297-301

10. Cross, M.J., Stewart, A., Hodgking, M.N., Kerr, D.J., Wakelam, M.J.O (1995) J.Biol.Chem., 270, 25352-25355

11. Thelen, M., Uguccioni, M., Bösinger, J. (1995) Biochem.Biophys.Res. Commun. , 217, 1255-1262
12. Baggiolini, M., Dewald, B., Schnyder, J., Ruch, W., Cooper, P., Payne, T.G.(1987) Exp.Cell.Res., 169, 408-418
13. King, W.G., Kucera, G.L., Sorisky, A., Zhang, J. and Rittenhouse, S.E. (1991) Biochem. J., 278, 475-480.
14. Mc Phail, L., C., Quallotine-Mann, D., Waite, K.A (1995) Proc.Natl.Acad.Sc.(USA), 92, 7931-7935
15. Ohta, H., Okajima, F., Ui, J (1985) J.Biol.Chem., 260, 15771-15780
16. Siddiqi, A.., Smith, J.L., Ross, A.H., Qlu, R.G., Symons, M. and Exton, J.H. (1995) J. Biol. Chem. 270, 8466-8473.
17. Ohguchi, K., Banno, Y., Nakashima, S., Nozawa, Y.(1996)J.Biol.Chem., 271, 4366-4372
18. Lambeth, D., Kwak, J.Y., Bowman, E.P., Perry, D., Uhlinger, D.J., Lopez, I.(1995) J.Biol.Chem., 270, 2431-2434
19. Lopez, I., Burns, D.J. and Lambeth, J.D. (1995) J. Biol. Chem., 270, 19465-19472.
20. Ptasnik, A., Traynor-Kaplan, A., Bokoch, G.M (1996) J.Biol. Chem., 270, 19969-19973
21. Li, R.Y, Gaits, F., Ragab, A., Ragab-Thomas, J., Chap,H. (1995) EMBO J., 14, 2519-2526
22.Stephens, L., Eguinoa, A., Corey, S., Jackson, T. and Hawkins, P.T. (1993) EMBO J., 6, 2265-2273.
23. Stephens, L., Smrcka, A., Cooke, F.T., Jackson, T.R., Sternweiss, P.C., Hawkins, P.T.(1994), Cell, 77, 83-93.
24. Ribbes, G., Cane, A., Planat, V., Breton, M., Chap, H., Béreziat, G., Record, M., Colard, O. (1996) J.Cell.Biochem., In Press.
25.Jiang, H., Luo, J.Q., Urano., T., Frankel, P., Lu, Z., Foster, D.A., Feig, L.A.(1995) Nature, 378, 409-412.
26. Liscovitch, M., Chalifa, V., Pertile, P., Chen, C.S. and Cantley, C. (1994) J. Biol. Chem. 269, 21403-21406.
27. Pertile, P., Liscovitch, M., Chalifa, V. and Cantley, L.C. (1995) J. Biol. Chem. 270, 5130-5135.

12(S)-HETE PRIMES A PHOSPHOLIPASE D PATHWAY IN ACTIVATED HUMAN BLOOD MONONUCLEAR CELLS

Alexia Zakaroff, Nadia Meskini, Catherine Joulain, Georges Némoz, Michel Lagarde, and Annie-France Prigent

INSERM U 352, Biochimie et Pharmacologie, INSA-Lyon, Villeurbanne, France

INTRODUCTION

Hydroxyeicosatetraenoic acids (HETEs) are major monohydroxylated arachidonic acid metabolites that are produced during inflammatory and immunological reactions[1]. Among the various lipoxygenase end-products, 12(S)-HETE, the main hydroxylated arachidonic acid derivative synthesized by activated macrophages and platelets, has been shown to either positively or negatively modulate mitogenic processes depending on the cell type considered. Thus, 12(S)-HETE has been shown to stimulate the proliferation of epidermal[2] and endothelial[3] cells in vitro. In contrast, opposite effects have been observed in lymphoid[4,5] as well as neuroblastoma[6] cells. For example, the treatment of human peripheral blood mononuclear cells (PBMC) with 12(S)-HETE induced a dose-dependent inhibition of lymphocyte proliferation, with 50% inhibition at about 20 μM[7,8], but the mechanisms involved remain poorly understood. The majority of 12(S)-HETE taken up by a wide array of cells is recovered in phospholipids[1,9] suggesting that it may affect cellular function through incorporation into membranes. One of the main biological changes associated with aging is an increase in free radical formation and lipid peroxidation with subsequent damage to cellular processes. Since the 12(S)-HETE level has been found higher in PBMC from aged subjects than in cells from young adults[10], we have hypothesized that this increased 12(S)-HETE level might be involved in the lowering of the immune defenses which accompanies aging.

On the other hand, several studies have highlighted the potential role of phosphatidic acid (PA) as a second messenger of growth factors[11] and its mitogenic properties for T cell clones[12] and fibroblasts[13]. During mitogenic activation of lymphoid cells, PA is mainly generated from diacylglycerol (DAG) through the DAG kinase activation[14]. Given the anti-mitogenic effect of 12(S)-HETE for lymphocytes, and its reported inhibitory properties toward DAG kinase[3,15], we asked whether an accumulation of 12(S)-HETE in blood cells might result in an altered production of PA upon mitogenic activation of PBMC or in the synthesis of 12(S)-HETE-containing PA species.

SYNTHESIS OF DAG AND PA IN CON A-ACTIVATED HUMAN PBMC.

As we have previously observed in rat thymocytes[14], the stimulation of [^{3}H]arachidonate-prelabeled PBMC by the mitogenic lectin concanavalin A (5μg/10^6cells) did not induce significant variations in the relative amount of [^{3}H]arachidonate-labeled DAG (0.4% of the total labeled lipids) whatever the length of the stimulation period between 0 to 30 min. In contrast, Con A activation significantly increased the [^{3}H]arachidonate-labeled PA from 0.6% of the total labeled lipids to about 2%, at 5 min. PA remained at this high level until 10 min, and then returned to base line values.

SYNTHESIS OF DAG AND PA IN CON A-ACTIVATED HUMAN PBMC PREVIOUSLY ENRICHED WITH 12(S)-HETE.

Since PA has been described as a DAG kinase inhibitor[3,15], a first hypothesis to explain the antiproliferative effect of 12(S)-HETE on human lymphocytes might be a decrease of PA synthesis through DAG kinase inhibition. The treatment of human PBMC with three

Table 1. Synthesis of PA in control and 12(S)-HETE-enriched human PBMC.

	PA[1] (ng/10^6 cells)	
PBMC	control cells	12(S)-HETE-enriched cells
without addition	10.4 ± 1.5 (n=14)[2]	17.2 ± 3.4 (n=8)
+ Con A	37.3 ± 5.1 (n=14)	66.6 ± 8.8 (n=8)†
+ Con A + EtOH	43.9 ± 14.8 (n=3)	24.6 ± 6.6 (n=3)*
+ Con A + suramin	35.6 ± 1.3 (n=3)	30.2 ± 6.4 (n=3)*

[1] PA was quantitated on TLC plates by videodensitometry (BIOPROFIL) after Coomassie blue staining.

[2] n indicates the number of separate experiments.

† significantly different from PA level measured in control cells stimulated with Con A (ANOVA, $P<0.01$)

* significantly different from PA level measured in 12(S)-HETE-enriched cells stimulated with Con A (ANOVA, $P<0.01$)

successive 30 min pulses of 1 μM unlabeled 12(S)-HETE induced a substantial enrichment of cell phospholipids in 12(S)-HETE, mainly phosphatidylcholine (PC) and to a lesser extent phosphatidylinositol (PI)[8]. In 12(S)-HETE-enriched cells further labeled with [^{3}H]arachidonic acid, the basal production of radiolabeled DAG was not altered, whereas a significant decrease was observed upon Con A stimulation (0.61 ± 0.07 % vs 0.40 ± 0.04 %). These data clearly indicate that 12(S)-HETE does not inhibit DAG-kinase in human PBMC.

In resting conditions, [^{3}H]arachidonate-labeled PA levels were similar in control and 12(S)-HETE-enriched cells. Con A activation significantly increased the [^{3}H]arachidonate-labeled-PA production in both control and 12(S)-HETE-treated cells, the increase being slightly higher in 12(S)-HETE-treated than in control cells. In contrast, Con A increased more potently the mass of PA in 12(S)-HETE-treated than in control cells (Table 1). Thus, the Con A-induced PA mass was significantly higher in cells treated with 12(S)-HETE (66.6 ± 8.8 ng/10^6cells) than in control cells (37.3 ± 5.0.ng/10^6cells).

EVIDENCES FOR A PHOSPHOLIPASE D PATHWAY IN THE 12(S)-HETE-INDUCED INCREASE OF PA SYNTHESIS.

Influence of ethanol on the PA mass

The fact that, in 12(S)-HETE-enriched cells, the Con A-induced rise in PA mass was significantly higher than in untreated cells might suggest that 12(S)-HETE was able to trigger an additional pathway of PA formation which was not activated by Con A alone in control cells. To test the possible involvement of a phospholipase D activity, the Con A-induced PA synthesis was monitored in the presence of 1% ethanol. In control cells, the Con A-induced synthesis of PA was not altered by the presence of ethanol (Table 1). In contrast, in 12(S)-HETE-enriched cells, the Con A-induced rise of the PA mass was significantly lowered (-63%) by ethanol. Similar lowering effects of ethanol were obtained when control cells were stimulated by 100 nM tetradecanoylphorbol acetate (TPA), a well-known activator of phospholipase D (not shown).

Synthesis of [^{3}H]arachidonate-labeled phosphatidylalcohol

Since the decrease in PA synthesis occurring during the transphosphatidylation reaction is accompanied by phosphatidylalcohol formation, experiments were designed to evaluate the agonist-induced synthesis of [^{3}H]arachidonate-labeled phosphatidylalcohol in both control and 12(S)-HETE-enriched cells. In the presence of the primary alcohols ethanol and 1-butanol, a significant 2.5-3-fold increase of [^{3}H]arachidonate-labeled phosphatidylalcohol was observed in 12(S)-HETE-enriched cells whereas 2-butanol was ineffective. In marked contrast, no phosphatidylalcohol was synthesized in control cells upon Con A stimulation in the presence of alcohol.

Influence of suramine

The polyanionic naphtalen-sulphonic acid derivative suramin is a known inhibitor of phospholipase D activity[16]. To get further insight into the potential mechanism responsible for the 12(S)-HETE-increased PA production upon Con A activation, we measured the accumulation of phosphatidic acid in response to Con A in the absence or presence of suramin. Our results show that suramin (1 mM) did not alter the Con A-induced increase of the PA mass in control PBMC but markedly decreased (-55%) that induced in 12(S)-HETE-enriched cells. Collectively, the above results suggest that Con A was actually able to stimulate a phospholipase D activity in 12(S)-HETE-enriched PBMC but not in control PBMC.

SYNTHESIS OF 12(S)-HETE-CONTAINING PA SPECIES IN 12(S)-HETE-ENRICHED CELLS

To know whether 12(S)-HETE-enriched PBMC were able to synthesize 12(S)-HETE-containing PA species, cells were treated with 12(S)-[^{3}H]HETE and stimulated or not with Con A. In unstimulated cells, 2.66% of the radioactivity associated with total phospholipids were found in PA. Taking into account the specific activity of 12(S)-[^{3}H]HETE and the measured mass of PA, 12(S)-HETE-containing PA species can be estimated at about 0.4% of total PA. Upon Con A stimulation, the mass of PA was by far more potently increased (+300%) than the radioactivity (+54%), so that the relative amount of 12(S)-HETE-containing PA species was decreased to 0.15% of total. Although we cannot exclude a selective accumulation of 12(S)-HETE-containing PA species in strategic compartments of the membrane bilayer, their formation in relatively low amount is very unlikely to be responsible for the antimitogenic properties of 12(S)-HETE.

FATTY ACID COMPOSITION OF PA IN NON-ENRICHED OR 12(S)-HETE-ENRICHED HUMAN PBMC STIMULATED WITH CON A.

As we have previously observed in rat thymocytes[14], saturated fatty acids accounted for more than 50% of the total PA fatty acids in resting human PBMC. Upon Con A-stimulation of control cells, the relative amount of palmitic acid (16:0) was significantly decreased while in the same time stearic (18:0) and arachidonic (20:4) acid proportions tended to increase (Figure 1). In 12(S)-HETE-enriched PBMC, the fatty acid variation pattern observed upon Con A stimulation was substantially different. Especially, the 16:0 proportion was not decreased any more by Con A whereas a significant rise of 18:0 was observed.

PBMC were incubated with three successive 30 min pulses of 1 µM 12(S)-HETE or with the vehicle (control cells). After removal of the medium and washing, cells were further incubated for 5 min at 37°C in the absence or presence of Con A (5 µg/10^6 cells).

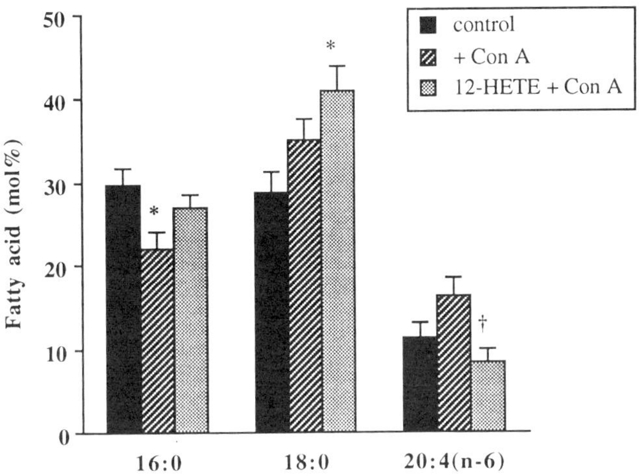

Figure 1: Variations in the palmitic, stearic and arachidonic acid content of PA from unstimulated control cells and from Con A-stimulated control or 12-HETE-enriched cells.

At the end of the incubation period, lipids were extracted from cells plus medium and PA was separated on TLC. PA fatty acids were transmethylated by BF_3/MeOH and analyzed by gas chromatography. The values are the means ± SE obtained in seven independent experiments. Data were compared by the paired Student's t test.* different from the value in control cells at $P \leq 0.02$, † different from the value in Con A-stimulated non-enriched cells at $P \leq 0.02$.

Similarly, 20:4 was no longer increased by Con A in 12(S)-HETE-enriched cells but even lowered. Thus, the fatty acid composition of PA synthesized upon Con A stimulation markedly differs whether PBMC have previously been enriched with 12(S)-HETE or not. In Con A-stimulated 12(S)-HETE-enriched cells PA exhibited a lower content in arachidonic and in total n-6 polyunsaturated fatty acids than in non-enriched cells and a higher content in total saturated ones (not shown).

CONCLUSION

Results from the present investigation demonstrate that 12(S)-HETE not only does not block the Con A-induced synthesis of PA in human PBMC, as it might be expected from its reported inhibitory properties toward DAG kinase, but on the contrary, induces a strong potentiation of PA production. This enhancing effect results from the activation of a phospholipase D whereas such a pathway is lacking in normal PBMC stimulated by Con A alone. Data obtained with primary alcohols and suramin clearly support this hypothesis. However, the mechanisms involved in the 12(S)-HETE-induced activation of phospholipase D in the presence of Con A remain presently unknown. It can be hypothesized that the esterification of 12(S)-HETE in cell phospholipids modifies the membrane properties in the vicinity of the Con A binding sites and allows a coupling which is impeded in normal cells. Alternatively, 12(S)-HETE might act as an extracellular ligand in its free fatty acid form. Indeed, 12(S)-HETE effect was antagonized by suramin, a membrane-impermeant agent known to interfere with interaction between peptide growth factors and their receptors[13,16,17]. Although the basal PA level tended to be higher in 12(S)-HETE-pretreated PBMC than in control cells, activation of the phospholipase D pathway required the simultaneous presence of 12(S)-HETE and Con A. Thus, Con A might act to mobilize 12(S)-HETE stored in the phospholipid pools and to increase the free 12(S)-HETE level available in the extracellular compartment. Indeed, we have previously observed a sustained increase of free 12(S)-[^{3}H]HETE in 12(S)-[^{3}H]HETE-labeled PBMC upon Con A stimulation[8]. A more detailed study supporting the above hypotheses has been published elsewhere[18].

Although 12(S)-HETE-containing PA species were effectively produced in 12(S)-HETE enriched PBMC, their relative amount was probably too low to significantly impair the mitogenic process. As demonstrated by GC analyses of PA fatty acids, the antimitogenic properties of 12(S)-HETE is very likely to be related to its capacity to stimulate the synthesis of saturated PA species with a low mitogenic potential, through the activation of a phospholipase D acting on a peculiar pool of phospholipids containing a low amount of arachidonic acid.

REFERENCES

1. A.A. Spector, J.A. Gordon, and S.A. Moore, Hydroxyeicosatetraenoic acids (HETEs), *Prog. Lipid Res.* 27:271 (1988).
2. P. Arenberger, L. Kemény, and T. Ruzicka, Characterization of high-affinity 12(S)-hydroxyeicosatetraenoic acid (12(S)-HETE) binding sites on normal keratinocytes, *Epith. Cell Biol.*, 2:1 (1993).
3. B.N.Y.Setty, J.E. Graeber, and M.J. Stuart, The mitogenic effect of 15- and 12-hydroxyeicosatetraenoic acid on endothelial cells may be mediated via diacylglycerol kinase inhibition, *J. Biol. Chem.* 262:17613 (1987).
4. J.W. Hadden, Transmembrane signals in the activation of T-lymphocytes by lectin mitogens, *Mol. Immunol.* 25:1105 (1988).
5. C.E. Low, M. Pupillo, R.W. Bryant, and J.M. Bailey, Inhibition of phytohemagglutinin-induced lymphocyte mitogenesis by lipoxygenase metabolites of arachidonic acid : structure-activity relationships, *J. Lipid Res.* 25:1090 (1984).
6. E.J. Wemer, R.W. Walenga, R.L. Dubowy, S. Boone, and M.J. Stuart, Inhibition of human malignant neuroblastoma cell DNA synthesis by lipoxygenase metabolites of arachidonic acid, *Cancer Res.* 45:561 (1985).
7. C. Hadjiagapiou, J. Travers, R. Fertel, and H. Sprecher, ß-oxidation of 12(S)-hydroxy-5,8,10,14-eicosatetraenoic acid by MOLT-4 lymphocytes, *Arch. Biochem. Biophys.* 292:112 (1992).

8. C. Joulain, N. Meskini, G. Anker, M. Lagarde, and A.F. Prigent, Esterification of 12(S)-hydroxy-5,-8,-10,-14 eicosatetraenoic acid (12(S)HETE) into phospholipids of human peripheral blood mononuclear cells. Inhibition of the proliferative response, *J. Cell. Physiol.* 164:154 (1995).
9. W.F. Stenson, M.W. Nickells, and J.P. Atkinson, Esterification of monohydroxy fatty acids into the lipids of a macrophage cell line, *Prostaglandins* 26:253 (1983).
10. N. Meskini, G. Némoz, P. Chapuy, P. Haond, J.F. Pageaux, E. Véricel, M. Lagarde, and A.F. Prigent, Glutathione peroxidase activity and metabolism of arachidonic acid in peripheral blood mononuclear cells from elderly subjets, *Clin. Sci.* 85:203 (1993).
11. W.H. Moolenaar, W. Kruijer, B.C. Tilly, I. Verlaan, A.J. Bierman, and S.W. de Laat, Growth factor-like action of phosphatidic acid, *Nature* 323:171 (1986).
12. E. Cano, M.A. Munoz-Fernandez, and M. Fresno, Regulation of interleukin-2 responses by phosphatidic acid, *Eur. J. Immunol.* 22:1883 (1992).
13. E.J. Van Corven, A. Van Rijswijk, K. Jalink, R.L. Van der Bend, W.J. Van Blitterswijk and W.H. Moolenaar, Mitogenic action of lysophosphatidic acid and phosphatidic acid on fibroblasts. Dependence on acyl-chain length and inhibition by suramin, *Biochem. J.* 281:163 (1992).
14. S. El Bawab, O. Macovschi, M. Lagarde, and A.F. Prigent, Time-course changes in content and fatty acid composition of phosphatidic acid from rat thymocytes during Concanavalin A stimulation, *Biochem. J.* 308:113 (1995).
15. G. Friedlander, C. Le Grimellec, J. Sraer, and C. Amiel, 12-HETE modulates Na-coupled uptakes in proximal tubular cells: role of diacylglycerol kinase inhibition, *Am. J. Physiol.* 259:F816 (1990).
16. C. Gratas, and G. Powis, Inhibition of phospholipase D by agents that inhibit cell growth, *Anticancer Res.* 13:1239 (1993).
17. A. Yayon, and M. Klagsbrun, Autocrine transformation by chimeric signal peptide-basic fibroblast growth factor: Reversal by suramin, *Proc. Natl. Acad. Sci. USA* 87:5346 (1990).
18. N. Meskini, A. Zakaroff, C. Joulain, G. Némoz, M. Lagarde, and A.F. Prigent, Triggering of a phospholipase D pathway upon mitogenic stimulation of human peripheral blood mononuclear cells enriched with 12(S)-hydroxyicosatetraenoic acid, *Eur. J. Biochem.* 233:907 (1995).

PHOSPHOLIPASE D ACTIVITIES AND PHOSPHATIDYLCHOLINE TURNOVER ARE DIFFERENTIALLY RELATED TO EXPRESSION OF PROTEIN KINASE C ISOFORMS AND MARCKS IN CONTROL AND TRANSFECTED NEURAL CELLS

Harold W. Cook, Stephen C. Van Iderstine, Sherry C. Morash, Sergio D. Rosé, Neale D. Ridgway and David M. Byers

Atlantic Research Centre
Departments of Pediatrics and Biochemistry
Dalhousie University
Halifax, NS B3H 4H7
Canada

INTRODUCTION

Phosphatidylcholine (PtdCho) and phosphatidylethanolamine (PtdEtn) are major membrane phospholipids constituting 70-80% of the total phospholipids in most cells (Vance, 1991; Kent, 1995). Accordingly, they have major roles in maintaining the structural integrity of membrane boundaries between cells and within subcellular organelles and compartments. Both PtdCho and PtdEtn are potential contributors to the generation of lipid second messengers following activation of phospholipases A_2 (PLA_2), C (PLC) and D (PLD) (Kiss et al., 1991; Kiss et al., 1994; Exton, 1990). Following activation of PLA_2, release of polyunsaturated fatty acids may result in bioactive oxygenated products including eicosanoids, leukotrienes and other derivatives. There is increasing evidence that fatty acids also may activate kinases or alter gene expression through receptors and nuclear response elements (Ailhaud et al., 1994; Graber et al., 1994). Release of diacylglycerol (DAG) and phosphatidic acid by PLC and PLD, respectively, also is important in signal transduction (Exton, 1990; Nishizuka, 1992). Initial production of DAG is part of a dual release of second messengers (DAG and inositol tris-phosphate) when phosphatidylinositol bisphosphate is hydrolyzed by PLC. Subsequent, sustained production of DAG appears to involve hydrolysis of PtdCho and PtdEtn (Nishizuka, 1992).

DAG, phosphatidic acid and some unsaturated fatty acids are activators of specific members of a large protein kinase C (PKC) family of isoforms. All isoforms share some common features within the catalytic domain; however, they differ within two conserved regions of the regulatory domain allowing PKC to be classified in three major groups (Dekker and Parker, 1994; Newton, 1995; Nishizuka, 1992). Classical or conventional PKCs (PKC-α, βI, βII, and γ) are activated by DAG and Ca^{2+}. In contrast, the new PKCs (PKC-δ, ϵ, μ and θ) do not require Ca^{2+} for activation. A third group of atypical PKCs (including PKC-ζ and λ) require neither DAG nor Ca^{2+} for activation. Accordingly, this latter group is not dependent on lipid hydrolysis for activation. The phorbol ester, 4β-12-*O*-tetradecanoyl-phorbol-13-acetate (TPA) mimics the effects of DAG on the classical and new isoforms of PKC. As TPA is relatively stable and effective at low concentrations (10-200 nM), it is frequently used in assessment of effects of PKC stimulation.

In earlier studies (Cook et al., 1989; Sproull et al., 1995), we investigated the synthesis and degradation of PtdCho in response to activation of PKC by TPA in four neural cell lines – C6 rat glioma, N1E-115 mouse neuroblastoma, and SK-N-MC and SK-N-SH human neuroblastomas. We reported the relative abundance of six PKC isoforms based on quantitation of protein by Western blot analysis using isoform specific antibodies and mRNA expression based on Northern and S1 nuclease protection assays. We also quantitated relative amounts of myristoylated alanine-rich C kinase substrate (MARCKS) in the four cell lines (Rosé et al., 1994). Levels of PKC-α and MARCKS were correlated positively with stimulation of PtdCho synthesis, measured as incorporation of radiolabelled choline (Cho) into PtdCho, in response to PKC stimulation by TPA. All cell lines had substantial basal PKC activity in cytosolic and particulate fractions and had similar binding constants with radiolabelled phorbol esters (Sproull et al., 1995). Synthesis of PtdCho in each cell line was stimulated by 100-200 μM oleic acid (18:1n-9) indicating that all cells have a capacity for enhanced synthesis of PtdCho. However, only C6 glioma and SK-N-SH neuroblastoma cells responded to TPA and contained substantial PKC-α and MARCKS; the two non-responding cell lines (N1E-115 and SK-N-MC) contained little detectable PKC-α or MARCKS. Thus, PKC-α and/or MARCKS appear to be essential for TPA-initiated stimulation of PtdCho synthesis.

Based on an observed lag of 15-30 min between the time of TPA addition and a detectable increase in PtdCho labelling, it appeared that stimulation of synthesis might be secondary to some earlier lipid turnover event. PKC activation occurred within seconds of TPA application. We postulated that the initial action of PKC on phospholipid turnover might involve PLD since PLD activity is enhanced by PKC (Conricode et al., 1994; Boyano-Adánez et al., 1994). To further evaluate mechanisms involved, we have now assessed PLD activity in the four cell lines using five different types of PLD assays and have determined the influence of TPA and bryostatin as activators and staurosporine (STS) and *bis*-indolylmaleimide (BIM) as inhibitors of PKC activity.

MATERIALS AND METHODS

Suppliers of activators or inhibitors of PKC and of radiolabeled substrates were identified previously (Sproull et al., 1995). Conditions for the culture of rat glioma (C6; line CCL-107), mouse neuroblastoma (N1E-115) and human neuroblastoma (SK-N-MC and SK-N-SH) cells from the American Type Culture Collection as well as pulse and chase conditions were described previously (Sproull et al., 1995). Sources of cDNA probes and antibodies for MARCKS and PKC isoforms, and preparation and culturing of transfection clones of SK-N-MC cells also have been described (Rosé et al., 1996).

RESULTS

Ethanolamine (Etn) Release from Prelabelled PtdEtn

C6 glioma cells were prelabelled with. Etn for 24 h and the release of label to the extracellular medium was monitored in response to TPA treatment (Figure 1). There was a time-dependent increase in the amount of labelled Etn recovered in the medium. Phosphoethanolamine (PEtn) accounted for less than 20% of medium label. Etn added as a "cold trap" during the chase enhanced accumulation of labelled Etn compared to controls. When TPA and Etn were added together, Etn release to the medium was substantially enhanced. Accordingly, PLD activity appeared to be primarily responsible for the hydrolysis of PtdEtn in response to TPA stimulation and this activity was detectable within 10 min.

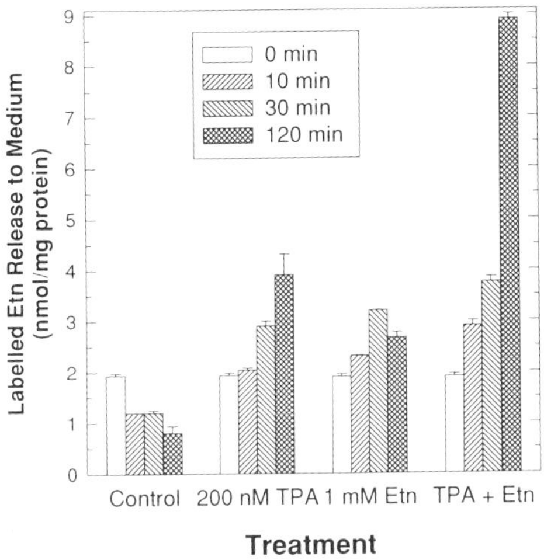

Figure 1. Time-dependent release of ethanolamine to the medium of glioma cells following phorbol ester treatment. C6 glioma cells were prelabelled for 24 h with [1,2-^{14}C]Etn and the medium was replaced with fresh medium. Control cells were treated with dimethysulfoxide. The combined treatments contained 200 nM TPA and 1 mM Etn. The ratio of Etn to PEtn (approximately 5:1) was identified by combinations of thin layer and column chromatography (Cook and Wakelam, 1989). Values are mean ± SEM for n=8 in four experiments.

Influence of PKC Activators and Inhibitors

PLD activity in C6 glioma cells was further evaluated by a combination of the Etn release assay and the formation of phosphatidylbutanol (PtdBuOH) by transphosphatidylation in cells prelabelled in the PtdEtn component (>76%) with [1-^{14}C]20:5n-3. Trends in PLD stimulation obtained with different combinations of agents were similar for Etn release and PtdBuOH formation (Figure 2), although the magnitudes of increase detected were different. TPA stimulated PLD activity in both assays. The non-specific protein kinase inhibitor, STS, was stimulatory, particularly at 2 μM. By contrast, the more specific PKC inhibitor, BIM (Toullec et al., 1991), was not itself stimulatory even at 5 μM. Low concentrations (200 nM) of STS that were not stimulatory partially inhibited the effect of TPA but even 2 μM STS did not eliminate the TPA effect. Some TPA-dependent activity was observed in the presence of 2 μM STS or 5 μM BIM. Independent measures of PKC under these conditions indicated that >98% of the PKC activity was eliminated. Accordingly, it appears that TPA caused stimulation of PLD hydrolysis of PtdEtn in a manner both dependent and independent of PKC activity. Based on the specificity of PtdEtn labelling in both types of experiment, it appears that all measured hydrolysis by PLD was specific to PtdEtn.

When C6 glioma cells were labelled in a similar manner with myristic acid so that a higher proportion of label appeared in PtdCho, 100 nM TPA and 1 μM STS stimulated PLD hydrolysis and PtdBuOH formation. At high concentrations of STS, the TPA effect was only partially inhibited such that approximately 20% of TPA-stimulated activity remained.

PLD Stimulation in Four Cell Lines

Using Cho release from cells prelabelled for 24 h with [methyl-^{3}H]Cho as a measure of PLD activity, effects of TPA, bryostatin (a non-phorbol stimulator of PKC activity) and STS were measured in each of the four cell lines (Figure 3). Observations in this assay with C6 glioma cells

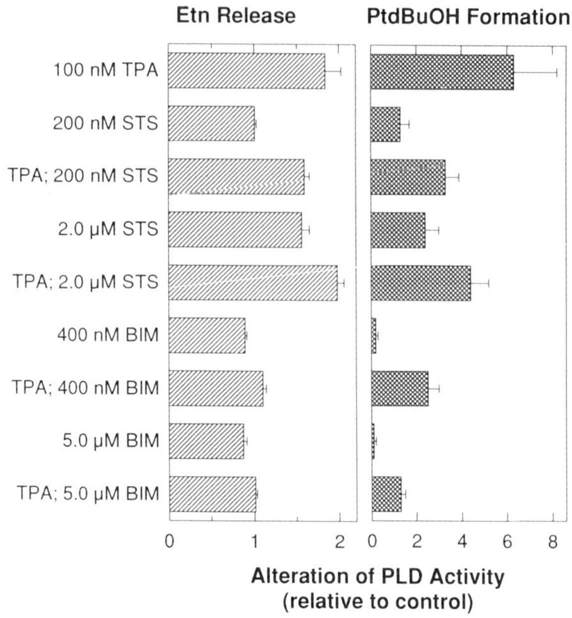

Figure 2. Effects of phorbol ester, staurosporine and *bis*-indolylmaleimide on PLD activity in glioma cells. For Etn release assays, cells were prelabelled for 24 h with [1, 2-^{14}C]Etn. For assay of PtdBuOH formation, cells were prelabelled for 24 h with [1-^{14}C]20:5n-3. After changing the medium, cells were pretreated for 15 min with STS or BIM as indicated. This was followed by 1 h incubation in the presence or absence of 100 nM TPA. For PtdBuOH formation, 1% 1-butanol was present for the last h in all assays. Values are mean ± SEM for n=6-11 in 3-6 experiments.

in response to TPA in the presence of STS closely resembled results reported for other measures of PLD activity. In contrast, neither β-TPA nor bryostatin stimulated PLD activity in N1E-115 cells; 100 nM STS appeared to inhibit basal PLD activity by approximately 40% both in the presence and absence of activators of PKC. Similarly, in both the SK-N-MC and SK-N-SH cell lines, neither β-TPA, bryostatin nor STS greatly influenced PLD activity. Thus, in contrast to C6 cells, TPA did not appear to activate PLD in SK-N-SH cells, even though PtdCho synthesis was stimulated (Sproull et al., 1995).

Overexpression of PKC-α and MARCKS in Clones of SK-N-MC Cells

To further assess the lack of response to TPA in SK-N-MC cells, clones transfected with PKC-α and MARCKS and with vector alone as controls (Rosé et al., 1996) were examined. PKC-α was overexpressed 30-80 fold based on detected protein and mRNA in two clones and MARCKS was overexpressed by 5-9 fold based on protein and mRNA levels in five others. Surprisingly, clones overexpressing PKC-α still did not respond to TPA in terms of enhanced PtdCho synthesis. However, cells overexpressing MARCKS were enhanced 1.9-2.3 fold in response to 100 nM TPA. In these responding cells, endogenous PKC-α was increased along with the overexpression of MARCKS. Thus, this PKC substrate appears to be essential for connecting TPA stimulation and enhanced PtdCho turnover.

As shown in C6 glioma and SK-N-SH cells containing adequate PKC-α and MARCKS, stimulation by TPA in MARCKS transfected clones involved a delay of 20-30 min, again suggesting that an earlier event such as PLD stimulation may precede PtdCho regeneration. In

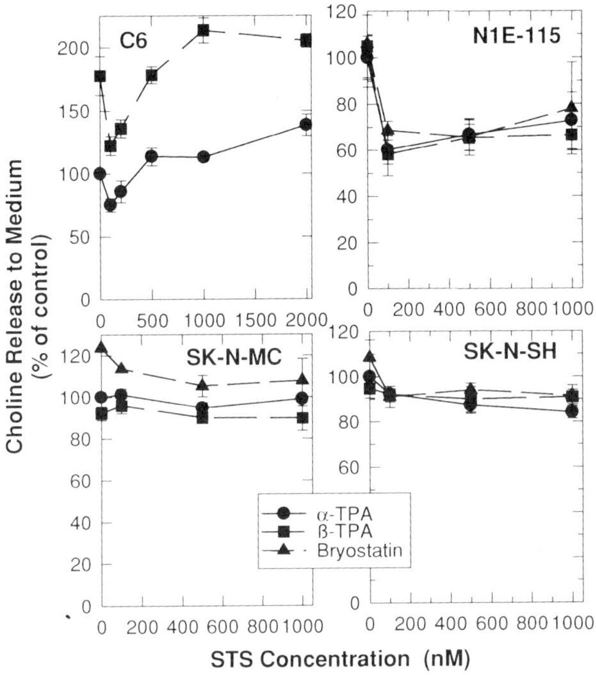

Figure 3. Effects of phorbol ester, bryostatin and staurosporine on stimulation of choline release from prelabelled PtdCho by phospholipase D. All cells were prelabelled for 24 h with [methyl-^{3}H]Cho. Cells were then incubated with dimethylsulfoxide vehicle (DMSO), 100 nM α-TPA, 100 nM β-TPA, or 100 nM bryostatin in the presence of varying concentrations of STS. After 1 h, release of labelled Cho to the medium was determined. Note the differences in the X- and Y-axis scales for individual cell types. Values are mean ± SEM for n=4-8 in 2-4 experiments.

a clone transfected with MARCKS, formation of labelled PtdBuOH in response to TPA following prelabelling of cells with either myristic acid or 18:1n-9 indicated that a low level of PLD activity was stimulated by TPA in contrast to no response in vector controls . Thus, MARCKS appeared to mediate stimulation of PLD after PKC activation. When the MARCKS transfected clone was assessed in relation to vector control using the Cho release assay, there was more label in the medium with TPA treatment during the first 30 min of chase (Figure 4; right panel). Interestingly, loss of label from the lipid fraction appeared to be greater in untreated compared to TPA-stimulated cells. Thus, the effect of TPA on these MARCKS transfected clones could not be attributed to stimulation of PLD mediated hydrolysis alone.

DISCUSSION

Turnover of membrane phospholipids is vitally linked to processes of lipid second messenger generation in response to a variety of cell surface stimulators. At least one, and in some cases both, of the products of phospholipase hydrolysis (PLA_2, PLC or PLD) can be a direct activator of downstream events including Ca^{2+} release, kinase activation and gene expression. To maintain the integrity of structural functions that phospholipids serve in cell surface and intracellular membranes, degraded phospholipids must be replaced rapidly by newly regenerated phospholipid molecules. Cycles involving degradation and synthesis have been described for inositol and choline phosphoglycerides (Pelech and Vance, 1989; Vance, 1991).

Given the variety of PKC isoforms and variations in their activation by different lipid products of phospholipid hydrolysis (Nishizuka, 1992), connections between specific PKC isoform expression

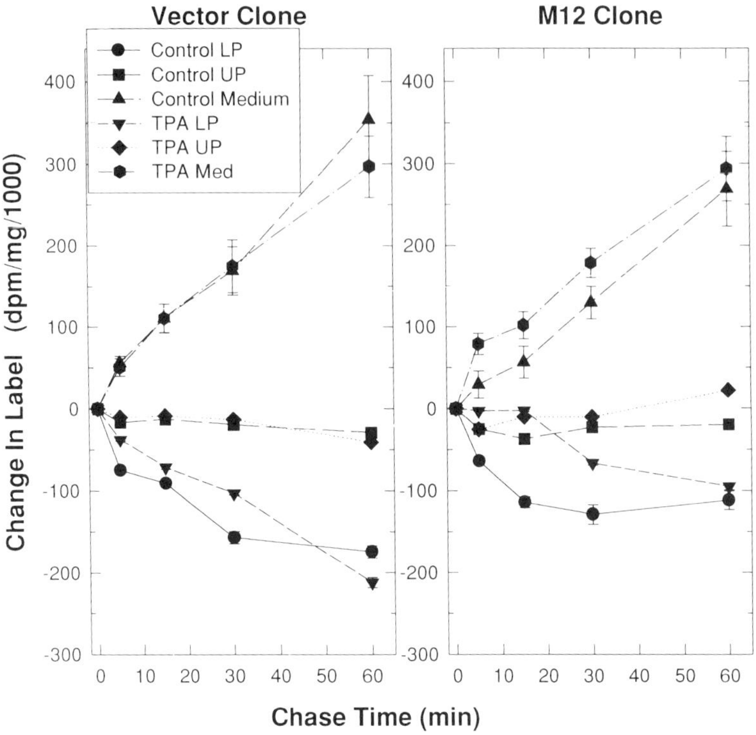

Figure 4. Phorbol ester treatment alters release of choline from prelabelled phosphatidylcholine only in a MARCKS-transfected (M12) clone. MARCKS-transfected and vector control clones of the SK-N-MC cell line were preincubated for 24 h with [methyl-^{3}H]Cho and the medium was changed. Cells were subsequently incubated in fresh medium containing 1 mM unlabelled Cho with or without 100 nM TPA. At various times, radioactivity in the chase medium and in the extracted upper phase (UP; H_2O soluble components) and lower phase (LP; lipids) of the cells was monitored. Values for the vector clone are mean ± SEM for n=6 in 3 experiments and in the M12 clone are mean ± SEM for n=2-6 in 1-3 experiments.

and phospholipase activation seem reasonable. Based on correlations of relative expression of PKC-α and MARCKS in four cell lines of neural origins (Rosé et al., 1994; Sproull et al., 1995), we previously postulated that both PKC-α and MARCKS may be essential for phorbol ester-stimulated turnover of PtdCho. Others also have provided evidence that PKC-α is required for regulation of various aspects of lipid metabolism in different cells (Balboa et al., 1994; Conricode et al., 1994; Slack et al., 1995). However, until our studies, there has been no direct evidence of a requirement for MARCKS in the regulation of PKC-mediated membrane phospholipid turnover. Involvement of MARCKS has been implicated in cytoskeletal mediated events, neurosecretion and mitogenic signal generation (Aderem, 1995; Hartwig et al., 1992) and a PKC- dependent role for MARCKS in vesicular trafficking has been described (Allen and Aderem, 1995). Our observations are the first to indicate that MARCKS is essential for stimulated PtdCho synthesis. Firstly, only in cells normally expressing MARCKS was PtdCho synthesis enhanced in response to TPA. More convincingly, in SK-N-MC cells normally lacking TPA-initiated PtdCho turnover, overexpression of MARCKS, but not of PKC-α alone, resulted in stimulated PtdCho synthesis with TPA treatment. In distinct contrast, vector controls were unaffected by TPA treatments. Thus, in this system MARCKS is an essential component of phorbol ester stimulated synthesis of PtdCho, mediated by PKC. It is not yet clear whether another PKC isoform (eg., PKC-β) inherent to SK-N-MC cells or the slightly enhanced PKC-α in transfected clones mediates the TPA effect but it is apparent that overexpression of PKC-α alone is insufficient to elicit TPA responsiveness measured by these parameters.

A temporal delay in response to TPA, measured as labelledCho incorporation into PtdCho, and its elimination by preincubation with TPA but not by preincubation with Cho, strongly implicate activation of events prior to the initiation of synthesis. PKC isoforms responsive to TPA are activated within seconds with maximal translocation occurring within 5 min with 20-200 nM TPA (Sproull et al., 1995). Activation of PLD might be one event preceding measurable synthesis, the latter serving as a replacement reaction for hydrolysed phospholipid. In C6 glioma cells, this could be the case as PLD measured with several different assays is rapidly activated by TPA before PtdCho synthesis can be detected. However, in SK-N-SH neuroblastoma cells, we could not detect enhanced PLD activity even though TPA did stimulate PtdCho synthesis. With MARCKS-transfected SK-N-MC clones, enhanced PLD activity was seen in response to TPA treatment suggesting that PLD activation may be the initial event. However, in these transfected clones, this can not be the complete explanation since loss of label from prelabelled PtdCho in the MARCKS clones actually was less in TPA stimulated cells. It seems that both enhanced degradation by PLD and stimulated resynthesis from intracellular Cho and PCho pools must be occurring simultaneously. More specific evaluation of effects of MARCKS transfection and TPA stimulation on individual steps in the biosynthetic pathway is ongoing but it appears that some components of the biosynthetic pathway also may be stimulated. Studies have indicated that the rate limiting enzyme, cytidylyltransferase, is not directly phosphorylated by PKC (Cook and Vance, 1985; MacDonald and Kent, 1994). Interactions of this or other enzymes involved in PtdCho synthesis with phosphorylated MARCKS has not been assessed.

Incomplete reversal of TPA-stimulated PLD activation in C6 glioma cells when all PKC activity (>98%) was inhibited by STS or BIM was unexpected. While approximately 80% of the TPA effect on PtdCho synthesis apparently was dependent upon PKC activity, the remaining 20% due to TPA treatment seems to occur by a mechanism independent of PKC activity. Others have proposed that PKC mediated activation of PLD may not require phosphorylation of a substrate (Singer et al., 1996; Conricode et al., 1992).

Overall, these studies have indicated that PLD activation is part of the response to PKC activation by phorbol ester in C6 cells containing PKC-α and MARCKS and in SK-N-MC clones transfected with MARCKS. An initial PLD activation, however, does not appear to explain all of the enhanced PtdCho synthesis when a signal is received for generation of lipid second messengers. It is apparent that both MARCKS and PKC-α may be essential components interacting in the regulation of PtdCho turnover. Thus, coordinated and temporal interactions of PKC isoforms and substrates have important implications in generation, turnover and balance of lipid second messenger molecules and in the maintenance of membrane integrity dependent upon regulation of a balanced phospholipid composition.

ACKNOWLEDGEMENTS

These studies were supported by a Program Grant (PG-11476) from the Medical Research Council of Canada. The technical assistance of Anne Murphy, Susan Powell and Robert Zwicker is gratefully acknowledged.

REFERENCES

Aderem, A., 1995, The MARCKS family of protein kinase-C substrates, *Biochem. Soc. Trans.* 23:587-591.

Ailhaud, G.P., Abumrad, N., Amri, E.-Z., and Grimaldi, P.A., 1994, A new look at fatty acids as signal-transducing molecules, *World Rev. Nutr. Diet.* 75:35-45.

Allen, L.-H., and Aderem, A., 1995, Protein kinase c regulates MARCKS cycling between the plasma membrane and lysosomes in fibroblasts, *EMBO J.* 14:1109-1121.

Balboa, M.A., Firestein, B.L., Godson, C., Bell, K.S., and Insel, P.A., 1994, Protein kinase C_α mediates phospholipase D activation by nucleotides and phorbol ester in Madin-Darby canine kidney cells. Stimulation of phospholipase D is independent of activation of polyphosphoinositide-specific phospholipase C and phospholipase A_2, *J. Biol. Chem.* 269:10511-10516.

Boyano-Adánez, M.D., Rodríguez, D., Aradottir, S., Alling, C., and Gustavsson, L., 1994, Protein kinase C-mediated phospholipase D activity is increased by linolenic acid supplementation in NG 108-15 cells, *Biochim. Biophys. Acta*,. 1214:263-271.

Conricode, K.M., Brewer, K.A., and Exton, J.H., 1992, Activation of phospholipase D by protein kinase C. Evidence for a phosphorylation-independent mechanism, *J. Biol. Chem.* 267:7199-7202.

Conricode, K.M., Smith, J.L., Burns, D.J., and Exton, J.H., 1994, Phospholipase D activation in fibroblast membranes by the α and β isoforms of protein kinase C, *FEBS Lett.* 342:149-153.

Cook, H.W., Byers, D.M., Palmer, F.B.S.C., and Spence, M.W., 1989, Alterations of phospholipid metabolism by phorbol esters and fatty acids occur by different intracellular mechanisms in cultured glioma, neuroblastoma and hybrid cells, *J. Biol. Chem.* 264:2746-2752.

Cook, H.W., and Vance, D.E., 1985, Evaluation of possible mecchanisms of phorbol ester stimulation of phosphatidylcholine synthesis in HeLa cells, *Can. J. Biochem. Cell Biol.* 63:145-151.

Cook, S.J., and Wakelam, M.J.O., 1989, Analysis of the water-soluble products of phosphatidylcholine breakdown by ion-exchange chromatography. Bombesin and TPA (12-*O*-tetradecanoylphorbol 13-acetate) stimulate choline generation in Swiss 3T3 cells by a common mechanism, *Biochem. J.* 263:581-587.

Dekker, L.V., and Parker, P.J., 1994, Protein kinase C--A question of specificity, *Trends Biochem. Sci.* 19:73-77.

Exton, J.H., 1990, Signaling through phosphatidylcholine breakdown, *J. Biol. Chem.* 265:1-4.

Graber, R., Sumida, C., and Nunez, E.A., 1994, Fatty acids and cell signal transduction, *J. Lipid Mediat.* 9:91-116.

Hartwig, J.H., Thelen, M., Rosen, A., Janmey, P.A., Nairn, A.C., and Aderem, A., 1992, MARCKS is an actin filament crosslinking protein regulated by protein kinase C and calcium-calmodulin, *Nature* 356:618-622.

Kent, C., 1995, Eukaryotic phospholipid biosynthesis, *Annu. Rev. Biochem.* 64:315-343.

Kiss, Z., Rapp, U.R., Pettit, G.R., and Anderson, W.B., 1991, Phorbol ester and bryostatin differentially regulate the hydrolysis of phosphatidylethanolamine in Ha-*ras*- and *raf*-oncogene-transformed NIH 3T3 cells, *Biochem. J.* 276:505-509.

Kiss, Z., Tomono, M., and Anderson, W.B., 1994, Phorbol ester selectively stimulates the phospholipase D-mediated hydrolysis of phosphatidylethanolamine in multidrug-resistant MCF-7 human breast carcinoma cells, *Biochem. J.* 302:649-654.

MacDonald, J.I.S., and Kent, C., 1994, Identification of phosphorylation sites in rat liver CTP: Phosphocholine cytidylyltransferase, *J. Biol. Chem.* 269:10529-10537.

Newton, A.C., 1995, Protein kinase C: Structure, function, and regulation, *J. Biol. Chem.* 270:28495-28498.

Nishizuka, Y., 1992, Intracellular signaling by hydrolysis of phospholipids and activation of protein kinase C, *Science* 258:607-614.

Pelech, S.L., and Vance, D.E., 1989, Signal transduction via phosphatidylcholine cycles, *TIBS* 14:28-30.

Rosé, S.D., Cook, H.W., Palmer, F.B.S.C., Ridgway, N.D., and Byers, D.M., 1994, Differential expression of MARCKS and other calmodulin-binding protein kinase C substrates in cultured neuroblastoma and glioma cells, *J. Neurochem.* 63:2314-2323.

Rosé, S.D., Morash, S.C., Ridgway, N.D., Byers, D.M., and Cook, H.W., 1996, Overexpression of MARCKS, but not protein kinase C-α, increases phorbol ester-stimulated synthesis of phosphatidylcholine in human SK-N-MC neuroblastoma cells, *J. Neurochem.* 66:1766-1769.

Singer, W.D., Brown, H.A., Jiang, X., and Sternweis, P.C., 1996, Regulation of phospholipase D by protein kinase C is synergistic with ADP-ribosylation factor and independent of protein kinase activity, *J. Biol. Chem.* 271:4504-4510.

Slack, B.E., Breu, J., Livneh, E., Eldar, H., and Wurtman, R.J., 1995, Phorbol ester stimulates choline uptake in Swiss 3T3 fibroblasts following introduction of the gene encoding protein kinase Cα, *Biochem. J.* 305:621-626.

Sproull, S.A., Morash, S.C., Byers, D.M., Palmer, F.B.S., and Cook, H.W., 1995, Phorbol ester stimulation of phosphatidylcholine synthesis in four cultured neural cell lines: Correlations with expression of protein kinase C isoforms, *Neurochem. Res.* 20:1397-1407.

Toullec, D., Pianetti, P., Coste, H., Bellevergue, P., Grand-Perret, T., Ajakane, M., Baudet, V., Boissin, P., Boursier, E., Loriolle, F., Duhamel, L., Charon, D., and Kirilovsky, J., 1991, The bisindolylmaleimide GF 109203X is a potent and selective inhibitor of protein kinase C, *J. Biol. Chem.* 266:15771-15781.

Vance, D.E., 1991, Phospholipid metabolism and cell signalling in eucaryotes, *Biochemistry of Lipids, Lipoproteins and Membranes* (D.E. Vance and J.E. Vance eds.), Elsevier, New York. pp. 205-240.

INDEX